Geographic Information Systems for Geoscientists: Modelling with GIS

WITHDRAWN

COMPUTER METHODS IN THE GEOSCIENCES
Daniel F. Merriam, Series Editor

Volumes in the series published by Pergamon (Elsevier Science Ltd):

Geological Problem Solving with Lotus 1-2-3 for Exploration and Mining Geology: G.S. Koch, Jr (with program on diskette)
Exploration with a Computer: Geoscience Data Analysis Applications: W.R. Green
Contouring: A Guide to the Analysis and Display of Spatial Data: D.F. Watson (with program on diskette)
Management of Geological Data Bases: J. Frizado (Editor)
Simulating Nearshore Environments: P.A. Martinez and J.W. Harbaugh

*Volumes published by Van Nostrand Reinhold Co. Inc.:

Computer Applications in Petroleum Geology: J.E. Robinson
Graphic Display of Two- and Three-Dimensional Markov Computer Models in Geology: C. Lin and J.W. Harbaugh
Image Processing of Geological Data: A.G. Fabbri
Contouring Geological Surfaces with the Computer: T.A. Jones, D.E. Hamilton, and C.R. Johnson
Exploration-Geochemical Data Analysis with the IBM PC: G.C. Koch, Jr (with programs on diskettes)
Geostatistics and Petroleum Geology: M.E. Hohn
Simulating Clastic Sedimentation: D.M. Tetzlaff and J.W. Harbaugh

*Orders to: Van Nostrand Reinhold Co. Inc., 7625 Empire Drive, Florence, KY 41042, U.S.A.

Related Pergamon/Elsevier Science Publications

Books

GAAL & MERRIAM (Editors): Computer Applications in Resource Estimation: Prediction and Assessment for Metals and Petroleum

HANLEY & MERRIAM (Editors): Microcomputer Applications in Geology I and II

MACEACHREN & TAYLOR (Editors): Vizualization in Modern Cartography

TAYLOR (Editor): Geographic Information Systems (The Microcomputer and Modern Cartography)

Journals

Computers & Geosciences

Full details of all Pergamon publications/free specimen copy of any Pergamon journal available on request from your nearest Elsevier Science office.

Geographic Information Systems for Geoscientists: Modelling with GIS

Graeme F. Bonham-Carter
Geological Survey of Canada,
Ottawa,
Ontario,
Canada

PERGAMON

U.K.	Elsevier Science Ltd, The Boulevard, Langford Lane, Kidlington, OX5 1GB, U.K.
U.S.A.	Elsevier Science Inc., 660 White Plains Road, Tarrytown, New York 10591-5153, U.S.A.
JAPAN	Elsevier Science Japan, Tsunishima Building Annex, 3-20-12 Yushima, Bunkyo-ku, Tokyo 113, Japan

---

Copyright ©1994 G.F. Bonham-Carter

All Rights Reserved. No part of this publication may be reproduced, stored in a retreival system or transmitted in any form or by any means: electronic, electrostatic, magnetic tape, mechanical, photocopying, recording or otherwise, without permission in writing from the publishers.

First edition 1994. Reprinted 1996.

Library of Congress Cataloging in Publication Data

Bonham-Carter, Graeme.
Geographic information systems for geoscientists:
modelling with GIS / Graeme Bonham-Carter. -- 1st ed.
p. cm. -- (Computer methods in the geosciences)
Includes bibliographical references and index.
1. Geographic information systems. I. Title. II. Series.
G70.2.B66 1994 910'.285--dc20 94-28315

British Library Cataloguing in Publication Data

A catalogue record for this book is available from the British Library.

ISBN 0 08 041867 8 Hardcover
ISBN 0 08 042420 1 Flexicover

Printed in Canada by Love Printing Service Ltd, Ontario

Contents

Series Editor's Foreword

GIS, the magic words in today's computer world. GIS, or Geographic Information Systems, are the basis for understanding our world through spatial information. Although geographers, and others working with maps and geographic data, have long been involved with GIS, geologists have been rather slow to adapt to their use. This situation is now changing, and geoscientists are now embracing this technology for a variety of applications.

This book is an introduction to this important aspect of geology written by a geologist for geologists and other geoscientists. The book takes the reader from a basic introduction through to advanced applications. It will find use by students, academicians, and practitioners alike. Once GIS is defined and explained, the author delves into spatial analysis based on a GIS. Chapters on spatial data models and structures, visualization - graphics - he goes into spatial data transformations, and tools for analyzing single maps, pairs of maps, and multiple maps. Anyone involved in spatial analysis will find these chapters enlightening and instructional.

The author states in the Preface that he wrote this book to fill a void on the subject, and indeed that is true. He also notes that the book was to be an introduction to the subject, and that too is true. For anyone interested in the subject, this is the place to start - with definitions, explanations, examples, and references to pertinent literature. Graeme Bonham-Carter has many years experience in the analysis of geological data and modelling, and he is well qualified to introduce other geoscientists to this important subject.

The book fits nicely into the series on *Computer Methods in the Geosciences*. It complements the other books concerned with mapping and spatial analysis. GIS will become an increasingly used tool, as geological databases of all kinds are expanded, and added to the growing volume of geographic data in digital form. GIS will ease the manipulation, display, and combination of geographically-distributed data, save time and effort, and allow new insights into the understanding and interpretation of spatial problems.

D.F.Merriam

Preface

The idea for writing this book occurred to me in 1990, when planning a course on Geographic Information Systems (GIS) to be given in the Department of Geology at the University of Ottawa. At that time, although the GIS literature was already quite voluminous and rapidly expanding, there were no books suitable as a textbook for the kind of course that I envisioned. The textbooks on the subject did not deal, for the most part, with data or problems relevant to geologists. Furthermore, few GIS books devoted much space to methods of spatial analysis and modelling.

The goal of this book is to introduce the ideas and practice of GIS to students and professionals from a variety of geoscience backgrounds. The digital manipulation of spatial data is now an essential part of many geological and environmental studies. Students doing thesis projects very readily see the application to their work, and familiarity with GIS technology can be a significant factor in their search for employment. Many geoscientists, particularly those working in the areas of mineral resources and environmental studies, are finding that GIS technology is increasingly important in their business. Furthermore, GIS is fun to work with, and stimulates many users to become more analytical in their approach to spatial data. This in turn leads to a curiosity and heightened interest in the methods of data processing and mathematical analysis that underlie GIS. The book therefore aims to introduce some of the mathematical framework for defining, transforming and analyzing spatial data.

Many of the applications discussed in this book deal with the integration of multiple spatial datasets for mineral exploration. The types of data and the general approach used for evaluating mineral potential are similar to those used in a number of other geoscience applications, such as the analysis of landslides, and the assessment of impacts in environmental studies. The emphasis in the book is to show how spatial data from a variety of sources (principally paper maps, digital images and tabular data from point samples) can be captured in a GIS database, manipulated and transformed to extract particular features in the data, and combined together to produce new derived maps, that are useful for decision-making and for understanding spatial interrelationships.

The book introduces the reader to some basic ideas of spatial analysis and modelling. For example, some simple methods of exploring the spatial associations between spatial objects are introduced, to answer such questions as "are the mineral deposits spatially related to fold axes", or "which bedrock units are most closely associated to the presence of till". Towards the end of the book, some methods for combining several datasets together for prediction are discussed, such as the use of fuzzy logic and Bayesian statistics for selecting the best location for a landfill, or for ranking areas according to mineral potential.

The book is intended as both a textbook for a course on GIS, and also for those professional geoscientists who wish to understand something about the subject. Readers with a mathematical bent will get more out of the later chapters, but relatively non-numerate individuals will understand the general purpose and approach, and will be able to apply methods of map modelling to clearly-defined problems.

At the time of writing, several new textbooks have appeared that provide useful supplementary reading. For example, the book by Legg (1992) discusses a variety of geoscience applications, with emphasis on the analysis of raster images. The book published by ESRI (1993) is mainly directed towards users of the ARC/INFO system, but gives a very readable introduction to the principles of GIS, particularly the digitizing, manipulation and production of maps. Tomlin (1990) provides an interesting introduction to map modelling. These books are recommended, because they emphasize aspects of GIS covered only briefly by the present volume. In addition the books by Aronof (1989), Burroughs (1987), and Maguire et al. (1991) are to be recommended. Many of the GIS papers relevant to earth science applications are dispersed in the literature. There are a number of proceedings volumes from GIS conferences that contain useful papers. The International Journal of Geographical Information Systems contains both theoretical and applied papers. Computers & Geosciences and several other journals, such as Photogrammettry and Remote Sensing, publish special issues on GIS.

The GIS field is diverse, and this book makes no attempt to cover the field as a whole. For example, cartographic applications of GIS are not emphasized, although digital capture and production of maps is the primary use of GIS by many geoscientists. Image processing is discussed only briefly, because although image processing systems are becoming increasingly used as GIS tools, there are a number of excellent textbooks on the subject, written mostly for remote sensing, but directly applicable to other image types. The database aspects of GIS are treated only briefly, despite the importance of creating large custodial spatial databases for supporting mineral exploration and environmental studies with GIS. The emphasis of the book is on how Geographic Information Systems work and how they can be applied to practical problems that involve spatial data analysis and modelling.

As all GIS users know well, the technical difficulties of assembling a large number of spatial datasets from diverse sources are challenging. Many users become so involved in the technical problems of data transfer, data merging and data visualization, that they lose sight of the more substantive reasons for carrying out a GIS project in the first place, namely the characterization, understanding and prediction of spatial phenomena, often as the basis for making decisions. Having assembled the datasets in a GIS (and computer systems are becoming increasingly good at aiding this process) the question is often "What do I do next?". This book attempts to give the reader some ideas about possible lines of attack, but it must be emphasised that the topics covered are far from exhaustive.

One of the problems in writing a book of this type is that it is difficult to be specific without using the language and functional organization of a particular commercial GIS. As far as possible, I have written the book in a generic form, so that it is system-independent. For my own work, I mainly use the system called SPANS, developed by an Ottawa-based company, Intera-Tydac Inc. Although a general-purpose GIS, SPANS provides the kinds of functionality and data structures that are well-suited for problems in mineral-potential mapping, my primary field of research. For this reason, the quadtree data structure, one of the structures used in SPANS, is discussed in some depth, the map modelling syntax is close to that used by SPANS, and many of the illustrations were created using SPANS or the companion desktop-mapping package, SPANSMAP. There should be no difficulty, however, in using this book with other systems, because the basic principles are emphasized, not the system-specific details of actual implementations. There are many excellent GIS on the market, for both workstations and personal computers. It is suggested that where the book is to be used as a textbook for a course with an associated laboratory for "hands-on" experience, that it be augmented by exercises that are tailored for the particular GIS being employed.

Many of the tools and modelling methods needed for analysis of GIS data are simply not found in commercial packages. There are several reasons for this. The primary one is that there is no universal set of methods that applies to a wide variety of applications fields. Another reason is that the GIS discipline is still developing, and as the subject matures, a consensus will develop about the content and organization of the ideal GIS "toolbox". Furthermore, many of the potential applications of GIS are so specific that it is unlikely that a general-purpose GIS will ever be adequate for all needs. The solution adopted by many GIS users at present, and perhaps this will always be the case to some degree in the future, is to couple a commercial GIS to other specialized programs. For example, specialized packages for spreadsheet analysis, multivariate statistics, mathematical morphology, expert systems, models of groundwater flow, contaminant dispersal in rivers, and many other areas, can be interfaced with the GIS. The GIS then becomes a kind of spatial-data server, as well as a visualization tool and data transformation aid. For this purpose, the GIS must be able to provide data files and "hooks" for inter-communication with other software.

BOOK OUTLINE

The outline of the book follows a progression of topics, becoming more specialized in the later chapters which deal with analysis and modelling. The general philosophy is to give some background for understanding how GIS work, rather than dwelling on the kind of practical details that can be learned from GIS manuals, and tend to be system-specific anyway. Chapter 1 defines the meaning, purpose and functions of GIS, and illustrates a typical GIS application.

Chapters 2 and 3 go together in that they introduce the methods for organizing spatial data in a GIS. In Chapter 2, the idea of data models is developed in some detail, covering the vector and raster models for spatial data objects, followed by some discussion of the relational data model for nonspatial data attributes. Data structures are discussed in Chapter 3, showing how the vector and raster data models are represented. For the vector model, the topological organization of data is the key concept for efficient organization and handling of area objects. Topological attributes are those geometrical properties of spatial objects, such as adjacency and containment, that are unaffected by scaling, rotation or shearing transformations. Data structures for the raster model are described, with particular emphasis on quadtree and octree structures, the latter being of particular interest for geological applications in three spatial dimensions.

Chapters 4 and 5 are devoted to data input and data visualization, respectively. Geographic projections, data digitizing, and methods of geographically registering vector and raster data are introduced as important concepts for spatial data input. Discussion of the details of individual digitizing procedures and data formats are omitted, because these topics are best covered in a GIS-specific laboratory manual. In the visualization chapter, the ways in which computers handle colour are described, both for colour monitors and output displays on paper. In addition, the special effects used to enhance digital images with shading and perspective are briefly introduced.

One of the important functions of any GIS is to transform spatial data from one data structure to another, as discussed in Chapter 6. For example, a raster form of a geological map may need to be converted to a vector form, so that the outlines of geological units can be superimposed on a geophysical image, or further transformed to show distance to selected contacts. Or a spatially continuous geophysical surface, known only at sample point locations, may require interpolation to a grid format, to allow combination with other gridded datasets.

Chapters 7, 8 and 9 deal with the combination, analysis and modelling of maps in both raster and vector formats. The operations normally available in a GIS for manipulating single maps, and their associated attributes, are discussed in Chapter 7. Some of these operations, such as map re-classification on the basis of attributes in a table, are independent of topological factors. Others, such as image filtering, explicitly use data from a spatial neighbourhood, and spatial objects are considered in relation to adjacent objects. Operations on pairs of maps are treated in Chapter 8, dealing with overlay and modelling. Methods of characterizing spatial associations between pairs of maps are discussed, For example, given the locations of geochemical anomalies, what is their spatial correlation with shear zones?

Finally, Chapter 9 is a long one and deals with some methods of combining multiple maps, using models. For many readers, this may be the most useful and interesting part of the book, because a variety of map modelling approaches are discussed and illustrated in some detail with respect to a landfill site selection problem and a mineral potential mapping problem.

ACKNOWLEDGEMENTS

I would like to acknowledge some of the people and organizations that have helped me to prepare this book. I am indebted to the Geological Survey of Canada for allowing me time to write the book, and for the technical support used to prepare the illustrations. My colleagues in the Geomathematics Section have helped in a number of ways. Danny Wright has worked on a number of the examples illustrated in the book, particularly the mineral potential study described in Chapter 1, and has provided technical assistance on numerous occasions. Frits Agterberg reviewed the whole manuscript and suggested several changes in the final chapters. Andy Rencz supplied useful comments on the Visualization chapter. Many thanks are due to Giulio Maffini, Wolfgang Bitterlich and others at Intera-Tydac Ltd. for infecting me with their enthusiasm for GIS, and for their technical advice. Giulio and Wolfgang also reviewed the whole book and suggested many improvements. I am grateful to Tony Fowler, Chairman of the Department of Geology, University of Ottawa, who encouraged me to teach a course in GIS, because this incubated the whole book project. Phil O'Regan prepared many of the diagrams, and some were made by Charles Logan and Brian Eddy. Not least, I thank my wife, Gwendy, and youngest daughter, Gemma, who put up with my many long evenings on the computer.

Graeme Bonham-Carter
Ottawa. August 1993.

REFERENCES

Aronoff, S., 1989, *Geographic Information Systems: A Management Perspective*: WDL Publications: Ottawa, Canada, 294 p.

Burroughs, P.A., 1987, *Principles of Geographical Information Systems for Land Resource Assessment*: Clarendon Press, Oxford, 193 p.

Environmental Systems Research Institute, Inc., 1993, *Understanding GIS*: Longmans Scientific & Technical, Harlow, 534 p.

Legg, C.A., 1992, *Remote Sensing and Geographic Information Systems: Geological Mapping, Mineral Exploration and Mining*: Ellis Horwood, New York, 166 p.

Tomlin, C.D., 1990, *Geographic Information Systems and Cartographic Modeling*: Prentice Hall, Englewood Cliffs, New Jersey, 249 p.

Maguire, D.J., Goodchild, M.F., and Rhind, D.W. (editors) , 1991, *Geographical Information Systems. Volume 1 – Principles. Volume 2 – Applications*: Longman Scientific & Technical, London.

NOTE FOR SECOND PRINTING

Several errors and omissions have been corrected in the second printing. I am grateful to my colleagues who have pointed out mistakes in the original. The location of a recently-discovered VMS deposit has been added to Figure 1-10.

Graeme Bonham-Carter
Ottawa, October 1995

NOTE

In common with general usage, I have employed the acronym GIS to refer to either a single geographic information system, or to several systems, or to the field of geographic information systems as a whole. For the plural case, I have not added an "s" or "es" to the basic three letters. For example, the acronym is used in the following ways:

* "This GIS has a powerful analytical module..." (used as a singular noun), or

* "These GIS operate on personal computers..." (used as a noun in the plural), or

* "GIS is a rapidly growing field..." (used as a singular noun referring to the whole technology), or

* "The GIS approach to the solution..." (used as an adjective referring to the whole technology).

CHAPTER 1

Introduction to GIS

A minority of geologists and other geoscientists have been using computers for manipulation of spatial data since the 1960s. During the 1980s, advances in computer hardware, particularly processing speed and data storage, catalyzed the development of software for handling spatial data. The emerging capabilities for graphical display played an important role in this development. One of the significant products of this period of rapid technological change was GIS. The impact of GIS has been widely felt in all fields that use geographic information, in resource management, land-use planning, transportation, marketing, and in many applications in the geosciences and elsewhere. The majority of large geological organizations are now using GIS, and the handling of spatial data of all types by computer is widespread.This chapter introduces GIS by describing what GIS means, the purposes and functions of GIS, how GIS relates to other kinds of software for spatial data handling, and presents a typical geological application.

WHAT IS GIS?

A geographic information system, or simply GIS, is a computer system for managing spatial data. The word geographic implies that locations of the data items are known, or can be calculated, in terms of geographic coordinates (latitude, longitude). Most GIS are restricted to data in two spatial dimensions, although some systems of particular interest to geologists have true three-dimensional capabilities and can represent objects such as recumbent folds. The word **information** implies that the data in a GIS are organized to yield useful knowledge, often as coloured maps and images, but also as statistical graphics, tables, and various on-screen responses to interactive queries. The word **system** implies that a GIS is made up from several interrelated and linked components with different functions. Thus, GIS have functional capabilities for data capture, input, manipulation, transformation, visualization, combination, query, analysis, modelling and output. A GIS consists of a package of computer programs with a user interface that provides access to particular functions. The user may control GIS operations with a graphical user interface, commonly called a **GUI**, or by means of a **command language**, consisting of program statements that dictate the sequence and type of operations.

GIS are computer tools for manipulating maps, digital images and tables of **geocoded** (geographically located) data items, such as the results of a geochemical survey. GIS are designed to bring together spatial data from diverse sources into a unified database, often employing a variety of digital data structures, and representing spatially varying phenomena as a series of data layers (such as bedrock geology, depth to water table, Bouguer gravity anomaly, and so on), all of which are in spatial **register,** meaning that they overlap correctly at all locations.

GIS is filling a very real need in the face of the rapid growth of digital spatial data in the geosciences. Many spatial datasets are now being generated by government agencies, private companies and university researchers, and they would be ineffectively used and result in wasted resources without good systems of data management. Satellite images are a prime example of this data explosion. Without digital systems for the processing and display of images, the enormous volumes of remote sensing data collected daily would simply remain on computer storage devices, and the wealth of information they contain would remain unrevealed and indigestible. Data collected from geophysical instruments, airborne, ship-borne, ground-based, down-hole and others, also make up enormous volumes of numbers that reveal little until they are properly organized and displayed. Similarly, geochemical surveys of rocks, soils, water, sediments and plants, often with thirty or more elements determined from each sample, yield huge amounts of spatial data whose information content cannot be assessed without efficient spatial data systems.

GIS has made a tremendous impact in many fields of application, because it allows the manipulation and analysis of individual "layers" of spatial data, and it provides tools for analyzing and modelling the interrelationships between layers. Geoscientists need to understand the spatial relationship between all the various kinds of spatial data that they collect. For example, mineral exploration requires the simultaneous consideration of many kinds of spatial evidence for mineral deposits, such as the geology, structure, geochemical and geophysical characteristics of a region, as well as the locations and type of past mineral discoveries. In environmental problems, the interaction of many processes must be considered, and the simultaneous analysis of multiple datasets is imperative.

Before widely-available commercial GIS appeared on the scene in the late 1980s, most geoscientists working with multiple spatial datasets were doing their work on light tables. A relatively small proportion of the geoscience community was computer-orientated, working mostly with mainframe computers and locally-developed software. GIS technology is still in its infancy (1993), with the majority of applications being carried out by specialists. However, GIS has the potential to change the geological workplace drastically. In a few short years, the personal computer has virtually eliminated the typewriter and the hand calculator from the office, and GIS is likely to replace the light table and the map cabinet. Like many fields where computers are employed, GIS has the potential to free the user from the technical, slow, laborious problems of data handling, and to enhance the capability for creative data analysis and interpretation.

The acronym "GIS" has come to mean much more than a type of computer program. GIS implies the science of geographic information management and analysis. In many countries there are now academic institutes devoted to all aspects of GIS. In the United States, the National Center for Geographic Information and Analysis (NCGIA) is headquartered at the University of California, Santa Barbara. In the United Kingdom, a group of regional laboratories has been established for GIS research. GIS courses are now part of the academic curriculum in many universities and technical colleges. There are a large number of GIS periodicals, many of them commercially-orientated, but some entirely for the publication of scientific papers, such as the International Journal of Geographical Information Systems. Numerous conferences are devoted to GIS, catering to the needs of those working in GIS in academia, government and business. GIS is enjoying a period of popularity because it is new and topical. Over time, GIS may simply be absorbed into the various disciplines that it affects, like some other computer applications. It seems more likely, however, that the discipline of GIS will become permanently established, with its own agenda of research, to meet the needs of the many fields that deal with spatial data.

PURPOSE OF GIS

The ultimate purpose of GIS is to provide support for making decisions based on spatial data, as illustrated by a few geological examples. The exploration manager may use GIS to assemble data in the form of a mineral potential map to decide priorities for future exploration; the mining geologist may evaluate the effects of acid mine drainage with GIS to decide the kinds of remediation that would be cost-effective; the engineering geologist may evaluate slope stability conditions with GIS to decide the best route for a new road. Sometimes the purpose of using a GIS is to support general research. For example, a geochemist may use GIS to investigate the spatial association between the distribution of selenium in surface water, and the distributions of local rock type, pH of the water and the local vegetation; or the geophysicist may employ GIS to study the spatial factors related to earthquakes. Of course, GIS is invaluable for collecting, maintaining and using spatial data in a database management role, as well as for producing both standardized and customized cartographic products. Application of GIS achieves these major goals through one or more of the following activities with spatial data: organization, visualization, query, combination, analysis and prediction.

Organization

Anyone who has collected a large mass of data for a particular purpose knows the importance of data organization. Data can be arranged in many different ways, and unless the organization scheme is suitable for the application at hand, useful information cannot be easily extracted. Schemes for organizing data are sometimes called **data models**, as discussed in Chapter 2. The principal characteristic for organizing GIS data is spatial location. A table of geochemical data may be interesting for analyzing the relationships between geochemical elements on scatterplots, but without knowing the *locations* of samples, the interpretation of spatial patterns and relationships with other spatial data, such as rock type, cannot be established. GIS data are also organized according to non-spatial characteristics. For example, the interpretation of geochemical data may depend on recognizing spatial patterns of element ratios, or of groups of observations based on the method of analysis or the year of collection. Data models must, therefore, organize observations both by spatial and non-spatial attributes. The efficiency and type of data organization effects all the other five activities and is therefore of fundamental importance.

Visualization

The graphical capabilities of computers are exploited by GIS for visualization. Visual display is normally carried out using the video monitor, but other output devices such as colour printers are used for hardcopy displays. Humans have an extraordinary ability to understand complex spatial relationships visually, whereas the same information may be quite unintelligible when presented as a table of numbers. For example, given a table of geochemical analyses, a geologist is normally unable to recognize the spatial distribution of highs and lows in the data, but when the same table is converted to an effective map display, spatial patterns are immediately revealed. Visualization is achieved in GIS with colour and symbols, and by specialized methods using perspective, shadowing and other means.

Spatial Query

Visualization reveals spatial pattern amongst collections of organized data items. However, visualization is not so helpful for answering questions about special instances in the data, such as the value of particular data items. Spatial query is a complementary activity to data visualization. For example, a particular display of a combination of mineral deposit points and a geochemical map might suggest the existence of a spatial relationship in some areas of the display, but not in other areas. Spatial query permits the user to find out the special circumstances of each case, by finding out the name and other details of individual mineral occurrences, and the characteristics of individual geochemical samples in the selected neighbourhoods of interest. This often helps to find the reason behind the spatial pattern. GIS provides tools for two types of interactive query. The first type is the question "What are the characteristics of this location?" The second type is "Whereabouts do these characteristics occur?"

For example, suppose an aeromagnetic map is displayed on the video monitor, one might wish to know in detail the rock formation, the distance to the nearest road, the topographic elevation, the value of the Bouguer anomaly and the location and analysis of the nearest geochemical sample for any specified location. Many GIS allow the user to generate a summary table of selected characteristics (appearing in a display window) pertaining to a specific location. The location is often identified interactively with the cursor, and the table is instantly updated as the cursor is moved to each new location. Clearly this requires that data is efficiently organized by spatial location to permit fast retrieval.

As an example of the second type of question, one might wish to know all the locations on the map where a particular set of conditions is satisfied. In a simple case, it might be useful to know all the locations where arsenic in soil is greater than 250 parts per million (ppm). Many GIS use a query language for constructing specific questions. The user types in one or more statements, causing a search, from which the result may be a table or a map with the identified locations. The query might be more complex, asking for all locations within 250 m of a lake, at elevations greater than 1000 m above sea level, underlain by Carboniferous limestone. Efficient searches of this type require organization of data both by spatial and non-spatial attributes.

Other kinds of questions may be related to distance, orientation, and conditions of adjacency or containment like "Find all instances of granite in contact with a limestone?" Such questions require not only efficient search of data items, but also the capability for deriving their geometric and topological attributes. The term topology refers those characteristics of a data object, like adjacency and containment, that are not affected by spatial transformations. Thus Africa is *adjacent to* the Mediterranean Sea, no matter what coordinate projection is used to represent the boundaries of these spatial objects.

Combination

The ability to merge spatial datasets from quite different sources and display and manipulate combinations can often lead to an understanding and interpretation of spatial phenomena that are simply not apparent when individual spatial data types are considered in isolation. For example, by superimposing a digitized geological map on to a satellite image, it may become clear that a particular lithology has a distinctive texture or spectral response on the image. The process of combining layers of spatial data is sometimes called data integration, and can be carried out either by visualizing composite displays of various kinds, or with integration models that effectively create a new map from two or more existing maps. Integration models are symbolic mathematical models, using arithmetic and logical operations to combine layers of data together. A simple example is the combination of a map of lakes, derived by digitizing a topographic base map showing drainage features, with a digital map of radiometric data, derived from an airborne survey. The radiometric map contains, say, 100 colours, each colour representing the intensity of the radioelement measurement, whereas the lake map contains only 2 colours, lake present and absent. The algebraic statement that combines these two maps might indicate that the new map be "equal to the class (in this case colour) of the radiometric map except where lakes are present, class 0 otherwise". Such statements are usually written in a programming language, sometimes known as "**map algebra**", specific to the GIS. The combined map can now be treated as a single map, revealing the spatial relationship of water bodies to radiometric patterns.

One of the really powerful features of GIS is the ability to link several map algebra statements together to form more complex algorithms. Several maps and tables of attribute data can be combined in a single processing step. The process of combining maps together is often called **map** or **cartographic modelling.**

Analysis

Analysis is the process of inferring meaning from data. Analysis is often carried out visually in a GIS, as already indicated. Analysis in a GIS can also be carried out by measurements, statistical computations, fitting models to data values and other operations. For example, an analysis of areas on a map may lead to a table (or histogram) summarizing the proportions of a region underlain by classes of surficial materials, or an area cross-tabulation summarizing the overlap relationships of bedrock classes and surficial classes. A statistical summary might be used to compare the mean and standard deviation of airborne uranium measurements by rock type. Or a regression model might be fitted to bivariate geochemical data and the resulting predicted values displayed as a map to see if deviations from the model were related to other known environmental factors. Sometimes a classification analysis based on clustering together locations with the same multivariate characteristics can lead to useful interpretation. Often the tools for specific analyses are not found as part of a GIS, but require that data be exported to other computer programs. Analysis is carried out either on data organized as maps, or on data organized as tables.

Spatial analysis in a GIS sense means simply the analysis of spatial data. For example, the area cross-tabulation of two maps may lead to useful conclusions about the relationship between the two maps, although spatial coordinates play no direct role in the statistical summary. In the statistical literature, however, spatial analysis often implies analysis specifically involving spatial location. For example, trend surface analysis is a method of fitting a mathematical surface to observed data values, and explicitly uses the spatial coordinates in the calculations. The book by Unwin (1981) is a good introduction to this more specific type of spatial analysis; see also the advanced book by Cressie (1991).

Prediction

The purpose of a GIS study is often for prediction. For example a number of data layers indicative of gold deposits might be combined together to predict the favourability for gold as a new map. Such a map may then be used as a basis for making exploration decisions, or land-use decisions such as "Is this region suitable as a national park?" Prediction in a GIS involves the use of map algebra for defining symbolic models that embody the rules for combining data layers together. Prediction is sometimes a research exercise to explore the outcome of making a particular set of assumptions, often with the purpose of examining the performance of a model. For example, one might be interested in the number and area of sites predicted as unstable with a slope stability model, as a function of changes in slope and soil saturation, simply to evaluate model performance and sensitivity. Alternatively, the purpose might be to use the results for choosing building sites or planning road construction. The modelling tools of GIS provide the means to apply spatial data in problem solving, and take spatial data beyond simply the retrieval and display of information.

GIS AND RELATED COMPUTER SOFTWARE

There are a number of close relatives in the family of computer software products developed for handling spatial data, as illustrated in Figure 1-1. Many of these are computer programs that are similar to GIS in some, but not all, respects. Even amongst the products that qualify as full-fledged GIS, there is a great range in functional capabilities, some excelling at making cartographic products, some being good for map modelling, others offering superior database management, and so on. The following paragraphs provide a cursory survey of some of the software categories. This helps to clarify what is, and what is not, a GIS, and also illustrates some of the factors that have influenced the development of GIS.

Because of the focus of this book, GIS has been placed at the centre of the diagram in Figure 1-1, but in fact any one of the boxes could be at the hub. If the book were about database management, for example, the DBMS box would be the central focus. An alternative representation is to show a series of overlapping boxes, because in reality many of the programs (and systems of programs) overlap in functionality. Furthermore, the situation is

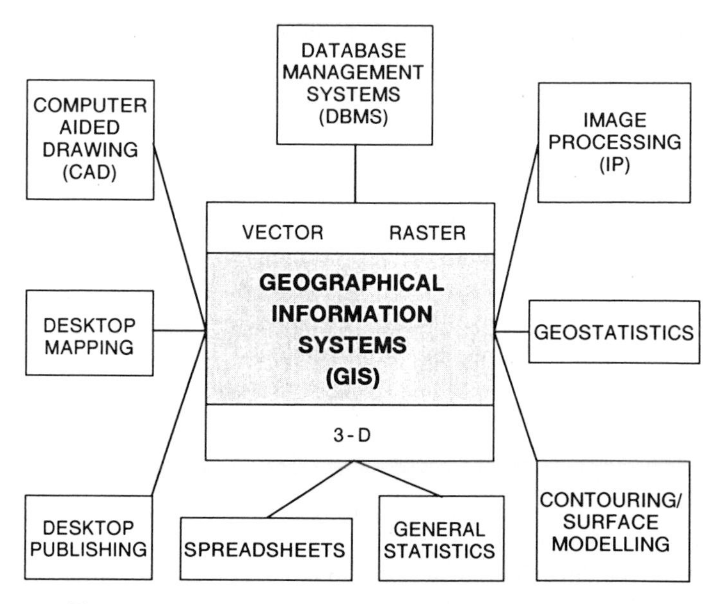

FIG. 1-1. GIS and other related software systems, many of them dealing with spatial data and including some GIS functionality.

always changing, because the commercial systems release new versions with expanded capabilities on a regular basis. This leads to a convergence of functionality, with for example, desktop mapping systems offering advanced functions that make them quite close to GIS, or computer-drawing packages that include the GIS projection transformation functions. This fuzziness between different kinds of systems, and the steady evolution of software systems, also make it difficult to choose a program suitable for a particular application. There are no specific recommendations here, just a note of caution that the spectrum is broad, the categories not very well-defined, and the boundaries between categories move with time. It is rather like buying hardware; you must base your decision on data available at time x, although you know that at time $x+t$ the data and your decision will change.

Two important close relatives of modern GIS are computer aided drawing (CAD) systems and image processing (IP) systems. Both kinds of system deal with spatial data, but they are based on quite different data models and functions. However both CAD and IP have played an important role in the development of GIS.

Computer Aided Drawing

CAD systems were originally developed for engineering drawings. They employ a **vector** data structure for representing points, lines and graphical symbols. The use of *vector* means that a point in a drawing is defined by a pair of spatial coordinates (rather than a single number or scalar), and that lines are built up by a series of ordered points. Areas are represented by boundary lines, also held digitally as strings of connected points. The digital data defining a drawing consists of a large number of coordinate pairs. CAD systems are excellent for the digital representation, spatial transformation and display of drawings, including maps. However CAD systems on their own are not designed to handle non-spatial attributes, except in a basic manner, and are unsuitable for manipulating digital images. They are therefore not good for handling data tables or gridded data, nor are they able to provide the analytical functions of a GIS. Many of the original GIS were developed with vector data structures for handling spatial data, but later added database management and functional capabilities for analysis and modelling.

Image Processing Systems

On the other hand IP systems were developed for manipulating and visualizing digital images in **raster** format, originally images generated principally from satellite sensors and also medical imaging sensors. A *raster* is simply a lattice of **pixels** (*pic*ture *el*ements) similar to those on a TV screen. Each pixel is like a cell in a rectangular grid. The terms raster and grid are used interchangeably in the GIS literature. Spatial locations of pixels are not stored explicitly with each pixel value, but are stored implicitly by the sequence in which the pixel data is held digitally. IP systems are very strong for the display and analysis of digital images, and the gridded data structure makes the overlap and combination of spatial datasets simpler to compute than in vector systems. However pure IP systems are often weak at handling vector data and they lack the linkages to tables containing non-spatial attribute data. They are not generally suitable for producing high quality maps, because the boundaries of map units are jagged, although some high-resolution digital images are of map quality. Some of the original GIS were based on a raster data structure and treated all spatial data as a series of gridded layers, like an IP system. When the data being analyzed is spatially distributed over earth's surface (as opposed to images of thin section of rocks, or images of human insides) an IP system is the same as a GIS. Many of the more advanced IP systems have improved their vector handling, can add cartographic annotation to the hardcopy of images, and possess analytical and modelling capabilities.

3-D GIS

Most commercial GIS have been developed for applications in a variety of fields, and are not specifically orientated towards the geosciences. GIS designed specifically for geological work, particularly for mining and oil exploration, need to be fully three-dimensional, so that each data object is characterized by its location in space with three spatial coordinates (x,y for horizontal and z for vertical position). This allows two or more objects that occur at the same (x,y) location to be distinguished by their z value. For geoscientific data where objects at one (x,y) location only have a single z value, two-dimensional GIS is adequate. Most 2-D GIS have facilities for perspective display of surfaces that are single-valued (not more than one z value per location). This is often referred to as a two-and-a-half dimensional capability. The "height" of the surface can be any attribute, not simply elevation. For projects that involve data such as multiple points down boreholes, seismic sections, recumbent or complex faulted structures, or three-dimensional geotechnical and geophysical data, three-dimensions are needed. This book does not devote much space to 3-D topics, and those interested in the field are referred to the books edited by Raper (1989), Turner (1992) and Pflug and Harbaugh (1992) as an introduction. There are a number of 3-D GIS available commercially, and there is clearly some overlap of functionality with 2-D GIS. However, the specialized data structures, visualization and analysis needs associated with 3-D problems make 3-D systems a rather different breed from regular 2-D GIS, at least at the time of writing in 1993.

Database Management Systems

Another important relative of GIS is DBMS, which stands for database management system. DBMS are computer systems for handling any kind of digital data. Some form of database management system lies at the heart of any GIS, and many commercial GIS are explicitly linked to a particular DBMS. Many of the data collected by geologists are stored as tables of numbers and text. For example, geochemical data are often recorded in tables with the rows being sample sites, and the columns being the chemical elements. When the location of each site is recorded by a pair of spatial coordinates (i.e. a vector), such tables comprise one of the most important inputs to a GIS. Where a CAD system is used in conjunction with a DBMS, many of the data handling and vector functions of a GIS can be implemented, although the data structures are usually not sufficiently complex for the more advanced analysis and modelling operations.

Desktop Mapping Systems

These are systems mainly for display and query of spatial data. Often they use databases that have already been assembled in a GIS or DBMS. They are less expensive and are easier to learn than a full GIS. They are suitable for users of spatial data that need to visualize and explore an existing database, without getting into database creation or advanced analysis and modelling.

Contouring and Surface Mapping Packages

Where data are organized as a table, with each row being the record of data sampled at a geographic location, the immediate goal of data analysis is often to make contour maps of one or more of the variables, stored as columns of the table. For example, the depth to formation tops held as columns in a file of well data are often contoured to show the shape of a stratigraphic surface in plan view. Surface mapping is a more general term for the estimation of surface characteristics from irregularly-spaced point data, not simply expressing the result as contour lines, but also as gridded or triangulated data structures. Some surface mapping systems offer sophisticated analysis, such as the handling of faults in surfaces, permit operations between multiple mapped surfaces, provide visualization and modelling tools, and approach full GIS functionality. Surface mapping packages have been widely used by geologists, particularly in the oil industry. Several chapters of Davis (1986), one of the standard textbooks for statistical methods in geology, are devoted to contouring and statistical analysis of geological surfaces.

Geostatistics Programs

Geostatistics is the branch of statistics dealing with **regionalized variables**. A regionalized variable is a quantity whose value changes with spatial location, and whose behaviour is somewhere between a truly random variable and one that is deterministic. The behaviour of regionalized variables is studied with variograms, that depict the average squared difference between data values as a function of distance and orientation between data locations. Variograms are used in kriging, a method of estimating the value of a regionalized variable from scattered data points. Kriging is a popular method of spatial interpolation for contouring and surface mapping. Geostatistics program packages are normally designed to handle point location data, similar to contouring and surface mapping programs, and can generate co-variograms and co-kriging estimates based on the spatial co-variation of pairs of variables. A good introduction to geostatistics is the book by Isaaks and Srivastava (1989).

Mathematical Morphology Programs

Mathematical morphology refers to a branch of spatial analysis that deals with the shapes and sizes of geometrical objects in images, Serra (1982). It is an important field for the extraction and analysis of features in images, and has been applied geologically mainly to images of rocks in thin or polished sections, see for example Fabbri (1984) . The extraction of features on satellite images and the analysis of features on maps is a methodology of importance to GIS, although at present few IP or GIS incorporate mathematical morphology routines.

Other software

Other programs often used in association with GIS for specialized tasks are **spreadsheets, statistical analysis** programs, particularly for multivariate analysis, and **expert system** shells.

CUSTODIAL VERSUS PROJECT-RELATED GIS

A major proportion of GIS resources, particularly in large organizations, is devoted to the development and maintenance of large **custodial** databases that serve as data sources for a large group of users over an extended period of time. At the opposite extreme is the category of GIS activity that is **project-related**, involving datasets that are assembled for a particular purpose for a small group of users, and are not maintained after the duration of the project. There are some important differences between the typical activities and concerns of those working with custodial databases and those engaged in project-related work.

The size of a custodial database is likely to be much larger than the data gathered for a project-related study. A custodial database must employ data standards that are widely accepted and stable with time, whereas a project-related set of data need maintain standards appropriate only for the purpose at hand. A typical custodial database might be a national geochemical database, or regional geological map database for a province or state. On the other hand, a project-related GIS study might be to map the landslide potential for a small region for the purpose of planning the location of a new housing subdivision, road or pipeline, using data generated just for that application.

In order to create a custodial database, data standards must be established for data entry and data maintenance that can be applied by different groups of people over time. Thus both the geochemical and the geological map database require detailed data definitions, and data models that establish exactly how data items are to be structured and organized. The GIS chosen for custodial use must excel at providing the functions needed for updating and editing of spatial data. The project-related GIS need not be as rigorous on the data standards issue (although the data model must still be unambiguously defined), nor so strong on the data maintenance side, but must be suitable for data analysis and modelling, providing a flexible computing platform for research rather than production work.

Note that we have used the word *database* for the custodial GIS, whereas for the project-related data the words *collection of spatial datasets* are more appropriate. The very notion of a database usually implies long-term custody, maintenance and access by multiple users. (In practice, however, the term spatial database is applied by those using GIS for both custodial and project-related types). In project-related studies there are seldom the resources for long-term maintenance, and once the immediate objectives of the project are complete, the datasets are backed up on to mass-storage devices, such as tapes or disks, for long-term "mothballing". In practice, the data collected together for geological GIS projects are generally obtained in part from custodial databases (often the topographic data, and data from regional geochemical and geophysical surveys) and in part from less formally organized sources and from custom-digitized maps.

GEOLOGICAL APPLICATION OF GIS

Mineral potential mapping has been selected to illustrate a typical geological application of GIS. However the operations that are carried out for mineral potential mapping are in many respects similar to those employed for a variety of other GIS applications, and the aim here is to illustrate the general approach rather than to dwell on the particular details of mineral deposits. Search for other kinds of geological resources, evaluation of hazards, environmental impact and site selection studies also require the simultaneous appraisal of spatial data from several sources. Some other earth science applications of GIS that could be used to illustrate GIS methodology are:

1. Hazard mapping related to slope stability and landslides, earthquake damage zonation, volcanic eruption impacts, flood damage from rivers and tsunamis, coastal erosion, impacts of pollution as a result of mining or industrial activity, global warming.

2. Site selection for engineering projects, such as waste disposal (municipal landfills, nuclear waste in disposal wells), pipeline, road and railway routing, dams, building developments.

3. Resource evaluation for a variety of geological commodities besides metallic minerals, such as water, sand and gravel, building stone, petroleum, natural gas, coal, geothermal energy.

4. Investigation of possible cause and effect linkages of environmental interest between different spatial datasets, such as the incidence of disease (in plants, animals or man) in relation to geochemical patterns in rocks, or soils, or water. Diseases may also be related to complex combinations of spatial environmental factors.

5. Exploratory investigations of spatial inter-relationships between datasets during the course of geological research, such as understanding the regional geochemical and geophysical signatures of I- and S-type granites, or the evaluation of spectral signatures from satellite images in relation to lithology and vegetation.

Mineral Potential Mapping

Mineral exploration is a multi-stage activity that begins at a small scale and progresses to a large scale, ultimately leading to the selection of sites as targets for drilling for buried deposits. At a small scale, exploration companies must delineate general zones that may be of potential interest for mineral deposits of a selected type, usually based on broad geological considerations. At a medium scale, parts of these general zones are selected for more detailed follow-up exploration, based on evidence from geological mapping, regional geochemical and geophysical surveys and the locations of known mineral occurrences. Having identified those more specific zones of favourability, targets may be selected directly, or a further stage of detailed survey work undertaken. Ultimately, this process leads to a large-scale map showing the locations and ranking of potential sites. Of course, actual drilling decisions are based also on other considerations such as physical access, and economic factors. The decision making process must consider many types of spatial data together.

Exploration is similar to mineral resource assessment, an activity often undertaken by governments in order to compare the mineral value of a tract of land with its value for other activities such as forestry, agriculture, or recreation. Like mineral exploration, mineral resource assessment involves the subdivision of land into zones according to mineral favourability. However, in order to assign a monetary value to the mineral resources (needed for making comparisons with competing land uses), the number of estimated deposits, plus their size and grade, must be attached to the favourable tracts. Such information can then be used to build an inventory of mineral resources, and to facilitate land use decisions like the siting of parks, or the settling of land claims.

Both mineral resource estimation and mineral exploration utilize spatial data from a variety of sources. In the past, the selection, evaluation and combination of evidence for mineral deposits was undertaken with the aid of a light table, the various maps being physically superimposed on one another to determine the overlap relationships between anomalies. GIS has greatly improved the efficiency with which this process can be carried out, and expanded the possibilities for specialized data processing and spatial data analysis.

The application described in this section deals with evaluation of mineral potential of a small area in Manitoba, Canada. This work was undertaken as part of a research project at the Geological Survey of Canada dealing with new technology applied to exploration for base metals, focused on volcanogenic massive sulphide (VMS) deposits in the Snow Lake greenstone belt. An early phase of the work is described in Reddy et al. (1992), and the illustrations shown here come from a second phase of study due to be published by the Geological Survey of Canada in 1994.

Conceptual model

Mineral deposits can be grouped or classified into different types, depending on their characteristics. No two deposits of a single type are completely identical, and sometimes one class may include a range of variation. Each class can be represented by an idealized mineral deposit, known as a **mineral deposit model**, one that has all the typical characteristics of the group. Mineral deposit models are conceptual models, usually described in words and diagrams. They describe the typical characteristics of a group of deposits, are accompanied by an interpretation of the processes of deposit formation, and are useful for providing criteria for mineral exploration (Hodgson, 1990). Mineral deposit models are important for providing the theoretical framework to guide GIS studies of mineral potential. They help in data selection and data modelling, in deciding which features to enhance and extract as evidence, and for deciding how to weigh the relative importance of evidence in estimating mineral potential. Conceptual models of different types are important for all kinds of GIS applications.

Volcanogenic massive sulphide deposits are formed at volcanic vents on the ocean floor. For a description of the characteristics and genesis of VMS deposits in general, see Lydon (1984, 1988). The Chisel Lake VMS deposit occurring in the Lower Proterozoic Snow Lake greenstone belt is described by Bailes and Galley (1989), who summarize the geology and mineral deposits of the Anderson-Chisel-Morgan Lake area. In this region (the area of study), the principal characteristics of the deposits useful for potential mapping are as follows.

1. The deposits occur within thick sequences of subaqueous volcanic rocks. Deposits are found mainly in felsic lava flows, but also in other volcanic rock types. Proximity to the contact between felsic flows and volcanoclastic units may be important.

2. The volcanic rocks are associated with large felsic intrusions, believed to act as a subvolcanic heat source driving a hydrothermal circulation. Seawater is thereby pumped through the volcanic rocks, from which metals are dissolved and re-deposited at or near vents to the seafloor.

3. The vent areas are associated with dykes and vertical fractures that cut through the volcanic rocks, localizing the upward flow of hydrothermal fluids. Heat from the intrusive dykes also helps to drive the circulation.

4. The hydrothermal fluids react with the volcanic rocks to produce alteration minerals. Silicification and the presence of amphibole minerals occurs in large zones, semiconformable with the volcanic stratigraphy. Ankerite occurs in more restricted zones.

5. Weathering and erosion of deposits produces a dispersion halo of metallic elements in drainage sediments and till.

6. Regional magnetic surveys show variations in magnetic susceptibility of the rocks, mainly due to differences in rock type. Sharp gradients in the magnetic field measurements usually reflect changes in rock type at geological contacts.

7. Geophysical data can be particularly important evidence for VMS deposits, which have distinctive electrical characteristics. Airborne and ground EM data are widely used to explore for these deposits, particularly in association with magnetic data.

These criteria constitute part of the deposit model used in the GIS study, providing a framework for data selection and analysis.

Brief Description of GIS study

Most GIS projects can be boiled down to three major steps or stages, as illustrated in Figure 1-2 for mineral potential mapping. The first step is to bring all the appropriate data together into a GIS database. The second step is to manipulate the data to extract and derive those spatial patterns relevant to the aims of the project, which in this case are the patterns critical as evidence for VMS deposits. The third step is to combine the derived evidence to predict mineral potential. Consider these steps in more detail for the VMS example, with reference to Figure 1-3.

Step 1

The initial database-building step was the most time-consuming phase of the project. This is typical of GIS studies in general. It involved establishing the spatial extents of the study area, deciding an appropriate working projection, and assembling the various spatial data to be used in the study in digital form, properly registered so that the spatial components overlap correctly.

The most important source of data was a recent geological map published by Bailes and Galley (1992). The paper map was digitized, transformed from table coordinates into a working projection, converted from vector into raster mode, and simplified by grouping some

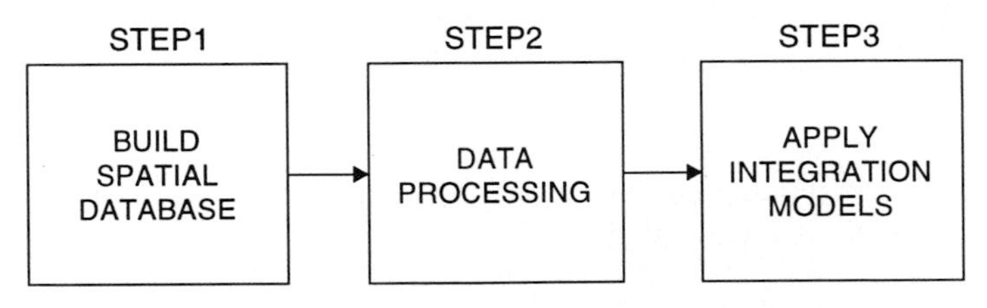

FIG. 1-2. Mineral potential mapping with a GIS as a 3-step process.

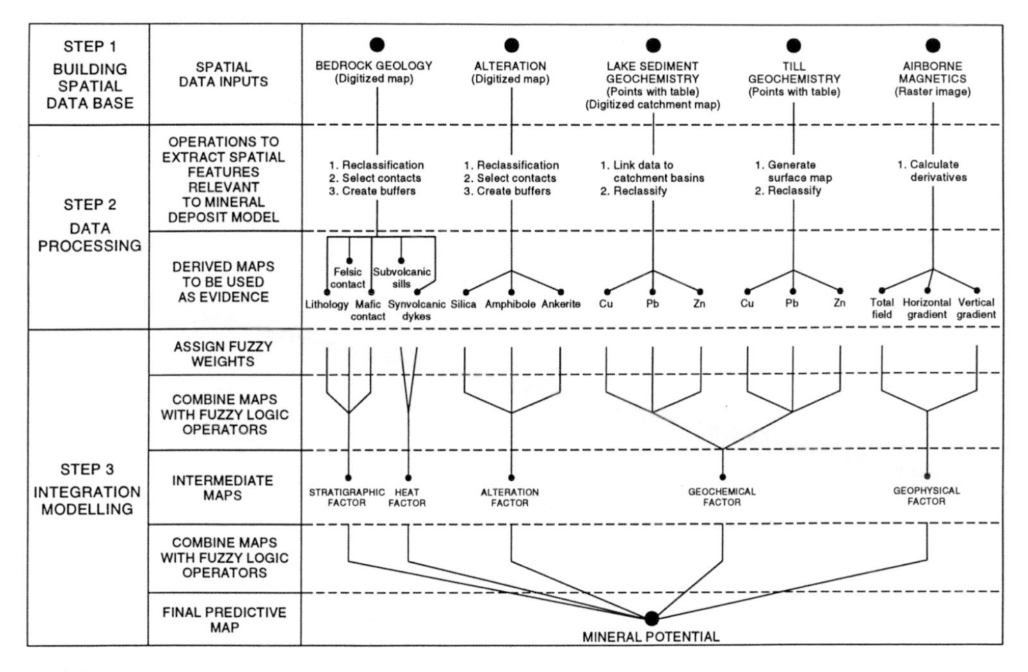

FIG. 1-3. Flowchart for the 3-step mineral potential study of part of the Snow Lake greenstone belt, Manitoba. The three steps in the study are: 1) A database phase of assembling the available spatial data in a GIS; 2) Extracting and enhancing those features of the primary datasets important for predicting volcanogenic massive sulphide deposits; and 3) Integrating the spatial evidence together using modelling to predict mineral potential. The final result is a map showing regions ranked according to favourability.

units together, see Figure 1-4. The alteration map was also digitized from the same source, and subjected to a similar treatment, Figure 1-5. Geochemical data were available from a regional survey of lake sediments. Samples taken from the many lakes in the area were analyzed for a variety of metallic elements. The data were obtained in digital form as a table, one row per sample, with columns containing spatial and non-spatial attributes. The spatial locations were recorded as (latitude, longitude) coordinates, and the chemical attributes as numerical fields. Each sample was treated as being representative of the local drainage catchment basin in which it occurs, so the data table was applied to basins instead of to the sample points, see Figure 1-6. The catchment basins were digitized from a topographic basemap, as were the drainage features of the area. Aeromagnetic data were brought into the database in a gridded raster format, and converted to the working projection, Figure 1-7. These geophysical data had already undergone several processing steps before being obtained as a digital file for the project. The original flight-line data had been transformed from a series of point measurements along lines and interpolated on to a regular grid. Likewise, the mineral deposit and geochemical sample locations were already in digital form when they were obtained.

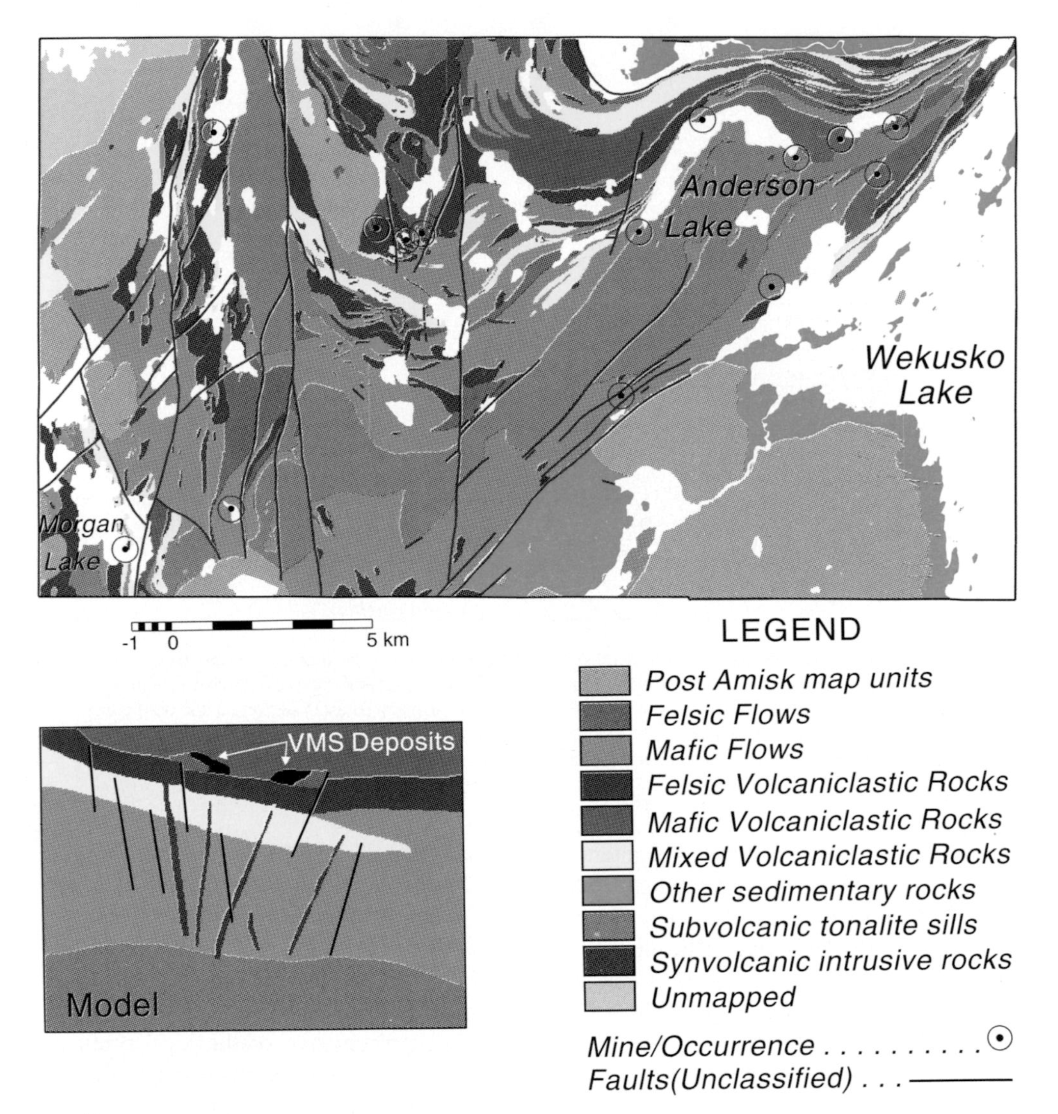

FIG. 1-4. Simplified geology of the Morgan-Chisel-Anderson lakes area, digitized and generalized from a 1: 15 000 scale map by Bailes and Galley (1992). A thick sequence of submarine volcanic rocks (green and red units) are underlain by large tonalite sills (orange), believed to have acted as a heat source for a hydrothermal system. Fracturing and intrusion of dykes (purple) helped to focus the transport of metal-rich fluids during various phases of submarine volcanism. The small model diagram shows a typical cross-section, with sulphide deposits often associated with felsic flows, being deposited at or close to the paleoseafloor. The points show the known VMS deposits, some of which are mines, from Fedikow et al. (1989).

Step 2

The second step was to process the input layers to extract the evidence critical to the prediction of VMS deposits. The geological map was generalized into a smaller number of map units or classes, preserving the information believed to be significant for mineral prediction. The contact between felsic flows (rhyolites) and volcanoclastic units was selected from the geological map and dilated, or buffered, to produce a proximity map, because some

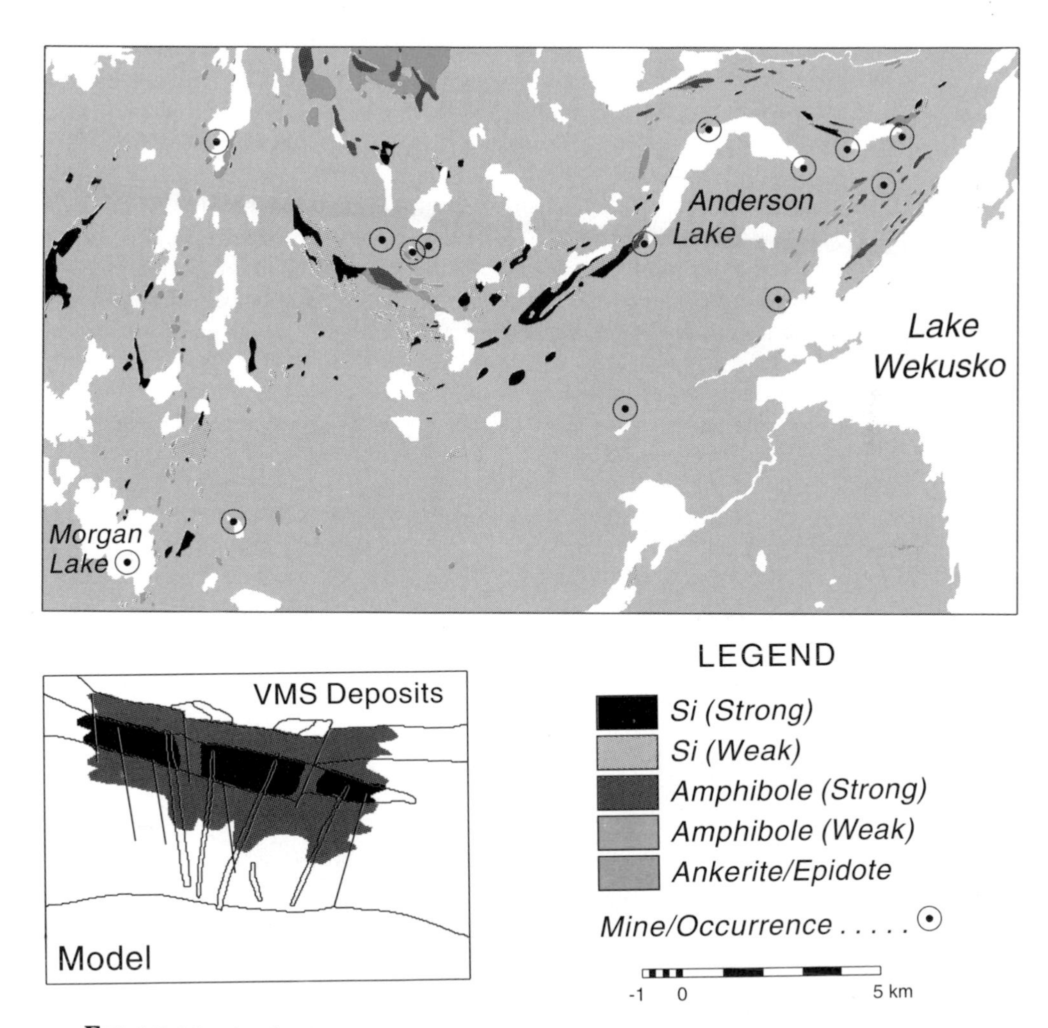

FIG. 1-5. Map showing the extent of alteration zones, also digitized from the Bailes and Galley (1992) map. Large zones of semi-conformable alteration, mainly silicification and amphibole alteration. Alteration minerals indicate the occurrence of past chemical reactions between volcanic rocks and hydrothermal fluids.

VMS deposits show a spatial association with this type of contact. Likewise, proximity maps were derived from the geological data showing distance to dykes and sills, used as evidence of proximity to a heat source, as shown in Figure 1-8. The different zones of alteration were separated and buffered, allowing the presence, and proximity to, alteration of various types to be modelled. The geochemical data from elements believed to be important indicators of base-metal deposits were classified on a scale designed to accentuate and enhance those areas with anomalously high values, and converted from the data table to catchment basin maps, Figure 1-6. The magnetic data was enhanced by choosing a suitable classification scheme, and a derivative map was constructed, showing the areas where rapid vertical change occurs in the local magnetic field, Figure 1-7.

Step 3

The third step consisted of combining together the various maps that provide evidence for VMS deposits. This was carried out in stages, with the ultimate product being a predictive map showing the relative favourability, or VMS potential. The combination process involves the weighting and fusion of evidence and can be carried out in a number of different ways, as discussed at some length in Chapter 9. In this case, fuzzy logic was employed as the

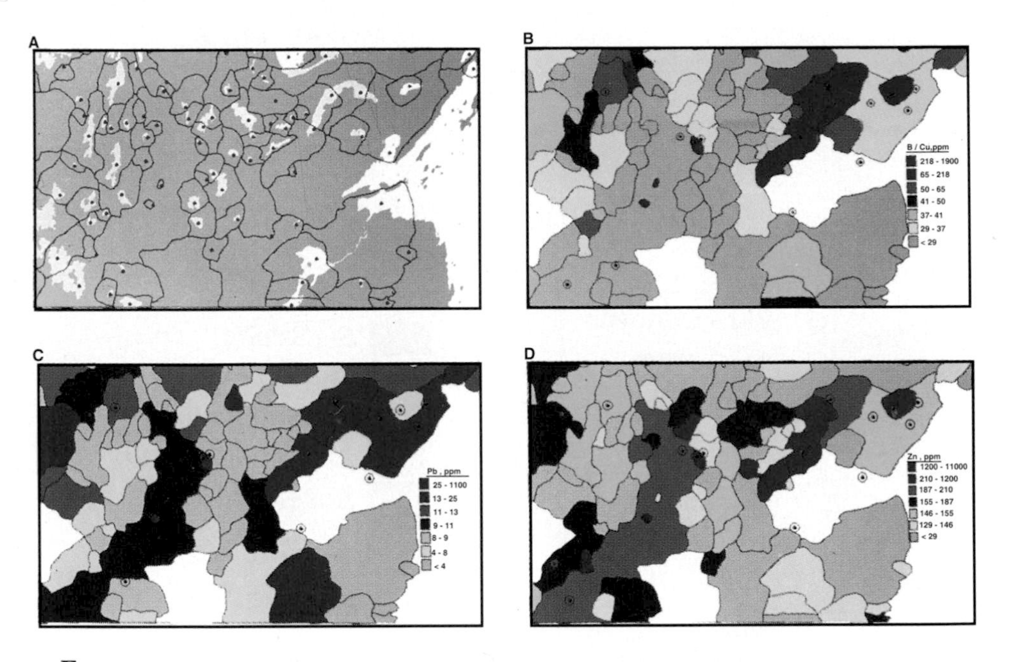

FIG. 1-6. Geochemical maps based on element concentrations in samples of lake sediment. **A.** Sample locations, and boundaries of digitized catchments used to make element maps. **B.** Basins classified according to the copper (Cu) content. **C.** Lead (Pb) content. **D.** Zinc (Zn) content. The dots enclosed by circles are the deposit locations.

A

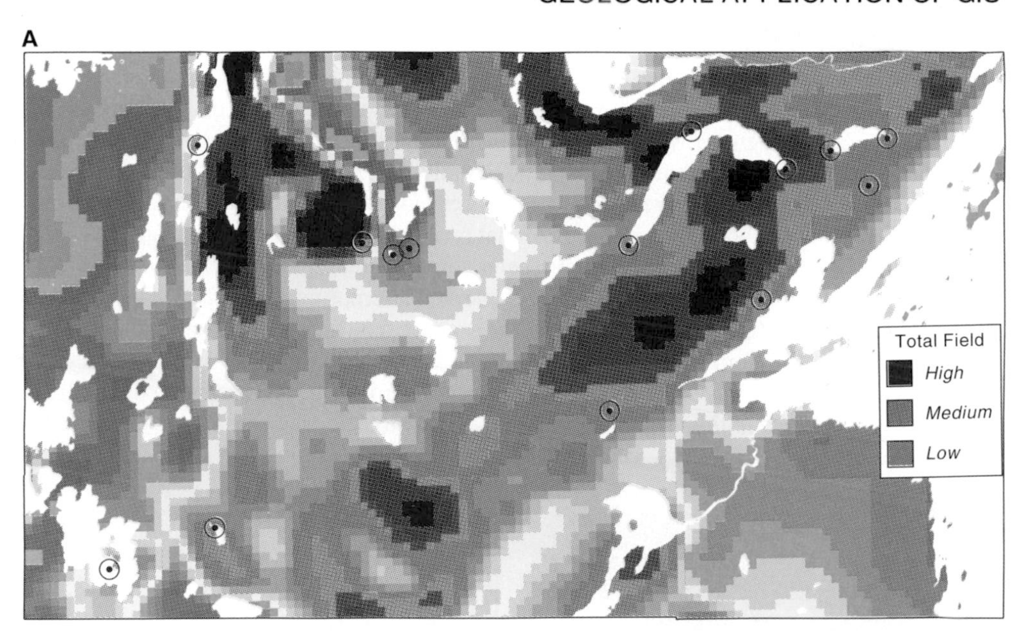

B

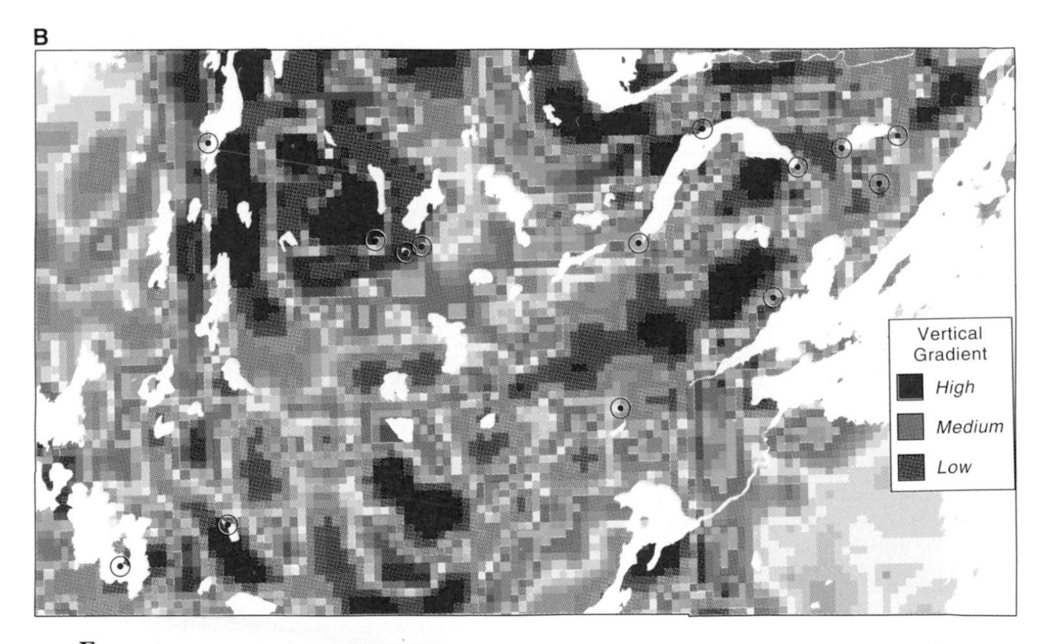

FIG. 1-7. Geophysical maps, from an airborne magnetic survey. The original data were measured at points along flight lines, but now have been interpolated on to a grid and coloured by intensity. **A.** Total field data. **B.** Vertical gradient data. The dots are locations of deposits.

method to represent the evidence and to propagate the evidence from input maps, via intermediate factor maps, illustrated in Figure 1-9, to the ultimate output map, Figure 1-10. Fuzzy membership functions were assigned subjectively to each type of evidence, thereby controlling the relative weighting of data. Alternatively, the evidence could have been weighted statistically (see Chapter 9), using the observed associations of the evidence maps with the known mineral occurrences. The potential map represents only one of many possible solutions, because the evidence could have been selected, manipulated and weighted in a variety of alternative ways. The ability to generate alternative maps by changing the assumptions is one of the important aspects of this approach. In this situation, the potential map was calculated without using the locations of the known deposits directly in the analysis. As can be seen in Figure 1-10, the deposits nearly all occur in zones of elevated potential, as might be expected. In addition, a number of highly favourable zones occur where no deposits are known at present.

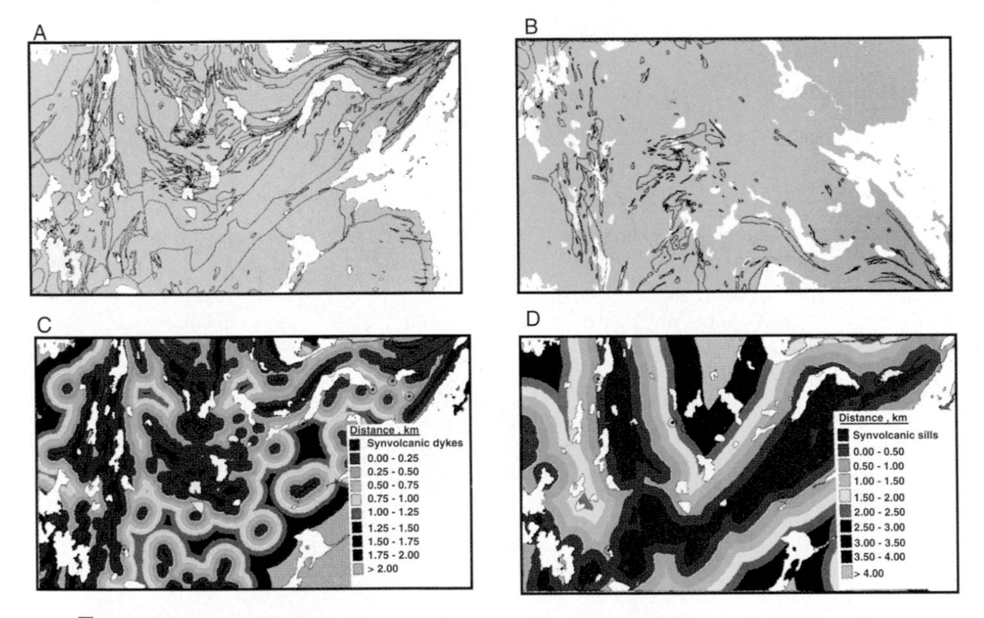

FIG. 1-8. Maps showing processing steps for some of the geological data. **A.** Geological map showing boundaries of map units. **B.** Boundaries of dykes, selected from the geological contacts. **C.** Dykes buffered (dilated) to show distance to nearest dyke. **D.** Distance to nearest tonalite sill, also derived from the geological map by selection of contacts and buffering. Maps showing distance to dykes and sills are employed as evidence of proximity to a heat source.

FIG. 1-9 (opposite). Maps of three of the five intermediate factors generated as part of the integration modelling process. **A.** The stratigraphic factor, based on the presence of favourable lithologies and proximity to favourable contacts. **B.** The heat factor, combining the effects of proximity to dykes and tonalite sills. **C.** The alteration factor, due to the combined effects of the presence, and proximity to, alteration zones. The numbers in the legend refer to the relative favourability on a scale between 0 and 1.

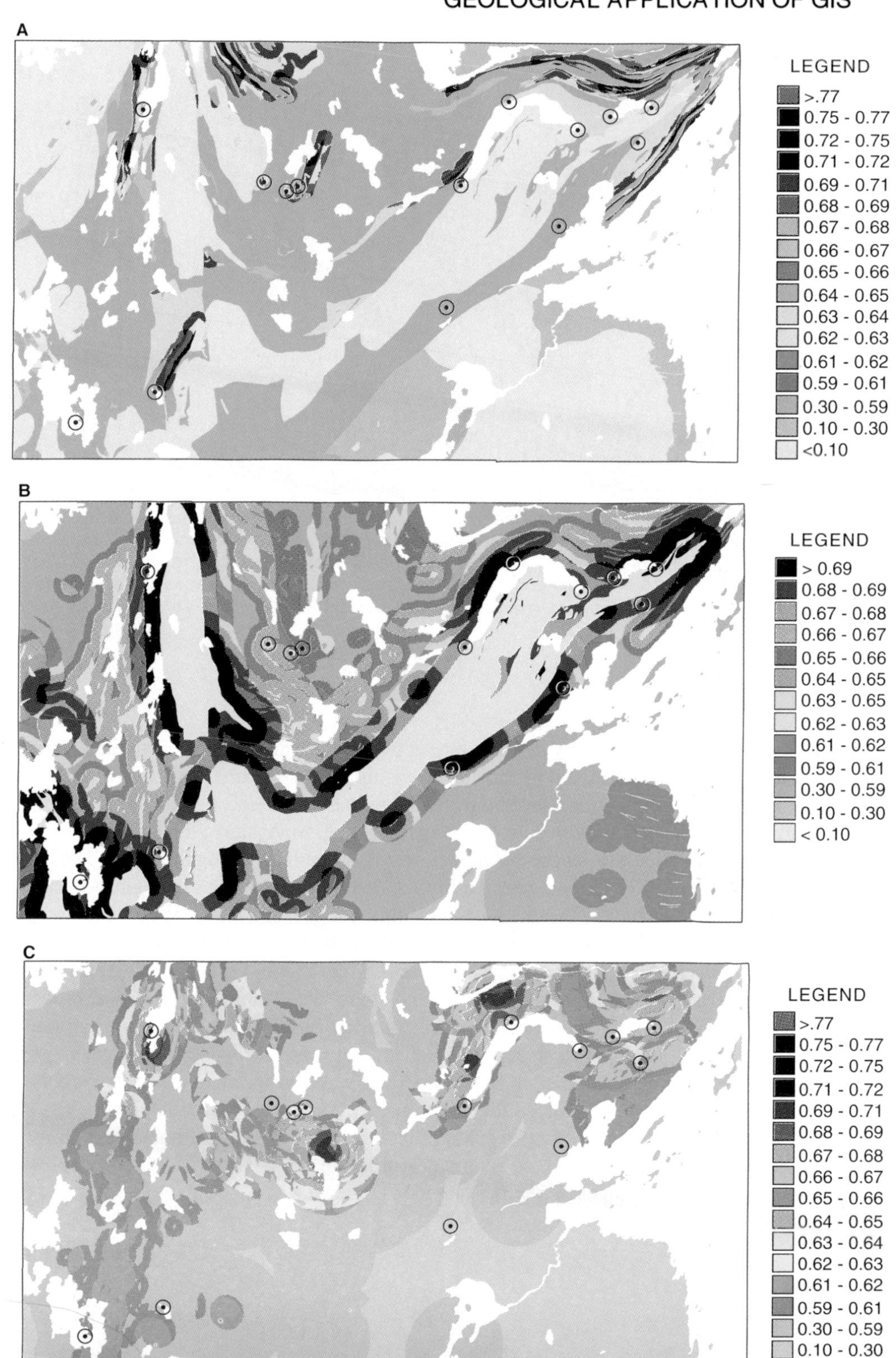
A
LEGEND
>.77
0.75 - 0.77
0.72 - 0.75
0.71 - 0.72
0.69 - 0.71
0.68 - 0.69
0.67 - 0.68
0.66 - 0.67
0.65 - 0.66
0.64 - 0.65
0.63 - 0.64
0.62 - 0.63
0.61 - 0.62
0.59 - 0.61
0.30 - 0.59
0.10 - 0.30
<0.10
B
LEGEND
> 0.69
0.68 - 0.69
0.67 - 0.68
0.66 - 0.67
0.65 - 0.66
0.64 - 0.65
0.63 - 0.65
0.62 - 0.63
0.61 - 0.62
0.59 - 0.61
0.30 - 0.59
0.10 - 0.30
< 0.10
C
LEGEND
>.77
0.75 - 0.77
0.72 - 0.75
0.71 - 0.72
0.69 - 0.71
0.68 - 0.69
0.67 - 0.68
0.66 - 0.67
0.65 - 0.66
0.64 - 0.65
0.63 - 0.64
0.62 - 0.63
0.61 - 0.62
0.59 - 0.61
0.30 - 0.59
0.10 - 0.30

FIG. 1-10. Map showing the favourability for volcanogenic massive sulphide deposits, derived by combining evidence from the five intermediate factors. This is the mineral potential map that can be used for exploration decision-making, for example to decide priority areas for additional detailed surveys. The known deposits (black dots) occur in highly favourable zones, as expected. A new previously unknown deposit (ca. 1 m tonnes) was discovered in 1994 within the large favourable zone that intersects the top of the map.

Not shown here is the ability to make interactive spatial queries of the data and results at specific locations as determined by the cursor position. This greatly helps to understand the reason why particular locations have high VMS potential. Also not shown here is the kind of specialized output, such as perspective displays, that help to visualize the results.

It is emphasised that the conceptual model was essential for guiding the study. The choice of data, the kinds of information extracted from it, and the assignment of weights to evidence were all dependent on the deposit model. For example, vegetation maps, airborne radiometric maps and Landsat imagery were not employed, because they were not directly usable as VMS evidence. Unfortunately, some evidence, defined in the model as being valuable, such as airborne EM data, was also not available for the study. The extraction of critical evidence, employing transformation methods to select contacts, buffer lines, calculate derivatives, interpolate from points to surfaces, generalize maps by reclassification, was greatly influenced by the conceptual model. Most importantly, the assignment of fuzzy weights and the choice of fuzzy logic operations for combining evidence were due to the deposit model.

In short, analysis and modelling of spatial data in a GIS is not simply a matter of throwing the data layers into a "black box" computer program. A conceptual model, preferably formulated in the early stages of a study, is used to guide the various stages of GIS processing.

REFERENCES

Bailes, A.H. and Galley, A.G., 1989, Geological setting of and hydrothermal alteration below the Chisel Lake massive Zn-Cu sulphide deposit: *Manitoba Energy and Mines, Minerals Division, Report of Field Activities*, p. 31-37.

Bailes, A.H. and Galley, A., 1992, Chisel-Anderson-Morgan Lakes Preliminary Map 1992 S-1: *Manitoba Energy and Mines*.

Cressie, N.A.C., 1991, *Statistics for Spatial Data*: John Wiley & Sons, Inc., New York, 900 p.

Davis, J.C., 1986, *Statistics and Data Analysis in Geology*: Second Edition, John Wiley and Sons, New York, 646 p.

Fabbri, A.G., 1984, *Image Processing of Geological Data*: Van Nostrand Reinhold, New York, 244 p.

Fedikow, M.A., Ostry, G., Ferreira, K.J. and Galley, A.G., 1989, Mineral deposits and occurrences in the File Lake area, NTS 63K/16: *Manitoba Energy and Mines, Mineral Deposit Series*, 277 p.

Hodgson, C.J., 1990, Uses (and abuses) of ore deposit models in mineral exploration: *Geoscience Canada*, v. 17 (2), p. 79-89.

Isaaks, E.H. and Srivastava, R.M., 1989, *Applied Geostatistics*: Oxford University Press, New York-Oxford, 561 p.

Lydon, J.W., 1984, Ore deposit models - 8: Volcanogenic massive sulphide deposits, Part 1: A descriptive model: *Geoscience Canada*, v. 11, p. 195-202.

Lydon, J.W., 1988, Ore deposit models - 14. Volcanogenic massive sulphide deposits, Part 2: genetic models: *Geoscience Canada*: v. 15, p. 43-65.

Pflug, R. and Harbaugh, J.W. (eds.), 1992, *Computer Graphics in Geology: Three-Dimensional Computer Graphics in Modeling Geologic Structures and Simulating Geologic Processes*: Lecture Notes in Earth Sciences, Springer-Verlag, Berlin, v. 12, 298 p.

Raper, J., (editor), 1989, *Three Dimensional Applications in Geographical Information Systems*: Taylor and Francis, London-New York, 190 p.

Reddy, R.K., Bonham-Carter, G.F. and Galley, A.G., 1992, Developing a geographic expert system for regional mapping of volcanogenic massive sulphide (VMS) deposit potential: *Nonrenewable Resources*, v.1(2), p. 112-124.

Serra, J., 1982, *Image Analysis and Mathematical Morphology*: Academic Press, London-New York, 610 p.

Turner, A.K., (editor), 1992, *Three-Dimensional Modeling with Geographic Information Systems*: Kluwer Academic Publishers, Dordrecht, 443 p.

Unwin, D., 1981, *Introductory Spatial Analysis*: Methuen, London-New York, 212 p.

CHAPTER 2

Spatial Data Models

INTRODUCTION

Computers and GIS cannot directly be applied to the real world: a data gathering step comes first. Digital computers operate on numbers and characters held internally as binary digits. The real-world phenomenon of interest must, therefore, be represented in symbolic form. The abstraction process of representing the geology, structure, geophysical or another property of Earth's surface in a computer-accessible form involves the use of **symbolic models**.

Models are simplifications of reality. A geological map is a symbolic model, because it is a simplified representation of part of the real world, as seen through the eyes of a field geologist. The components of the model are spatial objects, approximating spatial entities of the real world; they are represented on the map by graphical symbols. The key for understanding the symbols is recorded, at least in part, in the map legend. On the other hand, a computer file containing a digital representation of the geological map is also a symbolic model of the same piece of the real world, with a further process of abstraction from graphical symbols to digital codes. The key for communicating the meaning of the digital data is an essential part of the dataset itself. No two geologists will produce identical maps of the same area. They observe different objects, the objects that they see convey different meanings, and the interpretation of the underlying geology depends on their experience and philosophy. In turn, the digital representation of the map involves choices concerning the objects to be represented, and how they are to be organized. The value of a spatial dataset is not simply whether the digital format is understood by the software. Unless the rules for measurement and organization of data are clearly established, the data are of limited use, except possibly to the individual or organization who generated them.

Take a more straightforward example of a digital computer file containing a grid of elevation values, often called a DEM or digital elevation model. A DEM is also an example of a symbolic model of the real ground surface. The cells of the grid are the spatial objects, whose values are symbolized by numbers in the data file. In this case, the data definition and data organization requires less interpretation than a geological map, yet unless the organizational scheme is known, the data are worthless. Another example is a digital dataset containing well data from drill holes in a coalfield, which can also be regarded as a symbolic model of a real coalfield. The spatial entities are the sample locations in the wells at which formation tops, and other attributes, are recorded. In this case, the definition and organization of the nonspatial and spatial attributes are critical for data interpretation.

The importance of the conceptual step in organizing data can be illustrated by considering the interpretation of an aerial photograph. Suppose that a structural geologist, a geomorphologist and a geobotanist all wish to make a map from the same scene. The structural geologist records lineaments, bedding attitudes and fold axes; the geomorphologist focuses on raised beaches and the distribution of outwash gravels; the geobotanist maps vegetation units, and relates their distribution to lithology. Clearly, the conceptual model in the minds of these individuals is very different, even if they are looking at the same area. A skilled photointerpreter may in fact record all of the above features as a map. However, if the map is now to be recorded digitally, the manner of defining and organizing the data will differ depending on the purpose to which the data is to be put. Thus the question of data content and organization arises at both the primary step of mapping, and at the secondary step of transforming a map into a digital product.

The process of defining and organizing data about the real world into a consistent digital dataset that is useful and reveals information is called **data modelling,** see Peuquet (1984), Goodchild (1992) and other authors. The logical organization of data according to a scheme is known as a **data model**. Peuquet recognizes several steps in the data modelling process. Real world data must be described in terms of a **data model**, then a **data structure** must be chosen to represent the data model, and finally a **file format** must be selected that is suitable for that data structure. For example, spatial data about the height of the ground surface above sea level might be organized according to a raster (grid) model, the raster in turn is organized in a run-length encoded data structure, and the data written on a digital storage device in a file format such as a *.CUT file. Alternatively, the ground surface might be organized according to a vector model, expressed as polygons bounded by contour lines; the data being arranged in a topological structure called POLYVRT (see Chapter 3) and written on a storage device in a DLG (Digital Line Graph) file format. The TIN (Triangulated Irregular Network) model is yet another data model that could be equally well used for elevation data. There is, therefore, no unique way of organizing data, and numerous alternative representations are possible. Each spatial data model has several alternative data structures, and each structure can be stored digitally with many file formats. Simply knowing the file format is, therefore, only part of the story, and unless the data model and data structure are understood, the user cannot utilize the data properly.

Data can be defined as verifiable facts about the real world. **Information** is data organized to reveal patterns, and to facilitate search. Spatial information is difficult to extract from spatial data, unless the data are organized primarily by spatial attributes. Also, because of the nature of digital computers, data items must be identified as discrete spatial objects in order that they can be manipulated digitally. Geographical space must, therefore, be represented discretely, because spatially continuous fields, such as electrical or temperature fields, cannot be stored digitally as continua. All spatial data models make use of discrete spatial data objects, such as points, lines, areas, volumes and surfaces. Spatial objects are characterized by attributes that are both spatial and nonspatial, and the digital description of objects and their attributes comprise spatial datasets.

Spatial data can be organized in different ways, depending on the way they are collected, how they are stored, the amount of interpretation added to them, and the purpose to which they are put. The **vector** and **raster** models are the commonly recognized schemes for spatial data organization in a GIS. The vector model divides the world into points, lines, and areas bounded by lines, whereas the raster model uses cells or pixels as spatial units.

A **database** is a collection of inter-related data and everything that is needed to maintain and use it (Bisland, 1989). A **database management system (DBMS)** is a collection of software for storing, editing and retrieving data in a database. Most GIS manage spatial and nonspatial data separately. Some GIS use an internal DBMS to manage nonspatial data, others provide linkages to external DBMS.

This chapter compares the two principal spatial data models, the vector and raster models. In addition, the relational data model, widely used in database management systems, is discussed principally as it pertains to the organization of nonspatial attribute data. The details of specific data structures for the vector and raster models are described in Chapter 3. File formats for spatial data are not treated in this book, and readers are referred to GIS manuals for this information.The chapter begins with a classification and discussion of spatial objects, the building blocks for spatial data models.

SPATIAL OBJECTS

The real world exhibits properties that are either spatially continuous, like temperature, or discontinuous like the states of matter--solid, liquid and gas. Thus, gravity measured from place-to-place is a continuous **field** variable, whereas many rock bodies are discontinuous, having sharp boundaries at faults, intrusive contacts and unconformities. Where discrete spatial entities occur in the real world, they can be treated as natural **spatial objects,** generally irregular in shape, in a data model. For variables that are, in reality, spatially continuous, space must be divided into discrete spatial objects that can either be irregular or regular in shape. A lattice of sampling points (or equivalently a grid of square cells or pixels) is a collection of regularly-shaped imposed objects. Continuous fields can also be divided into naturally-shaped irregular objects bounded by contour lines.

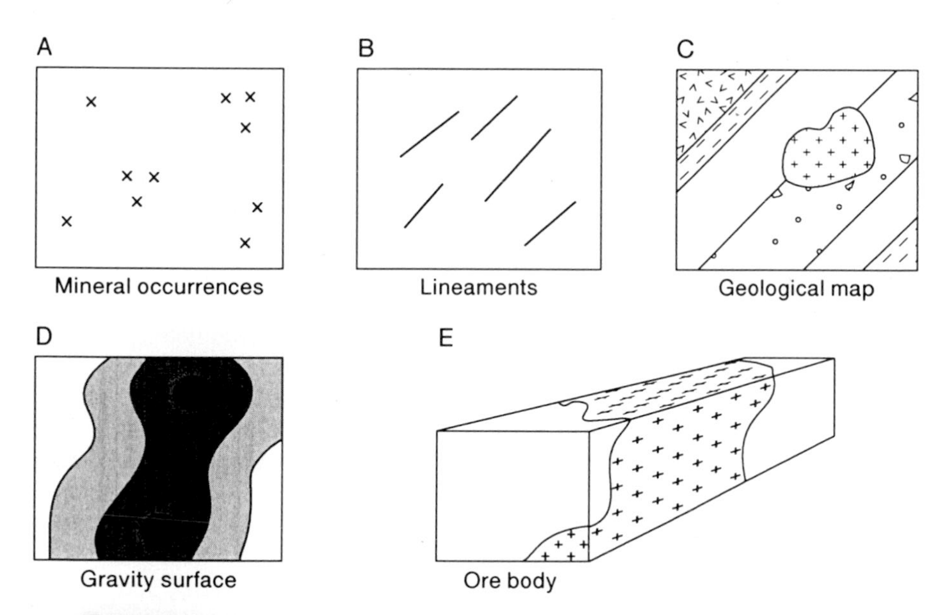

FIG. 2-1. Geological examples of spatial objects classified according to their spatial dimensions. **A.** Mineral occurrences can be regarded as points (0-D) at small map scales. **B.** Lineaments are lines (1-D). **C.** Areal patterns on a geological map are areas (2-D). **D.** A spatially continuous gravity surface is 2.5D, as depicted on a contour map. **E.** An ore body is 3-D.

SPATIAL DATA MODELS

Spatial objects can be grouped into points, lines, areas, surfaces and volumes, varying in spatial dimension as illustrated in Figure 2-1. Points, lines and areas have zero, one and two spatial dimensions, respectively. Volumes are three-dimensional, and so are the complete bounding surfaces of a volume object. Some surfaces are called 2.5-dimensional, because they are restricted to only one value at every horizontal location. Examples include topographic elevation of the ground surface (with the exception of vertical or overhanging cliffs), the elevation of the upper surface of a relatively undistorted coal-seam, or the conceptual iso-surface of a continuous field variable, like the magnetic field, as measured at or near the Earth's surface. The complete bounding surfaces of rock bodies, or the upper surfaces of beds in overturned folds or reverse faults require full three-dimensional representation, because at any (x,y) location, the position of the surface can occur at more than one value of z. Line networks can also be regarded as spatial objects, although this type of object seems to be of limited interest to geologists, being applied mainly to communications networks, such as rail, road, telephone, electrical power lines and other utilities.

In reality, many spatial objects in the real world have fractional spatial dimensions. When a natural line object is measured at progressively larger scales, its length increases as greater detail becomes apparent. The classic example is the length of a coastline that gets longer as the length of the measuring stick decreases, Mandelbrot (1983). If the logarithm of the total length (L) is plotted against the logarithm of the length of the measuring stick (d), a straight line relationship is found which can be modelled as

$$\log L(d) = a + (1 - D) \log d \qquad (2\text{-}1)$$

The slope of the line is $(1 - D)$, where D is the known as the **fractal** dimension, and a is a constant. Values of D are typically about 1.25, although the deeply incised coast of Norway has a value of $D = 1.52$ (Feder, 1988). Similarly, the fractal dimension of a natural surface is found to be between 2 and 3.

Many natural objects have the property of looking similar when enlarged, of being self-similar under changes of scale, as implied by this model. Fractal geometry provides a system for describing complex natural objects, which can only be approximated with difficulty using ordinary Euclidean geometry. A variety of authors have used fractal geometry for characterizing rock pores, caves, fragments in crushed rock, meander loops, coral reefs, earthquakes, floods, oil fields, ore bodies and others (Goodings and Middleton, 1991). In this book, however, spatial objects are treated as having the conventional Euclidean dimensions, i.e. 0, 1, 2, and 3, with 2.5 being applied to single-valued surfaces.

Besides their dimensionality, spatial objects can also be grouped according to whether they are "natural" or "imposed". **Natural spatial objects** correspond to discrete spatial entities recognizable in the real world, like a river or an orebody. **Imposed spatial objects** are artificial or man-made entities, like a property boundary, or a pixel. Unwin (1981) has written clearly about a typology of spatial objects in his book on spatial analysis. Some geological examples of spatial objects are shown in Figure 2-2 and Table 2-1, and are discussed in the following paragraphs.

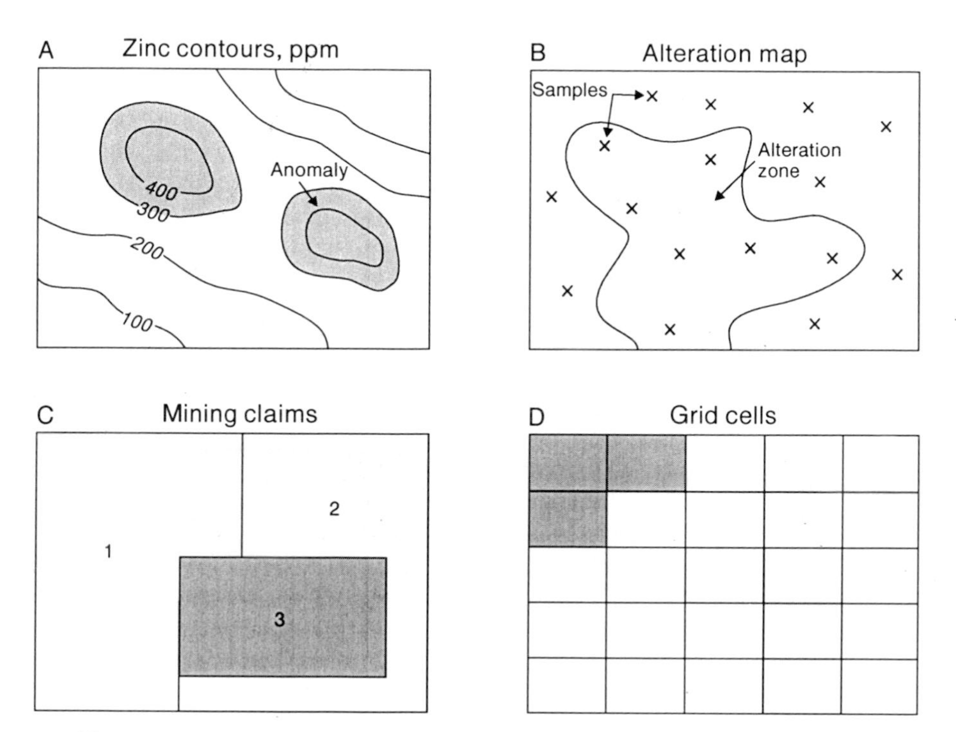

FIG. 2-2. Geological examples of spatial objects classified according to whether they are natural or imposed, and whether they are regular or irregular. **A.** The zinc anomalies are definition-limited, because their size and shape changes as the cutoff is changed. **B.** The alteration pattern is sampling-limited because the size and shape is determined by the sampling density. **C.** A map of mining claims is made up of irregular imposed spatial objects. **D.** A grid map consisting of regular imposed area objects.

Sampling-Limited Spatial Objects

A dyke is a naturally-occurring spatial entity which on a map is usually represented as an area object. Raper (1989) calls such objects sampling-limited, because information about their shape and extent is limited only by the amount of available sampling, not by their definition. Examples of sampling-limited objects are: the surface-trace of an unconformity (line), the top of a coal-seam (surface), and an oil-pool bounded by water below and cap-rock above (volume).

Definition-Limited Spatial Objects

A metallic orebody is a good example of a definition limited volume object. The size and shape of the orebody is defined by the cutoff grade. If the cutoff is reduced, the orebody definition changes and the orebody becomes larger. Seismic epicentres are examples of point objects defined by the sensitivity of the seismometer or by an intensity threshold that

Table 2-1. Some geological examples of spatial objects, organized by type and by spatial dimension. Objects that are naturally-occurring can be organized according to whether they are sampling limited or definition limited. Objects that are imposed are either regular (like the pixels in a raster image), or irregular, like the polygons in a Voronoi diagram. Naturally occurring spatial objects of 1 or more dimensions are virtually always irregular in shape (exception: columnar basalt).

TYPE		SPATIAL DIMENSION 0-D POINT	1-D LINE	2-D AREA	2.5-D SURFACE*	3-D VOLUME
NATURALLY OCCURRING	SAMPLING LIMITED	lineament intersection	inferred contact	flood zone	top of coal seam in subsurface	salt dome
	DEFINITION LIMITED	seismic epicentre	contour line	geochemical anomaly	thermocline	ore body
IMPOSED	IRREGULAR	soil sample locations	drill hole in vertical cross-section	mining claim	non-planar cross-section	3-D excavation
	REGULAR	sample locations on grid	flight line traverse	grid cell	planar cross-section	voxel

*Single-valued surfaces, such that any (x,y) location has only a single value of z. A folded surface would therefore not qualify, because multiple values of z occur at given (x,y) locations.

distinguishes true seismic events from noise. An elevation contour is an example of a definition-limited line object and a geochemical anomaly is typical of a definition-limited area object, because as the upper limit of the "background" is changed, the size and shape of the anomaly changes. Definition-limited objects are more common than sampling-limited objects, although in practice many objects are limited by both definition and sampling.

Irregular Imposed Spatial Objects

An administrative zone and a mining claim are irregularly-shaped area objects imposed by man in subdividing a region. They need bear no relationship to naturally occurring spatial entities. Triangulated Irregular Networks (TIN) , used to subdivide surfaces into triangular facets, and Thiessen polygons, also used to partition space into a specialized type of mosaic, also fall into this category. Irregularly distributed sample locations are irregular imposed point objects; a road is an example of an imposed line object and an underground tunnel is an example of a volume object of this type.

Regular Imposed Spatial Objects

Any regular subdivision of space produces regularly shaped area objects, such as square pixels in a raster image, or rasters based on hexagons or equilateral triangles. A straight line of section and a planar cross section are examples of line and surface objects of this type. A voxel or cubic volume element is a typical three dimensional example.

A key difference between the raster and vector data models is that the raster model uses regular imposed spatial objects, whereas the vector model uses irregular spatial objects that can be either natural or imposed. The vector model employs a boundary representation of area objects, because each object is different from the next, whereas the pixels or cells in a raster

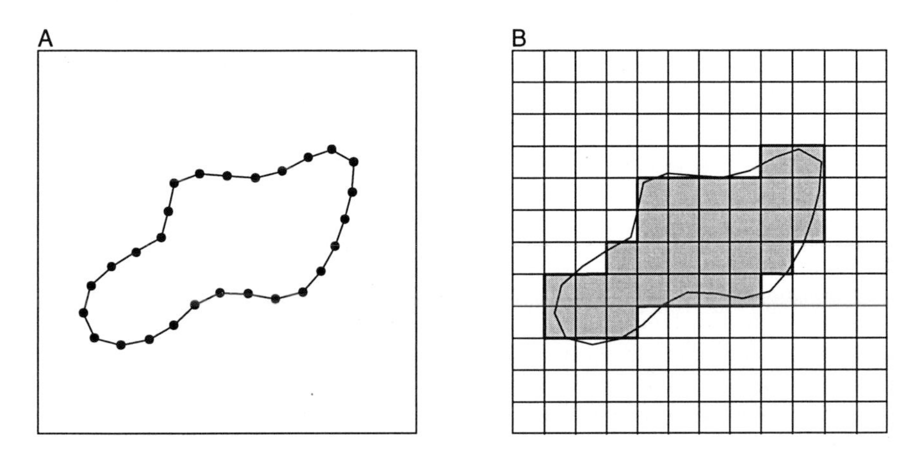

FIG. 2-3. A. Boundary representation of an area objects in a vector model. **B.** Spatial enumeration of the same area object in a raster model.

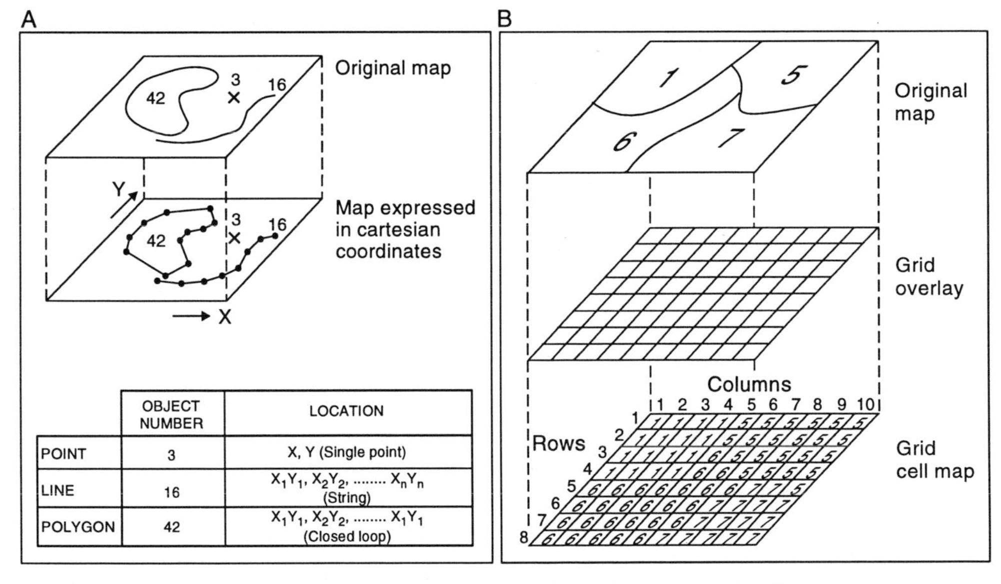

	OBJECT NUMBER	LOCATION
POINT	3	X, Y (Single point)
LINE	16	$X_1Y_1, X_2Y_2, \ldots\ldots X_nY_n$ (String)
POLYGON	42	$X_1Y_1, X_2Y_2, \ldots\ldots X_1Y_1$ (Closed loop)

FIG. 2-4. A. Digital representation of spatial objects on a map in a vector model (spaghetti model). **B.** Representation of area objects in a raster model.

are constant in size and shape and do not require individual boundary definitions, see Figures 2-3 and 2-4. In either model, a spatial object is assumed to have properties that are homogeneous, so that a single object is described by attributes that are constant for the whole object.

Spatial objects may also be simple or compound. A compound object is composed of two or more simple objects. For example, a system of interconnected lakes may be regarded as a compound object, composed of a collection of lakes that are simple objects. An irregular object such as a polygon on a map may be a simple object in a vector representation. However, in raster representation, the same polygon may be composed of several pixels, so the object is now compound, composed of simple objects (pixels) having attributes in common.

RASTER AND VECTOR SPATIAL DATA MODELS

The raster model is particularly well suited for subdividing spatially continuous variables, like the Earth's gravity field, for digital computers. The raster can be represented as a rectangular matrix of numbers, see Figure 2-4B, similar to a two-dimensional array in FORTRAN or BASIC, and can be stored on a disk with a simple file structure, with straightforward addressing of pixels by sequence in the file. Digital scanning devices and video-digitizers produce data in raster form. Many output devices are based on rasters, such as video display monitors, line printers and inkjet plotters. The raster model is used for digital images; digital image processing and analysis are well-established disciplines with a broad range of applications in remote sensing, medical imaging, computer vision and other areas.

On the other hand, the vector model is well-suited for representing maps. Points, lines, polygons and symbols on maps are difficult to capture with fidelity in a raster without making the pixels very small, resulting in high storage costs. In vector mode, the lines surrounding polygonal areas are made by linking sequences of points, or **vertices**, each vertex being an ordered pair of spatial coordinates, hence the name vector, see Figure 2-4A. If vertices are placed very close together, and the coordinates are expressed in numbers with sufficient precision, curved lines can be represented precisely. The vector model fits cartographic needs well. Digitizers that use line following principles produce digital data in vector form. Digital pen plotters are relatively inexpensive output devices that produce drawings and maps with smooth curves. However, the data structures needed for storage of vector data are considerably more complex than their raster counterparts and algorithms for overlaying maps in vector form are more involved. Display of vector data on video monitors and raster plotting devices requires a conversion to raster mode.

Raster and vector models can be differentiated on the basis of how they represent space, as well as by the type of spatial objects they use. Whereas the raster model employs areal or volumetric **enumeration**, the vector model uses boundaries or surfaces to represent areas or volumes. The raster describes areal or volume elements directly; the vector model stores the boundaries of objects, and uses a labelling scheme to keep track of their attributes. This labelling scheme may involve the notion of **topological** attributes, that are spatial attributes that define adjacency and containment relationships between spatial objects.

Raster Model

One of the advantages of the raster model is that spatial data of different types can be **overlaid** without the need for the complex geometric calculations required for overlaying different maps in the vector model. Each layer of grid cells in a raster model records a separate attribute. The cells are constant in size, and are generally square, although rectangles, hexagons and equilateral triangles have also been used. The locations of cells are addressed by row and column number. Spatial coordinates are not usually explicitly stored for each cell, because the storage order does this implicitly. Information about the number of rows and columns, plus the geographic location of the origin are saved with each layer.

The spatial **resolution** of a raster is the size of one of the pixels on the ground. At 100 m resolution, a square area 100 km on a side requires a raster with 1000 rows by 1000 columns, or one million pixels. At 10 m resolution, the same area requires 10,000 columns by 10,000 rows or one hundred million pixels. If one computer byte (requiring eight bits or binary digits of computer storage) is used per pixel (and this is usually a minimum, because a byte is limited to integer numbers between 0 and 255), the storage needed is 100 megabytes. This is a considerable amount of space for a single map or image. Because storage requirements increase geometrically with increasing resolution, raster models are often represented with data structures that conserve storage space by compression methods, as discussed in Chapter 3. In choosing the spatial resolution in a raster model, there is a trade-off between maximizing spatial fidelity and minimizing storage and processing costs.

In raster mode, points are represented as single pixels and lines by strings of connected pixels. This is often unsatisfactory, because the pixel size is too coarse to resolve closely-spaced objects. Having fixed the size of a pixel at the time of creating a raster, levels of detail that require greater resolution are lost. On geological maps, thin stratigraphic units like dykes or thin sedimentary formations may require a spatial resolution that makes a raster prohibitively large. Raster structures that offer data compression, such as run-length encoding or quadtrees (Chapter 3) can alleviate this problem to some extent.

Processing raster data is efficient for some tasks, such as neighbourhood query, operations such as spatial filtering that carry out calculations on a square window of adjacent pixels, and overlay operations for combining two or more images together. Raster processing is not so efficient for operations like finding all the adjacent pixels belonging to one compound object that are completely contained by pixels belonging to another compound object.

The raster organization is well suited for modelling spatial continua, particularly where an attribute shows a high degree of spatial variation, such as data on satellite images. The regular spacing of pixels in a lattice is ideal for calculating and representing spatial gradients.

Vector Model

The basic type of vector model has come to be known as the "**spaghetti model**", see Figure 2-5A. Points are represented as pairs of spatial coordinates, lines as strings of coordinate pairs, and areas as lines that form closed loops or polygons. If each point, line and polygon on a geological map is digitized with a line-following digitizer, the resulting strings

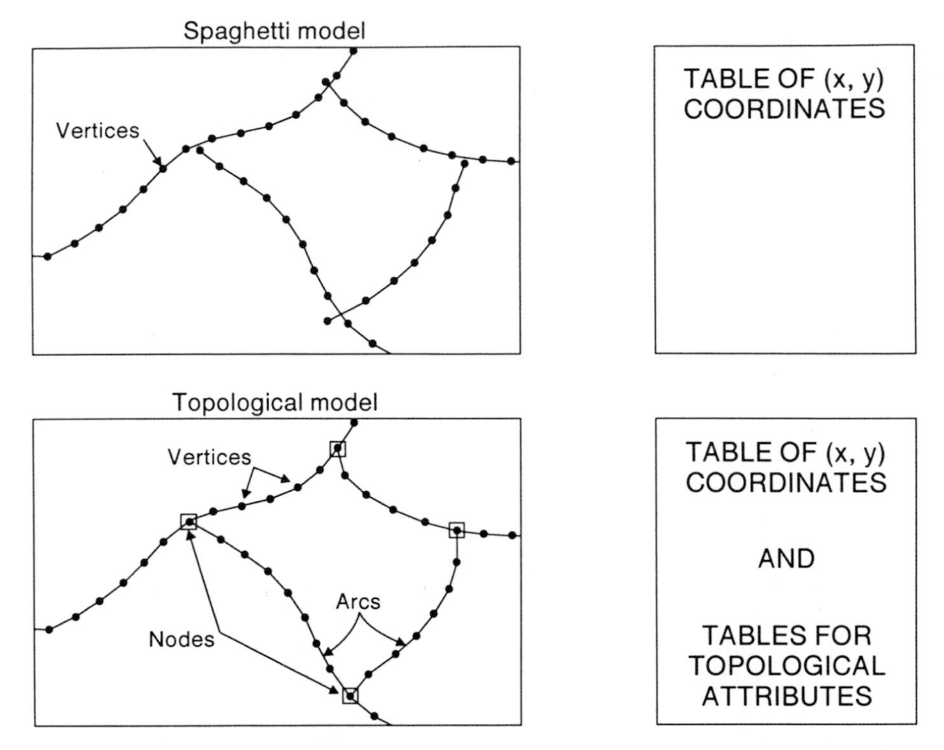

FIG. 2-5. A. In the spaghetti model, lines are defined by vertices whose spatial locations are recorded in tables of geometric coordinates. **B.** The topological model has additional information about the adjacency, containment and connectivity of arcs and nodes. Each arc is composed of one or more straight line segments defined by vertices.

of coordinate pairs can be stored in a relatively unstructured form. This straightforward model, and the equivalent data structure, is ideal for inexpensive graphical systems. The spatial objects can be regarded as graphical elements. Digital drawings can be readily scaled, transformed to various projections and output to pen-plotters or displayed on a video screen. Video display involves reconversion to raster mode, but this operation is fast and is carried out "on the fly".

If spatial objects are stored with attribute information, usually with spatial and non-spatial data in the same file, points can be plotted with different symbols, lines can be given different colours and weight, and polygons can be filled with patterns and colour, depending on the values of associated attributes. With even minimal database management of the spatial objects and their attributes, spatial objects can then be selected for display based on prescribed attribute criteria. Scale and projection changes to vector data are straightforward, so that specific zones can be enlarged or modified for a variety of purposes. Many inexpensive commercial packages with this kind of functionality are available, and locally-developed computer programs with these capabilities can be written at low cost. For many applications in geoscience, such programs are useful and represent cost-effective solutions for basic mapping needs.

Some **computer-aided drawing packages (CAD)** offer impressive functionality, both for digitizing maps and for display of selected map features. These packages are not limited to two-dimensional objects, allowing surface representation and display of solid objects with advanced visualization techniques. However, the key to whether such vector-based systems offer true GIS functionality is whether they use **topological** data to describe the spatial relationships between the spatial objects. In a vector model, topological attributes are essential for efficient overlay and modelling operations. The structuring of vector data according to topological criteria is the key difference between the spaghetti data model and the topological data model.

In the **topological model**, the boundaries of polygons are broken down into a series of **arcs** and **nodes**, and the spatial relationships between arcs, nodes and polygons are explicitly defined in attribute tables, Figure 2-5 B. In the spaghetti model, the boundary between two adjacent polygons is stored twice, once for each polygon. This is wasteful of storage, and leads to double boundaries that do not match exactly. In the topological model, the polygon to the left and right of each arc is explicitly defined, so polygon boundaries are never repeated.

Another difference between spaghetti and topological models is whether the polygons completely fill the space on the map. In the spaghetti model, polygons need not form an interlocking mosaic that is spatially exhaustive and non-repeating. For example, if a region is subject to regular flooding and a map showing the extent of floods of various ages is digitized, in a spaghetti model the polygons for floods may overlap each other, and need not cover the whole region. For a small database, this is a convenient data organization for simple spatial queries, such as determining which, if any, of the flooded areas contain the spatial location in question. For large spatial databases with thousands of polygonal areas, however, and for analyses that demand complete map coverage with no null areas, the spaghetti model for polygons is inefficient.

This situation is overcome in a topological model by **planar enforcement**, which results in a the creation of a set of polygon objects that completely fill the plane of the map. A point selected at any location on a planar-enforced polygon dataset occurs in one, and only one, area object. Both raster and topological vector models satisfy this condition. Figure 2-6 illustrates the concept of planar enforcement for a map showing the locations of overlapping surveys. Now suppose that this map is to be searched to find which surveys have been carried out at a particular location. Before planar enforcement, and using a spaghetti data model, the fact that the survey polygons overlap is immaterial, because every survey polygon is searched in turn to test whether or not the test location is included within it. However, after planar enforcement is applied, the number of polygons increases from 3 to 7, and the polygons form an interlocking mosaic with no overlaps and no null space. A search at any point location results in finding the presence of a single polygon. The associated attribute table is then used to find out which surveys occur in that polygon, with a lookup operation. Note that raster data is always planar-enforced, with no overlap between pixels.

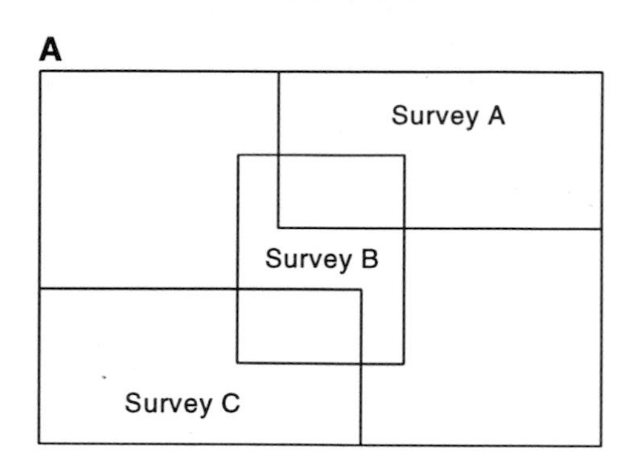

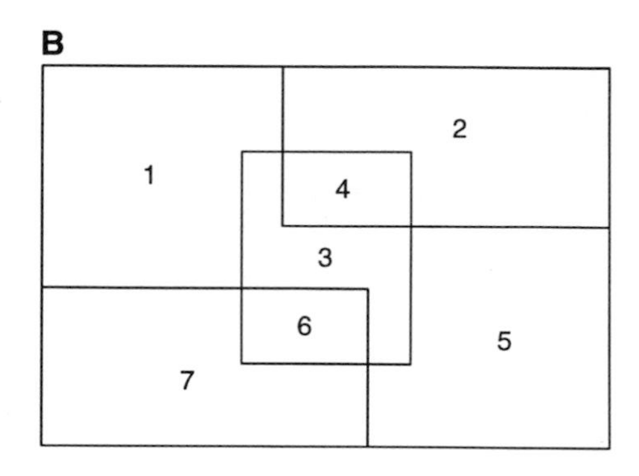

POLYGON	SURVEY: NONE	A	B	C
1	1	0	0	0
2	0	1	0	0
3	0	0	1	0
4	0	1	1	0
5	1	0	0	0
6	0	0	1	1
7	0	0	0	1

FIG 2-6. A. A map showing boundaries of three surveys carried out in different years. Note that the survey polygons overlap in some areas, and are completely absent in other areas. In a spaghetti data structure, this is legal. **B.** The same surveys after **planar enforcement**, forcing the creation of area objects (polygons) that are mutually exclusive and exhaustive. An attribute table is now required to relate polygons to surveys.

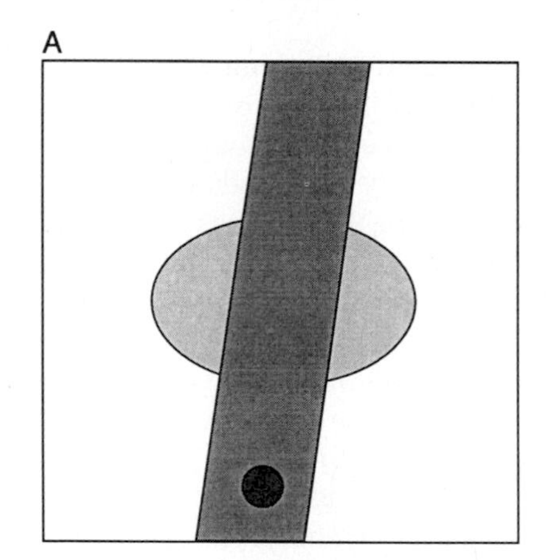

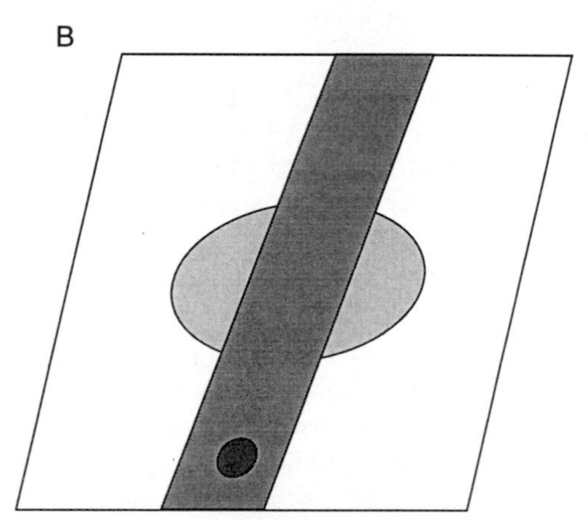

FIG. 2-7. A. Area objects on a geological map. **B.** The same objects after applying a shearing transformation.

Topological attributes of spatial objects are those spatial characteristics that are unchanged by transformations such as translation, scaling, rotation and shear. Spatial coordinates, and some geometric attributes such as area, perimeter and orientation **are** affected by such changes. But **contiguity** (pertaining to spatial adjacency), **containment** and **connectivity** are topological characteristics that remain unaltered, Figure 2-7. In the topological model, the topological attributes of spatial objects are defined in addition to the spatial coordinates of the spaghetti model. In GIS parlance, **building topology** means adding topological structure to a spaghetti file, and ensuring planar enforcement. For example, suppose a spaghetti map of polygons is created for use with CAD software, by digitizing a map, the topology can be added or "built" in a GIS before carrying out spatial analytical operations that require topological attributes. Once spatial data is represented in a topological model, changing, adding or deleting polygon boundaries not only affects the spatial coordinates of vertices, but also affects the topological attributes of arcs and nodes. Overlay of two or more map layers using the topological model involves creating a new layer of planar enforced polygons, as if the maps were overlaid on a light-table and all the boundaries from each map were traced on to a new sheet.

Digital files of spatial data organized according to the vector model usually take up less storage space than equivalent raster files, even using compressed raster data structures like quadtrees (see Chapter 3). The vector model is well-suited for representing graphical objects on maps, because smooth curves are better approximated by vertices that are not constrained by the coarse cartesian coordinates of a lattice, and points and small polygons can be represented precisely. On the other hand, cartographic fidelity is often more apparent than real, partly due to inaccurate positional information, but also due to inadequate sampling and fuzzy definition of objects. The topological model results in efficient computer processing where topological information is required. For example, finding all arcs which "have granite on one side" is an operation that requires only topological attributes; spatial coordinates can be ignored completely. On the negative side, building topology is a computer-intensive task, and must be carried out not only at the time of data entry, but also whenever editing of a map is required.

The vector model is not well suited for handling raster images, and because vector data structures are more complex than raster, operations involving the overlay of one map by another are simpler in raster mode. Most modern GIS support both raster and vector, taking advantage of the good characteristics of both data models.

Also of interest are the data models that use irregular polygons to model surfaces. The **Triangulated Irregular Network (TIN)** model is used mainly to represent digital elevation surfaces. The **Voronoi** model subdivides a region into **Thiessen** or **Voronoi** polygons. In both models, the data are derived from attributes of points. The surface to be modelled is the "height" of an attribute measured at the point locations. The objective of modelling is to convert the point objects into a mosaic of area objects that approximate a surface. The resulting surface is usually treated as a vector representation.

In the TIN model, point locations form the vertices of triangles that are as close to being equilateral as possible. The preferred method is to use **Delaunay** triangulation, which produces a unique set of triangles. Other triangulation methods produce solutions that are not unique, but depend on the starting point, (McCullagh, 1988). The resulting triangular facets form a mosaic of planar surfaces whose geometry in space are defined by the heights of the corners, Figure 2-8. Topographic relief of the ground surface is rather well approximated by sets of triangles. A triangular network is advantageous in that triangle size automatically adjusts to the point density, with small triangles where points are close together, and large triangles in areas with few points. Where the points originate from digitized contour lines, point density increases with contour line density, yielding smaller, more dense, triangles on steep slopes than on gentle slopes. The resulting piecewise surface is an efficient data model, with relatively low storage costs compared with a raster. The TIN model is also advantageous for allowing discontinuities, such as cliffs, faults, coastlines and valley bottoms to be explicitly represented using **breaklines**, (McCullagh, 1988). In converting TIN models to raster, interpolation can either be carried out directly from the planar surfaces (linear interpolation), or smooth patches can be used to produce a surface that blurs the sharp discontinuities between triangular facets (nonlinear interpolation).

Thiessen models consist of a mosaic of polygons, one polygon per data point, see Figure 2-8. The polygons have the property that the sides are always at right angles to a line joining adjacent points, and that any location within a polygon is closer to the point in that polygon than to the points of neighbouring polygons. Thiessen polygons have a variety of

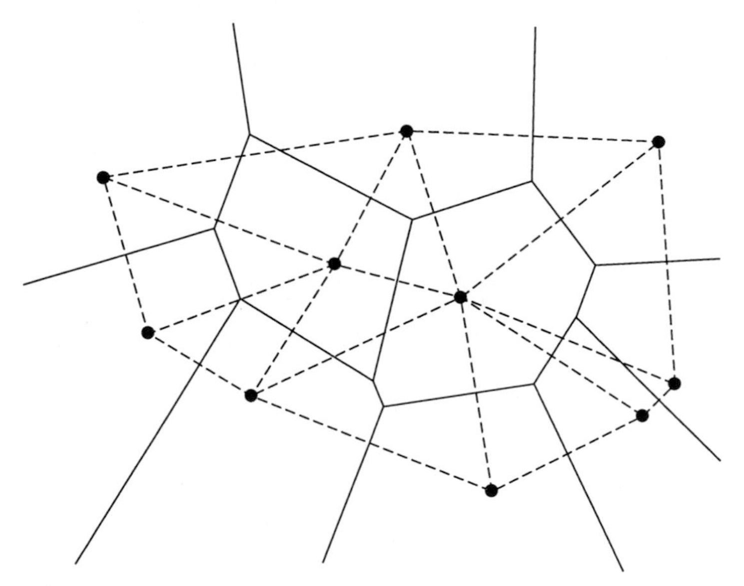

FIG. 2-8. Relationship between Delaunay triangles (dotted lines) with data points at the vertices, and Thiessen or Voronoi polygons (solid lines) that each contain a single point. The triangles form a Triangulated Irregular Network (TIN), a type of vector model. Thiessen polygons are often used as imposed spatial objects where point data are to be treated as areas, but interpolation is not suitable between points.

uses, but they can be used to model a surface as a series of plateaux, where the height of each plateau is constant and equal to the value of a selected attribute of the enclosed point. Whereas TIN triangles are sloping, Thiessen polygons are flat.

We now turn to the subject of describing and managing the attribute data in a GIS, touching on scales of measurement and the use of the relational model for data organization.

ATTRIBUTE DATA

The attributes of objects to be recorded in a database can be divided into three types: **spatial**, **temporal** and **thematic**. Most objects can be characterized by attributes of all three types, although in practice spatial and/or temporal attributes are sometimes neglected because they are constant or simply not important for the tasks at hand. For most GIS purposes, temporal and thematic attributes are lumped together as being nonspatial. Attribute values can be either based on primary observation or measurement, or they can be derived by secondary processing and calculation. Thus the values measured at gravity stations are observed values of a thematic attribute of point objects, whereas the calculated Bouguer anomaly is a derived thematic attribute.

Spatial Attributes

Spatial attributes record data about the location, topology and geometry of spatial objects. These are the characteristics that separate GIS from other kinds of database management systems. Spatial location of objects is recorded either in latitude and longitude coordinates, in coordinates of one of the standard cartographic projections, or in arbitrary rectilinear coordinates with a local origin. Geographic projections are discussed further in Chapter 4. One of the useful functions of GIS is the ability to transform spatial data from one coordinate system to another, so that maps in different projections can be compared. In general, geoscientific data (unlike some socio-economic data) are not spatially referenced to systems like postal code, or street address. However, national coordinate systems such as the National Grid in U.K. or the National Topographic System (NTS) in Canada are widely used for indexing maps and for providing geographic references in books and scientific papers.

Humans generally do not think about the positions of spatial objects in terms of locational coordinates, but rather in terms of their spatial (often topological) relations to other better-known objects. For instance, a mineral occurrence may be "close to the road, near the northwest corner of the lake", see Figure 2-9; or the granite contact "follows the river valley"; or "the glacial erratics can be found within 5 km of the coast, straddling the boundary between two geological provinces". Given our geographical and geological knowledge, we can understand a great deal about spatial location of an object from statements of this kind, whereas a string of geographic coordinates are, for human cognition, almost meaningless. How many individuals know the precise coordinates of their home or office? In order to capture spatial coordinates automatically from such information for use in a GIS, expert

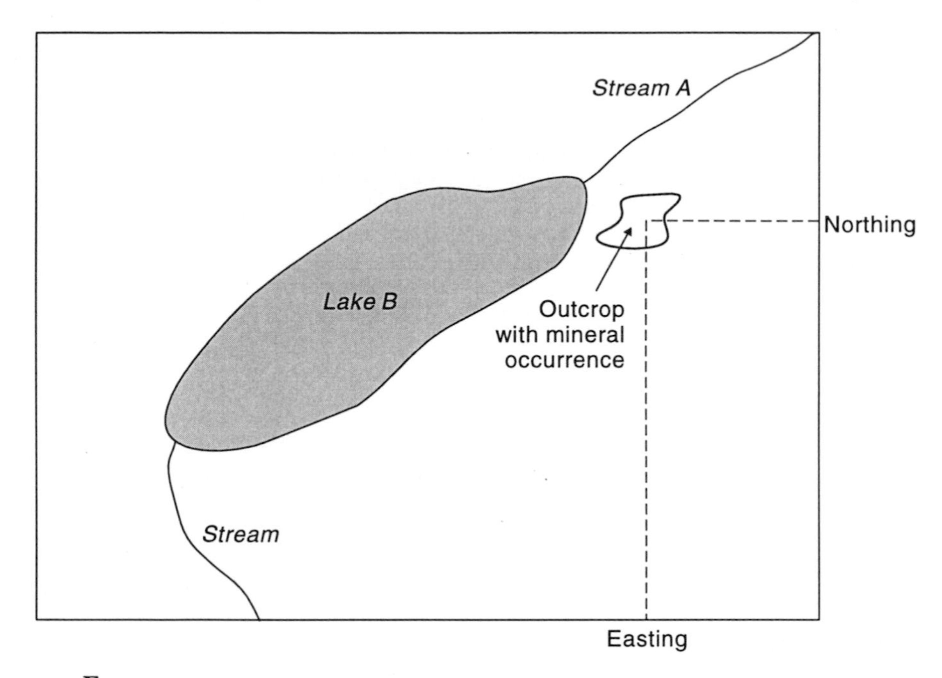

FIG. 2-9. The difference between a spatial reference in terms of absolute geographic coordinates "the mineral occurrence is at (easting, northing)", and in relative terms "located just SE of the place where stream A flows into lake B".

systems that store, manipulate and learn spatial knowledge will be required. Reasoning with topological characteristics will be of particular importance in future GIS that attempt to capture and locate spatial objects from textual description.

In current GIS, topological attributes generated automatically by "building topology" from digitized spaghetti are explicitly recorded in tables, as discussed in Chapter 3. Geometric attributes of spatial objects like length, area, shape and orientation apply to objects in vector mode, or to compound objects in raster mode.

Nonspatial Attributes

Temporal and thematic attributes of spatial objects are usually treated similarly in GIS. Temporal attributes refer either to the age of objects (e.g. geological age), or to the time of data collection or measurement. Thematic attributes refer to other kinds of properties of objects that are neither locational nor temporal, such as rock type, annual rainfall, the presence of minerals or fossil taxa. Measurement of attributes is made according to various scales or levels of measurement. These are briefly reviewed here, because they have an important bearing on the organization and treatment of attribute data.

Measurement Scales

Measurement is the process of assigning a score to an observed phenomenon according to an operational definition. The four commonly recognized levels of measurement are ratio, interval, ordinal and nominal, as illustrated in Table 2-2.

Measurements on a **ratio** scale are those such as "zinc content of soil" or "distance to granite contact". Zero on a ratio scale is absolute; zero zinc means no zinc, whether the units of measure are ppm or grams per ton. **Interval** scale measurements are similar to ratio measurements, in that the interval between units on the scale are equal, but there is no true zero. Temperature is a common example, because zero temperature differs according to the units of measurement. A hot spring at 150 degrees Fahrenheit is not twice as hot as a spring at 75 degrees Fahrenheit, because the ratio of these values is different using Celsius. Ratio and interval scale measurements produce numbers that can be manipulated by arithmetic operators; most importantly, the arithmetic mean is a suitable measure of central tendency for such data. The values of attributes depicted on maps commonly do not use ratio or interval scales. Even where an attribute is a continuous variable, like temperature, the range of the variable is divided, or quantized, into discrete intervals (depicted by contour lines or colour), that are not necessarily equally-spaced. However, the values of actual measurements or the results of calculations are often stored in attribute tables, linked to point, line or area objects.

Ordinal level measurements involve observations that are ranked according to relative position on a scale, with unequal intervals between units. The categories on maps that depict the values of a geochemical element according to a percentile scale, see Figure 2-10, are an example. In this particular case, the intervals on the concentration scale are set at arbitrary levels, leading to ordinal values that form a ranked sequence, but Zn values in class 4 are not twice those found in class 2. Ordinal scale numbers can be compared with "greater than" and "less than" operators, but may not be meaningfully added, subtracted, multiplied or divided. The median (or central value) is used as a measure central tendency for a set of ordinal observations, because the arithmetic mean requires using the addition and division operators.

Nominal scale, also known as **categorical** scale measurements, are simply numerical labels without quantitative content. The index values or pointers that link a spatial object to the records of an attribute table are an example of numbers on a nominal scale. Numbers assigned arbitrarily to rock types on a map are a common geological example. The logical

Table 2-2. Examples of different measurement scales. Nominal scale data can also be sub-divided according to whether the scale is binary or multi-state.

Scale Type	Geological Example	Legal operations*	Appropriate average*
Nominal or categorical	Rock type	=	Mode
Ordinal or ranked	Relative age	> <	Median
Interval	Temperature	+ - * /	Mean
Ratio	Distance		

* In any row, the operations for the preceding rows also apply. For example, ordinal scale data can use equality and mode as well as greater than, less than and median.

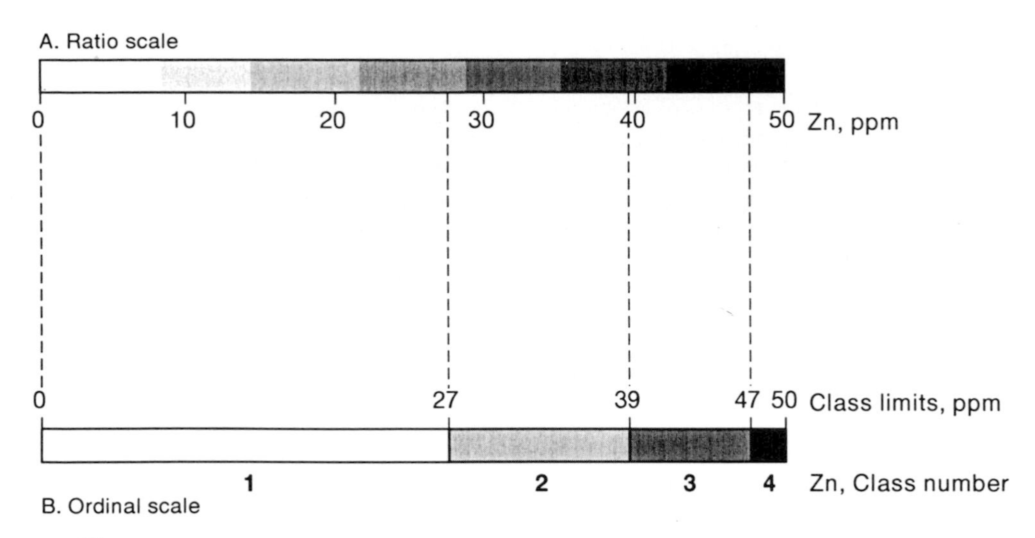

FIG. 2-10. Zinc values measured in parts per million converted to an ordinal scale by grouping the original data values into four unequal classes.

operation of equality is valid and the mode, or most frequently occurring value, is an appropriate measure of central tendency for a set of categorical observations. This is a very common level of measurement on maps.

Dichotomous or **binary** measurements are sometimes distinguished from **polychotomous** or **multi-state** measurements. Polychotomous refers to an attribute with multiple states that are usually nominal but may be ordinal. Binary attributes have two states, and can imply relative magnitude, such as the "background" versus "anomalous" states of a geochemical attribute, or simply be labels that act as symbols for the presence of one state versus the presence of alternative states. Binary attributes are often applied to map data for conveying the truth (=1) or falsity (=0) of a proposition. Thus areas on a map might be described according to the proposition "this location is less than 1000 m above sea level", or "this area is underlain by granite or granodiorite", leading to a map where the area objects are binary 1s or 0s. Such maps are then readily combined by overlay operations and the use of Boolean operators, AND, OR, NOT and exclusive OR. Sometimes polychotomous attributes are subdivided into those that are **free** and those that are **monotonic**. Monotonic implies at least an ordinal level of measurement, so that successive numerical values either increase or decrease in a consistent direction, whereas free implies nominal. The word **free** is used because categories of attributes can be grouped together into new classes, without considering their relative location on a monotonic scale.

Another scale of interest for some kinds of mapping is the probability scale from 0 to 1. This is useful for indicating the degree of certainty of an observation, or the degree of belief of a proposition. Thus an observation about the presence of granite might be 1 for definitely present, 0 for definitely absent, or 0.8 for probably present, and so on. A value between 0 and

1 can also be used to indicate the degree of "fuzzy membership" of an observation with respect to a class of objects, and fuzzy membership values can be combined with "fuzzy logic", as opposed to Boolean, operators. Some expert systems, such as the PROSPECTOR system for evaluating mineral prospects (Campbell et al., 1982) represent uncertain evidence this way. This topic is developed in more detail in Chapter 9. Although probability values may be stored in attribute tables to any reasonable precision, they are often converted to ordinal (or interval) scale classes for display as maps.

In general, attribute values are represented internally in computers either as integers, real numbers or as text. Most digital images use 8-bit integers (byte), with nonnegative numbers between 0 and 255, although image processing systems often support the use of 16-bit integers. One-byte storage is inadequate for maps with more than 256 spatial objects, where an index or pointer to an attribute table might require tens of thousands of values.

Attribute Tables

Attributes of spatial objects are usually organized into lists or tables. Whatever the internal computer organization of tables, they can be thought of as flat files, two-dimensional arrays of numbers with rows being the entities or objects and columns being the attributes. Any system that contains a large variety of different kinds of data, GIS included, requires organization and management of data files.

In the following section, some aspects of data modelling that pertain to general DBMS are discussed. This is important firstly because many GIS projects involve the manipulation of huge files of attribute data and their organization is crucial for efficient data handling, and secondly because operations on attribute data are an integral part of spatial query, spatial analysis and modelling.

Database management systems are based on either the hierarchical, network or relational models, but in recent years the relational model has become the most widely used. Although hierarchical and network principles are important for some geological databases, they are not discussed here. The following discussion of the relational model is drawn from Bisland (1989).

THE RELATIONAL MODEL

A relation is a two-dimensional structure that contains data. It is an abstract concept that corresponds in practice to a table. A row of the relation is a **tuple**, and column is a **field** or attribute. The tuple is analogous to a **data record** in a **flat file**, and contains a collection of data items that describe an object or entity. A **key**, or **keyfield**, is an attribute that uniquely identifies tuples and provides a link between one relation and another. Table 2-3 is an example of a relation that pertains to geochemical data, part of a large geochemical database. Each row or tuple is a sample. The fields are attributes that are properties of the samples. The sample number cannot be used as a keyfield on its own as it is not unique (although in the extract shown here it looks unique). Thus, the survey year and sample number are treated together as a composite key. Keyfields, whether single or composite, are also known as identifying attributes. The properties of a true relational database, as outlined by Codd (1970) are:

Table 2-3. Geochemical attribute table organized as a "flat file". Each sample is recorded on a separate line or record. Each attribute occupies a column or "field". This table can also be a "relation" in a relational database. Each row is called a "tuple". One or more columns can be "keyfields", because they link one relation to another. Here, the two fields "year" and "sample #" are combined and used as a composite keyfield.

Latitude	Longitude	Sheet	Year	Sample	Rktyp	Zn	Cu	Pb	Ni	Co
50.29557	-127.71470	' 92L'	88	3058	'BSLT'	95	80	1	37	37
50.29905	-127.71861	' 92L'	88	3059	'BSLT'	90	100	1	45	35
50.33650	-127.91139	' 92L'	88	3236	'ANDS'	117	43	2	26	17
50.33464	-127.83808	' 92L'	88	3093	'ANDS'	130	46	9	30	23
50.35596	-127.85528	' 92L'	88	3235	'ANDS'	112	57	4	43	23
50.37004	-127.84960	' 92L'	88	3082	'ANDS'	240	41	13	32	19
50.30198	-127.80173	' 92L'	88	1206	'BSLT'	160	48	2	32	24
50.31566	-127.75929	' 92L'	88	3060	'SLSN'	204	157	2	53	37
50.35135	-127.80686	' 92L'	88	3083	'ANDS'	230	62	7	41	24
50.35385	-127.73848	' 92L'	88	3056	'ANDS'	60	46	1	30	42
50.35642	-127.74000	' 92L'	88	3057	'ANDS'	74	27	1	23	22
50.35970	-127.69873	' 92L'	88	3126	'QRZD'	24	8	4	3	35
50.36629	-127.69994	' 92L'	88	3127	'QRZD'	82	38	1	24	23
50.37561	-127.92401	' 92L'	88	3265	'ANDS'	102	41	1	28	20
50.37772	-127.92280	' 92L'	88	3266	'ANDS'	135	43	3	55	23
50.39767	-127.94715	' 92L'	88	3263	'GRCK'	152	39	3	34	18
50.50899	-128.01807	'102I'	88	3040	'ANDS'	98	24	1	32	21
50.49847	-127.96268	' 92L'	88	1266	'ANDS'	101	30	1	19	14

1. ***All data must be represented in tabular form, as opposed to hierarchies or networks.***
2. ***All data must be atomic. This means that any cell in a table can contain only a single value. Thus in Table 2-3, each sample can have only a single zinc value. Replicate analyses require more tuples.***
3. ***No duplicate tuples are allowed.***
4. ***Tuples can be rearranged without changing the meaning of the relation. For example, samples can be in any order, and attributes can be in any order, without altering the contents of the table.***

An essential concept in the design of a relational database is **normalization**, which is the process of converting complex relations into a larger number of simpler relations that satisfy relational rules. The normalization process is best described by a simple example. Suppose that a geological map has been digitized and an initial table has been created to link polygons (the spatial objects or entities) to a series of attributes describing the lithology and age, Table 2-4 A. In relational database conventions, this relation can be referred to by

POLYGON(poly#,Fm_name,lithology,age)

where POLYGON is the name of the relation, poly# is the keyfield (underlined), Fm_name, lithology and age are attributes that in this case are text strings. "Fm_name" is the name of the geological formation, "lithology" is the primary rock type and "age" is the relative geological age. Because textual attributes can cause difficulties (definitions change over time and between individuals), the first step is to add new additional numerical attributes that can

Table 2-4. A. Polygon relation before normalization. Note that several formations are repeated more than once. A large table organized in this fashion would be difficult to edit, because any changes would need to be repeated, possibly many times.

POLYGON #	FORMATION NAME	LITHOLOGY	AGE
1	Shelly Fm.	Limestone	Pennsylvanian
2	Grit Fm.	Sandstone	Pennsylvanian
3	Slab Fm.	Shale	Pennsylvanian
4	Mount Fm.	Granite	Cretaceous
5	Mount Fm.	Granite	Cretaceous
6	Volcano Fm.	Tuff	Triassic
7	Mount Fm.	Granite	Cretaceous
8	Shelly Fm.	Limestone	Pennsylvanian
9	Slab Fm.	Shale	Pennsylvanian
10	Shelly Fm.	Limestone	Pennsylvanian

B. Same table after adding numerical codes, the first step in the normalization process.

Polygon #	*Formation#*	Name	*Lithology #*	Lithology	*Age #*	Age
1	*2*	Shelly Fm.	*7*	Limestone	*5*	Pennsylvanian
2	*3*	Grit Fm.	*6*	Sandstone	*5*	Pennsylvanian
3	*4*	Slab Fm.	*5*	Shale	*5*	Pennsylvanian
4	*1*	Mount Fm.	*2*	Granite	*8*	Cretaceous
5	*1*	Mount Fm.	*2*	Granite	*8*	Cretaceous
6	*5*	Volcano Fm.	*3*	Tuff	*7*	Triassic
7	*1*	Mount Fm.	*2*	Granite	*8*	Cretaceous
8	*2*	Shelly Fm.	*7*	Limestone	*5*	Pennsylvanian
9	*4*	Slab Fm.	*5*	Shale	*1*	Pennsylvanian
10	*2*	Shelly Fm.	*7*	Limestone	*5*	Pennsylvanian

substitute for textual attributes. Besides Fm_name, the numerical field Fm# is added, as are the fields lith# and age#. Thus the new expanded table shown in Table 2-4B can be referred to by the statement

POLYGON(<u>poly#</u>,Fm#,Fm_name,lith#,lithology,age#,age).

Notice that the formation numbers are repeated, because more than one polygon can belong to the same geological formation. Even in this very simple example, there is considerable redundancy in the basic table, because polygons 4,5 and 7 all contain the Mount Formation, a granite of Cretaceous age, and this identical information is repeated in three tuples. The first normalization step eliminates repeating attributes. This leads to the creation of a new FORMATION table and a simplification of the POLYGON table as follows:

POLYGON(<u>poly#</u>,Fm#), and
FORMATION(<u>Fm#</u>,Fm_name,lith#,lithology,age#,age).

Note that the formation uniquely determines lithology and age. In a large map with thousands of polygons, but only a small number of map classes, this step alone can save considerable storage, and more importantly, can make editing more efficient. The formation number now becomes the linking attribute between the two relations, being the keyfield in the FORMATION table. This allows the FORMATION table to be edited independently of the POLYGON table, and vice-versa. However, the FORMATION table still contains repeating groups, because formations 2, 3 and 4 have the same ages. This is rectified by simplifying the FORMATION table and creating a separate AGE table:

FORMATION(Fm#,Fm_name,lith#,lithology,age#), and
AGE(age#,age).

Now each tuple in the POLYGON, FORMATION and AGE relations contain no repeating groups, and the AGE relation can be separately edited and updated from the other tables. In a real database, the AGE table might have one hundred or more tuples, breaking down the geological age into many divisions, whether or not they are actually referred to in a particular FORMATION table. Relations without repeating groups of attributes are said to be in **First Normal Form (1NF)** in relational DBMS parlance.

The second normalization step is to ensure that each nonidentifying attribute is functionally dependent on the whole key. Any attribute not being used as a key in a particular relation is called nonidentifying, and keys can be based on multiple attributes. For example, suppose that a geological sample has a sample number, but the sampling scheme starts from 1 on each new survey, then a table might use both the survey number and the sample number as identifying attributes, in a paired or composite key; neither the survey number nor the sample number is unique, but only when used together. In the present example, this rule does not apply, so the relations as specified are said to be in **Second Normal Form (2NF)**, as well as 1NF.

The third step is to ensure that nonidentifying attributes are mutually independent. This clearly is not the case in the FORMATION table, because lith# and lithology are "dependent" upon one another in a 1:1 sense; given lith#, one can always specify the lithology and vice-versa. Formation name and number are also dependent in the FORMATION table, but Fm# is an identifying attribute, being the keyfield, so requires no normalization. To rectify the problem with lithology, the FORMATION table is further simplified and a new table for LITHOLOGY is defined:

FORMATION(Fm#,Fm_name,lith#,age#) and
LITHOLOGY(lith#,lithology).

This completes the normalization process, and has resulted in decomposing the original single table into four simplified tables that are now said to be in Third Normal Form (3NF). The simplified tables are, as shown in Table 2-5, summarized as

POLYGON(poly#,Fm#),
FORMATION(Fm#,Fm_name,lith#,age#),
LITHOLOGY(lith#,lithology), and
AGE(age#,age).

Table 2-5. Relations from Table 2-4 after normalization to "third normal form". Note that repetition is avoided, making editing much easier as shown in Table 2-6.

A. POLYGON relation

Polygon #	Formation #
1	2
2	3
3	4
4	1
5	1
6	5
7	1
8	2
9	4
10	2

B. FORMATION relation.

Formation #	Formation Name	Lithology #	Age #
1	Mount Fm.	2	8
2	Shelly Fm.	7	5
3	Grit Fm.	6	5
4	Slab Fm.	5	5
5	Volcano Fm.	3	7

C. LITHOLOGY relation.

Lithology #	Lithology
2	Granite
3	Tuff
5	Shale
6	Sandstone
7	Limestone

D. AGE relation.

Age #	Age
5	Pennsylvanian
7	Triassic
8	Cretaceous

The normalization process can be carried still further, to at least a fifth normal form (Bisland, 1989, chapter 15). The primary purpose of this progressive decomposition is to avoid what are known as anomalies, that are undesirable side effects that can occur if relations are not in normal form. These anomalies arise when tables are modified, either by inserting new tuples, deleting tuples or modifying existing tuples. As a simple illustration, suppose that the original POLYGON table in the example required modification to change the lithology associated with Fm# 1, the Mount Formation from granite to gabbro (see Table 2-6). Because this occurs as a repeating group in three tuples, the same changes need to be made three times in separate places in the original POLYGON table. However in the final form, the change is simpler. A new tuple for gabbro with a unique lith# is added to the LITHOLOGY table, say 4, and the lith# for Fm# 1 in the FORMATION table changed from 2 (the old granite) to 4 (for the new gabbro). This is illustrated in Table 2-7. This may seem trivial in a small example, but in a large database requiring constant maintenance, the requirement for speedy and accurate insertion, deletion and editing is paramount.

The paradox is that by decomposing tables into normal form, speed of search and retrieval is sacrificed. In order to speed up a search, tables must often be joined together again, and sorted and indexed to optimize efficient use. For large, constantly updated and complex databases, the normal form is desirable, both for editing and for ease of understanding the data model. For smaller databases, particularly ones that are stable and no longer being

Table 2-6. Editing the relation shown in Table 2-4 B before normalization. It has been discovered that the Mount Fm. is a gabbro not a granite (!). Even in this very trivial example it can be seen that the change must be repeated exactly the same in three places in the unnormalized table. Change lithology # and lithology each time the Mount Fm. occurs, see items in bold italics.

Polygon #	Formation #	Formation Name	Lithology #	Lithology	Age #	Age
1	2	Shelly Fm.	7	Limestone	5	Pennsylvanian
2	3	Grit Fm.	6	Sandstone	5	Pennsylvanian
3	4	Slab Fm.	5	Shale	5	Pennsylvanian
4	1	Mount Fm.	***4***	***Gabbro***	8	Cretaceous
5	1	Mount Fm.	***4***	***Gabbro***	8	Cretaceous
6	5	Volcano Fm.	3	Tuff	7	Triassic
7	1	Mount Fm.	***4***	***Gabbro***	8	Cretaceous
8	2	Shelly Fm.	7	Limestone	5	Pennsylvanian
9	4	Slab Fm.	5	Shale	1	Pennsylvanian
10	2	Shelly Fm.	7	Limestone	5	Pennsylvanian

Table 2-7. After normalization, the POLYGON and AGE relations remain unchanged, and two small changes are needed in the FORMATION and LITHOLOGY relations. Suppose this change is made in a real situation for a geological map containing several thousand polygons. The editing changes would still be the same as shown here, because the POLYGON relation requires no modification.

FORMATION relation (1 change shown in italics)

Formation #	Formation Name	Lithology #	Age #
1	Mount Fm.	***4***	8
2	Shelly Fm.	7	5
3	Grit Fm.	6	5
4	Slab Fm.	5	5
5	Volcano Fm.	3	7

LITHOLOGY relation
(1 insertion shown in italics)

Lithology #	Lithology
2	Granite
3	Tuff
4	***Gabbro***
5	Shale
6	Sandstone
7	Limestone

expanded, normalization is less desirable. Clearly there is a trade-off between maintenance and use. In many cases, the archival form of a database is held in normal form, and periodically it is "un-normalized" for distribution of selected subsets, or possibly for direct user access.

The relational data model is not widely used for the spatial coordinates of vertices in the vector model because the sequence of the vertices is vital for defining a line. This violates the last of Codd's properties of a relational database, that the tuples in a relation can be re-arranged without affecting the data. There are ways around this, as discussed in Chapter 3, but the relational form does not lead to a data structure that is the most efficient for operations like drawing a line, where the ordering of vertices is critical.

In Chapter 4 , we will return to the subject of spatial query and access to nonspatial attributes in relational databases. Whether or not data are organized in a full relational database, the relational principles, briefly introduced here, are useful for organizing data according to the relational data model. In the later chapters of this book, in discussing **map**

modelling, operations are carried out on sets of spatial objects linked to attribute tables. In the process of carrying out such operations on maps and their attributes, interrelated tables may be linked together, and to spatial objects, with keyfields. Relational principles are therefore important both in data modelling and map modelling.

Before leaving the topic of attribute tables, it should be noted that they form a unifying link between the raster and vector models. Suppose that a map showing soil type has been digitized, see Figure 2-11. The individual soil polygons are the spatial objects in a vector model, linked to soil classes in a polygon attribute table. Furthermore, assume that this same map has been transformed to the raster model, by vector-to-raster conversion, Figure 2-11B. The spatial objects are now the pixels, but the pixels belong to composite objects, comprising the groups of spatially contiguous pixels that make up the soil polygons in Figure 2-11A. The raster model records the polygon number as the pixel attribute, so both the raster and vector models share exactly the same polygon attribute table, Figure 2-11C. This example illustrates the duality of the raster and vector views. It also shows that pixels as spatial objects in raster mode can be regarded as members of composite spatial objects in vector mode, in this case the mapped polygons, and that attribute tables represent common ground between the two approaches.

A

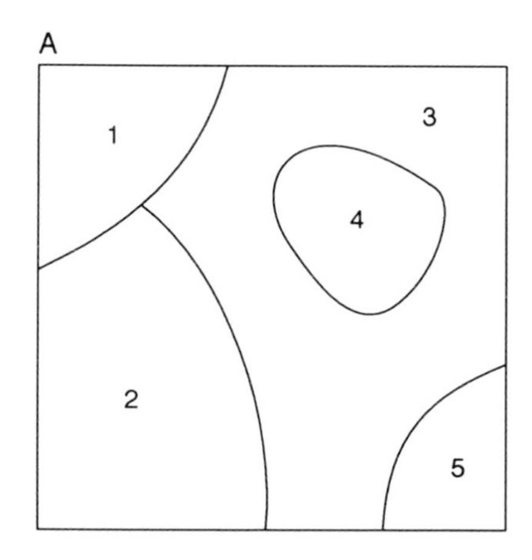

B

1	1	1	1	1	3	3	3	3	3	3	3
1	1	1	1	3	3	3	3	3	3	3	3
1	1	1	1	3	3	4	4	4	3	3	3
1	1	1	3	3	3	4	4	4	4	3	3
1	2	2	2	3	3	4	4	4	4	3	3
2	2	2	2	3	3	3	4	4	4	3	3
2	2	2	2	2	3	3	3	3	3	3	3
2	2	2	2	2	3	3	3	3	3	3	3
2	2	2	2	2		3	3	3	3	5	5
2	2	2	2	2	2	3	3	3	5	5	5
2	2	2	2	2	2	3	3	3	5	5	5
2	2	2	2	2	2	3	3	3	5	5	5

C

Polygon	Class	Rock type	Age	Name
1	15	sandstone	Late Pennsylvanian	Andrews Formation
2	6	limestone	Early Silurian	Barry Formation
3	3	shale	Middle Silurian	Clinton shale
4	14	granite	Devonian	Delta granite
5	14	granite	Devonian	Delta granite

FIG. 2-11. A. Soil map in vector model. **B.** Same map in and raster model. **C.** Both models utilize the same polygon attribute table. Note that the attribute value in the raster is a pointer to the polygon number.

REFERENCES

Bisland, R.B., 1989, *Database Management: Developing Application Systems Using ORACLE*: Prentice-Hall International, 441 p.

Campbell, A.N., Hollister, V.F. and Duda, R.O., 1982, Recognition of a hidden mineral deposit by an artificial intelligence program: *Science*, v. 217 (3), p. 927-929.

Codd, E.F., 1970, A relational model of data for large shared databases: *Communications of the ACM*, v. 13 (6), p. 377-387.

Feder, J., 1988, *Fractals*: Plenum Press, New York-London, 283 p.

Goodchild, M.F., 1992, Geographical data modeling: *Computers & Geosciences*, v. 18(4), p. 401-408.

Goodings, D. and Middleton, G.V., 1991, *Fractals and fractal dimension: Non-Linear Dynamics, Chaos and Fractals*: Geological Association of Canada, Short Course Notes: Editor, Middleton, G.V., v. 9, p. 13-22.

Mandelbrot, B.B., 1983, *The Fractal Geometry of Nature*: W.H. Freeman & Co., New York, 468 p.

McCullagh, M.J., 1988, Terrain and surface modelling systems: theory and practice: *Photogrammetric Record*, v. 12(72), p. 747-779.

Peuquet, D., 1984, A conceptual framework and comparison of data models: *Cartographica*, v. 21 (4), p. 66-113.

Raper, J., 1989, A 3-dimensional geoscientific mapping and modelling system: a conceptual design: In *Three Dimensional Applications in Geographical Information Systems*, Editor, Raper, J., Taylor and Francis, London-New York, p. 11-19.

Unwin, D., 1981, *Introductory Spatial Analysis*: Methuen, London-New York, 212 p.

CHAPTER 3

Spatial Data Structures

INTRODUCTION

Spatial data structures refer to the organization of spatial data in a form suitable for digital computers. As discussed in Chapter 2, a data structure can be regarded as being intermediate between a data model and a file format. For example, run-length code is a data structure applicable to the raster model, and can be written to a digital file in a variety of file formats. The difference between data *model* and data *structure* is poorly defined. For example, it is

Table 3-1A. Attribute table for points on a regular lattice, with values of lead (Pb) from soil samples collected every 100 m.

X, m	Y, m	Pb, ppm
100	100	6.7
200	100	7.2
300	100	8.9
400	100	7.1
500	100	6.2
100	200	7.3
200	200	7.8
300	200	10.1
400	200	9.7
500	200	8.3
100	300	7.6
200	300	9.2
300	300	12.3
400	300	10.6
500	300	9.2
100	400	7.8
200	400	9.5
300	400	10.7
400	400	9.4
500	400	8.1
100	500	5.1
200	500	8.4
300	500	8.6
400	500	7.6
500	500	3.1

Table 3-1B. Two-dimensional array of the same data as Table 3-1A, except with the spatial locations specified implicitly by row and column index. DX and DY are the spacings between points in the X and Y directions, both equal to 100 m.

		X Direction, (columns)				
		1	2	3	4	5
Y Direction (rows)	1	6.7	7.2	8.9	7.1	6.2
	2	7.3	7.8	10.1	9.7	8.3
	3	7.6	9.2	12.3	10.6	9.2
	4	7.8	9.5	10.7	9.4	8.1
	5	5.1	8.4	8.6	7.6	3.1

not clear at what stage a topological model becomes a topological structure, because the word "topological" is used in both instances. The distinction is not important, as long as the organizational scheme is clear.

Raster structures are widely used in image processing systems and in raster GIS. Vector structures predominate in CAD systems and GIS with strong cartographic capabilities. Several alternative structures have been developed for both the raster and vector data models. Choice of an optimal data structure depends on the nature of the data and how they are used. Most GIS employ both raster and vector models, with several alternative data structures for the same data. Different structures are used for different tasks, depending which are the most efficient and most suitable. For example, a dataset of sampling points and associated attributes might exist 1) as an attribute table with spatial coordinates, 2) as a raster containing values interpolated from the points for one of the attributes, 3) as vectors, showing the boundaries of polygons on a contour map created from the raster, and 4) as a TIN structure. Each of these structures has advantages and disadvantages, and the ability to transform from one structure to another is valuable. Some of the common raster and vector data structures are introduced in this chapter. A short section is devoted to three-dimensional structures for surfaces and volumes.

RASTER STRUCTURES

If a region is densely sampled at point locations on a square lattice, a table containing spatial coordinates and nonspatial attributes is the simplest structure for organizing the data, as illustrated in Table 3-1A. Similarly, if the points are regarded as lying at the centres of square cells or pixels, a table with explicit geographic coordinates for each cell, similar to Table 3-1A, can be used to define the data structure. In the normal raster structure, the spatial coordinates are omitted altogether, because the sequence of pixel values provides an implicit spatial address, once the "scan order" or sequence of rows and columns is established. The most common organization is a rectangular array of pixel values, in which the row and column coordinates define a particular location, similar to the index values for a two-dimensional array in FORTRAN, as shown in Table 3-1B. For three dimensional problems, 3-D cells (**voxels** or *vo*lume *el*ements) are organized as a cube with an additional coordinate for the vertical location. Given information about the size of the array (e.g. the numbers of rows, columns, and possibly levels for the 3-D case), and the ordering convention, arrays can be stored as one-dimensional lists. The array structure for the raster model is sometimes called a **full raster** or an **expanded raster**.

Although full rasters are convenient for many spatial data processing functions, being particularly efficient for the overlay of multiple data layers (with each layer being a raster image depicting a separate attribute), they occupy large amounts of storage space, particularly for high-resolution images. Several schemes are used to compress raster data. The most common are **run-length encoding** and hierarchical data structures called **quadtrees** and **octrees**. Boundaries and other lines can be represented in raster as chains of pixels, sometimes compressed using **Freeman codes**. Lines can also be modelled with special kinds of hierarchical data structures such as strip trees, arc trees, line quadtrees and edge quadtrees (Samet, 1990b), but discussion of these topics is beyond the scope of this book.

Full Raster Structure

Most digital image processing systems use full raster structures. The structures differ slightly from one another mainly in the way that attribute data are organized and represented. The simplest and most popular structure is to restrict each grid layer to a single attribute, and to limit the number of values of the attribute to the integer range 0-255 (i.e. one byte of storage per pixel). In image processing terminology, attributes are often called **bands**, referring to the bandwidths of the electromagnetic spectrum measured by satellite imagers like Landsat. The sequencing of pixels in a full raster is usually by row-order, starting in the upper left and scanning left-to-right, top-to-bottom. A number of alternative orderings is possible, and this becomes relevant for compression structures such as run-length codes and quadtrees.

The full raster structure can be organized as **band sequential (BSQ)**, with the values of a single band or attribute arranged in row order, as above. If there is more than one attribute, the second band starts where the first band finishes. Multi-band images can also be stored as **band interleaved by line (BIL)** or **band interleaved by pixel (BIP)** structures. For BIL, each row of pixels is repeated **m** times where **m** is the number of bands, before moving to the next row. For BIP, the band values for each pixel are stored together, so that for a 7-band image the first seven values refer to the first pixel, followed by the next seven values for the second pixel, and so on. Clearly, BIP and BIL formats are advantageous for operations involving the combination of images, because the physical addresses of attribute values for the same pixel are close together; but for very rapid display of single attributes from a large multi-band dataset, BSQ is more efficient.

The most common raster image structures store only positive integers in the range 0-255 (one byte). One byte (8 binary digits or bits) can hold nonnegative integers up to 2^8, but two bytes per pixel can hold signed integers up to 2^{15} with one bit for the sign. Some image systems allow character or real variables to be stored as an alternative to one or two byte integers. Gridded geophysical image data often require a range greater than 0-255, because high spatial

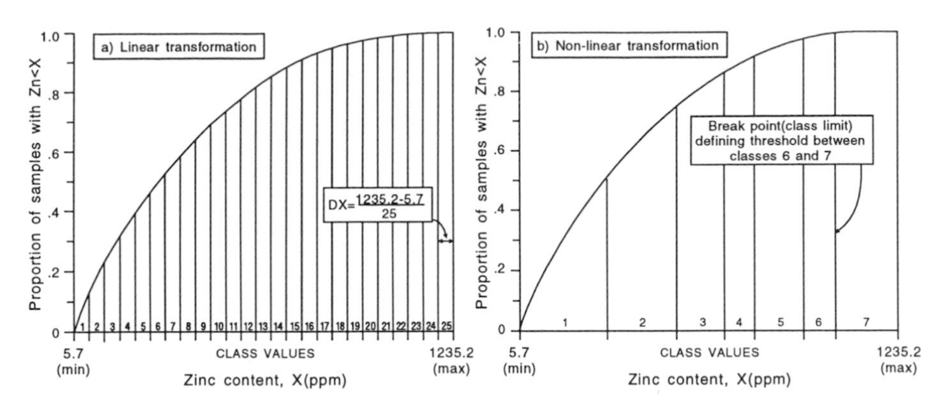

FIG. 3-1. Recoding of attributes according to a classification. **a**. Linear transformation producing equally-spaced classes. **b**. Class intervals irregular, with cutoff limits either specified by a nonlinear transform or by a direct lookup operation. Recoding with a binary classification, where the selected class can be in mid-range, is usually called **density slicing** in image processing.

frequencies (such as the steep gradients of magnetic values that occur over geological contacts or dykes) are important. High-frequency information is lost by quantizing into a small numerical range, so that geophysical images often require more than one byte per pixel.

As mentioned in Chapter 2, the attribute values of an image can act as pointers to an attribute table, instead of storing each attribute as a separate grid layer. This is particularly advantageous for digitized maps, in which the attributes of map units, or polygons, are stored as a table. In such cases, one byte integers are usually insufficient, because the attribute table often has more than 256 unique records. Storage space of two or more bytes are therefore required for pointer values.

Attribute Classification

A short digression is in order here, to touch upon the classification of single attribute values into groups or classes. Classification of attributes is often required, whether the vector or raster model is being used, for data display and for data generalization. Suppose, for example, that zinc values have been determined from soil samples in a survey. In order to visualize the spatial variation of zinc values, point symbols of different colour or shape are to be used (or colour and/or contour lines are to be employed for a representation of a zinc surface). In either case, a relatively small number of discrete zinc classes are needed, certainly no more than 255, but often fewer than 10. For display, classification is needed because the eye and mind are confused by too many symbols or colours.

Attributes that refer to variables measured on ratio, interval or ordinal scales are subdivided into classes using a series of cutoff values that define the class limits. This is similar to grouping a variable into **bins** for drawing a histogram. The series of ordered cutoff values can itself be called a "classification". Once grouping has taken place, an integer number indicating the class is used instead of the original value. This can obviously save storage space, but degrades the precision of the attribute measurements, and often degrades the level of measurement from ratio or interval to ordinal scale. Ideally, the original untransformed values are retained in an attribute table, so that alternative classifications can be applied and the precision of the original data retained. Attributes that are nonquantitative (nominal scale) can also be re-coded into groups, but because the attribute values are "free" and not ordered; the grouping is not constrained by the numerical sequence. This topic is revisited later in Chapter 7 under the topic of map reclassification.

The classification process is illustrated in Figure 3-1. The geochemical abundance of zinc in soils might range from a minimum of 5.7 parts per million (ppm) to a maximum of 1235.2 ppm within a particular survey. Suppose that two classifications are to be compared, one with 25 classes, the other with 7 classes:

1) The intervals are to be constant in size, using a linear transformation, see Figure 3-1A.

2) The intervals are to be unequal, with cutoffs related to the percentile distribution of the data, see Figure 3-1B.

There are two common methods for carrying out these binning transformations. The first is to specify the functional form of the transform and the necessary parameters. The second is to supply either the class limits as a table, or to supply a complete **lookup table** relating each untransformed value to its transformed equivalent. Whichever method is used, the transformed data is a **quantized** version of the original, and reconversion does not recover the original precision, except in the case where all the original attribute values are assigned to a unique class.

Although attribute data are usually classified for the purpose of generalization, classification also produces data compression, reducing fractional numbers to one or two-byte integers. We now turn to another type of data compression for reducing the space requirements of a full raster.

Run-Length Encoding

The storage requirements for full raster images increase geometrically with the decreasing size of a pixel, becoming unmanageably large for some high-resolution data, particularly for three-dimensional data. Space problems can be mitigated to some extent by compression

Table 3-2. Run-length encoding of a raster with 5 rows (or lines) and 10 columns (or pixels). The coding is shown line-by-line. Each encoded data pair consists of (run length in pixels, and run class). Usually runs terminate at the end of each line. Raster images of maps can often be greatly compressed by this method. The letter symbols used here would be replaced by numbers in an actual digital file.

		1	2	3	4	5	6	7	8	9	10	Run-length encoding
Rows	**1**	A	A	A	A	B	B	B	A	A	A	(4,A),(3,B),(3,A)
	2	A	A	A	B	B	B	A	A	A	C	(3,A),(3,B),(3,A),(1,C)
	3	A	A	B	B	B	A	A	A	C	C	(2,A),(3,B),(3,A),(2,C)
	4	A	B	B	B	A	A	C	C	C	C	(1,A),(3,B),(2,A),(4,C)
	5	A	A	A	A	A	A	C	C	C	C	(6,A),(4,C)

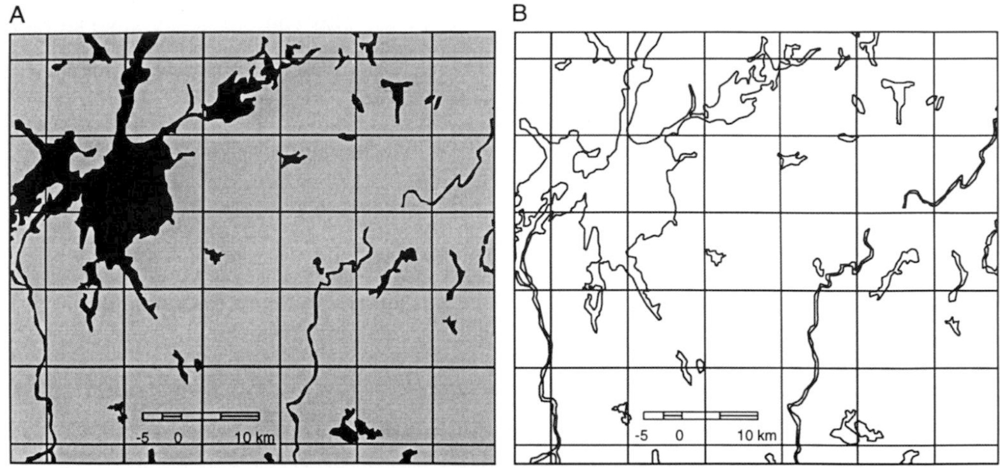

FIG. 3-2. Lakes in a raster image with 794 rows and 893 columns. The full raster, **A**, (1 byte per pixel) uses 709,042 bytes of storage. The run-length encoded version requires 21,903 bytes. A quadtree version requires 19,473 bytes. A vector version, **B**, requires 17,890 bytes.

methods. Run-length encoding is a simple data structure that can reduce the space requirements of some images drastically. It is also an efficient structure for image display and for some processing algorithms. For other processing operations it is not suitable, and encoded data must be unscrambled to a full raster, for example in neighbourhood analysis.

The principle of run-length encoding is shown in Table 3-2. Adjacent pixels having the same value are combined together as a run, represented as a pair of numbers. Each run pair consists of a number for the length of the run in pixels, followed by a second number (shown in the example as a letter) for the attribute (or class) value of the run. Each new row starts with a new run. The number of bits used for run length is dependent on the number of columns in the image. For example, an image with 1024 columns needs 10 bits for runs that can be up to 2^{10} long, or 12 bits for an image with 4098 columns. The number of bits needed for the run class depends on the maximum class value in the image. Typically two bytes are used for run length, one byte for the run class, for a total of three bytes per run pair. Alternatively, the run length can be restricted to a small fixed size, such as one byte (255), as long as consecutive runs are allowed to have the same class.

Some further compression is possible by allowing runs to continue from one row or line to the next. An image in which all the pixels belong to one class would be a single run. A 1024 by 1024 image of this type all having the class value of 1 is represented by the run pair (1048496,1). Note however that this particular length of run is too large to store in two bytes, and would need more storage space. Run-length encoding of raster digitized maps can produce dramatic savings in space, because long runs are typically present. Figure 3-2A shows a 2-class black and white image of lakes. The full raster has 794 rows and 893 columns and is saved as a file with storage of one byte per pixel (although 1 bit per pixel would actually suffice in this case). The file size is therefore 709,042 bytes. The run-length encoded file,

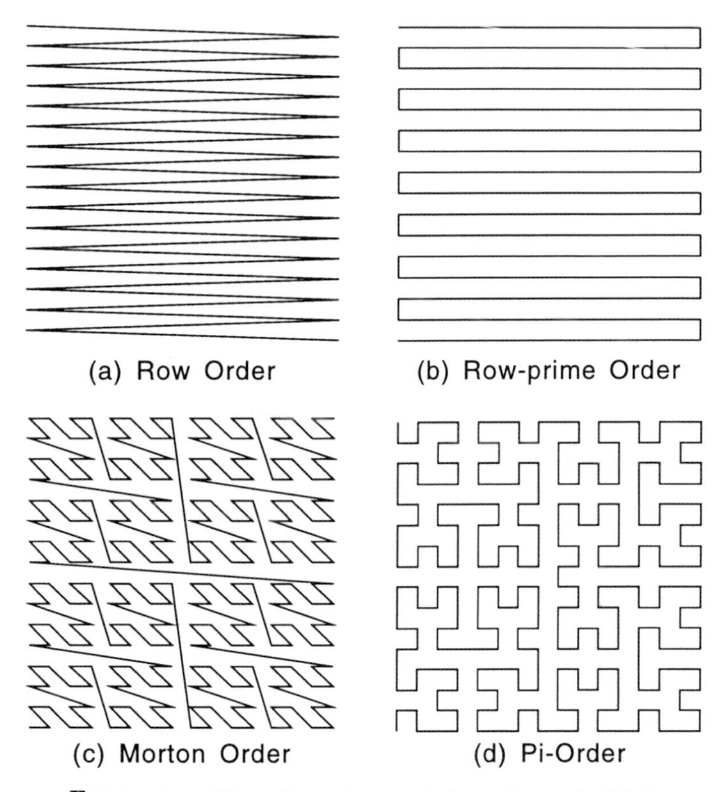

FIG. 3-3. Four different kinds of raster ordering. **a**. Row order (this is the most comon method). **b**. Row-prime order. **c**. Morton order (used in quadtrees). **d**. Hilbert-Peano order. Figure adapted from Goodchild and Grandfield (1983).

each row starting a new run, with 3 bytes per run pair, occupies a total of 21,903 bytes, and is therefore smaller by a factor of 33 times. Even with satellite image data, in which the length of runs is on average quite small, a small saving in space is usually achieved.

Scan Order for Rasters

The ordering of a full raster can affect the efficiency of run-length encoding. Figure 3-3 shows four orderings: **row order**, **row-prime order**, **Morton order** and **Hilbert-Peano order**. Row prime order simply alternates the scan direction between each row. Morton and Hilbert-Peano orderings are both types of **space-filling** curves that visit all the pixels in one block before moving on to the next block in a repetitive manner. The Morton order is particularly interesting, because the blocks are arranged into quadrants, each quadrant itself being subdivided into four more quadrants, and this process continues recursively. This arrangement is at the heart of the quadtree, to be discussed next. However, Goodchild and

Table 3-3. Morton matrix for an 7 by 8 image. A. As decimal numbers, B. As base 4 numbers. The row and column indices are shown starting at 0 rather than 1, with the origin in the SW corner. The Morton numbers for (row 1, column 3) are highlighted, see text.

A

7	42	43	46	47	58	59	62	63
6	40	41	44	45	56	57	60	61
5	34	35	38	39	50	51	54	55
4	32	33	36	37	48	49	52	53
3	10	11	14	15	26	27	30	31
2	8	9	12	13	24	25	28	29
1	2	3	6	**7**	18	19	22	23
0	0	1	4	5	16	17	20	21
	0	1	2	3	4	5	6	7

B

222	223	232	233	322	323	332	333
220	221	230	231	320	321	330	331
202	203	212	213	302	303	312	313
200	201	210	211	300	301	310	311
022	023	032	033	122	123	132	133
020	021	030	031	120	121	130	131
002	003	012	**013**	102	103	112	113
000	001	010	011	100	101	110	111
0	1	2	3	4	5	6	7

Grandfield (1983) show that the efficiency of run-length encoding is not greatly affected by the order, except that row-prime and Hilbert-Peano orders are slightly more efficient than row and Morton orders. This is due to the tendency for adjacent pixels to belong to the same class, and in general this increases with the degree of **autocorrelation** present in the data. Autocorrelation is a term meaning the degree of correlation of the data with itself, as measured by the cross-correlation of a map with a replicate map under translation. Auto and cross-correlation are discussed further in Chapter 6.

Morton Order and Morton coordinates

The Morton order and Morton system of spatial coordinates are discussed here, because they lead naturally into the subject of quadtrees, to be covered in the following section. Quadtrees are raster data structures that are based on the recursive subdivsion of a square image into quadrants. The quadtree structure permits data compression with a kind of two dimensional run-length encoding. The quadtree structure is used in some GIS, and the principle of Morton addressing can be applied to any system that employs recursively-divided square **tiles** for partitioning a region.

The trajectory formed by linking pixels in an image according to the Morton order forms a repetitive reverse Z-shaped pattern, as shown in Figure 3-3C. The pixels in a square image, with the length of side being a power of 2, can be spatially indexed with a Morton address. Each Morton coordinate is a single number denoting position along the Morton trajectory. Morton coordinates are convenient for spatial search operations, because instead of having to search by row and by column separately, the two coordinate values are combined into a single number. Morton indexing can greatly improve the efficiency of operations such as finding the class value of a map at, or close to, a specified location.

Table 3-3 shows a **Morton matrix** for an 8 by 8 image. This simply labels the pixels in Morton sequence according to a base 10 or decimal number (Table 3-3A) and in base 4 (Table 3-3B). Base 4 is convenient for indexing the addresses of quadtrees because of the recursive fourfold partition of quads. Note that the origin is in the SW corner; some authors use the NW corner as the origin, e.g. Samet (1990a). In Morton order, the pixels form a sequence that

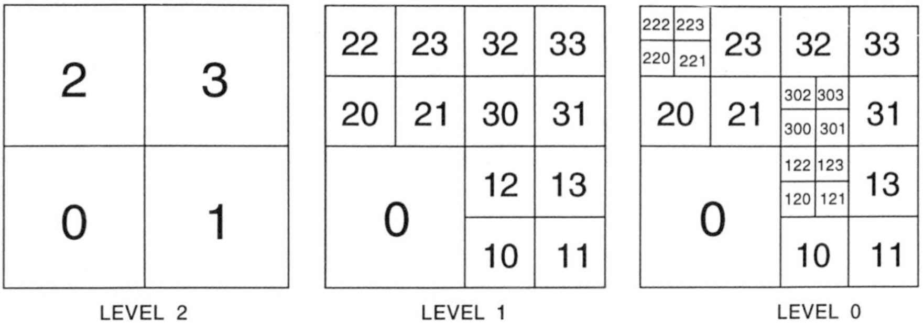

FIG. 3-4. Morton indexing of recursively subdivided square blocks. Base 4 numbers are used. At level 2, the square block on the left is divided into 4 quadrants. At level 1, each level 2 block is subdivided into 4 for a maximum total of 16 blocks. However, blocks are not subdivided unless they are homogenous. At level 0, the process is repeated. The numbering sequence follows the Morton order shown in Figure 3-3C.

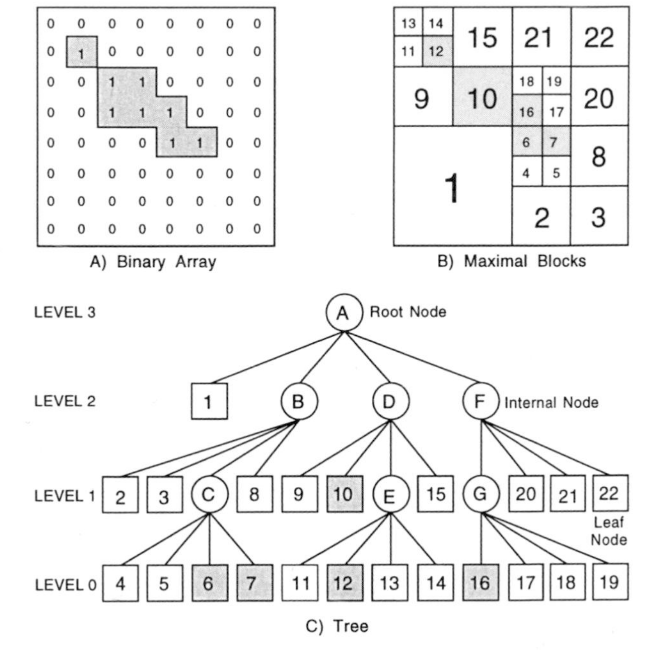

FIG. 3-5. A. Binary raster array showing a region of 1s surrounded by 0s. **B**. Decomposition into maximal blocks, numbered in Morton order, starting at 1 in the SW corner. **C**. Tree structure showing internal nodes as circles, and leaf nodes as squares. The shaded leaf nodes are the areas occupied by 1s in the original array.

is area-filling, never crossing itself, and in which the number of large jumps is minimized. The trajectory is built up from a basic 2 by 2 pattern into a series of levels, repeating the same pattern at each level. This is illustrated in Figure 3-4, where the blocks for each level are indexed by a base 4 number, with the four quadrants of a block being 0 (SW), 1 (SE), 2 (NW) and 3 (NE). At level 2, there are four square blocks of equal size. At level 1, three of the level 2 quadrants have been subdivided, and at level 0, three of the level 1 blocks have been subdivided. The base 4 number is known as the **Morton coordinate**, or **Morton key**. The number of digits in the Morton coordinate are inversely related to the level and therefore to the size of the block.

The Morton coordinate is used as a locational code. It is calculated from the conventional raster address by **bit-interleaving** the row and column coordinates at a binary level, (Abel, 1986). For example, in Table 3-3A, take the base 10 (decimal) Morton number of 7 (row 1, column 3). The

Table 3-4. Quadtree from Figure 3-5 as a basic table with numerical pointers to leaf nodes and alphabetical pointers to internal nodes.

Node (P)	Father (P)	Son (P,1)	Son (P,2)	Son (P,3)	Son (P,4)	Class (P)
A	-	1	B	D	F	Grey
1	A	-	-	-	-	White
B	A	2	3	C	8	Grey
2	B	-	-	-	-	White
3	B	-	-	-	-	White
C	B	4	5	6	7	Grey
4	C	-	-	-	-	White
5	C	-	-	-	-	White
6	C	-	-	-	-	**Black**
7	C	-	-	-	-	**Black**
8	B	-	-	-	-	White
D	A	9	10	E	15	Grey
9	D	-	-	-	-	White
10	D	-	-	-	-	**Black**
E	D	11	12	13	14	Grey
11	E	-	-	-	-	White
12	E	-	-	-	-	**Black**
13	E	-	-	-	-	White
14	E	-	-	-	-	White
15	D	-	-	-	-	White
F	A	G	20	21	22	Grey
G	F	16	17	18	19	Grey
16	G	-	-	-	-	**Black**
17	G	-	-	-	-	White
18	G	-	-	-	-	White
19	G	-	-	-	-	White
20	F	-	-	-	-	White
21	F	-	-	-	-	White
22	F	-	-	-	-	White

row and column numbers are converted to binary, and interleaved starting with the row. Binary of decimal 1 is also 1, and we add a leading 0 to make 01. The binary for decimal 3 is 11. Interleaving these binary digits, starting with the row is shown as follows:

ROW	0 1	(1 in base 10)
COLUMN	1 1	(3 in base 10)

Interleaving results in the digits 0-1-1-1, which is binary 0111 or (0*8)+(1*4)+(1*2)+(1*1)=7 decimal. The binary number 0111 can also be converted to base 4 by grouping successive binary digits in pairs, 01 is 1 and 11 is 3. Therefore the base 4 Morton

Table 3-5. Same as Table 3-4, but retaining leaf nodes only, omitting columns for father and sons, and adding a base 4 locational code and level. The last two columns comprise the basic structure of linear quadtrees, requiring only an ordered list of (level, class) pairs.

Node (P)	Location (P)[1]	Level (P)[2]	Class (P)
1	000	2	White
2	100	1	White
3	110	1	White
4	120	0	White
5	121	0	White
6	122	0	**Black**
7	123	0	**Black**
8	130	1	White
9	200	1	White
10	210	1	**Black**
11	220	0	White
12	221	0	**Black**
13	222	0	White
14	223	0	White
15	230	1	White
16	300	0	**Black**
17	301	0	White
18	302	0	White
19	303	0	White
20	310	1	White
21	320	1	White
22	330	1	White

1 Morton number for lower left corner of block.
2 Size of block is related to its level by the relation
size $= 4^{level}$

coordinate of this block is 13. The number of quad levels needed for an 8 by 8 raster is 3 (levels 0, 1 and 2) , and at level 0 the number of digits in the Morton coordinate is 3. Therefore the proper base 4 address is 013, with a leading 0 to make up the right number of digits.

Alternatively, given a base 4 Morton coordinate of 023, or 23 after removing the leading 0, the process is reversed to produce row and column number. Base 4 23 is 1011 in binary. Taking the first and third bits forms the row number, binary 11 or decimal 3, and bits 2 and 4 forms the column number, binary 01 or decimal 1. Note that the row and column numbering of the raster starts from zero for this interleaving method to work.

Region Quadtrees and Octrees

Quadtrees and octrees are hierarchical data structures based on successive subdivision of blocks into 4 quadrants, or 8 octants, for pixels or voxels, respectively. The word **region quadtree** is applied to hierarchical structures for regions. There are many different kinds of quadtrees, some of which are used for points and lines (Samet, 1990b). **Bintrees** are similar to quadtrees, except successive division is made into 2 rather than 4, and division in the x-direction alternates with division in the y-direction. In this book, the term quadtree is used to refer to region quadtrees unless stated otherwise.

Suppose that an 8 by 8 raster with a binary attribute that is either black (=1) or white (=0), as shown in Figure 3-5A, is subdivided into four blocks or quadrants. Each block is in turn subdivided, but only if it contains a mixture of black and white pixels. If the block is homogenous, no further subdivision is carried out. At the third stage, level 0, the blocks are the size of the pixels in the original raster. In Figure 3-5B, these **maximal blocks** are shown, being labelled in a decimal sequence according to their Morton order, starting at 1 in the SW corner. This image can also be shown as a tree, Figure 3-5C. The **leaf nodes** are shown as boxes, labelled by the sequence number. **Internal nodes** (non-leaf) are shown as circles, labelled by letter.The top node is the **root node**. Each node represents an areal block of varying size, depending on the **level** in the tree. The lowest level (greatest depth) is level 0, in this case depth 3. At this level, the blocks are equivalent to the pixels in the full raster. At level

Table 3-6. Similar to Table 3-5, but retaining only locational codes and levels for blocks belonging to a single class (black). this scheme is only suitable for binery images.

Node	Location	Level
6	122	0
7	123	0
10	210	1
12	221	0
16	300	0

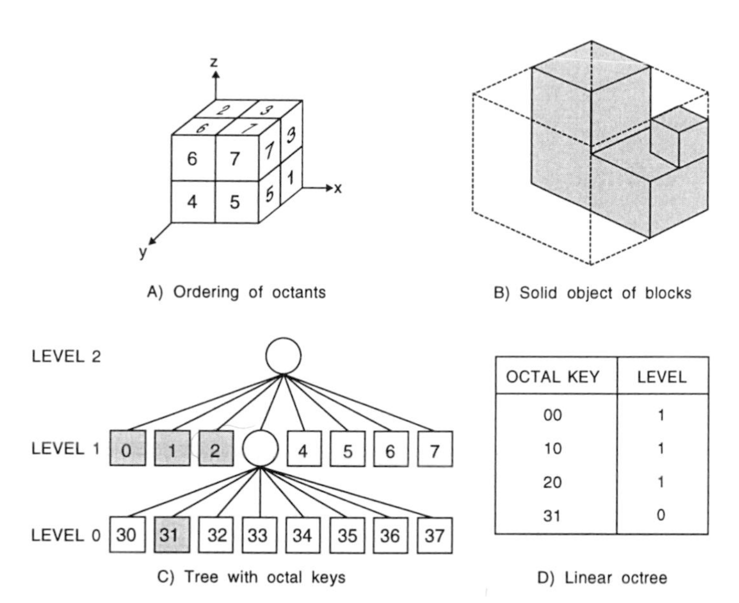

FIG. 3-6. A. Ordering of blocks subdivided into octants, numbered 0 through 7. **B**. Solid object as blocks in a volume with 2^4 voxels. **C**. Octree representation of B., labelled by octal locational code. **D**. Linear octree for black leaf nodes only.

1, the blocks are four times larger than level 0, level 2 blocks are four times the size of level 1 blocks, and so on. Note that some authors use the term *"level"* or "quadlevel" to refer to the "depth". The actual name is not significant, but it is important that depth and level be distinguished. In this book the conventions of Samet (1990a) are used, so that level 0 implies that the blocks are the same size as the pixels of the raster. The spatial origin and extent of the square region represented by the quadtree is known as the **universe**.

In the example in Figure 3-5, leaf nodes can only be either black or white; internal nodes are a mixture of black and white, or grey. Of course, most maps represent attributes with multiple classes, and in these cases leaf nodes contain only pixels (or level 0 blocks) belonging to a single class value. A very basic table structure of the quadtree, Table 3-4, contains a row for each node (both internal and leaf), in this case using the same letters and number pointers shown in Figure 3-5. For the P-th node (row), the first column is the **father** node or Father(P); the second column is the node of the first **son**, or Son(P,1); columns 3, 4 and 5 are the other three sons, Son(P,2), Son(P,3) and Son(P,4); and the final column is the node class or type, in this case either black, white or grey. Leaf nodes have no sons and their type is either white or black; internal nodes are grey, being a mixture of classes. The sons of internal nodes are either more internal nodes or leaf nodes. This table represents a structure that clearly is not very efficient because the leaf nodes all point to NIL, and go no further. There are several alternative structures for quadtrees that are more economic of space, either using pointers or

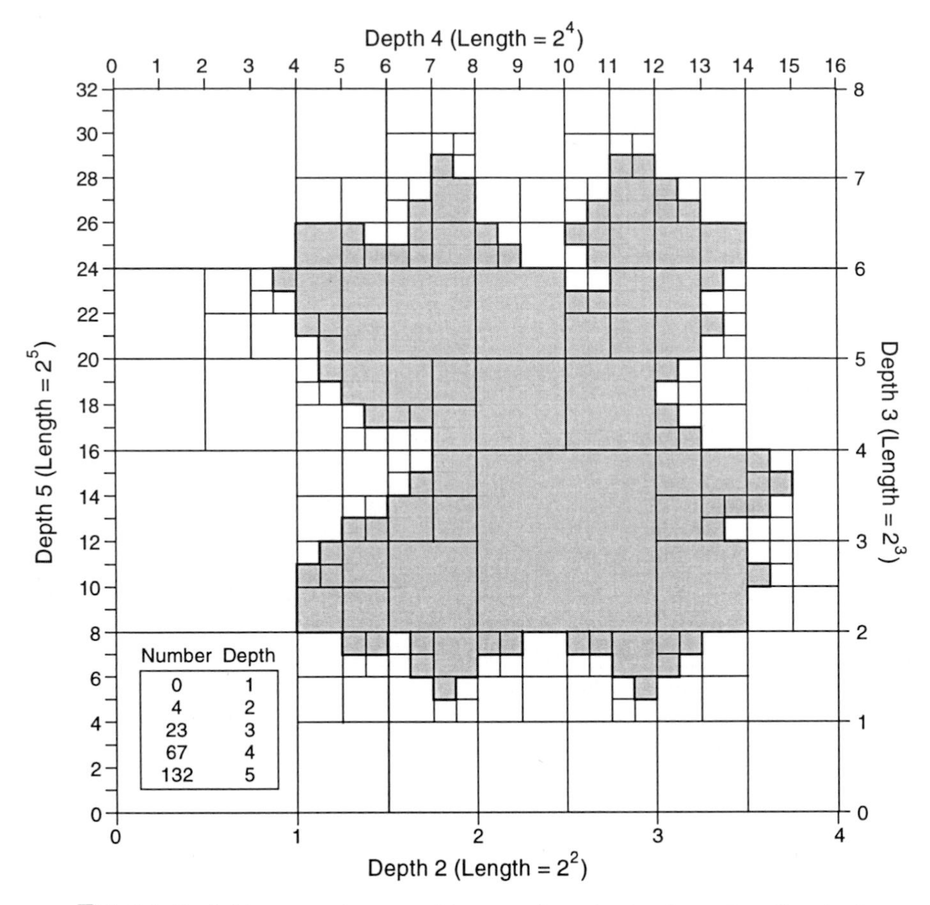

FIG. 3-7. Simple binary map decomposed into a quadtree, showing the number of maximal blocks at each level of the tree (inset). Note that the small blocks are only needed at the boundaries of the shaded object. The length of the side of the area is shown in terms of the number of quads for each depth.

collecting together leaf nodes in Morton sequence and omitting pointers, Samet (1990a). Without going into great detail, some alternative methods are briefly outlined using simplified versions of the tree in Figure 3-5 and Table 3-4.

Table 3-5 shows a representation that retains only the leaf nodes as rows. Pointers are omitted altogether. The first column is the leaf node number, and can be omitted as long as the sequence of rows is retained in Morton order. The second column is a locational code (base 4 Morton coordinate); the third column is the level of the leaf; and the fourth column is the leaf type or class. Notice that the difference in base 4 between Morton coordinates of adjacent rows is related to the level. Thus, the difference in locational codes between leaf node 1 and leaf node 2 is base 4 (100-000)=100, and the level of leaf node 1 is 2. The Morton difference between leaves 2 and 3 is 010 and the level of leaf 2 is 1. The Morton difference between leaves 4 and 5 is 001 and the level of leaf 4 is 0. The pattern is

Morton difference	level
100	2
010	1
001	0

due simply to the fact that block area increases by a power of 4 with unit increase in level. This fact leads to two possible further simplifications.

Given the level and class of blocks in Morton order, i.e. <level(1), class(1), level(2), class(2), ... level(k), class(k)> where k is the total number of leaf nodes in the quadtree, the whole tree is completely specified without pointers or locational codes, as proposed by Gargantini (1982). This is known as a **linear quadtree** (not to be confused with a line quadtree). The tree can be thought of as a list of ordered pairs, rather like run-length code except that runs follow the Morton order and are constrained by the boundaries of homogenous quadrants. The Morton coordinate for any block is found by summing 4^{level} for all the previous blocks in the tree in decimal and converting to base 4. Thus in Table 3-5, the tenth position in the Morton sequence of leaf nodes has the Morton key equal to $(4^2+4^1+4^1+4^0+4^0+4^0+4^0+4^1+4^1)$, which is decimal 36 or 210 in base 4. In practice, navigation through a linear quadtree can be accelerated by constructing an accompanying index table, also in Morton order, as discussed by Gahegan (1989). For a particular search, the closest position in the index is found; this acts as the entry point to the linear quadtree and reduces the search of the main tree to the interval between index points. Two dimensional run encoding in Morton order can also be useful for additional compression of the quadtree, as discussed by Lauzon et al. (1985).

The further simplification of Table 3-5 is to retain only the locational codes and levels of particular leaf classes. Suppose that only the black pixels are required, the list would contain just 5 pairs as follows <(122,0),(123,0),(210,1),(221,0),(300,0), as shown in Table 3-6. In other words, instead of using (level, class) pairs for the complete tree, (location code, level) pairs are used for the subset of leaf nodes that belong to a particular class. This scheme has been described by Gargantini (1989) for representing 3-D spatial objects with octrees. Octrees are similar in every way to quadtrees except each father has 8 sons instead of 4, and locational coordinates are base 8 (octal) numbers. Figure 3-6 shows a solid object and its octree representation both graphically and as a linear octree.

Kavouras and Masry (1987) use a linear octree for 3-D solid objects in which black nodes and grey (i.e.internal) nodes are retained, but not white ones. This allows the possibility of creating low-resolution generalized images by choosing an intermediate depth of the tree for display.

The basic advantage of using quadtrees and octrees instead of a full raster structure is to reduce the space requirements for raster data. Typically, a doubling of spatial resolution (e.g. of rows and columns) quadruples the size of the raster, but only doubles the size of the quadtree. Figure 3-7 shows a binary raster image with an arbitrary shape, and illustrates the subdivision of the image into quads (maximal blocks) at various depths of a quadtree. Depth 0 corresponds to the universe, containing a single large square. At depth 1, there are 4 possible quads, but all of them contain a mixture of classes, so none of them are "maximal" and they are all subdivided. At depth 2, there are up to 16 maximal blocks, but only 4 of them are homogenous and the other 12 are further subdivided at the next level. At depth 3, there are 23 maximal blocks, 67 at depth 4 and 132 at depth 5. At this level, there are up to a possible maximum of 32 times 32 blocks of the smallest size, so an ordinary raster with this resolution would occupy 32*32=1024 pixels. The total number of maximal blocks of all depths actually used in the quadtree is 226. Each quad requires a (level, class) pair of numbers in the data structure, so the number of computer words required is 226*2=452, as compared to 1024 for the full raster. This represents about a 50% saving of space, assuming one byte per pixel, and two bytes for a (level,class) pair. Now as the resolution of the image is increased, also increasing the depth of the quadtree, the difference between the storage requirements for the raster and the quadtree becomes more accentuated. For example, in Figure 3-2A, the full raster image of lakes occupies 794*893=709402 bytes, whereas a quadtree calculated to a depth at which the smallest quads are the same size as the original pixels required 19473 bytes, a compression factor of more than 36 times. Images with continuous variation between adjacent pixels sometimes do not compress at all. In the worst case of a chess-board pattern, a linear quadtree is double the size of a full raster, because each pixel is a leaf node at level 0, requiring a (level, class) pair instead of just a class number for a full raster. However in practice, even 8-bit Landsat images are usually found to compress by about 10% (depending on the scene content), and more if they are reclassified into a smaller number of classes. Just as important as the space-saving is the efficiency of some algorithms for processing quadtrees. Because quadtree locations can be addressed by a single Morton coordinate, (or found in a linear quadtree by cumulatively adding 4^{level} for each block as the tree is traversed in Morton order), searching for a block with a particular address is relatively fast, and can be further accelerated by using an additional system of locational pointers. Overlay operations between two or more images in quadtree form are also efficient. Even if the levels of two maps in a quadtree structure are different, the block boundaries are aligned as in a full raster, and what may be a leaf node in one tree can be readily matched to a spatially equivalent hierarchy of internal nodes and leaves in other trees. Operations like finding all the contiguous regions belonging to the same class (known as **connected component labelling**) take advantage of the tree structure, although the algorithm and its efficiency varies depending whether or not the structure uses pointers, Samet (1990a).

On the other hand, building quadtrees or octrees, either from a full raster or from vector data, can be time-consuming, particularly for trees with a large number of levels and with a great deal of spatial variation. Similarly some processing operations are slow with quadtree structures, compared with a full raster, particularly those that use 4-connected or 8-connected neighbours. Display of quadtree data is slower than display of run-length data. Quadtrees are not simple to translate, rotate or scale, operations that require the creation of a totally new tree. Quadtrees from one universe cannot be used directly with quadtrees from another universe. Quadtrees are, therefore, better suited for project-related rather than custodial GIS. The choice of whether or not to use quadtrees for raster data is generally a tradeoff between processing speed and storage constraints.

Lines and Points in Raster

Lines can be approximated in raster by chains of connected pixels. Points can likewise be approximated as single pixels. For many applications, raster resolution is too coarse for closely-spaced lines or points. In a full raster layer, it is generally undesirable to fill a whole pixel to represent a single point, that in reality has zero area. Similarly a string of connected pixels is a poor way to represent a line that in fact has zero thickness. This is particularly a problem when the scale is changed: points and lines either get too big or disappear, depending whether the scale is increased or decreased.

Tomlin (1990) illustrates a raster system which distinguishes between 47 types of cell containing line information. He calls each type a **lineal condition**. For example, type 0 is a single point at the cell centre; type 6 is a line segment from the cell centre to the upper right

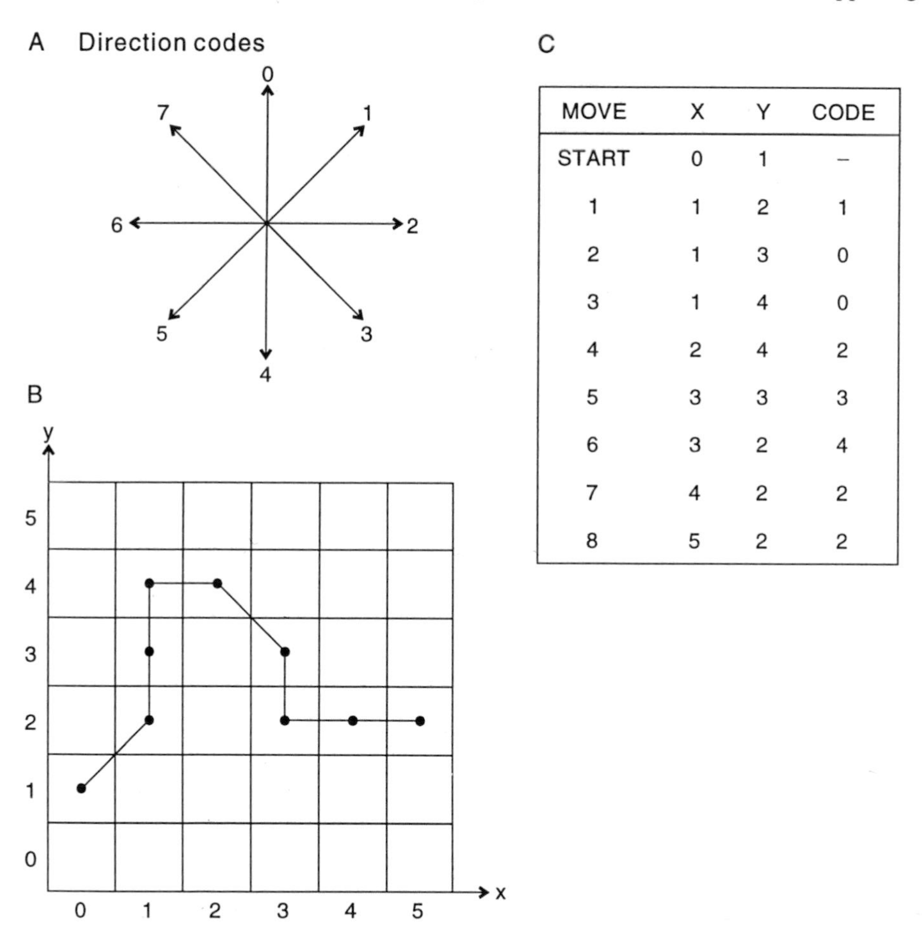

MOVE	X	Y	CODE
START	0	1	–
1	1	2	1
2	1	3	0
3	1	4	0
4	2	4	2
5	3	3	3
6	3	2	4
7	4	2	2
8	5	2	2

FIG. 3-8. Freeman code. **A**. Move directions and numbering system. **B**. Example of a line. **C**. Freeman code for line, plus associated (x,y) coordinates.

corner; type 39 contains 3 line segments leading from the cell centre, one to the upper left, one to lower right and one to lower left. Lineal characteristics, like a pipeline or a property boundary, can be represented in a data structure by the presence or absence of adjacent cells classified according to these lineal conditions. Such a structure carries important geometric information about the lineal characteristics of spatial objects, which is advantageous for some applications like the analysis of transportation or utility networks.

A well-known structure for lines is the **Freeman code**, (Freeman, 1961). Instead of representing a chain of pixels as a set of coordinate pairs, the incremental change in x and y directions is coded as a series of move directions, usually a maximum of eight possible directions, and assigned the integers 0 through 7, see Figure 3-8. Freeman code requires only one number per move instead of a coordinate pair. Further, the move direction requires only 3 bits (numbers up to 2^3), whereas spatial coordinates require a greater number of bits to give the requisite precision. Freeman code is not particularly useful for general applications, but is handy as an intermediate structure, mainly for vector data originating from a raster scanner or from an image that is vectorized.

In a GIS that uses quadtrees, points (and the vertices of vectors) can be addressed using Morton coordinates. This has the advantage that a single spatial coordinate is used instead of two coordinates, and that the point address can be used to find the attribute values at that location in any of the quadtree map layers. Of course, the spatial origin and extent of the quadtree universe must be compatible with the Morton coordinates for the points. Point locations are generally specified assuming a quadtree depth of at least 16 (equivalent to a raster with 2^{16} pixels per side). At depth 16, a Morton coordinate needs 32 bits of storage space, which can be accommodated by a 32-bit computer word, so precise positioning of points (and lines) may require more than one 32-bit word for each Morton coordinate. When Morton coordinates have been calculated for a table of points, such as sample locations, and the samples sorted into ascending Morton order, the spatial location is not only compactly represented, it is also in a convenient form for spatial search operations, and for merging point with area data.

We now move on to consider the data structures used for the vector data model. The spaghetti model, the topological model and the TIN model, introduced in Chapter 2, are associated with various spaghetti, topological and TIN structures. Although the distinction between data model and data structure is poorly defined here, the data structure is closer to being a "blueprint" for a digital file format than the data model.

VECTOR DATA STRUCTURES

Spaghetti Structure

In the spaghetti structure, tables of locational coordinates are associated with each of the basic spatial objects (points, lines, or polygons) as illustrated in Table 3-7. No topological attributes are used, so that navigating around a map must be accomplished by searching lists of spatial coordinates. This is costly for many search operations, but is efficient for display purposes. Separate tables are used for points, lines and polygons, and no attempt is made to

Table 3-7. Tables showing a very simple data structure ("spaghetti structure") for points, lines and closed polygons. Note that the spatial and non-spatial attributes are held in the same tables, and that no topological data accompanies the lines and polygons. Where polygons form an interlocking mosaic, all interior boundaries are defined twice, being part of two adjacent polygons.

A. Point table. X and Y are locational coordinates, A_1, A_2, ...A_n are thematic attributes. Each record or row is a single point object, such as a mineral deposit location, or geochemical sample site.

ID #	X	Y	A_1	A_2	..	A_n
1	x_1	y_1	a_{11}	a_{12}	.	a_{1n}
2	x_2	y_2	a_{21}	a_{22}	.	a_{2n}
3	x_3	y_3	a_{31}	a_{32}	.	a_{3n}
.	.	.	.	.	.	.
...	.	.	.	.	.	.
m	x_m	y_m	a_{m1}	a_{m2}	.	a_{mn}

B. Line table[1]. Many lines are held in the same table or file. Each new line begins with a header (one or more records), followed by the locational coordinates of the vertices or points defining the line. In this case the first field of the header record is the line ID#, the second field is the number of vertices, and the third and fourth (or more) fields are attributes, such as feature codes. There are *m* lines.

1	**5**	**2**	**7**	**Header for line 1**
	x_1	y_1		Coordinates of vertices for line 1
	x_2	y_2		
	x_3	y_3		
	x_4	y_4		
	x_5	y_5		
2	**2**	**4**	**7**	**Header for line 2**
	x_1	y_1		Coordinates for line 2
	x_2	y_2		
3	**15**	**2**	**8**	**Header for line 3**
.	.	.		.
m	etc	etc		etc

[1] The table is nonstandard, because it contains more than one kind of record.

C. Polygon table[1]. This is essentially the same as for lines, except that the last vertex has the same coordinates as the first vertex in each polygon. Therefore there must be a minimum of four vertices per polygon. Each polygon may have many attributes, in which case the attribute data are held in a separate table, linked by polygon number. One attribute must define priority for plotting, to take care of the presence of islands. There are *m* polygons.

1	**5**	**429**	**18**	**Header for poly 1**
	x_1	y_1		Coordinates of vertices for polygon 1
	x_2	y_2		
	x_3	y_3		
	x_4	y_4		
	x_5	y_5		
2	**4**	**39**	**12**	**Header for poly 2**
	x_1	y_1		Coordinates for polygon 2
	x_2	y_2		
	x_3	y_3		
	x_4	y_4		
3	**81**	**9**	**3**	**Header for polygon 3**
.	.	.		.
m	etc	etc		etc

[1] This table is also nonstandard, because it contains more than one kind of record.

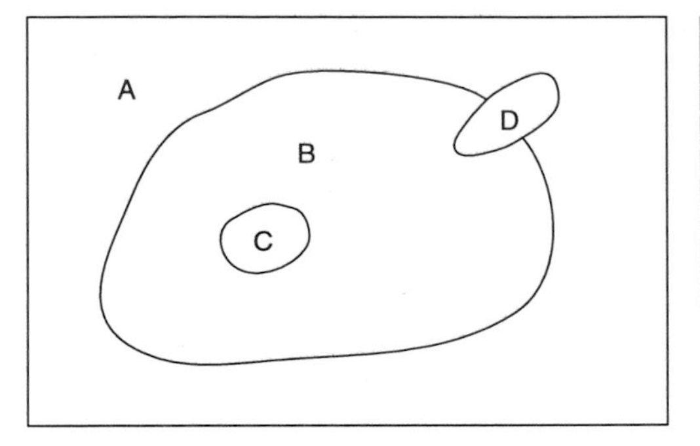

Rock	Polygon	Priority
shale	A	0
sandstone	B	1
granite	C	2
granite	D	2

FIG. 3-9. Using priority in a spaghetti structure to denote islands. Plotting sequence is determined by priority.

provide linkages between spatial objects. All such linkages are determined by computation from the spatial coordinates. Even on a map of moderate complexity, spatial coordinates can occupy large amounts of storage space, so that accomplishing tasks like finding whether a point is in a polygon, or locating the intersection of two lines, is time consuming and expensive. Consider the three types of tables:

1. Point tables are straightforward. Each point is a row of the table, with the locational attributes as columns, see Table 3-7A. Nonspatial attributes are often associated spatial coordinates in the same table.

2. Lines are strings of connected straight-line segments defined by ordered sequences of points or vertices. Because the number of vertices used to define a line is variable, line tables usually have two kinds of record: a record with line id, number of vertices and possibly other attributes, followed by coordinate records, one for each vertex, see Table 3-7B. ("Table" may be a poor word for line and polygon tables, because the number of rows and columns is not fixed).

3. Polygon tables are very similar to line tables, except that the last vertex is the same as the first vertex, see Table 3-7C. For drawing the boundaries of polygons, this simple structure is adequate. However, if the polygons are to be filled by colour or pattern for display, those polygons that are islands within other polygons must be recognized. The simplest way to handle islands is to add one "island" attribute to each polygon header indicating a priority. Low priority polygons are drawn first and filled. High priority polygons are plotted last, so that island polygons overwrite the earlier polygons and take precedence over them, see Figure 3-9. Contour maps require several levels of priority, because contours form a nested sequence of islands within islands.

The great advantage of spaghetti structures is their sequential organization for digital plotting. There are two main disadvantages of the spaghetti approach. The first is data redundancy, the second is computational expense due to the absence of topological attributes.

Redundancy comes about because polygon boundaries are repeated. A map consisting of a mosaic of polygons, with each polygon saved as a closed line, repeats each line twice, except where the line forms the edge of the map. This is simply because each line is on the boundary of two adjacent polygons. Sometimes, polygons require thousands of vertices, and the size of polygon spaghetti files can almost double the size of a file in which polygon boundaries are recorded only once, as in the topological model. In addition, if polygons are digitized as spaghetti, the second version of every line will not be identical to the first, leading to the creation of artificial gaps **(slivers)** and overlaps between adjacent polygons.

Spaghetti files are sometimes called **unstructured** because topological relationships must be derived through computation. To illustrate the importance of topological attributes, consider three tasks where spatial relationships between objects are used.

1. Find all granite contacts on a geological map that are also limestone contacts.

2. Remove all boundary lines between adjacent polygons that have the same classification. This is a common problem for maps that have been simplified by reclassification. A geological example is the reclassification of geological formations into new map classes on the basis of geological age or lithology.

3. Find points on a structure map where fault traces intersect.

With spaghetti data, for the first case, one might start by making a list of granite polygons and another list of limestone polygons. Then the vertices of each granite polygon must be matched with the vertices of every limestone polygon, clearly a very time consuming set of computations.The second case is similar to the first. All polygons belonging to the same class need to be matched with one another to find common boundaries. In the third case, each fault must be matched with every other fault, but this time pairwise comparison of vertices is not enough, because faults could intersect anywhere, not just at vertex points. Each adjacent vertex pair from one fault must be compared with every adjacent vertex pair of another fault to see if the lines cross. Again, this is expensive for large datasets.

With a topological data structure, these three examples become trivial. Spatial coordinates are actually unnecessary for these tasks, because all the computations can be carried out on tables of topological attributes.

Topological Data Structures

Some additional terminology is required here, to indicate some special kinds of points and lines. **Points** are either isolated, or linked to form lines, in which case they are **vertices.** A **line** is a sequence of ordered vertices, where the beginning of the line is a special vertex or **start node** and the end a special vertex called an **end node**. A **chain** is a line which is part of one or more polygons. It can have (left, right) polygon identifiers as well as (start,end) nodes. Chains are also called **arcs** or **edges**. A **node** is a point where lines or chains meet or terminate. A **polygon** consists of one outer **ring** and zero or more inner rings. A **ring** consists of one or more chains. A **simple polygon** has no inner ring, whereas a **nonsimple or complex polygon** has one or more inner rings and is said to have "holes" or "islands".

A Basic Topological Structure for a Map

There are many different topological structures that are in use or have been proposed. Structures differ in detail, but in general they conform to a pattern that can be recognized through a "jungle" of specific terminology. The first basic structure described here is one

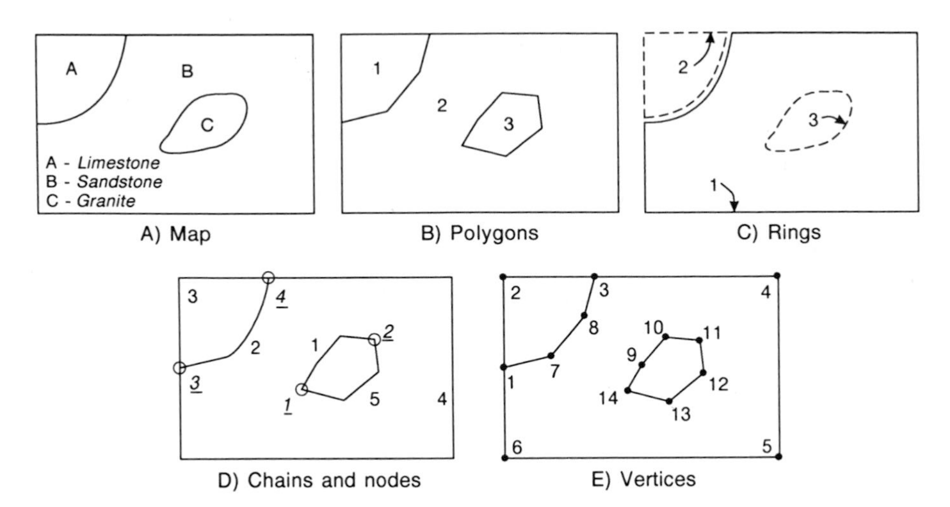

FIG. 3-10. Illustration of spatial components used in van Roessel's topological data structure. This figure should be examined in combination with the relational tables in Table 3-8. **A**. Geological map, showing three rock types. **B**. Polygons, one per rock type in this instance. **C**. Rings. Each polygon is defined by an outer ring and zero or more inner rings. **D**. Chains and nodes (italics, underlined). Each chain (arc, or edge) starts and ends at a node. **E**. Vertices. These are points that correspond either to nodes, or to locations that define the position of a chain between nodes. Spatial coordinates are associated only with vertices.

Table 3-8. Relational tables defining the topological structure of the geological map in Figure 3-10. The linkages between tables are summarized in Figure 3-11.

A. Polygon topology table

Polygon #	Ring #	Ring Sequence #
1	2	1
2	1	1
2	3	2
3	3	1

B. Ring topology table

Ring #	Chain #	Chain Sequence #
2	3	1
2	2	2
1	2	1
1	4	2
3	1	1
3	5	2

C. Chain topology table

Chain #	Start Node	Stop Node	Left Polygon	Right Polygon
1	1	2	2	3
2	3	4	1	2
3	4	3	1	0
4	4	3	0	1
5	1	2	3	2

D. Node-to-vertex table

Node #	Vertex #
1	14
2	11
3	1
4	3

E. Chain-to-vertex table

Chain #	Vertex #	Vertex Sequence #
1	14	1
1	9	2
1	10	3
1	11	4
2	1	1
2	7	2
2	8	3
2	3	4
3	3	1
3	2	2
3	1	3
4	3	1
4	4	2
4	5	3
4	6	4
4	1	5
5	14	1
5	13	2
5	12	3
5	11	4

F. Coordinates of vertices table (part only)

Vertex #	X	Y
1	x_1	y_1
2	x_2	y_2
.	.	.
.	.	.
14	x_{14}	y_{14}

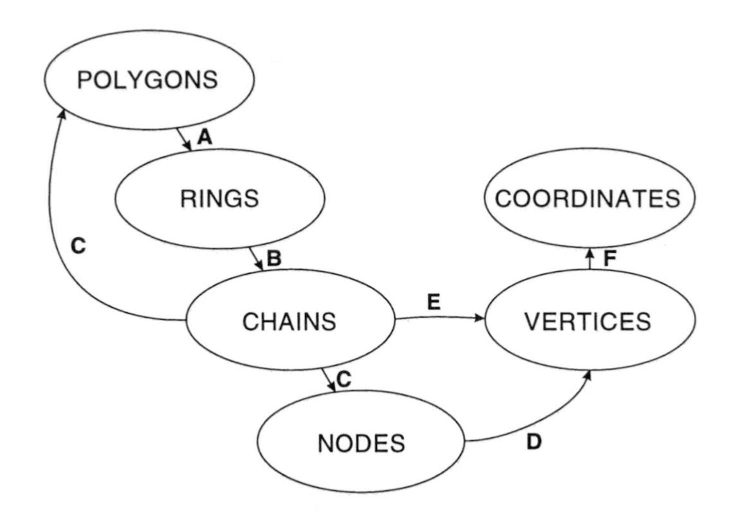

FIG. 3-11. Summary of linkages between van Roessel's topological tables. The arrows with letters denote the tables in Table 3-8. For example, Table 3-8C refers to the chain topological table that links chains to polygons, and chains to nodes.

described in relational normal form by van Roessel (1987), who intended it to be used as a basis for an interchange structure for changing from one vector structure to another. It is not an efficient structure for operational use, but provides a clear and unambiguous description of topological relationships. The structure is described using a simple geological example.

Consider the geological map in Figure 3-10A, which shows regions underlain by three rock types, limestone, sandstone and granite, occurring as three polygonal areas. The polygons are numbered in Figure 3-10B, and there is a 1:1 correspondence between polygon and rock type. Each polygon is made up from one **outer ring** and zero or more **inner rings**. From the figure and table, note that:

1. Polygon 1 is simple, and is defined only by ring 2 (an outer ring),

2. Polygon 2 is nonsimple, being circumscribed by ring 1 (an outer ring) and containing ring 3 (an inner ring) as an island, and

3. Polygon 3 is the island, and is defined by ring 3 only (an outer ring).

Polygons are defined in terms of rings by a polygon topology table, see Table 3-8A. Van Roessel uses a ring sequence number field. This keeps track of cases with more than one ring per polygon, and satisfies the normalization rules of the relational model. If there is more than one ring per polygon, the first one is the **outer ring**, and all the others are **inner rings**.

The second table (Table 3-8B) links rings to chains (Figure 3-10C and D). Ring 2 is made up from chain 2 and chain 3, ring 1 is composed of chains 2 and 4, and chains 1 and 5 form ring 3. This table is the ring topology table.

The third table (Table 3-8C) links chains to nodes and polygons. Thus chain 1 starts at node 1 and ends at node 2; it has polygon 2 on its left and polygon 3 on its right. This table is useful for searching for particular types of polygon contact, without resorting to an exhaustive coordinate search.

Tables 3-8D and E provide linkages from nodes and chains, respectively, to a table of vertices and coordinates, Table 3-8F. Notice that spatial coordinates are held in one table only, quite separate from topological attributes.

This set of six tables completely defines the spatial and topological relationships found on the map. Nonspatial attributes have not been added, but additional tables could be used to link any of the spatial objects to thematic and geometric attributes. The most likely in this case would be a table linking polygons to geological attributes, such as rock formations. But a table containing chain attributes might also be used to store the kinds of geological contact, such as faulted, unconformity, gradational, and so on. Thus, to summarize the relational linkages: polygons point to rings, rings to chains, chains to nodes and polygons, and both nodes and chains to vertices, as shown in Figure 3-11. Vertices are not directly linked to polygons or rings, nor is there a direct link between nodes and rings, nor between nodes and polygons. If necessary, these relationships can be derived from the other tables.

The advantages of this structure over the spaghetti structure are:

1. there is no repetition of spatial coordinates between one polygon and the next, except at nodes, so that repeat lines are eliminated, and

2. topological information is explicitly stored and is separated from the spatial coordinates, facilitating search that requires adjacency, containment and connectivity information.

For example, the problem of finding granite-limestone contacts reduces to a search of the chain topology table for (left,right) polygon pairs that are either (granite,limestone) or (limestone,granite). This is fast and efficient, requiring no reference to the locations of vertices. The problem of finding adjacent polygons belonging to the same class can also be rapidly solved from the chain topology table by looking for (left,right) polygon pairs where left and right have the same class. The third case of fault intersections requires network topology. Here the structure is not concerned with polygons but with intersection and connectivity of a set of lines, to be found in the node topology table. The node list is searched for nodes with at least 2 lines where the lines are classified as faults (as opposed to streams or roads).

The disadvantages of the topological structure are that:

1. topological tables must be generated in the first instance, which is computationally expensive and requires some overhead in storage space, and

2. some simple operations like graphic display are slow and cumbersome, because they require the spatial coordinates in the most accessible form and do not require the topology. For example, to draw the boundaries of all polygons that contain granite would require that chains be selected (from the chain topology table) by finding those records where either the left or right polygon was granite, determining the start and end nodes from the same record, obtaining the coordinates of these nodes from the vertex coordinate table, via the node-vertex table, and finding the coordinates of intervening vertices on the chain via the chain-vertex table.

The choice of whether or not to generate topology depends on whether the data is to be used for analysis, or simply for display. Using the relational form for topological tables has the advantage of being very clear and unambiguous. In addition, editing lines or inserting new lines is relatively simple, because the coordinates are kept separate, repeating groups of attributes are eliminated, each tuple is unambiguously associated with a unique key, and the sequence of tuples is immaterial. In practice, topological vector structures that are used operationally are streamlined, and do not satisfy relational criteria. For example, ring objects are usually omitted, and the spatial coordinates of vertices are directly associated with chains in the same table. Furthermore, the ordering of tuples, particularly the sequencing of vertices, is vitally important for fast access, so one of Codd's relational principles is sacrificed.

Some Operational Topological Structures

To give some flavour of the differences between topological structures for operational use, four structures are compared with the basic relational structure discussed above, as summarized in Figure 3-12. The operational structures are:

1. The POLYVRT structure developed by Peucker and Chrisman (1975),

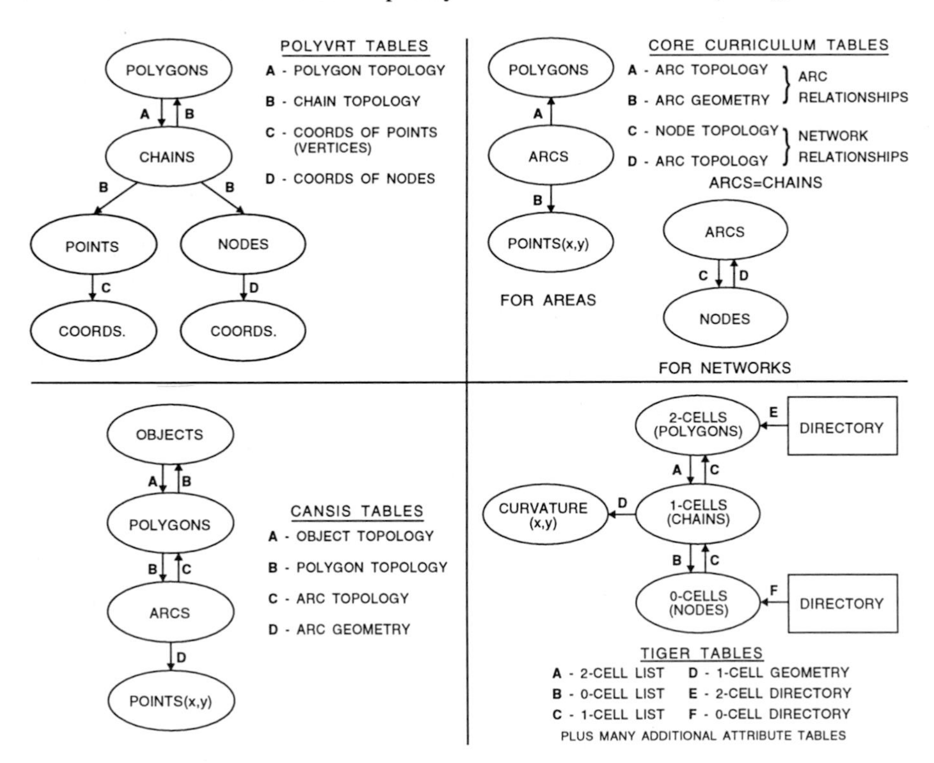

FIG. 3-12. Comparison of four operational topological data structures.

2. a pair of simple structures for area and network relationships described in the NCGIA core curriculum (Goodchild and Kemp, 1990),

3. the CANSIS structure developed by the Canadian Department of Agriculture in the 1970s, also described in the NCGIA core curriculum, and

4. the TIGER structure, a comprehensive data structure, originally developed for the 1990 U.S. census (Marx, 1986).

The POLYVRT structure uses polygons, chains (arcs), nodes and points (vertices). The polygon topology table specifies chains directly, without using intervening rings. Chains that form inner rings are flagged. The POLYVRT chain topology table contains (start,end) node pointers and (left,right) polygon pointers. Coordinate data are held separately from the topological data in two tables, one for vertices, the other for nodes. Within the chain topology table, two columns are used for accessing coordinates of vertices: one column is a pointer, the second indicates the number of vertices. The POLYVRT structure can be used for both areal and network objects. As opposed to the van Roessel relational tables, the sequence of records in the POLYVRT tables is important. For example, the boundary of a polygon can be traversed in the correct order by following the chain and coordinate sequences.

In the NCGIA core curriculum structures, the topological information is reduced to a parsimonious level. For area relationships, an arc (same as chain) topology table contains (left,right) polygons, and an arc geometry table holds the coordinate strings; for network relationships, an arc topology table specifies (start,end) nodes and a node topology table specifies a list of arcs. Notice that there is no polygon topology table, although one could be easily derived from the (left, right) polygon information in the arc topology. Also, node coordinates are not stored separately, but again they could be derived readily from the other tables.

In the CANSIS structure, there are tables linking objects to polygons, polygons to arcs and objects, arcs to polygons, and an arc geometry table containing coordinates of vertices. Here, the word "object" is used to mean a group of polygons (composite polygon) belonging to the same class. Thus a soil class might be an object with a number of polygons. Nodes and node topology are not used in this structure, because the design is for areal objects rather than networks. Sequencing of records is vital.

The TIGER structure is more complex (Marx, 1986), and is only briefly touched upon here. TIGER stands for Topologically Integrated Geographic Encoding and Referencing. TIGER uses some different terminology: "0-cells" are equivalent to nodes, "1-cells" are chains, and "2-cells" are polygons. The topological relationships between 0-, 1- and 2-cells are held in tables (called "lists"). The 0-cell and 2-cell lists both point to 1-cells, and are equivalent to node and polygon topology tables, respectively. The 1-cell list contains pointers to (left, right) 2-cells and (start,stop) 0-cells, just like the chain topology table. A 1-cell "curvature" list stores the coordinate data, in other words it contains the spatial coordinates of vertices. Besides the "list" files, 0-cell and 2-cell objects have "directory" files, allowing rapid and efficient access to particular records in the list files. For example, the 0-cell directory contains a column for a bit-interleaved address called the Peano key (equivalent to Morton number in binary). 0-cell records are sorted by Peano key, allowing efficient spatial search, with pointers leading to the list files which are simply in the order of data entry. Special features of the data structure allow "threading" from one table to another, minimizing data redundancy. Extensive nonspatial attribute data forms an integral part of the data structure, organized according to the 0-, 1- and 2-cell classification.

Vector Structures for Surfaces

Surface modelling has a long tradition in geology, particularly for subsurface problems in the oil and mining industries where observations are made from well and drill holes. Surfaces are either sampling-limited (e.g. stratigraphic surfaces), others are definition-limited (e.g. orebodies). Surface modelling methods are widely used to interpolate continuous variables from irregularly-spaced data points on to a regular grid. Once in the gridded or raster form, surfaces can be displayed, sectioned, combined and analyzed. Surfaces are usually single-valued and can be treated as 2.5-D. Multi-valued surfaces, like salt domes or orebodies, must be treated as fully 3-D.

Like the raster and vector models for 2-D, 3-D objects can be represented either as volumetric grids, also known as voxel models, or with boundary representations using surfaces. Three-dimensional hierarchical data structures (octrees) have been briefly touched on for modelling volumes, providing both data compression and efficient spatial addressing. In this section, vector structures for modelling surfaces are briefly introduced.

For 2.5-D data, where some continuous attribute is measured at a series of (x,y) locations, the value of the attribute can be treated as a vertical coordinate, z. In the case of elevation data, z is truly the position of the ground surface above some datum. Where z values are observed on a regular (x,y) lattice, the surface is simply treated as a mosaic of rectangular facets whose height are proportional to z. In a sense this is a boundary representation, a 2-D boundary lying on top of a 3-D volume that is subdivided into a series of rectangular pillars. This 2-D boundary can also be modelled as triangular facets, with data points at the nodes, as in the TIN model. In order to make planar facets (of any shape) into continuous curved surfaces, mosaics of surface patches may be employed, a technique that is popular in 3-D CAD systems.

The data structure for the TIN model is described here, because it fits in well with topological vector structures and because it is widely used in GIS. To close the chapter, the topic of vector structures for multi-valued surfaces is briefly introduced.

Triangulated Irregular Networks (TIN)

Irregularly distributed points on a surface may be joined together into a network of interlocking triangles. The nodes of the triangles are the original points; the triangles themselves are the polygons; and the sides of the triangles are a special case of chains--they are straight-line segments with the nodes being the only vertices. TINs are often used as a data structure for digitally representing topography, and sometimes for other single-valued surfaces. Each triangle, or "facet", can be treated as planar, with the geometry of the plane completely defined by the (x,y,z) values of the three nodes. The TIN has the advantage that

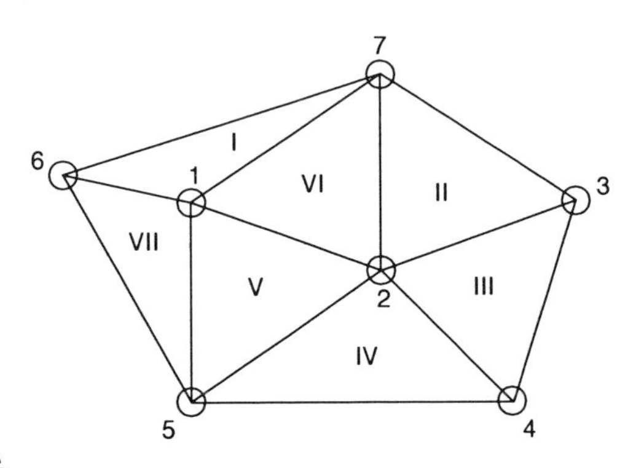

METHOD A

NODE COORDINATES

Node	X	Y
1	.	.
2	.	.
3	.	.
4	.	.
5	.	.
6	.	.
7	.	.

TRIANGLE TOPOLOGY

Triangle	Nodes	Adjacent Triangles
I	1,6,7	VII,O,VI
II	2,7,3	VI,O,III
III	2,3,4	II,O,IV
IV	2,4,5	III,O,V
V	5,1,2	VII,VI,IV
VI	1,7,2	I,II,V
VII	6,1,5	O,I,V

O=Outside

METHOD B

NODE COORDINATES

Node	X	Y	Pointer
1			1
2			6
3			12
4			.
5			.
6			.
7			.

NODE COORDINATES

Index	Connected Nodes
1	5
2	6
3	7
4	2
5	0 —
6	1
7	7
8	3
9	4
10	5
11	0 —
12	2
.	.
.	.

FIG. 3-13. TIN topology. **Method A.** The triangles are the basic spatial objects. Each triangle has topological attributes that describe the nodes and adjacent triangles. **Method B.** The nodes are the basic spatial objects. Each node is linked via a pointer to a table showing the directly-connected nodes, arranged in clockwise order. A zero (null) node is used to indicate the end of a string of connected nodes.

the density of triangles varies with the density of the data points (as opposed to the regular density of pixels in a raster). The topological structure of a TIN can be explicitly stored, leading to efficient processing. For example, algorithms for calculating slope and aspect of the surface, for automatic contouring, hidden-line removal in perspective plots and surface-shading can make effective use of the topological structure. TINs require less storage space than raster structures.

TINs can be generated using a Delaunay triangulation (McCullagh and Ross, 1980; Auerhammer, 1991). They are particularly useful structures for surfaces with discontinuities, because vertices placed at a linear discontinuity can act as a "breakline". For example, streams, cliffs and shorelines can act as different types of breaklines on topographic surfaces. Faults are an important type of breakline on a structure surface.

There are various ways of storing TIN topology. Probably the most common one uses the triangle as the basic spatial object, with topological links to adjacent triangles and a list of nodes. Alternatively, the nodes are the basic spatial objects, with links to all connected nodes.

In the first method, Figure 3-13 (Method A), each record of the triangle topology table lists the three adjacent triangles in clockwise order (triangles adjacent to an edge of the triangulated region use a special symbol for one of the triangles) and the three nodes, also in clockwise order. The spatial coordinates of each node (x,y,z), are saved in a separate file. This structure allows efficient processing where areal adjacency is used. In the second method (Method B), the node coordinate file is the same, except that an additional attribute is used as a pointer to a table of connected nodes (Peucker and Chrisman, 1975), as shown in Figure 3-13. Clearly this is efficient if connectivity along triangle edges is required. The relative merits of these two approaches depends on the processing algorithm.

Another structure, called the **doubly-connected list** (Preparata and Shamos, 1985), is also widely used for triangular networks in computational geometry.

Multi-valued Surfaces

Fully three-dimensional topological models can be developed using polyhedral tesselations. For example, Youngmann (1989) discusses the use of inter-linked tetrahedra for modelling sedimentary basins. This is analogous to the TIN model, but extended into the third dimension, with data points lying at the tetrahedral nodes.

Mallet (1989) shows how triangulated points can be used to model multi-valued surfaces. His solution involves the use of control nodes, which are the original data points, and new interpolated nodes that then form a triangulated surface which satisfies a minimum roughness criterion. Tipper (1978) also has discussed the multi-valued surface problem and uses a solution that involves polynomial patches.

Several authors have described 3-D systems that combine both surface and volumetric representations. For example, Bak and Mill (1989) use bounding surfaces and octrees. The bounding surfaces are made up from the faces of polygons, whose vertices are the known data points. The data structure is a coordinate table for the vertices and a connectivity table linking polygons to vertices. A conversion routine is used to build the octree from its enclosing boundary surface. The surface is relatively straightforward to construct, edit and display; the octree is superior for spatial search and for combining two or more solid images, for example in intersecting a mine excavation with an orebody.

Jones (1989) summarizes how octrees can be used for storing exact definitions of vectors, such as vertices, edges and faces. He calls these vector octrees, but they have also been called polytrees or extended octrees by other authors. Vector octrees appear to offer promise for integrating volumetric and boundary representations within a unified structure.

REFERENCES

Abel, D.J., 1986, Bit-interleaved database access keys in spatial data processing using tesseral methods: *Proceedings of the Tesseral Workshops, National Environment Research Council*, Swindon, Wiltshire, U.K., Eds., Diaz, B.M. and Bell, S.M.B., p. 163.

Auerhammer, F., 1991, Voronoi diagrams: a survey of a fundamental geometric data structure: *ACM Computing Surveys*, v. 23(3), p. 345-405.

Bak, P.R. and Mill, J.B., 1989, Three dimensional representation in a geoscientific resource management system for the minerals industry: In *Three Dimensional Applications in Geographical Information Systems*, Editor, Raper, J., Taylor and Francis, London-New York, p. 155-182.

Freeman, H., 1961, On the encoding of arbitrary geometric configurations: *Transactions on Electronic Computers*, v. EC10, p. 260-268.

Gahegan, M.N., 1989, An efficient use of quadtrees in a geographical information system: *International Journal of Geographical Information Systems*, v. 3, p. 201-214.

Gargantini, I., 1982, An effective way to represent quadtrees: *Communications of ACM*, v. 25, p. 905-910.

Gargantini, I., 1989, Linear octrees for fast processing of three-dimensional objects: *Computer Graphics and Image Processing*, v. 20, p. 365-374.

Goodchild, M.F. and Grandfield, A.W., 1983, Optimizing raster storage: an examination of four alternatives: *Proceedings AUTOCARTO* 6, v. 1, p. 400-407.

Goodchild, M.F. and Kemp, K.K. (Editors), 1990, *NCGIA core curriculum project*: National Center for Geographic Information and Analysis, University of California, Santa Barbara, California, variously paged.

Jones, C.B., 1989, Data structures for three-dimensional spatial information systems in geology: *International Journal of Geographical Information Systems*, v. 3 (1), p. 15-31.

Kavouras, M. and Masry, S.E., 1985, An information system for geosciences: design considerations: *Proceedings AUTOCARTO* 8, p. 336-345.

Lauzon, J.P., Mark, D.M., Kikuchi, L. and Guevara, J.A., 1985, Two-dimensional run-encoding for quadtree representation: *Computer Vision, Graphics and Image Processing*, v. 30, p. 56-69.

Mallet, J.L., 1989, Discrete smooth interpolation: *ACM Transactions on Graphics*, v. 8(2), p. 121-144.

Marx, R.W., 1986, The TIGER system: automating the geographic structure of the United States census: Reprinted in *Introductory Readings in Geographic Information Systems*, 1990, Editors: Peuquet, D.J. and Marble, D.F., Taylor and Francis, London, p. 120-141.

McCullagh, M.J. and Ross, C.G., 1980, Delaunay triangulation of a random data set for isorithmic mapping: *Cartographic Journal*, v. 17, p. 93-99.

Peucker, T.K. and Chrisman, N., 1975, Geographic data structures: *American Cartographer*, v. 2, p. 55-69.

Preparata, F.P. and Shamos, M.I., 1985, *Computational Geometry: An Introduction*: Springer-Verlag, Berlin, 390 p.

Samet, H., 1990a, *Applications of Spatial Data Structures*: Addison-Wesley Publishing Company, New York-Amsterdam, 507 p.

Samet, H., 1990b, *The Design and Analysis of Spatial Data Structures*: Addison-Wesley Publishing Company, New York-Amsterdam, 493 p.

Tipper, J.C., 1978, Computerized modeling for shape analysis in geology: In *Recent Advances in Geomathematics*: Editor: Merriam, D.F., Pergamon Press, Oxford, p. 157-170.

Tomlin, C.D., 1990, *Geographic Information Systems and Cartographic Modelling*: Prentice Hall, Englewood Cliffs, New Jersey, 249 p.

van Roessel, J.W., 1987, Design of a spatial data structure using the relational normal form: *International Journal of Geographical Information Systems*, v. 1, p. 33-50.

Youngmann, C., 1989, Spatial data structures for modelling subsurface features: In *Three Dimensional Applications in Geographical Information Systems*, Editor: Raper, J., Taylor and Francis, London-New York, p.129-136.

CHAPTER 4

Spatial Data Input

INTRODUCTION

A major proportion of the effort in any GIS project is tied up in assembling the data in digital form, and creating a spatial database in which all the maps, images and spatial data tables are properly geocoded and in spatial register. The topics covered in this chapter are related to various aspects of spatial data capture and conversion. They are **data sources, map projections**, **digitizing** and **coordinate conversion**.

Data sources can be broadly classified according to whether they are **primary** or **secondary** and digital or nondigital. Most GIS projects use secondary data, by which is meant data previously gathered, manipulated and stored by others. Secondary data for GIS input consists mainly of maps (nondigital or analogue), tables (often digital) and images (nearly always digital).

The subject of **map projections** is probably regarded by most GIS users as a tiresome red herring, a necessary obstacle to be overcome before the interesting part of a study can begin. But projections are significant for GIS both at the data input and data visualization stages. For example, errors in map registration, caused by incorrect specification of projection parameters at the time of map input, can make nonsense of subsequent analyses.

The process of **digitizing** maps using manual and automatic methods is the bottleneck for most GIS work. Improvements in scanning technology and automatic feature recognition may improve this phase of data capture in the future.

Digital data are often not in the correct form for use in a particular GIS and require data conversion. Many GIS contain routines for importing and converting data from other sytems, via a variety of interchange formats. These conversions concern changes in file format and are not discussed further. On the other hand, **coordinate conversions** are often required. These are more fundamental, and a section is devoted to some general principles. Typical coordinate conversions involve the transformation of table coordinates derived from a digitizer to the coordinates of a map projection, the forward and inverse transformations from geographic to projection coordinates, and the conversion of arbitrary raster coordinates to the coordinates of some known projection.

Prior to the availability of modern GIS, spatial data capture and conversion were sometimes difficult and frustrating tasks. One of the advantages of GIS is that computing tools have been packaged to make the process easier, making more time available for the main purpose of understanding and predicting spatial relationships in the data.

DATA SOURCES

Examples of some primary and secondary data sources for geosciences are shown in Table 4-1.

Table 4-1. Types of data sources for GIS, with some examples.

	PRIMARY	SECONDARY
NONDIGITAL	Field mapping Hand-recorded data Analogue well logs	Maps Tables
DIGITAL	Field digitizing Geophysical data Geochemical data Geotechnical data Remote sensing images	Digital databases

Primary data

Primary geophysical, geochemical and geotechnical data are now collected to a large extent by instruments that record digitally. These data are usually observations taken at points, either using *in situ* measurements or laboratory measurements on samples. Geographic locations of sample sites are, in many cases, established by hand from identifiable ground features shown on topographic base maps. Locations on base maps may be digitized on a digitizing table, or measured directly in the field with Global Positioning System (GPS) instruments using satellites. GPS will undoubtedly become very widespread for primary collection of spatial coordinates as the accuracy and portability of these instruments improve and their costs decrease.

Although geological mapping has traditionally been carried out by a multi-stage manual drafting and compilation process, portable field digitizing systems now offer computer drafting and spatial database management tools. The recent trend towards the adoption of computing tools for field mapping is a significant step towards the goal of building geological map databases. Usually the mapper still records observations on to an air photograph or topographic base map, but the data can be transferred on a regular basis to a digital base in field camp. Portable computers with digitizing facilities, usually equipped with low-cost CAD (computer aided drawing) software, are becoming commonplace field tools. Systems have been developed to aid geological mapping, not only for collecting the graphical data but also for systematic recording of lithology, structure and other observations in a digital format.

For example, FIELDLOG (Brodaric and Fyon, 1988; Thomas, 1991) is a computer-based field mapping and data storage system developed at the Ontario Geological Survey and Geological Survey of Canada. The system provides for data entry in either graphic or keyboard mode. Graphical data is captured using widely available CAD software. Attribute data are stored in a commercially-available relational database. The system allows field survey data to be stored, managed and retrieved in a variety of graphic and tabular formats. The data-gathering-to-map publication cycle is shortened substantially. Such a system can also be a "user-friendly" front-end to a GIS. Customized data definitions are created using a language designed specially for field geologists. In contrast, the digitizing modules of many full-featured GIS packages tend to be complex, unfriendly and use terms unfamiliar to non-specialists. The conversion of CAD drawings to topologically complete files can usually be carried out within a GIS, although the mode of CAD digitizing can greatly affect the ease of this conversion process.

Having improved the *mechanics* of digital capture of spatial data, there is now a great need for comprehensive data models to be developed for geological maps of all types. The data models need to be structured in such a way as to be flexible and modifiable as far as possible, because no two geologists will agree exactly on how data should be organized. Unless there can be some agreement on, and general adoption of, data models, it is difficult to make progress towards the ultimate goal of producing "seamless" databases, in which the join between two or more adjacent sheets is invisible.

Secondary data

When primary data are interpreted, edited and processed for use by others, they become secondary data sources. Many of the commonly-used sources of secondary spatial geoscientific data are not in digital form, such as maps, printed tables, geophysical well logs on paper, and geocoded information in journals and books. Digital capture of these data is usually only practical for specific projects, although in some cases large programs of digitization have and are being undertaken. For example, many of the older paper well logs have been converted to digital form by oil companies, and topographic maps are being converted to digital form in several countries. Outside the geoscientific world, utility companies are spending millions of dollars on data conversion.

There is an increasing amount of secondary digital data available for GIS projects, mostly lodged in custodial databases. The utility of such data varies depending on the data models, data structures and data standards that have been applied. The question of standards is vital for the sharing of GIS data. Standards pertain not only to data interchange formats, but also to the underlying data models and data structures.

The most difficult area for establishing standards is for geological mapping. Geological maps are interpretive and subjective. When a map is published, it represents an interpretation of one (or a few) geologists, at one moment in time. Even adjacent regions mapped by the same geologist, over a gap of a few years, may be incompatible, because new concepts have been developed that affect the underlying interpretation and the resulting map model. The data model also changes with the scale of mapping. For example, map units that are appropriate at a scale of 1:250,000 are too generalized for 1:20,000 scale mapping, and vice-versa. In addition to incompatible data models and structures, the question of standards for file structures compounds the difficulties.

Standards for data models, structures and files pertaining to laboratory or *in situ* measurements on point data are somewhat easier to establish than those for maps. Regional geochemical survey data are typical examples of point data, consisting of geochemical element determinations and site characteristics of samples. Standard methodologies are established for field sampling, laboratory procedures, methods of sample preparation, chemical decomposition and analysis.The accuracy and precision of geochemical levels are essential characteristics for later interpretive studies, particularly those involving regional comparisons. Control samples are inserted into each sample batch, some controls being standard reference materials whose chemical compositions have been accurately characterized, others being field duplicates. Control samples permit an evaluation of accuracy and precision of any geochemical element. GIS projects that involve environmental analysis, mineral exploration and other applications of geochemical data depend on such control data for information about data reliability.

Some kinds of secondary data are extensively pre-processed before being archived and distributed. For effective data sharing, standards for pre-processing must be applied. For example, airborne geophysical data undergo processing to convert digital signals to calibrated values of the variables being measured. Gravity data are combined with elevation data to produce free-air and Bouguer anomaly maps. Satellite images are processed to remove undesirable effects due to such factors as "striping" (caused by an error to a line of the image), atmospheric attenuation and motion of the spacecraft.

Standards describing the provenance of secondary data become even more critical for "derived" data sources. These are data that have been extensively manipulated and combined, often within a GIS. A derived GIS map layer may have undergone a straightforward transformation, like a reclassification, or it may have been modified from the original source

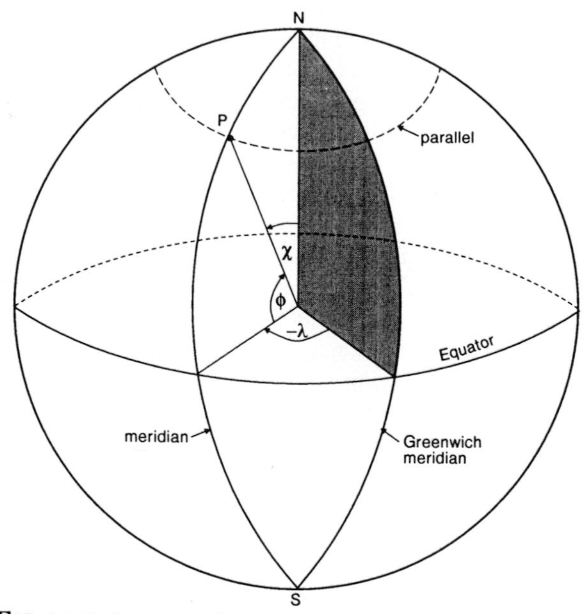

FIG. 4-1. Definition sketch for geographic coordinates. The plane of the Greenwich meridian is shown cross-hatched. Meridians are great circles, parallels are small circles. Geographic coordinates of point P are defined by the angles ϕ (latitude) and λ (longitude). The co-latitude $\chi = 90 - \phi$.

by a series of transformations and data combinations. The subsequent use of derived datasets by others is dependent on the availability of an informative "audit trail" that records all prior manipulations. Data about data are often referred to as **metadata**.

Many government and industrial organizations are attempting to establish and introduce standards for digital spatial data in the geosciences. This is clearly an issue whose outcome will have a bearing on the effective utilization of geoscientific data for GIS projects, particularly those that involve studies on a continental and global scale.

One of the most important steps in building a spatial database is the geometric conversion of spatial coordinates to a common geographic projection. The principles of map projections are discussed here, because it is helpful for a GIS user to understand: 1) the relative merits of different projections, 2) the parameters needed for defining a particular projection, and 3) the algebraic concepts for making transformations from one projection to another.

MAP PROJECTIONS

The location of a spatial entity on the earth's surface is defined in mathematical terms using either geographical (sometimes called global) coordinates, or planar coordinates according to some projection. It is possible for a GIS to store and manipulate all spatial data in geographical coordinates (latitudes and longitudes). Ultimately, however, all spatial data are visualized on paper, or film, or flat video monitors, with planar coordinates. Thus, most GIS use planar map projections for storing spatial coordinates, in order to avoid the repeated transformation from geographic to projection coordinates every time the data are viewed.

There are many kinds of map projection, devised for different purposes. Maling (1992) and Snyder (1987) provide authoritative treatments of the subject. This introduction briefly discusses coordinate systems, the figure of the earth, distortions introduced by projections and some examples of the equations used for transforming from geographic to projection coordinates.

Geographic coordinates

Geographic coordinates are expressed in terms of latitude and longitude. A line joining the N and S pole of the globe through some point on the surface, *P*, is called a **meridian,** see Figure 4-1. The latitude of *P* measures the angle, ϕ, between *P* and the equator along the meridian (although this definition does not strictly hold for an ellipsoidal shape); the longitude measures the angle, λ, between the meridian through *P* and the **central meridian** (through Greenwich, England) in the plane of the equator. A **great circle** is a line at the Earth's surface formed by a plane passing through the Earth's centre (again, this definition is strictly true only for a spherical Earth). Planes passing through the Earth but not intersecting the centre form **small circles** at the Earth's surface. Meridians are thus great circles, whereas lines of constant latitude, called **parallels**, are small circles, except the equator itself, which is a great circle.

The great circle distance between a pair of points on the earth's surface, *A* and *B*, can be calculated from the relationship

$$Dist(A,B) = R \arccos [\sin\phi_A * \sin\phi_B + \cos\phi_A * \cos\phi_B * \cos(\lambda_A - \lambda_B)] \quad (4\text{-}1)$$

where R is the radius, and the earth is assumed to be spherical. A **graticule** is a network of intersecting meridians and parallels as viewed in a projection. As rules of thumb, one second of latitude is about 30 m, or 1 degree is approximately 110 km. One second of longitude varies with latitude from a maximum at the equator to zero at the poles.

Plane coordinates

Locations on a plane are defined by **polar** or **cartesian** (rectilinear) coordinates. Given an arbitrary origin, as shown in Figure 4-2, the distance r and angle θ (with respect to a fixed direction usually pointing N) define the location of point *P* in polar coordinates. With the same origin, and two rectilinear coordinate axes, with *Y* pointing N and *X* pointing E, the familiar cartesian coordinates of *P* are the distances *x* and *y*. The ordered coordinate pairs (r,θ) and (x,y) are readily converted from one coordinate system to the other by the relationships:

$$x = r\sin\theta \quad ; \quad y = r\cos\theta \quad \text{and} \quad (4\text{-}2)$$

$$r = \sqrt{x^2 + y^2} \quad ; \quad \theta = \tan^{-1}(y/x) \quad . \quad (4\text{-}3)$$

Note that here the angle, θ, is defined as the azimuth angle, clockwise from North, not the more common trigonometric convention of being measured counterclockwise from East.

The Pythagorean distance between two points, *A* and *B*, on a plane is defined by

$$Dist(A,B) = \sqrt{(x_A - x_B)^2 + (y_A - y_B)^2} \quad , \quad (4\text{-}4)$$

where the points are located at (x_A, y_A) and (x_B, y_B).

Geometric distortions

Projection transformations from the globe to a plane introduce geometric distortions. A useful device for describing distortion is Tissot's indicatrix. Graphically, this can be represented as the shape of a tiny circle on the globe which becomes distorted to an ellipse in the projection process. Projections can be classified according to their geometric distortion characteristics into **conformal (equiangular)**, **equal area**, and **equidistant** types.

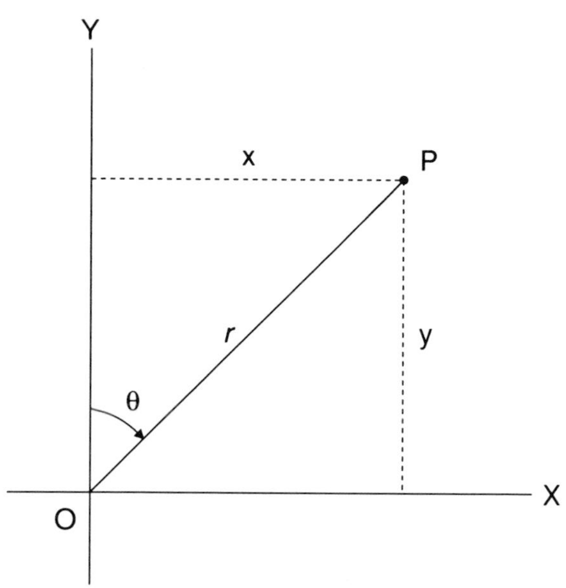

FIG. 4-2. Definition sketch for planar coordinates showing the relationship between polar and Cartesian types.

Conformal projections (e.g. Mercator, Figure 4-3 top) preserve angular relationships between features. Lines of constant orientation with respect to the N pole on the globe are straight on the map (rhumb lines). Parallels and meridians cross at right angles and Tissot's indicatrix is circular everywhere, but varying in size. Small areas remain relatively undistorted, but conformal projections are unsuitable for large regions, such as oceans and continents, because areas are distorted.

Equal area projections (e.g. Albers' equal area projection, Figure 4-3 middle) preserve areas but at the expense of angular relationships. Thus Tissot's indicatrix is constant in area, but varies in elliptical shape. Equal area projections are useful for representing point distributions over large regions, because point density is unaffected, whereas apparent density changes are introduced artificially by conformal or equidistant projections.

Equidistant projections (e.g. azimuthal equidistant projection, Figure 4-3 bottom) preserve neither angular nor area relationships, but distance relationships in certain directions are maintained. Equidistant projections are often used in atlases covering large regions because they are a compromise between the severe angular distortions of equal area maps and the areal distortions of conformal maps.

Figure of the Earth

In order to define projections mathematically, a geometrical model known as the **figure of the earth** is used to generate projections. The simplest geometrical models are the plane and the sphere; a more realistic model is the **spheroid**, a figure produced by rotating an ellipse about its minor axis. The plane is a suitable model only for small regions, such as those shown on vertical air photographs, and amounts to "no projection". The sphere was widely used in traditional map making for large regions, the size of a continent or greater, where the differences introduced due to the use of a sphere instead of a spheroid are negligible. The spheroidal model requires more complicated mathematical calculations for forward and inverse transformations, but must be used for accurate mapping, particularly at large scales. In GIS packages, the spheroid is routinely used at all scales, but because the complicated formulas are embedded in computer code, they are invisible to the user.

The radius of the earth is about 1 part in 300 shorter at the poles than at the equator. The spheroid is thus **oblate**, and can be precisely defined by the lengths of the semi-major (equatorial) and semi-minor (polar) axes, a and b, respectively. The flattening, f, and eccentricity, e, of the spheroid, terms used in some of the transformation equations, are defined by

$$f = \frac{a - b}{a} \quad \text{and} \quad e^2 = 2f - f^2 \quad , \tag{4-5}$$

respectively. Table 4-2 shows the parameters for four spheroids in common use. Maling (1992) lists thirty one determinations of the figure of the earth, and notes that twelve principal spheroids are in use. When datasets are imported into the working projection of a GIS, the correct spheroid must be specified for both the input data and for the working projection.

Developable surfaces

Projections can be classified into **planar (also known as azimuthal)**, **conic** and **cylindrical** types depending on the shape of the **developable surface**. These surfaces can be visualized as flat, cone-shaped or cylindrical, touching or cutting the globe in one of six basic ways, as shown in Figure 4-4. In the **tangent** case, the developable surface touches the globe along a great circle for a cylinder, or along a small circle for a cone, or at a point for a plane. For the **secant** case, the developable surface *cuts* the globe as illustrated in Figure 4-3 (middle).

For **planar projections** in the tangent case, points are projected from the surface of the globe to the plane. A commonly-used projection of this type is the stereographic conformal projection. To convert from geographic to polar coordinates, using a sphere (rather than the more accurate spheroid), and assuming that the surface is tangent at the pole, the following formulas can be used for this projection:

$$r = 2 \tan(\chi / 2) \quad \text{and} \quad \theta = \lambda \quad , \tag{4-6}$$

FIG. 4-3 (opposite). Diagrams to illustrate the distortion characteristics of projections. **Top.** Conformal (equiangular) projection, as illustrated by the ordinary Mercator projection. Note that the developable surface is cylindrical. **Middle.** Preservation of area relationships, illustrated by the Albers' equal area projection.The developable surface is conic. **Bottom.** Preservation of distance relationships, illustrated by the azimuthal equidistant projection, using a planar developable surface. Reproduced from a poster on geographic projections, published by the United States Geological Survey, Reston, Virginia.

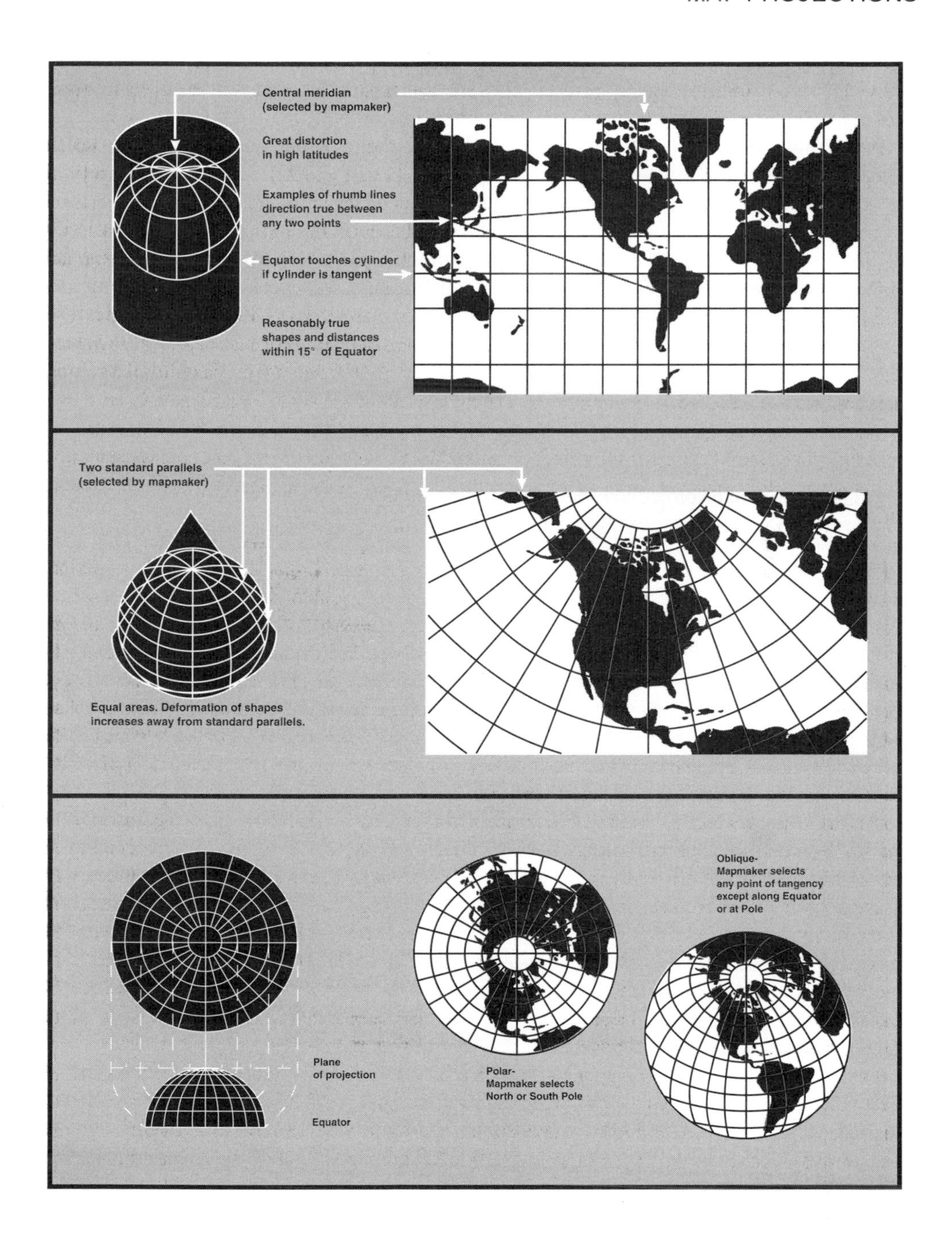
Central meridian
(selected by mapmaker)
Great distortion
in high latitudes
Examples of rhumb lines
direction true between
any two points
Equator touches cylinder
if cylinder is tangent
Reasonably true
shapes and distances
within 15° of Equator
Two standard parallels
(selected by mapmaker)
Equal areas. Deformation of shapes
increases away from standard parallels.
Plane
of projection
Equator
Polar-
Mapmaker selects
North or South Pole
Oblique-
Mapmaker selects
any point of tangency
except along Equator
or at Pole

where the quantity $\chi = (90 - \phi)$ is known as the **colatitude**. The resulting polar coordinates can then be readily converted to cartesian form by equation (4-2). Equation (4-6) is used for the **forward** transformation of geographic to planar coordinates. For data input to a GIS, the **inverse** transformation is usually required first, in order to transform from the plane coordinates of the digitizer to geographic coordinates. Other examples of planar projections are the gnomic, Lambert's equal area and orthographic projections.

For **conic projections**, points from the surface of the globe are projected to a cone that is either tangent at a small circle (one "standard parallel") or intersecting at two small circles (two "standard parallels"). The forward transformation equations for an equidistant conical projection with one standard parallel (at colatitude χ_0) assuming a spherical earth are

$$r = \tan(\chi_0) + \tan(\chi - \chi_0) \quad and \quad \theta = \lambda \cos(\chi_0) \quad . \tag{4-7}$$

Again, the polar coordinates are converted to cartesian coordinates for GIS purposes. A very popular conic projection is Lambert's conformal conic projection with two standard parallels.

Cylindrical projections involve the use of a developable surface that is a cylinder. For the ordinary Mercator projection, the axis of the cylinder passes through the poles, and the cylinder touches the globe at the equator. Equations that assume a spherical earth are

$$x = \lambda \quad and \quad y = \log_e \tan(\pi/4 + \phi/2) \quad . \tag{4-8}$$

One of the most widely-used system of cartesian coordinates is the Universal Transverse Mercator (UTM) system, established in 1936 by the International Union of Geodesy and Geophysics, and adopted by many national and international mapping organizations. The UTM grid utilizes the transverse Mercator projection, which results from wrapping the cylinder round the poles instead of round the equator, as for the ordinary Mercator projection. The **central meridian** is the meridian where the globe touches the sphere, and it changes as the sphere is rotated about the poles. The globe is subdivided into sixty UTM zones, see Figure 4-5, numbered from west to east, starting with zone 1 at 180°W. Each zone is thus six degrees of longitude wide, and extends from 84°N to 80°S. The origin of each zone is the intersection of the central meridian at the equator. Displacements in the *x* and *y* directions are called **UTM eastings** and **UTM northings**, respectively. Conventionally the origin of each

Table 4-2. Some reference spheroids[1] in use throughout the world, after Snyder (1987).

Name	*a*, metres	*b*, metres	*f*	Use
Australian 1965	6,378,160	6,356,774.7	1/298.25	Australia
Krasovsky 1940	6,378,245	6,356,863.0	1/298.3	Soviet Union
Clarke 1866	6,378,206	6,356,583.8	1/294.98	North America
Airy 1830	6,377,563	6,356,256.9	1/299.32	Great Britain

[1]Also called ellipsoids

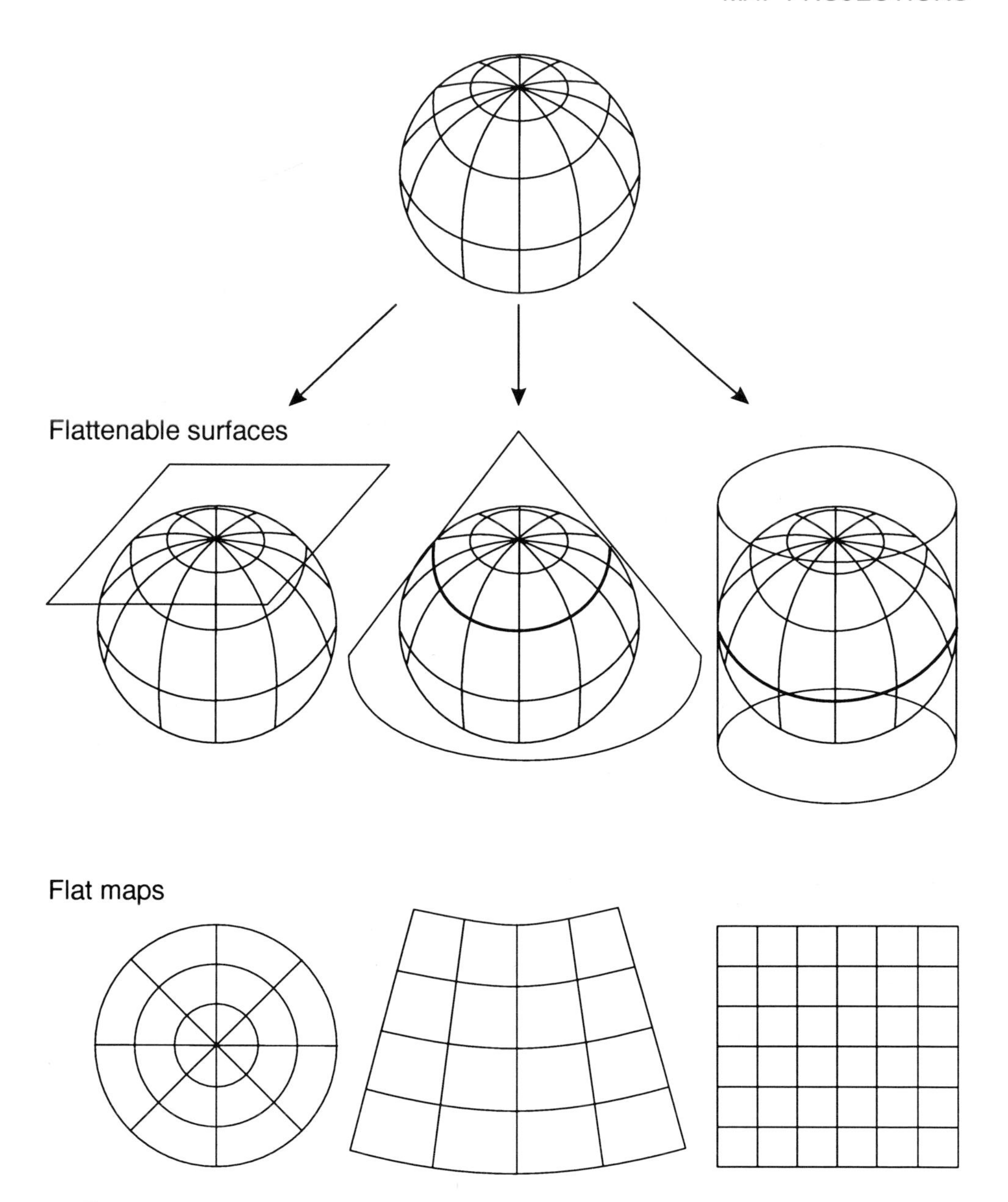

FIG. 4-4. The three principal types of developable (flattenable) surfaces: planar on the left, conic in the middle, and cylindrical on the right. These are all shown in the tangent case (touching the globe at a point or along a line).The secant case results when the conformable surface cuts the globe, instead of touching it.

zone is offset to the west, and assigned an easting of 500,000 m, so that, within that zone, eastings are always positive. The northing of the equator in the northern hemisphere is zero m, but in the southern hemisphere it is arbitrarily assigned a value of 10,000,000 m, in order to avoid negative northings. To minimize geometric distortion across each zone, the scale at the central meridian is reduced by a **scale factor** equal to 0.9996. This produces two parallel lines of zero distortion approximately 180 km either side of the central meridian.

The UTM is an excellent system for regions covered by maps at scales of 1:250,000 and larger. At smaller scales, distortions generally become unacceptable. As its name suggests, UTM coordinates have been widely and consistently applied. A UTM spatial reference requires three numbers, the easting, northing and either the zone number or central meridian. Problems can arise when using data from adjacent zones, because no simple relationship exists between the coordinates of one zone and the coordinates of the other. Thus eastings and northings from one zone must first be converted to geographic coordinates, then to the eastings

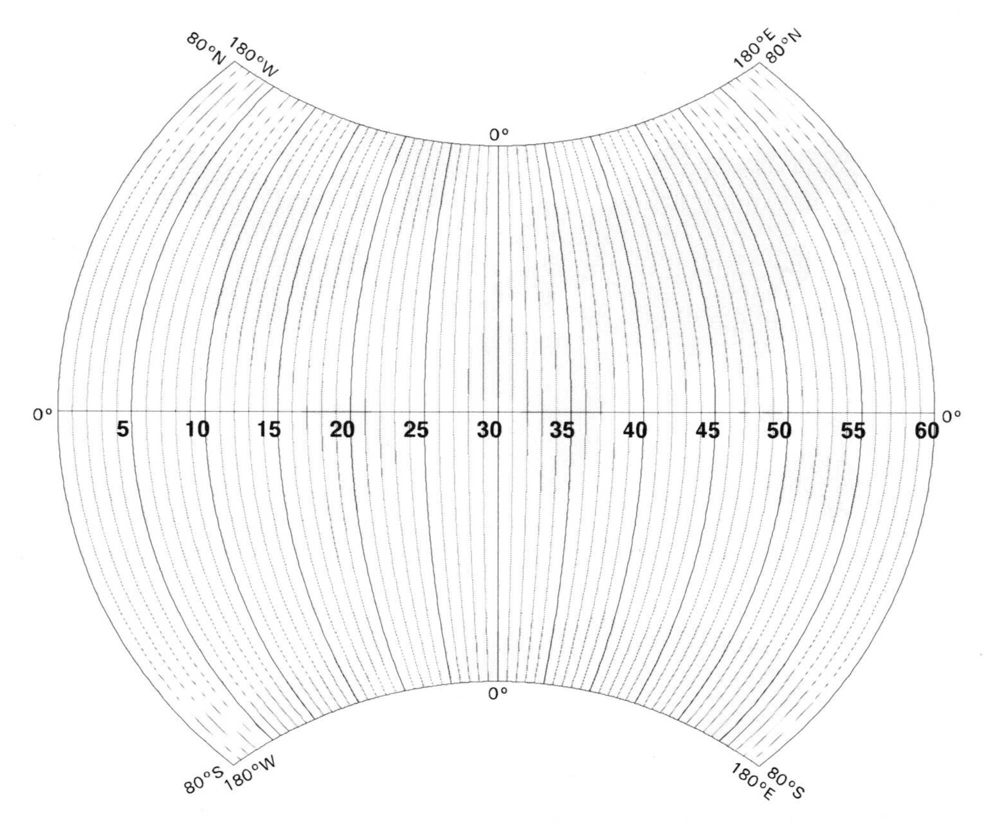

FIG. 4-5. Universal Transverse Mercator (UTM) zones, numbered from W to E. There are sixty zones, each six degrees of longitude wide, extending from 80°S to 84°N, although the figure only shows them extending to 80°N.

and northings of the other zone. Usually a Murphy's Law prevails that states that the region of interest always straddles two UTM zones. In that case, an ordinary transverse Mercator projection is still a good choice, because with the central meridian placed at the centre of the study area, and the scale factor set to 0.9996, the projection characteristics will be the same as UTM, but unrestricted by zone boundaries.

Equations (4-6) to (4-8) give the general flavour of the forward transformation relationships assuming a spherical earth. The full equations for the spheroid are more complex, and in some cases require numerical, rather than analytical, solutions (Snyder, 1987). Snyder is a good source of information about the forward and inverse transformations, as employed by GIS, to convert from geographic coordinates to planar coordinates, and back again.

In the following section, some aspects of digitizing are discussed. Digitizing maps generates planar coordinates in table units (inches or mm). Table coordinates must be transformed to the eastings and northings of the original map. Given the parameters of the map projection (the spheroid, the central meridian, and so on, depending on the projection type), the eastings and northings of the map can then be converted (with the inverse transformation equations) to geographic coordinates. Once in geographic coordinates, the forward transformation equations are employed to convert the (latitude,longitude) values to planar coordinates of the actual projection selected for the GIS database. This step also requires the definition of the necessary parameters for the working projection of the database. Coordinate conversions are discussed in more detail at the end of the chapter.

DIGITIZING

The digital capture of data from maps is carried out in one of two principal ways: 1) by manual digitizing and 2) by raster scanning using optical scanners.

Manual Digitizing

Manual digitizing uses a digitizing table, like a drafting table except that it is equipped with a stylus or cursor for tracing and electronically recording the positions of points and lines. The map is mounted on the table (or a smaller digitizing "tablet"), taped securely to prevent movement, and traced with the cursor, as shown in Figure 4-6. The electronics in the system convert the position of the cursor to a digital signal with a precision of about 5 to 50 points per mm. The signals are transmitted from the table to the computer where a digitizing program processes and stores the raw data. The cursor position is recorded in plane cartesian coordinates in mm or thousandths of inches. Most tables work on an electrostatic principle. Associated with the cursor is a "puck" containing buttons or switches that allow the user to enter commands without moving from the table to the keyboard.The resulting strings of spatial coordinates comprise the raw data for the vector representation of points, lines and polygons.

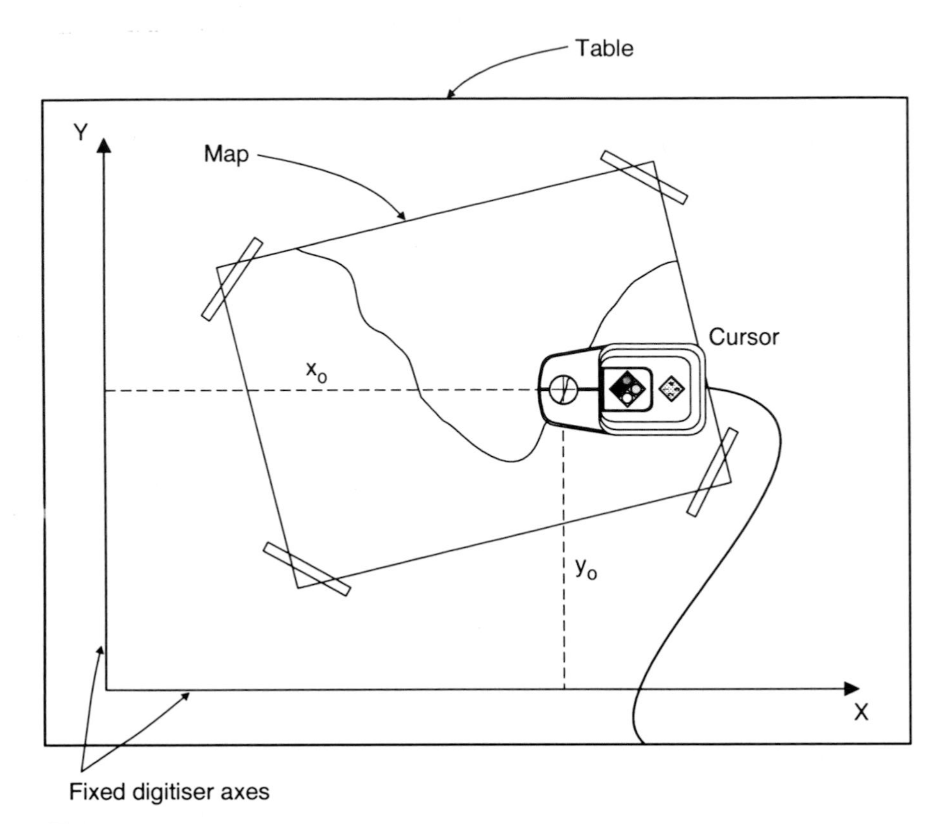

FIG. 4-6. Digitizing table setup. The spatial objects on the map are digitized in vector mode according to the Cartesian coordinates of the table.

The accuracy of digitizing is affected by the number and accuracy of **control points**, by the stretch of the source document, and by the skill of the operator. It is obvious that no amount of careful digitizing can compensate for a map that is not accurately positioned, and great care should be exercised in selecting and digitizing the control points. Paper is notorious for stretching with changes in humidity, by sometimes as much as 3%. Where possible, input documents composed of mylar or other stable-base materials should be used. The specific procedures for manual digitizing differ depending on the software in use.

In general, the first step is to digitize 3 **reference points** that lie outside the map area in the corners of the document. These define the position of the document with respect to the table. If the document is removed from the table and replaced later in a different position, only the three reference points need be re-digitized before continuing to collect new, or edit old, data. The resulting coordinates generated by the digitizing software are table coordinates (planar) that are consistent, although the source document may have been placed in different positions on the table in several digitizing sessions.

The next step is to define and digitize several control points. The eastings and northings of the points are entered in the projection coordinates of the map. The positions of control points are used to determine the transformation parameters for converting from table coordinates to projection coordinates of the input map (both planar coordinate systems), as will be described in the coordinate conversion section, below. Instead of eastings and northings, the positions of the control points can be specified in geographic coordinates, as long as the projection parameters and projection type are also known, therefore allowing the eastings and northings of control points to be calculated. This control point step is vital for **geocoding** the data, the process of assigning geographic locations to spatial objects. Geocoded data can be placed in the correct geographic location, which is essential for comparing maps from different sources with each other and with other data types.

Having completed these preliminaries, actual data gathering then proceeds either in **point**, **line** or **stream** mode. In point mode, individual locations (such as sample sites or well locations) are recorded by positioning the cursor over the point and pressing the appropriate button, generating a single table coordinate pair. In line mode, straight line segments are digitized by recording a point (vertex) at each end; curved lines are built up by a series of connected straight line segments. In stream mode, curved lines are defined by vertices generated automatically at specified time or distance intervals. Digitizing in stream mode can rapidly create very large files.

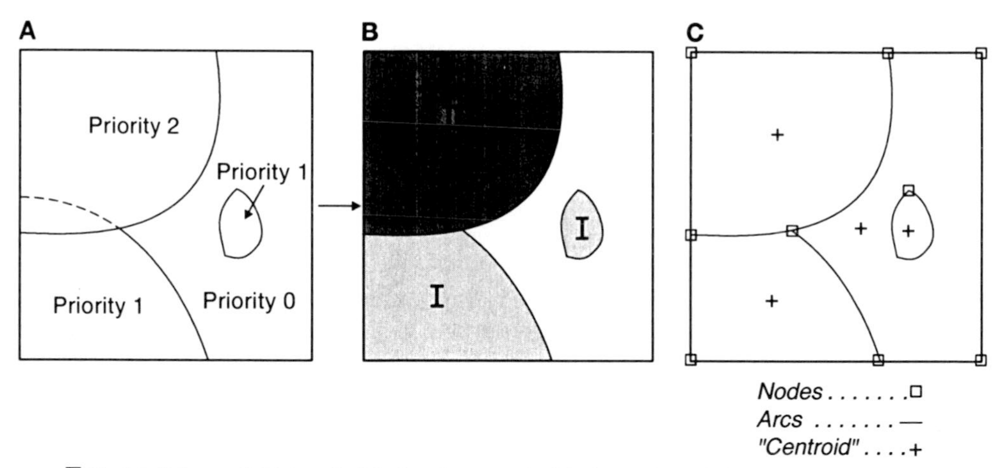

FIG. 4-7. Polygon digitizing. **A.** Whole polygon method. Polygons boundaries are digitized as closed loops and assigned priorities to establish plotting sequence. Notice that the boundary polygon is assigned priority 0; the two priority 1 polygons take precedence over priority 0, where priority 0 and 1 are together. Similarly the priority 2 polygon takes precedence over the priority 1 polygon where they occur together. **B.** Polygons plotted with fill patterns according to priority would result in the desired pattern. **C.** Topological method using nodes and arcs. Centroids are digitized for each polygon and are used to assign polygon attributes.

When points and lines are digitized, they are assigned numerical identifiers that are used to link the spatial coordinates to nonspatial attributes. For example, samples taken at points in a soil survey might be assigned the point identifiers 1,2,....N during digitizing, where the identifier is a numerical field in the point attribute table, allowing the spatial coordinates to be merged with other point attributes. The coordinates may alternatively be retained in a separate table, linked to an attribute table by a common keyfield. Where lines are to be treated as "spaghetti", as might be the case on a map of lineaments digitized from a satellite image, each line is tagged with a numerical identifier, allowing the linkage of graphical coordinates to other attributes, as for points.

Where lines are to be treated as polygon boundaries, the problem of digitizing and linking spatial coordinates to polygon attributes becomes more complex. There are three commonly-used approaches: 1) the whole polygon mode, 2) the manually-assigned topology mode and 3) the automatically-assigned topology mode, see Figure 4-7.

1. Whole Polygon Mode

The complete external boundary of each polygon is digitized as a separate spatial object. This has the advantage that the complete set of boundary coordinates are stored together for each polygon, which can be useful for rapid point-in-polygon search and for algorithms that automatically "fill" closed polygons with colour or pattern. However, there are a number of serious disadvantages of the approach from a GIS standpoint, as pointed out in Chapter 3.

The first is that polygon boundaries are digitized twice, except at the boundaries of the study area, because each line is common to two adjacent polygons. This leads to unnecessarily large files, and to discrepancies between the two versions of the same line. Inevitably, many "sliver" polygons are produced, leading to severe problems later where regions within these slivers either belong to no recognized polygon or belong to more than one polygon. Some digitizing systems overcome this by using a method of purposeful overlaps, as follows. Having digitized polygon A as accurately as possible, and having assigned it a priority, polygon B is digitized. Where B abuts A, a false digitized boundary is deliberately overlapped, and polygon B is given a lower priority than A, indicating that the boundaries of A take precedence over those of B. Polygons C,D,E... are digitized following the same rules. In a subsequent processing step, the redundant overlapping regions and boundaries are eliminated under the control of the priority assignments.

The second disadvantage of the whole polygon approach is that a system of priority assignments must also be used to deal with "island" relationships, see Figure 4-7A and B. On many maps, several levels of islands within islands are common. This means that a point lying within the enclosing boundary of one polygon may also be within the enclosing boundaries of other, larger polygons. This leads to ambiguities in spatial search where one location may be in several polygons simultaneously; also to display problems where islands may be obliterated by larger enclosing polygons, depending on the display sequence. Priorities assigned to polygons overcome these difficulties. Despite the disadvantages, whole polygon digitizing can be useful for simple maps, for on-screen digitizing of a few polygons, and for maps where polygons are not planar-enforced, as discussed in Chapter 2.

2. Manual topology-assignment mode

Instead of treating polygons as the primary digitizing objects, lines (arcs) and their points of intersection (nodes) become the primary objects in the digitizing process. With each digitized arc, a pair of numerical attributes are entered that are identifiers of the polygons to either side of the arc. In this manner the basic topological information is coded directly during the digitizing process. Each arc is digitized only once, avoiding the sliver problem. The island problem is automatically taken care of by the topological relationships between objects. The disadvantage of this approach is that the digitizing operator must constantly divert attention from tracing the lines with the cursor to recording the topological data, leading to a loss of concentration. The alternative is to use the automatic topology-assignment mode.

3. Automatic topology-assignment mode

Here the operator concentrates on digitizing the nodes and arcs, without assigning the left and right polygon identifiers. In a subsequent step, a point is digitized inside each polygon and tagged with a numerical identifier that provides a link to a polygon attribute table. Note that these points are called **centroids,** although they are not **geometric centroids**. A geometric centroid is a point lying at the centre of mass of a polygon. It may, in some cases, lie outside the polygon boundary, depending on the geometry. The software assigns the topology automatically, generating an arbitrary set of polygon identifiers that are subsequently linked to the centroid identifiers by point-in-polygon calculations.

Editing digitized vector data

Three common digitizing errors that require editing are caused by improper line-node joins, see Figure 4-8. The first is an **undershoot** so that two separate polygons remain connected at the gap between the end of the line and the node. The polygon that is open at an undershoot is called a **leaking polygon**. The second is to produce one or more superfluous small polygons close to the node, by not forming a clean join. The third is to **overshoot**, producing a **dangling arc**. These errors can usually be overcome by using the **snap** procedure, which results in automatic closure, and guarantees that the final vertex of an arc is a node. Many other errors are possible, such as not identifying a node where lines meet, omitting a line, inserting superfluous lines, improper positioning of a line, and so on.

Specialized vector editing tools are used for a variety of editing tasks, including the removal of **sliver polygons**, and **edge-matching** adjacent sheets. Sliver polygons are produced by overlay of two maps with the same polygons that do not have identical boundaries. A common instance of this type is the overlay of a geology map with another map (soils, Quaternary geology, etc), where shorelines occur on both maps. Slight differences between the two sets of shorelines produce a large number of sliver polygons that must be removed for the two maps to fit together properly. Edgematching between two adjacent sheets

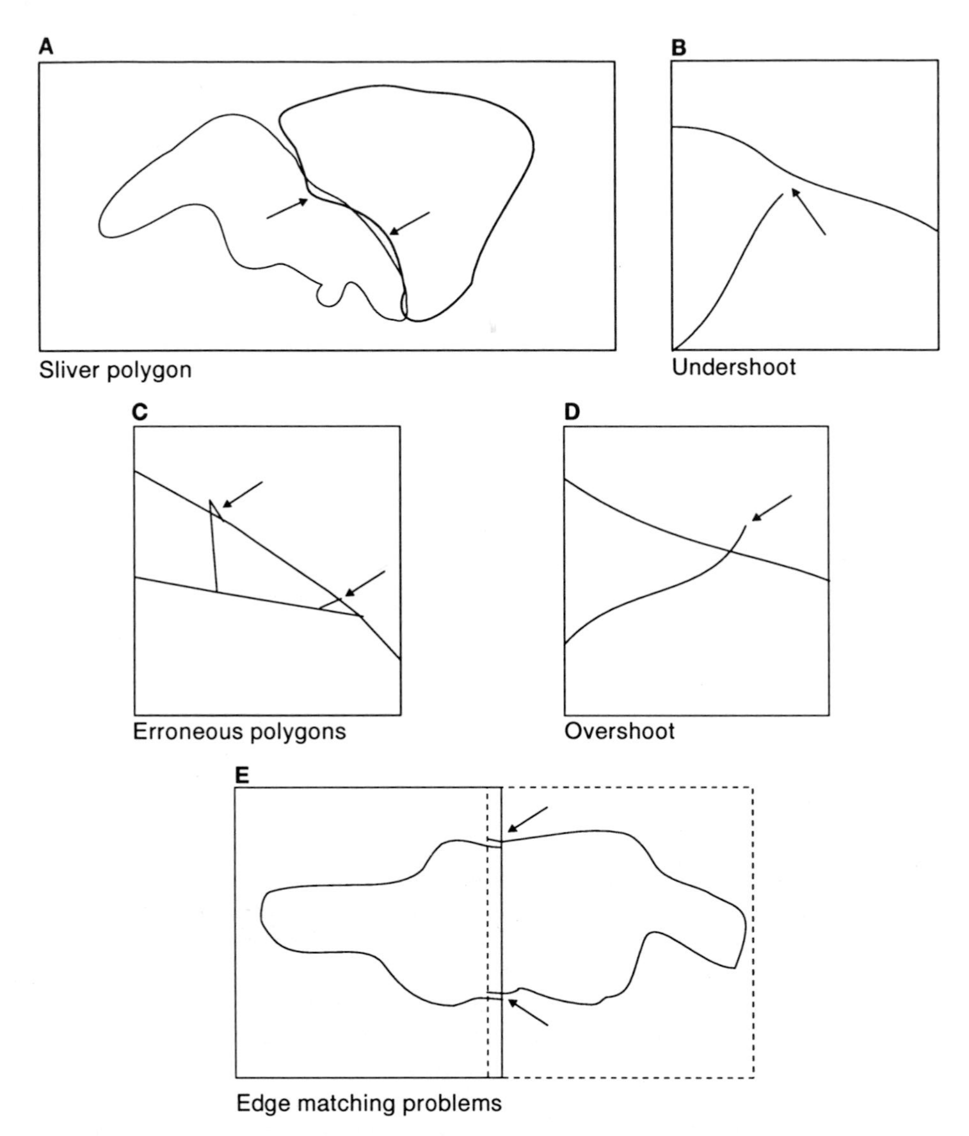

FIG. 4-8. Typical digitizing errors. **A**. Sliver polygons, occurring when attempting to draw a boundary between two polygons twice. Normally this does not occur in digitizing, because either the whole polygon or topological methods are used. Sliver polygons are often produced in the overlay of two polygon maps. **B**. An undershoot, caused by failing to make two lines meet exactly. The problem is avoided in the topological method by digitizing a node on the first line at the point of intersection, and "snapping" the second line exactly to the node. **C**. Erroneous polygons, often very small and hard to identify. Avoided by using nodes and snapping lines to them. **D**. Overshoot, producing a dangling arc. Can be avoided by snapping to nodes. **E**. Edge matching problems between adjacent map sheets.

often requires modification of boundaries to ensure a seamless join. **Conflation** is a term applied to a process by which maps are matched and merged by selecting the "best" features from each individual map to be added to the composite map.

The editing of vector data is a topic of great practical importance, but the actual details of efficient editing are specific to the particular GIS in use, and are not discussed further here.

Raster Scanning

Optical scanners, either of the rotating drum , as shown in Figure 4-9, or flatbed variety, have been used for cartographic data entry since the mid-1960s (Carstensen and Campbell, 1991). A raster scanner generates a large matrix of digital values, each pixel integrating the reflectance over a small portion of the original image. The radiometric resolution describes the range of values the device can discriminate per pixel. Most scanners are capable of 8-bit resolution, or 256 brightness levels. The spatial resolution varies from about 600 dots per inch, dpi, (the distance between dots is about 42 microns) in the more expensive devices, to about 200 dpi (125 micron distance) at the low end. The large drum scanners can accommodate documents up to about 1 m by 1 m; inexpensive desktop devices are usually restricted to the page sizes as used by office copiers.

Optical scanners can be used in binary or 8-bit mode. In 8-bit mode (some devices are capable of 8-bit resolution for each of the red, green and blue colours) a full raster image of a photograph, satellite image or map is scanned. This is then rectified to a geographic base with control points, as discussed under coordinate conversion below, for direct use as a

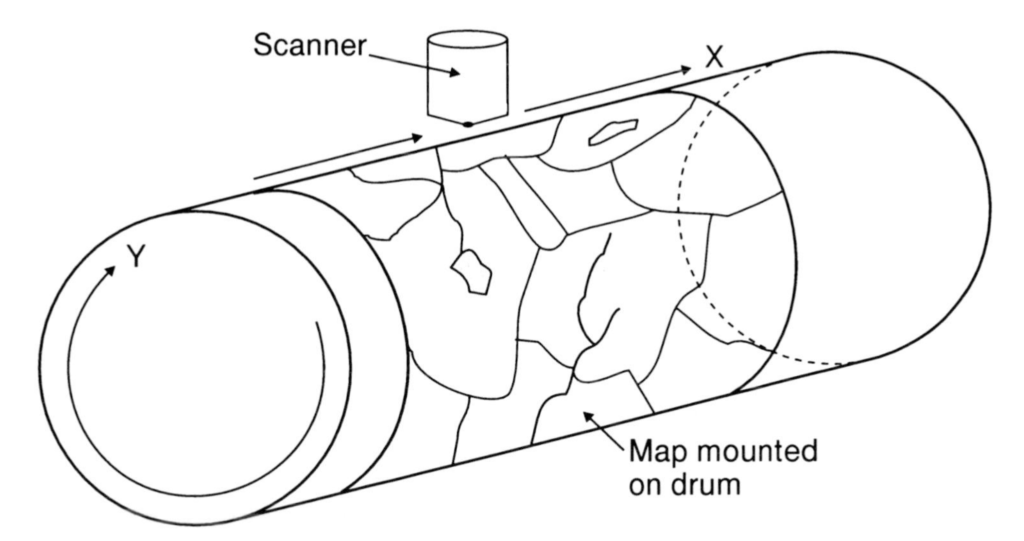

FIG. 4-9. Raster scanner, rotating drum variety. The scanning head moves in increments along the X direction, parallel to the long axis of the drum. The map moves in the Y direction with respect to the scanning head by drum rotation.

geocoded data layer. Images scanned from maps by this approach are sometimes used as "backdrops" on which other data are displayed, with no attempt to extract point, line and polygon objects, or to eliminate superfluous symbology. Two principal approaches have been used to extract vector features and build topological relationships from scanned data.

The first approach is to use the scanner in binary mode, combined with pre- and post-scanning steps. For example, the Canada Lands Data System (CLDS) successfully digitized thousands of maps since the mid 1960s using this method (Crain, 1984). The system is now somewhat dated, but the principles employed are common to more modern systems. Before scanning, each CLDS map was scribed or redrawn on to a stable base. The document to be scanned consisted, therefore, only of the boundaries of polygons, and a "neat" line at the map margins. After scanning, the resulting binary raster image was automatically processed and edited in various stages. The rasterized lines were **thinned** to one pixel wide, and nodes identified. Errors were flagged and corrected at a raster-editing workstation (corrections were made in a batch mode process prior to the workstation era). Rasterized lines were then **vectorized**, and **polygonized**, ultimately producing an arc-node data structure, complete with topological attributes. Polygons were tagged by manually digitizing centroids, to provide the link between the polygon objects and nonspatial attribute tables, as described under table digitizing methods above.

Such a system was reported to have the throughput of 30-50 times that of a digitizing table, and at the time of Crain's article, large drum scanners were roughly 10 times more expensive than large digitizing tables. The system required considerable manual intervention at both pre- and post-scanning stages. The high capital cost could only be justified if there was a continuous throughput of maps (Crain, 1984).

During the 1980s, an alternative approach became widely employed, that of semi-automatic line following. Instead of cleaning up the input document in a pre-processing stage, the **original** document is raster-scanned. Colour separations, where available, of a map can provide "cleaner" input documents. Colour separation can also be achieved digitally on the scanned document, separating out such features as drainage in blue, contours in brown, and so on. The raster image, or part thereof, is then displayed on a workstation, and a cursor controlled by a "trackball" is moved on to one of the lines to be captured. The line is then followed automatically until operator-intervention becomes necessary, such as at a gap, at a label, or at a node, whereupon the operator takes over until the obstruction is passed. The lines are automatically digitized during tracking, producing vector data. The resulting file is then edited, topology is built, and polygons are tagged as in other methods.The great advantage of this approach is that manual drafting is eliminated, and editing is reduced because, although feature extraction is not automatic, it is operator-controlled.

Low-cost table-top scanners appear to have remarkably good radiometric and spatial resolution, and will probably become increasingly important for input of small documents (Carstensen and Campbell, 1991). It is also likely that automatic feature extraction from raster scanned maps will play an increasing role for cartographic input to GIS (Ansoult et al., 1990). A combination of scanning and on-screen manual digitizing of the scanned document, is now sometimes called "heads-up" digitizing.

COORDINATE CONVERSION

Typical input data for a geoscientific GIS project might consist of:

1. A geological map (Albers equal-area conic projection) in digital form and occurring as a file of table coordinates and control point data,
2. a geochemical data table, with spatial coordinates in UTM eastings and northings, and
3. a satellite image in raster format, but not geocoded.

In order to bring these data from the coordinate systems in which they currently occur to a new uniform planar coordinate system within a GIS requires a sequence of **coordinate conversions**. A typical GIS project begins with choosing the geographical extents of the region to be studied, followed by selecting a suitable map projection as the **working projection**. Suppose, for example, that a Lambert conformal conic projection with standard parallels of 33°N and 45°N, central meridian at 160°W, and the Clarke 1866 spheroid have been chosen for the working projection.

The steps required to convert the digitized geological map into working projection coordinates are 1) the table coordinates are converted to Albers equal-area conic projection eastings and northings, 2) the Albers projection coordinates are transformed to geographic coordinates, and 3) the geographic coordinates are converted to eastings and northings in the Lambert conformal conic projection. For the geochemical point file, the UTM eastings and northings are 1) transformed to geographical coordinates and 2) transformed to Lambert eastings and northings. In the case of the satellite image, the input raster must be **resampled** to a new grid, where the coordinates are spaced according to the working Lambert projection. The **warping** or **rubber-sheeting** of one raster grid to another is carried out by selecting and applying **ground control points** that can be identified both on the untransformed image and at points on the ground.

These conversions are carried out within GIS input routines. The following sections describe in principle how such routines operate, although the details vary from one GIS to another. Although the various steps in the transformation process look confusing, the GIS user is usually shielded from the details. The following discussion is designed to give some insight into the process. It is divided into two parts, the first dealing with conversion of vector data, the second with raster data.

Vector conversion

Figure 4-10 illustrates diagrammatically the steps in converting three sources of vector input. Source A contains table coordinates from digitizing, source B is associated with input projection eastings and northings, and source C uses geographical latitudes and longitudes. The goal is to convert all three sources to eastings and northings in a GIS working projection.

Table to projection coordinates

For source A, where the projection is known, a small number of control points are used to link table coordinates to projection coordinates. The control points are digitized at known locations, such as "tic" marks or on the coordinate graticule printed on the map. The first step

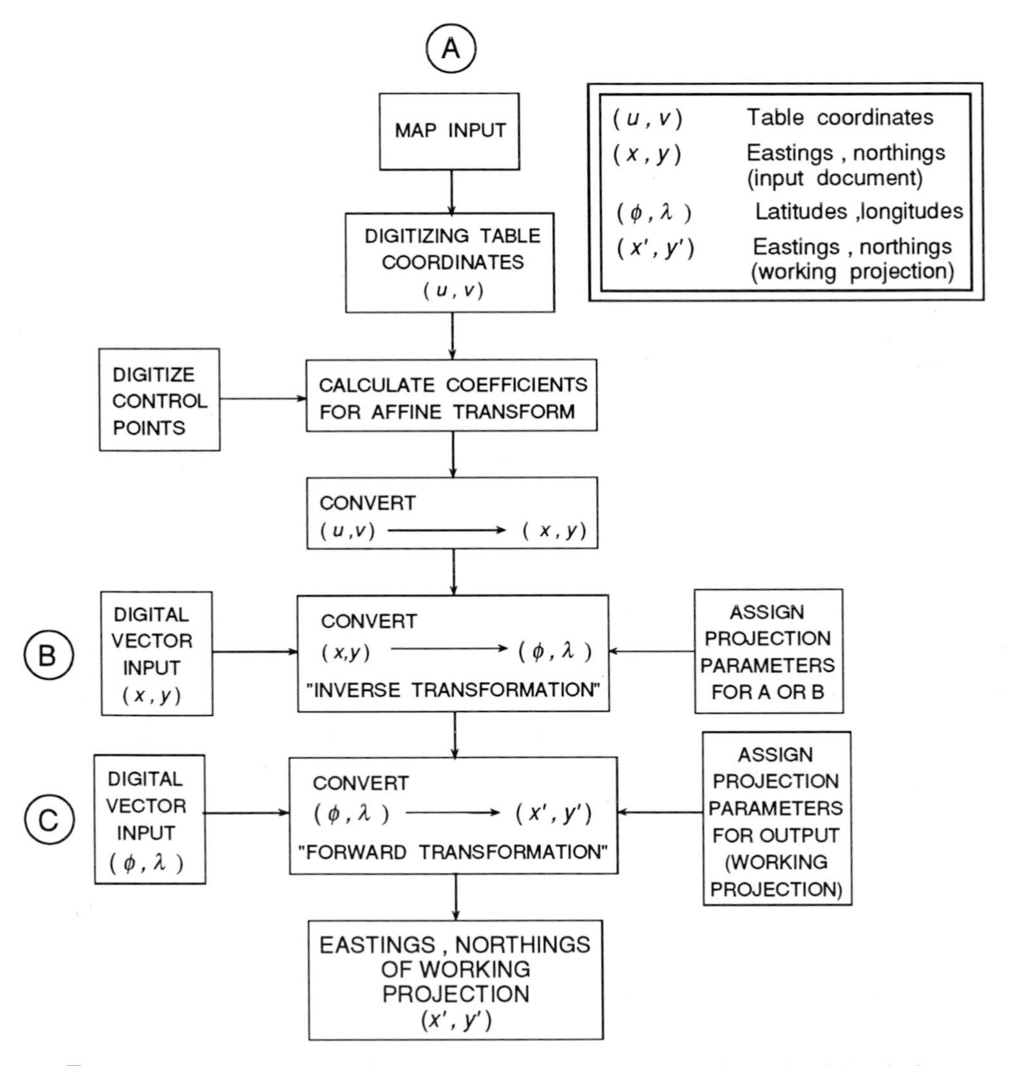

FIG. 4-10. Steps in converting vector data to the planar Cartesian coordinates of a working GIS projection. **Source A** consists of Cartesian coordinates from a digitizing table. **Source B** consists of eastings and northings in the projection of the input document. **Source C** is similar to **B**, except that the coordinates are already in geographic coordinates.

is to convert table coordinates to projection coordinates of the input map, using an empirical polynomial function that is fitted to the control points. At the control points, both the table coordinates (u,v) and the input projection coordinates (x,y) are known. The transformation accounts for changes in scale, for translation of the origin and for rotation, between the two sets of plane Cartesian coordinates. This type of transformation, which has the form of a first-order polynomial, is known as an **affine** transformation. The affine equations for converting from table coordinates (u,v) to eastings and northings (x,y) are:

$$x = a + bu + cv \quad and \tag{4-9}$$

$$y = d + eu + fv \quad , \tag{4-10}$$

where the coefficients a,b,c,d and e are determined from at least 3 control points, and (x,y) are the predicted eastings and northings. For three control points, with two coordinate values per point, the six unknown coefficients are obtained by solving the equations exactly. In this case, the eastings and northings of the control points are predicted without error, because the fit is exact. It may be desirable to use more than three control points, to reduce the chances of positional error. With more than three control points, the coefficients are determined by least squares fitting, and the predicted control point eastings and northings have residual errors. The residual error at a point is usually defined as the Pythagorean distance between the observed and predicted positions, calculated from the relationship

$$residual = \sqrt{(x-x_{obs})^2 + (y-y_{obs})^2} \quad . \tag{4-11}$$

A listing of residual values for control points shows whether some points are poorly located. Once satisfied with the fit of the relationship, possibly rejecting or redigitizing grossly misregistered control points and refitting, the affine equations are used to convert all the table coordinates in the file to eastings and northings.

In a situation where the map projection is unknown, as would occur if the vectors were digitized from lines marked on an airphoto or satellite image, or on a map with no information about the projection, then the conversion from table coordinates to planar Cartesian coordinates of a known projection requires more control points and a higher-order polynomial equation. The control points are often termed **ground control points** or GCPs, due to the fact that identifiable ground features, whose locations can be established from a topographic base map, or possibly by GPS measurements, are used. This is the same process employed for geocoding satellite images, discussed below, except that for raster data a resampling stage is necessary. The higher-order polynomial not only accounts for scaling, rotation and translation between the plane coordinate systems, but also takes warping effects into consideration. GCPs are chosen to give a uniform distribution over the map, at locations where both table coordinates and projection eastings and northings (for the output or working projection) are known. A polynomial function, either quadratic (6 coefficients per equation) or if necessary cubic (10 coefficients per equation), is fitted to the control points by least squares, as above.

Again predicted coordinates are calculated for each GCP and if one or two points have large residual values they can be rejected and the fit recalculated. Once satisfied with the fit, table coordinates are converted to eastings and northings. Under these circumstances, there is usually no attempt to identify the projection of the input document, and the conversion is made directly from the plane coordinates of the input document to the plane coordinates of the output or working projection, skipping the geographic coordinate step entirely.

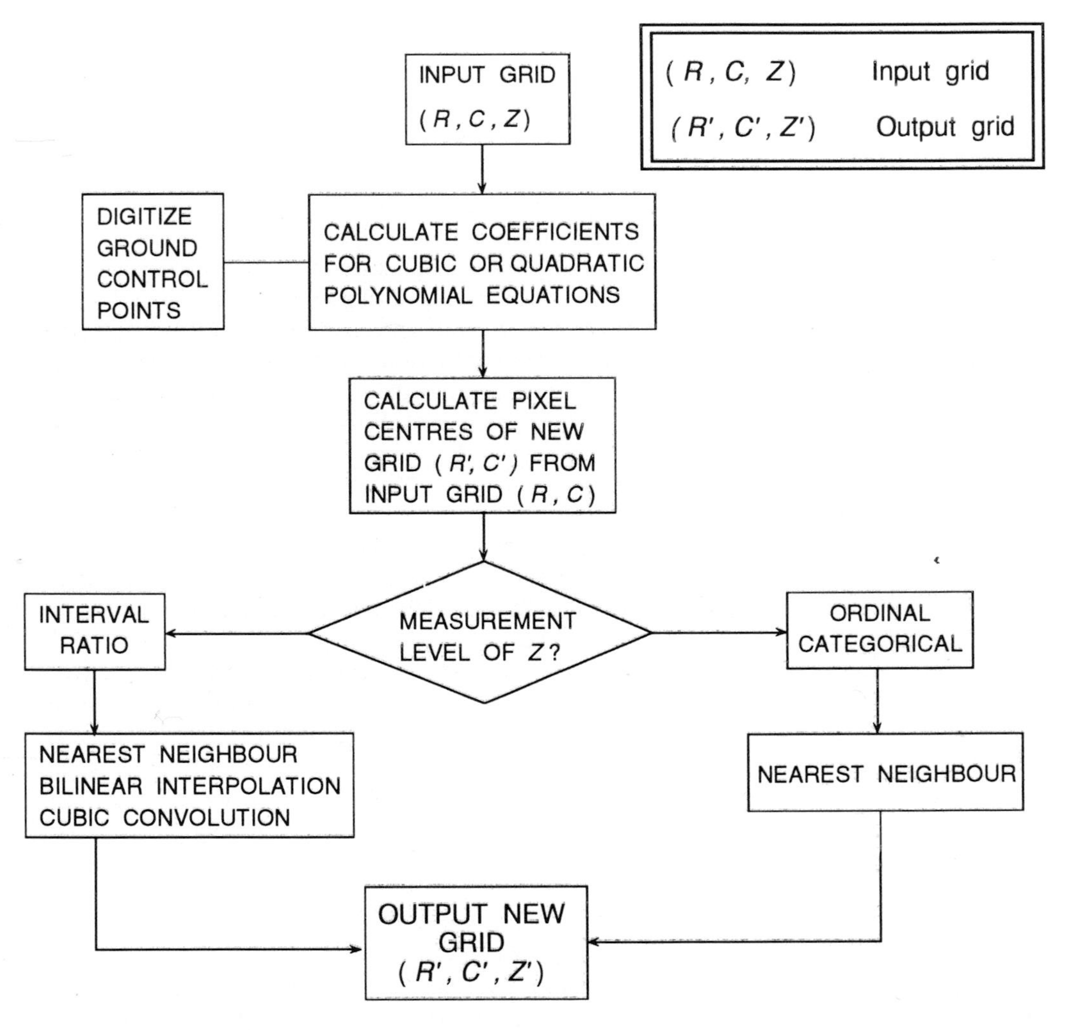

FIG. 4-11. Coordinate conversion of raster data from a raster image with unknown projection characteristics to a new raster in the working projection.

Projection coordinates to geographic coordinates

This step uses the **inverse transformation** equations to convert the eastings and northings of the input projection to latitudes and longitudes, (x,y) to (ϕ,λ). These are the mathematically-derived equations, whose form vary with the projection type and spheroid, as introduced above under map projections and discussed in detail by Snyder(1987). Note that for this step the input projection parameters are required.

Geographic coordinates to working projection coordinates

For this step the **forward transformation** equations are applied to convert latitude, longitude values to eastings and northings, where the projection parameters are now those of the working projection. Thus (ϕ,λ) are converted to (x',y'), ready for use in GIS analysis.

Raster conversion

In the case of raster input, see Figure 4-11, the conversion process involves establishing a new raster, whose coordinate axes and pixel coordinates conform to the GIS working projection, followed by **resampling** to determine the new pixel values from the old pixel values. In order to determine the locations of pixel centres in the "new" grid in terms of the spatial coordinates of the "old" input grid, a pair of transformation equations are used, as before. If the input raster is already in a known geographic projection, as will be the case for

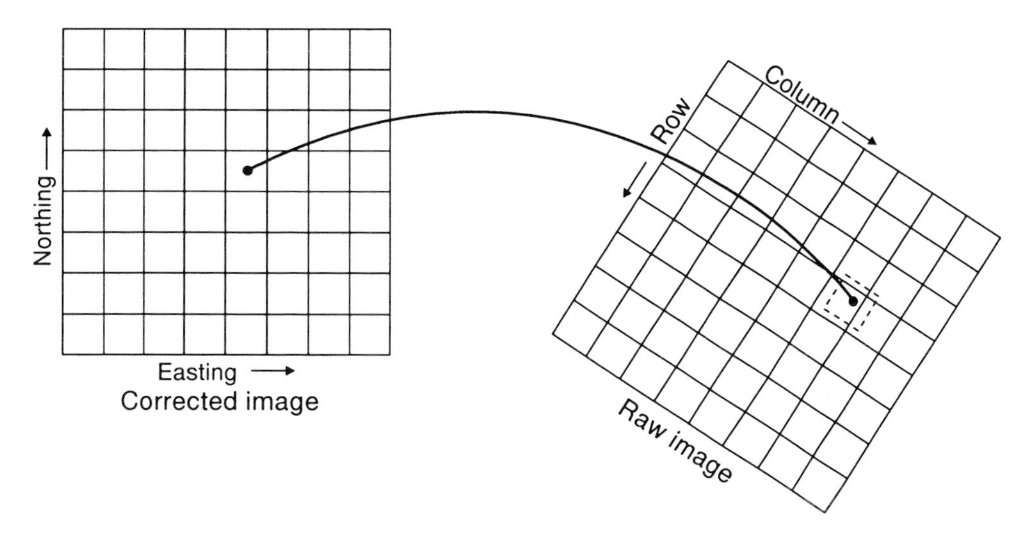

FIG. 4-12. Resampling to convert one raster to another. **A.** The "raw" input image is resampled to a new grid whose coordinates are known in relation to a working geographic projection.

geocoded and rectified satellite images, for example, the forward transformation equations, mentioned in the previous paragraph, are employed. If the old grid is not geocoded, then the equations are higher-order polynomials, fitted to GCPs.

The difference between the spatial conversion of rasters as opposed to vectors is that the actual attributes of the pixels must be converted from one grid to the other, not just the spatial coordinates. A functional relationship is still needed for transforming the grid coordinates. In addition, pixel values in the new grid are determined from one or more neighbouring pixels in the old grid. The process is illustrated diagrammatically in Figure 4-12.

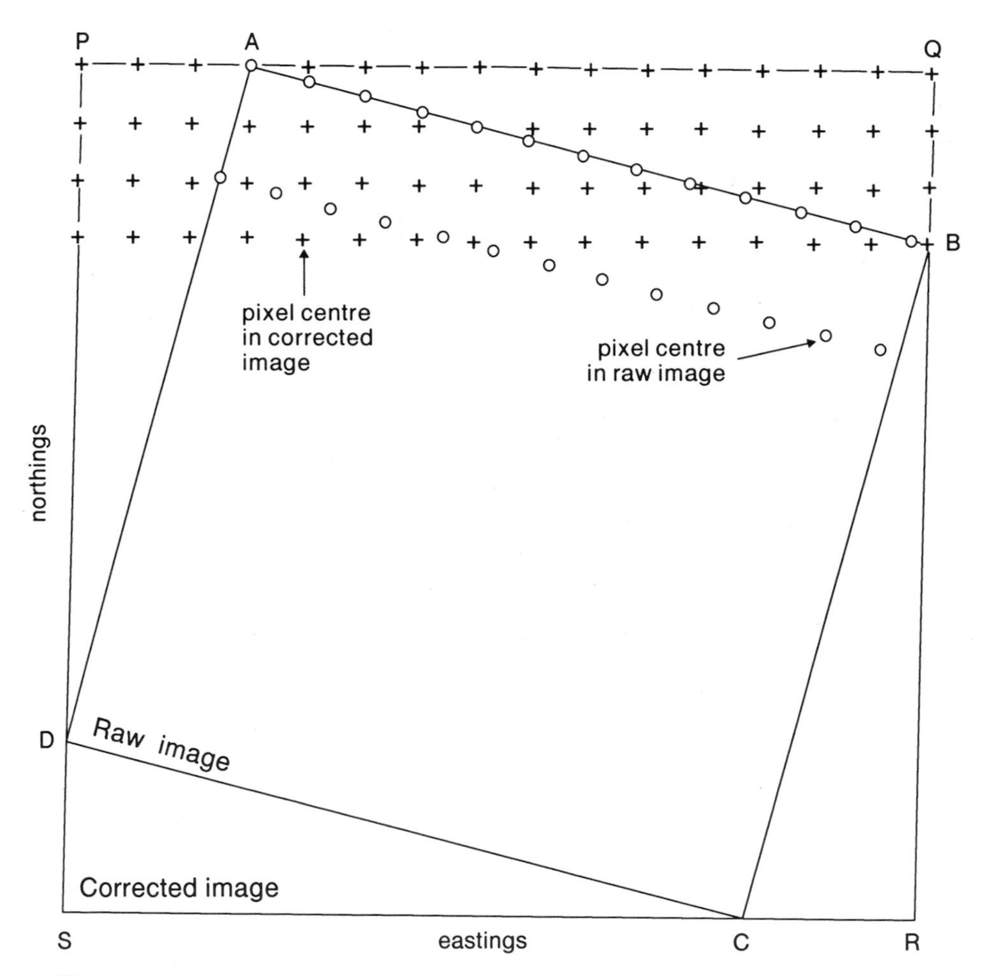

FIG. 4-12. B. The raw image superimposed on the corrected image to show the spatial relationship between the lattices of the pixel centres of the two images. Modified after figures in Mather (1987).

The collection and use of GCPs to fit a quadratic or cubic polynomial function is exactly the same as described for the vector case. When the input projection is unknown, which is often the case with satellite data, the transformation is direct from the input grid coordinates to the new grid in working projection coordinates, without the intervening geographic coordinate stage.

The resampling process is carried out by one of three methods: nearest neighbour, bilinear interpolation and cubic convolution. In the nearest neighbour method, each new pixel takes on the value of the closest pixel from the old grid. When the pixel attributes are categorical or ordinal scale measurements, the nearest neighbour method must be used because it makes no sense to calculate average values of attributes like rock type, or relative abundance. The nearest neighbour method can also be used with ratio and interval scale measurements, but much smoother surfaces are produced by interpolation methods. On the other hand, the interpolation methods are more demanding in terms of calculation, with the cubic convolution being the most costly in terms of computation. Mather (1987) provides a clear explanation of resampling methods.

REFERENCES

Ansoult, M.M., Soille, P.J. and Loodts, J.A., 1990, Mathematical morphology: a tool for automated GIS data acquisition from scanned thematic maps: *Photogrammetric Engineering & Remote Sensing*, v. 56 (9), p. 1263-1271.

Brodaric, B. and Fyon, J.A., 1988, OGS Field Log: A micro-computer-based methodology to store, process, and display map-related data: *Ontario Geological Survey Open File Report* 570, 73 p.

Carstensen, L.W. and Campbell, J.B., 1991, Desktop scanning for cartographic digitization and spatial analysis: *Photogrammetric Engineering & Remote Sensing*, v. 57 (11), p. 1437-1446.

Crain, I.K., 1984, A comparison of raster scanning and manual digitizing: *Canadian Lands Data Systems Division*, Lands Directorate, Environment Canada, Ottawa, Ontario, Report No. R001300, 10 p.

Maling, D.H., 1992, *Coordinate Systems and Map Projections*: Second Edition: Pergamon Press, Oxford, 476 p.

Mather, P.M., 1987, *Computer Processing of Remotely-Sensed Images*: John Wiley and Sons,Chichester-New York, 352 p.

Snyder, J.P., 1987, Map projections – a working manual: *United States Geological Survey Professional Paper 1395*, United States Government Printing Office, 386 p.

Thomas, C.W., 1991, OGS FieldLog: A micro-computer-based methodology to store, process, and display map-related data, a review: *British Geological Survey Unpublished Report*, 16 p.

Chapter 5

Visualization and Query of Spatial Data

INTRODUCTION

Visualization is a critically important function in GIS. The availability of low-cost computer graphics in the 1980s was one of the catalysts for the growth of GIS. The human ability to recognize spatial relationships on maps, images and other graphical displays is exploited in GIS, mainly through the use of video displays, but also through hardcopy output. Geoscientists as a group are accustomed to looking at maps, and their ability to recognize geological structure and pattern from maps is well-developed. Therefore, the capability of GIS to turn tables of data into pictures, and to visualize spatial associations, is immediately appealing to those working with data about the Earth.

On the other hand, the recognition of patterns and anomalies in visual displays often causes the need for examining specific data values in detail. Instead of ploughing through reams of paper tables, GIS, in common with other types of digital data systems, allow the user to make specific queries about the data. The ability to make rapid interactive spatial queries is a valuable GIS function that complements visualization. This chapter discusses some aspects of visualization and query, and deals with the following topics: the composition of cartographic images, display hardware, colour models, hardcopy plotting, visualization of surfaces, dynamically-linked data views and spatial query.

The visualization process in GIS mainly involves generating and looking at **cartographic images**, that are maps in the form of digital images. Visualization also involves other "views" of the data present in a GIS database. For example, instead of looking at a horizontal (x,y) plane, vertical profiles and 3-D perspective diagrams can be displayed. In addition, scatterplots, variograms, histograms, bar charts and other graphs allow visual appraisal of data spaces that are not necessarily geographic. Where more than one type of view is displayed simultaneously in a "windows" environment, with the objects on one view dynamically linked to objects on other views, the interpretation of anomalies and patterns becomes greatly enhanced. For example, maps, tables, graphs and scanned images (such as photographs) can, in some systems, all be displayed on a single screen and linked simultaneously. Dynamically linked ("hot-linked") data views are specially powerful if they are combined with interactive spatial query, permitting objects selected in one view to be simultaneously highlighted in other views.

Spatial query of a GIS database involves two principal types of question:

1) what conditions exist at some location? and

2) at what locations do certain conditions exist?

In the first case, the GIS user can point to a location with the cursor on a cartographic image displayed on the monitor, and retrieve information about the characteristics of selected data layers that occur at, or close to, that location. In the second case, the user can find those locations that meet certain criteria, such as "close to a road", "on granite", "arsenic in soil >300 ppm", and so on. The selected areas can either be highlighted on the monitor, or shown on a new view. In both cases, interactive visualization plays an important role.

We begin the chapter with a description of the basic process of composing a cartographic image, visually combining data about several spatial data layers and superimposing cartographic annotation.

DISPLAY OF CARTOGRAPHIC IMAGES

Any pictorial or spatial data can be represented as a digital image. Thus medical images, satellite images and scanned documents are examples of digital images. The images created with GIS differ from other types of digital images in that they are **geocoded**, so that pixels are correctly positioned according to a geographic projection. They are normally composite displays of several spatial data types, with cartographic annotation superimposed. At the display stage, both raster and vector data are represented in a raster format on video monitors, and hardcopy versions are made on colour plotters that are generally raster devices. Such displays can be called **cartographic images**, because they are digital images, yet have the characteristics of conventional maps. A **map** can be defined as "a graphic representation on a plane of selected features of a part or the whole Earth's surface" (Makower et al., 1990). The word "map" traditionally implies a paper document, on which the graphic representation is in a nondigital form.

The cartographic tools of GIS have reached a level at which, in the hands of a skilled draughtsperson, it is difficult to distinguish between cartographic images in hardcopy, and conventional paper maps. A cartographic image, generated as output from a GIS, is composed of graphical elements that are linked to several types of spatial objects and their attributes, recorded in the files of a spatial database. Because of this linkage to a database, the output need not be regarded as a one-time graphical representation of spatial data, but as one product out of many possible products, within the limitations imposed by the data model. Customized map products are straightforward to make from a well-structured database, but costly and difficult from a conventional map. One customer may want to see the geological map generalized by combining geological units according to lithology or stratigraphic age; another customer may want a map showing only a particular region at a large scale, on which the locations of selected mineral occurrences are shown. Making changes to a traditional map is impractical without a major revision, whereas maintenance of a spatial database, although still relatively expensive, is now straightforward. From a GIS, customized maps can be current, containing the latest revisions to the database.

On the other hand, much of the visual output generated by GIS is experimental, and not designed to be of professional cartographic quality. Displays are generated in order to try particular data combinations, to test hypotheses, and to browse through spatial data in a search for pattern and meaning. This experimental aspect of GIS is perhaps one of the technology's greatest strengths.

Some authors have used the term "map image" in the same sense that cartographic image is applied here (Rencz et al., 1993). The final cartographic (or map) image, whether it is displayed on a monitor, on a film transparency, or on a sheet of paper, can be regarded as the combination of two components: 1) a digital image composed of graphical elements, that are linked to spatial objects in a spatial database, and 2) cartographic annotation, such as labels, legend and scale. The form in which the cartographic image is stored digitally depends on the display device, although device-independent file structures are now in common use. To illustrate the steps in creating a cartographic image, a geological example is presented in Figure 5-1.

Components of a Cartographic Image

Graphical Elements Linked to Spatial Objects

Each type of spatial object, such as points, lines, irregular polygons and pixels, can be linked to graphical elements for display. Consider some of the possibilities for these object types.

1. *Point data* are represented on images with graphical symbols. The type, size and colour of symbols to be displayed at point locations can be set to constant values, or they can be allowed to vary according to one or more fields in the point attribute table. For example, in Figure 5-1, the towns (only two of them) are recorded by circles filled with the colour orange. The symbol type, colour and size are all constant, because, in this case, there are only two points. However, suppose that the points are the locations of seismic epicentres, and that several hundred epicentres occur within the study area. An effective display might then consist of circles whose diameter increases with earthquake magnitude, and whose colour changes with earthquake depth, controlled by the values recorded in a point attribute table. In the case of an attribute like earthquake magnitude, that is a continuous variable, a classification is usually necessary to divide the continuous measurements into discrete classes suitable for plotting. Although symbol size can change continuously within some reasonable range, symbol type and colour require discrete classes. Figure 5-2 illustrates some of the more exotic point symbols that can be used on geological maps, many of them for structures. However, for plotting point attribute data, circles, squares, diamonds, crosses and other simple point symbols are normally chosen. Some symbols may record vector rather than scalar data at points. For example, the strike and dip symbol, or the plunge of a fold axis, use the conventional geological symbols for these measurements, and record two quantities at each point: the azimuth direction and the angle with the horizontal. In some cases, one or more attribute fields of the point attribute table contains **feature codes**, which are used as numerical pointers to tables of symbol types.

2. *Line data* are represented on images either with continuous lines, that can vary in thickness and colour, and they can be displayed in different styles. Some styles are shown in Figure 5-2. Again, specified fields of line attribute tables may be used to control the graphical

characteristics of the lines. For example, in Figure 5-1, the roads are classified into three levels: the two-lane roads are shown as the thickest lines, the unclassified roads as thinner lines, and cart or winter roads as dotted lines.

3. *Irregular polygon data* are represented on images as closed polygons filled with colour or black and white (or grey) "fill", either as solid colour or as various kinds of pattern. The boundaries of polygons are displayed as lines that can vary in style, thickness and colour as with line data. The colour and pattern of the polygon fill is defined by values in one or more of the fields in polygon attribute tables, and classifications are often applied to the attributes to divide them into discrete classes. In Figure 5-1, the map unit polygons are filled with solid colour. The colours are defined using a **palette**, also known as **colour lookup table (LUT),** as will be discussed in greater depth in the next section. Fill patterns can either be discrete, or can form a more-or-less continuous sequence of patterns, such as point or cross-hatch patterns that change in density. Such fill patterns can symbolize variables measured on ordinal, interval or ratio scales of measurement, because they represent gradational change.

4. *Pixel data* are represented almost exclusively with colour, or shades of grey. In the case of a raster with no associated attribute table, the colour is determined from the pixel value using a colour LUT, as discussed below. On the other hand, the pixel value may be the pointer to an attribute table, in which case the value of a selected field is used to determine the colour, often after applying a classification. Up to three raster images can be displayed at one time on some systems, using a red-green-blue (RGB) colour LUT, or by determining a colour from another "colour space", such the intensity-hue-saturation (IHS) space. The lower left quadrant of the cartographic image in Figure 5-1 is a combination of two raster images, displayed in an IHS colour space. One image is from an airborne sensor which measures a radar intensity signal. The other image is the geological map in a raster format. In IHS space, the intensity is set equal to the radar value, the hue is controlled by the geological class, and the saturation is set to a constant. IHS and RGB colour representation is discussed below.

The sequence of displaying the various kinds of spatial objects represented in a cartographic image is important, because the objects that are displayed last overwrite the earlier graphical elements. The sequence followed is usually to plot the pixel data first, followed (optionally) by the polygon data, then the line and point data. Polygon data can sometimes be effectively plotted over the top of pixel data as cross-hatch or other patterns, allowing the colour of the pixel data to show through the superimposed patterns, or they can be simply displayed as boundary lines. One of the most common geological displays of this type consists of geophysical or geochemical image data with geological contacts superimposed as polygon boundaries.

FIG. 5-1 (opposite). Map of the Sudbury area in Ontario, illustrating the components of a cartographic image. **1.** Geological map units, as polygons surrounded by lines and filled with colour according to the classification shown in the legend. **2.** Locations of towns, as points whose attributes are recorded as a digital table. The symbols applied here to the towns are filled circles of constant radius and colour. **3.** Labels, linked to points with text attributes and represented with a choice of fonts, size and colour. **4.** Roads, represented as lines and linked to attributes such as road type. A variety of line styles and colour can be controlled by attribute values. **5.** Raster image in the inset. This is a combination of a radar image and geological units. **6.** Annotation, including labels for the geographic reference points, scale bar, titles, legend, and type of border. Modified from Rencz et al. (1993).

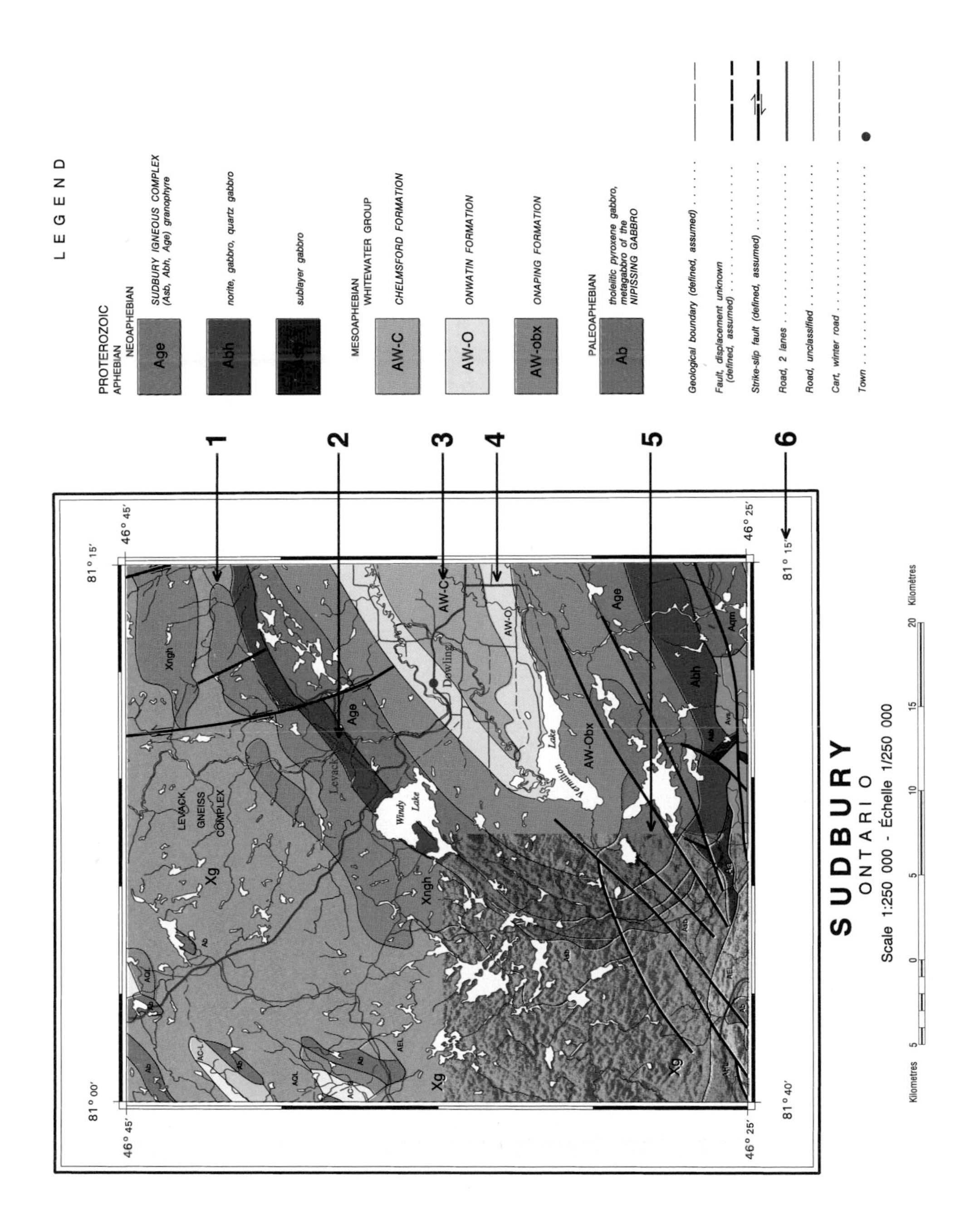
LEGEND
PROTEROZOIC
APHEBIAN
NEOAPHEBIAN
Age
SUDBURY IGNEOUS COMPLEX (Asb, Abh, Age) granophyre
Abh
norite, gabbro, quartz gabbro
sublayer gabbro
MESOAPHEBIAN
WHITEWATER GROUP
AW-C
CHELMSFORD FORMATION
AW-O
ONWATIN FORMATION
AW-obx
ONAPING FORMATION
PALEOAPHEBIAN
Ab
tholeiitic pyroxene gabbro, metagabbro of the NIPISSING GABBRO
Geological boundary (defined, assumed)
Fault, displacement unknown (defined, assumed)
Strike-slip fault (defined, assumed)
Road, 2 lanes
Road, unclassified
Cart, winter road
Town
1
2
3
4
5
6
81° 15'
46° 45'
46° 25'
81° 00'
81° 40'
Xngh
Xg
LEVACK GNEISS COMPLEX
Levack
Windy Lake
Dowling
AW-C
AW-O
AW-Obx
Age
Abh
Vermilion Lake
SUDBURY
ONTARIO
Scale 1:250 000 - Échelle 1/250 000
Kilometres
Kilomètres
5
0
5
10
15
20

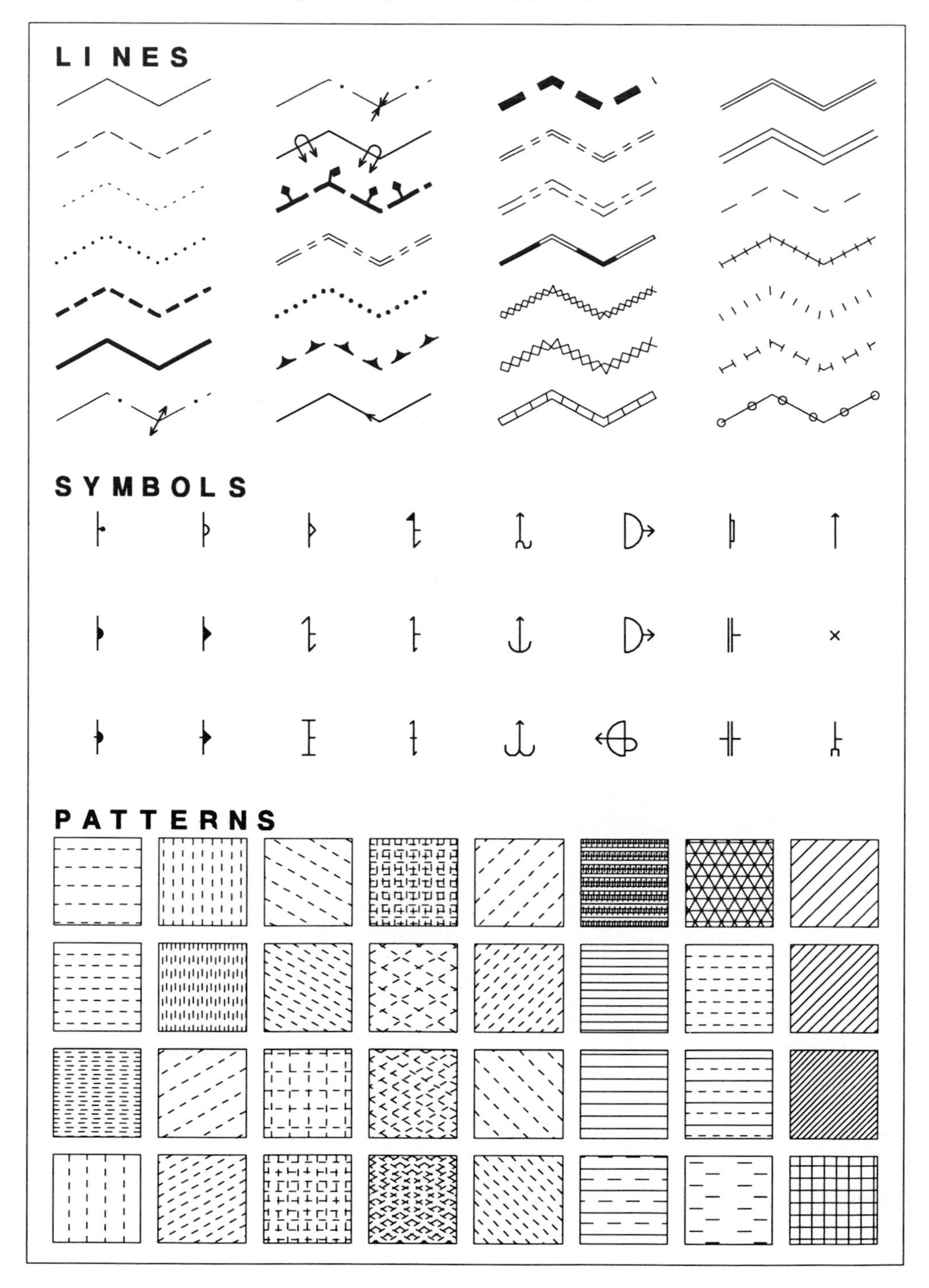
LINES
SYMBOLS
PATTERNS

Cartographic Annotation

The second major component to the cartographic image is the annotation. The annotation is generally plotted last, overwriting the graphical elements of the image. Annotation includes labels, titles, blocks of text, legends, a scale bar, a graticule and a north arrow. Labels are generally linked to point objects, so they can be placed at particular positions on the map. Fields in an attribute table can be used to define the text string, the font, the size, colour and orientation of labels. In some systems, label size can be treated as a variable that is automatically changed with an increase or decrease in the scale of the map. Title and text blocks can also be positioned at particular locations, and can have similar characteristics to labels. Tic marks or grid lines, or graduated lines that surround, or cross, the body of the map are used to indicate geographic or projection coordinates. The legend can, in some systems, be linked to the classes of spatial objects in special text files. In general, the design of cartographic annotation is made as flexible and interactive as possible, facilitating the production of good quality cartographic images.

The composition of cartographic images is either carried out in a series of interactive steps, or is controlled by a sequence of commands in a **command procedure**, or **script** file. The advantage of employing a command procedure is that complex sequences of operations are saved as a file that can be edited and used for multiple purposes.

Resolution, Scale and Metafiles

The hardware device used for the interactive display of cartographic images is a colour video monitor. The size of the image on the screen is governed by a combination of the physical dimensions of the screen, and the number of rows and columns of pixels on the screen, on the video memory board and in the digitally-held image. In most cases, the number of rows and columns on the screen is matched with the number of rows and columns of the video memory. The **resolution** of the image on the screen is the distance on the ground corresponding to a pixel. This is often unrelated to the resolution of the data in the database. For example, if a data layer in raster format with 4,000 rows and 4,000 columns is stored on the disk, but is displayed on a screen with 1,000 by 1,000 rows and columns, then the display uses only every fourth pixel in each direction. The image is said to be "decimated" by a factor of four in each direction. Conversely for a raster on the disk that is smaller than the screen raster, some pixels may be replicated to achieve a particular scale, and fill a greater part of the screen than would otherwise occur. Line and point data, and their associated symbols, are rasterized "on-the-fly" during the display process.

The scale of the display of a cartographic image can be altered by either a **hardware zoom** or a **software zoom**. In a hardware zoom, the effect is simply that of magnifying a portion of the screen, without changing the image resolution. Individual image pixels are replicated, and

FIG. 5-2 (opposite). Examples of some graphical elements that can be used to display lines, points and area patterns. The choice of line type, symbol type or pattern can either be chosen as a constant for one set of spatial objects, or can be made variable, depending on the value of one or more fields of an attribute table.

as the zoom factor increases, the image becomes progressively more blocky. A software zoom is often more useful, because the resolution is changed by redisplay from the database, as illustrated in Figure 5-3. A hardware zoom of a cartographic image containing vectorized lines, for example, produces a "staircase" appearance, whereas in the equivalent software zoom, the lines are as smooth as the scale of original digitizing will allow. Specific regions for zooming may be defined in most GIS as a **window**, whose geographic extents are saved internally. This allows the cartographic image to be redisplayed in the same window many times, without having to remember the precise coordinates of the corners. Images can be moved incrementally on the screen by "panning" or "roaming", usually under the control of the graphics mouse. These operations do not change the scale of the display.

Graphical output from GIS can be directed to devices other than video monitors, such as inkjet or laser plotters. In some cases, the contents of the video monitor (actually the display memory), are simply "dumped" to a raster file with the same number of pixels as the screen. In other cases, a digital plot file is created for a specific hardware device, such as a large-format raster or vector plotter. Alternatively, a device-independent file is created, known as a **graphics metafile**. A common format for graphics metafiles is a **PostScript** file. Besides being device-independent, graphics metafiles preserve the spatial resolution of the data, as opposed to raster images restricted to the resolution of the display memory, which can lose much of the fine detail of an image, a loss that is a serious disadvantage for plotting on large sheets of paper. Device-independent graphics files can then be directed to any available display or hardcopy device, assuming that the appropriate software drivers (programs that convert the metafile to specific plotting instructions) are available.

In the following section, some aspects of digital colour graphics are introduced, in order to give the GIS user an understanding of how colour is represented physically and symbolically on a computer.

DISPLAY HARDWARE FOR DIGITAL IMAGES

The typical setup for video-display hardware is shown diagrammatically in Figure 5-4. When displaying a data object, the computer program fills the appropriate locations in the display memory. This memory is logically subdivided into a number of **bit planes**, each one capable of storing a binary image with a fixed number of rows and columns. The number of bit planes determines the number of possible colours that can be displayed at any one time. For example, if there are 8 bit planes, then the number of colours is 2^8, or 256. This display memory usually resides physically on a single graphics "card" or "board". Also on the card is a hardware lookup table (LUT) that converts the values of the digital image into red, green, and blue (R,G,B) components. The digital R,G and B signals pass through a digital-to-analogue converter and finally to the display monitor. The value of each image pixel is converted to a voltage level, which governs the intensity of an electron beam at a specific location in the monitor (consisting of a cathode ray tube or CRT). The CRT **phosphors**, which occur as a lattice of individual dots, are excited to differing degrees by the intensity of the electron beam. On colour monitors, each pixel location has a trio of red, green and blue phosphors, and an associated trio of electron guns which separately excite the R,G and B phosphors to create colour images.

The amount and configuration of display memory on a particular graphics card varies considerably. In specifying the configuration, usually three number are used, such as (1024*768*8). This means that there are 1024 rows by 768 columns of pixels, and each pixel can have a colour value in the range from 0 to 2^8, or eight bits. Therefore a graphics card with the configuration (512*512*24) indicates an image with 512 rows by 512 columns and 2^{24} colour values at each pixel. Many image processing systems are setup to use 24-bit display memory, allowing three 8-bit images to be displayed simultaneously. For example, in order to display a LANDSAT image, a magnetics image and a gravity image simultaneously, eight bits (256 colour numbers) are used for each image. The LANDSAT values can be assigned

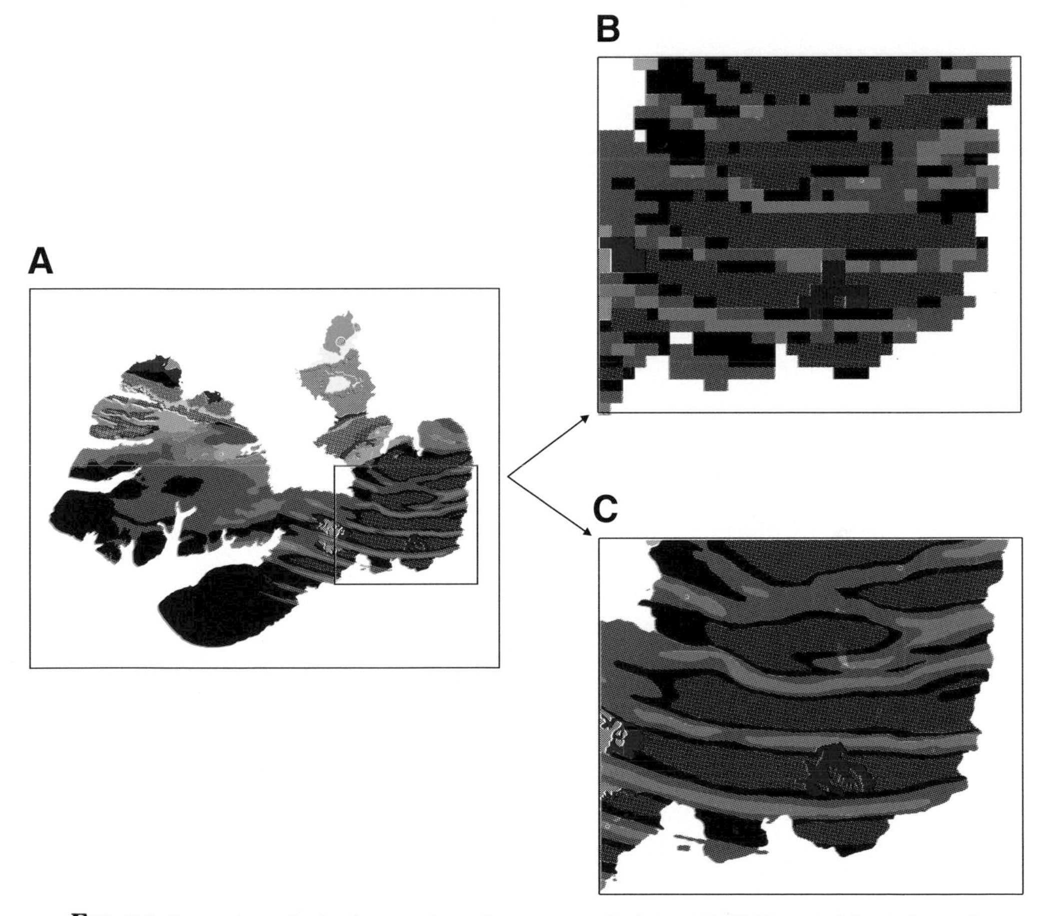

FIG. 5-3. Comparison of a hardware and a software zoom. **A.** A cartographic image of the geology of Melville Island in the Canadian Arctic Islands, showing the area to be enlarged. **B.** Region enlarged with a hardware zoom, resulting in pixel replication and a blocky appearance. **C.** Same area enlarged by a software zoom, with an improved definition of the geological contacts. Figure supplied by Brian Eddy, University of Ottawa.

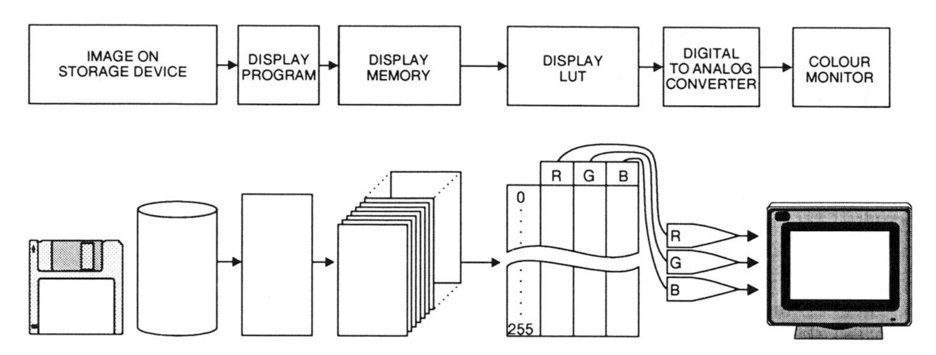

FIG. 5-4. Diagram to show a typical setup of display hardware. Display memory is physically stored on a graphics or video board.

to the red colour guns, magnetics to the green guns and gravity to blue guns, or some alternative combination, producing a **colour composite** image, a term widely used in image processing.

COLOUR

The use of colour for display of maps and images enormously enhances the ability to communicate spatial information. The human eye is able to perceive subtle colour differences and to recognize colour patterns. The photoreceptor cells on the retina of the eye are of two types, the rods and cones. The cones are believed to be associated with colour perception. According to the tristimulus theory of colour vision, the cones are of three kinds, each one responsive to one of the three primary colours of light, namely red, green and blue. Any colour can be produced by adding together red, green and blue light in various combinations. The theory is that humans perceive colour by the relative intensity of the stimuli on the red, blue and green cones. Colour television and video monitors work on the same principle. At each dot on the screen, a red, green and blue phosphor is present that can be excited in varying proportions. When only the red phosphor is excited, the dot appears red. When all three phosphors are excited together to the same intensity, the dot appears grey (or somewhere between white and black, depending on the intensity value). The screen is composed of a lattice of dots, which together create a visual image in colour.

In this colour model, the primary colours can be arranged at three corners of a colour cube, see Figure 5-5. In the lower corner of the cube, all three primary colours are of zero intensity, and the result is the colour black. The colour corresponding to any point in the cube can be described by the displacements on the red, green and blue (R,G,B) axes. The diagonal axis cutting across the cube is an intensity axis, ranging from black at (R,G,B) = (0,0,0) at the origin, to white at (R,G,B) = (100,100,100), where intensities are expressed in percent. Elsewhere in the cube, colours are defined by their R,G,B values; (50,50,50) is dark grey, (0,0,100) is bright blue, and so on. Red, green and blue colours are called **additive**, because

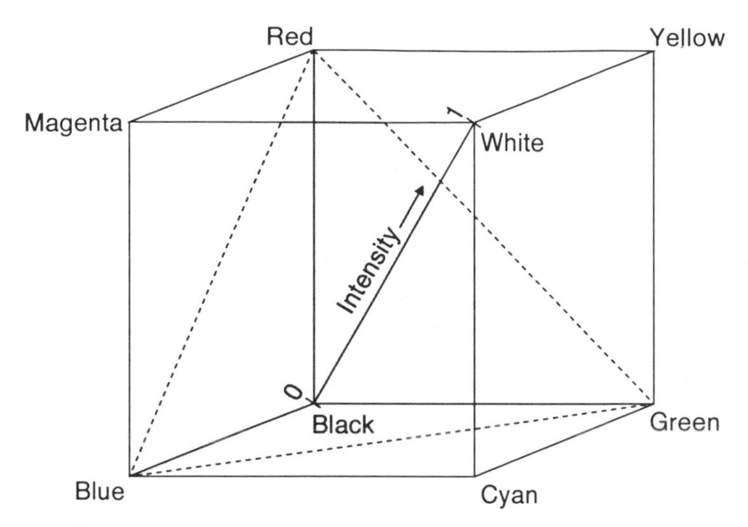

FIG. 5-5. Colour cube, showing the additive primaries (RGB) and subtractive primaries (CMY) plus black and white. A colour is defined as a point within this space either on the RGB axes (with black as the origin), or on the CMY axes (with white as the origin). In this case, the diameter of the cube is scaled to a length of 1, but a variety of other scales are used. The diagonal is the intensity axis, used in the IHS space.

new colours are obtained by adding them to black. The additive colours are used for video display devices, but **subtractive** colours are used in the dyes and inks used for many colour hardcopy devices.

The subtractive primary colours can also be represented in the colour cube. They are cyan, magenta and yellow (C,M,Y) as shown in the remaining corners, as shown in Figure 5-5. Cyan is formed by adding blue and green light, magenta by adding blue and red, and yellow by adding green and red. Dyes and printers ink use the subtractive primaries to produce other colours by subtracting their complementary colours from white. Colours are thus defined on an R,G,B scale for video displays, and on a C,M,Y scale for most printing and plotting purposes.

One of the problems with the RGB colour cube model is that linear changes of position within the colour cube do not lead to a corresponding linear change in colour perception by the human eye. Figure 5-6 shows that blue colours are relatively less sensitive than green and red, and this is correlated with the relatively small number of blue cones on the retina as opposed to red and green cones. An alternative colour model that attempts to overcome this problem is the intensity, hue, saturation (IHS, or HSI) formulation. The IHS model is geometrically related to the RGB colour cube as shown in Figure 5-7. The intensity axis is the same for both models, increasing from 0% (black) at the origin to 100% (white) at the top. The location of a point on any plane normal to the intensity axis is defined using polar coordinates, where hue is the angular distance and saturation is the distance from the centre.

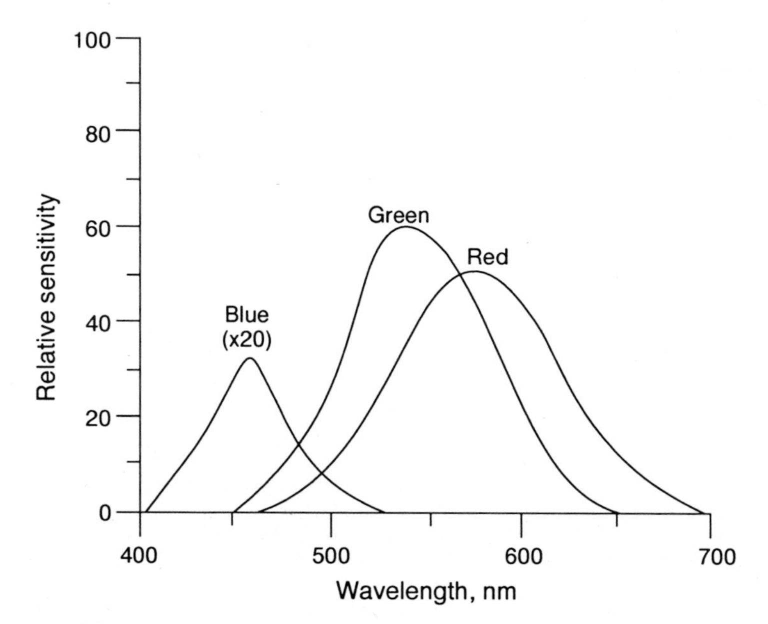

FIG. 5-6. Sensitivity of the eye to light in the red, green and blue wavelengths. Because the sensitivity varies with wavelength, linear change of position in the colour cube does not necessarily lead to a corresponding change of colour as perceived by the eye. Adapted from Mather (1987).

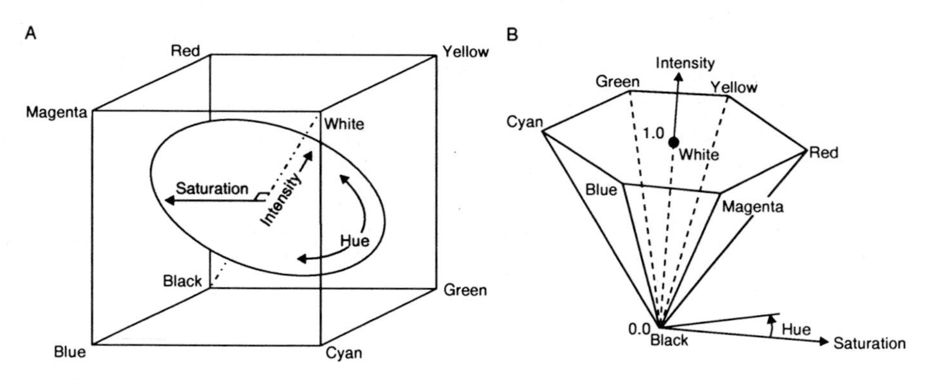

FIG. 5-7. Intensity, hue and saturation (IHS) hexone, and its relation to the colour cube. **A.** The colour cube showing the IHS axes. **B.** The IHS hexone.

Hue thus ranges in value from 0 to 360; the origin is arbitrary, but is normally either at pure red or pure blue, increasing counterclockwise. The IHS space can be pictured as a hexone, with pure colours lying round the top edge. Intensity is a measure of the brightness of the colour. The addition of white light produces less saturated, paler colours.

IHS coordinates can be calculated from RGB coordinates from the following algebraic relationships from Harrison and Jupp (1990):

$$I = R/\sqrt{3} + G/\sqrt{3} + B/\sqrt{3} \quad , \tag{5-1}$$

FIG. 5-8. An IHS image of the Coldwell complex, a carbonatite intrusion on the north shore of Lake Superior in the province of Ontario, Canada. The city of Marathon is just visible on the shore of Lake Superior on the east side of the image. The image combines radar reflectance (image intensity) with an airborne radioelement signature (image hue). The saturation is set to a constant. The IHS values are converted to RGB for display on the screen, or to CMY for hardcopy plots. Image supplied by David Graham, Canada Centre for Remote Sensing.

$$H = \tan^{-1}(\nu_2/\nu_1) \quad \textit{and} \tag{5-2}$$

$$S = \sqrt{\nu_1^2 + \nu_2^2} \quad , \tag{5-3}$$

where

$$\nu_1 = R/\sqrt{6} + G/\sqrt{6} - 2B/\sqrt{6} \quad ,$$

$$\nu_2 = R/\sqrt{2} - G/\sqrt{2} \quad .$$

Conversely, RGB values can be back-transformed from IHS values with the equations:

$$R = I/\sqrt{3} + \xi_1/\sqrt{6} + \xi_2/\sqrt{2} \quad , \tag{5-4}$$

$$G = I/\sqrt{3} + \xi_1/\sqrt{6} - \xi_2/\sqrt{2} \quad \textit{and} \tag{5-5}$$

$$B = I/\sqrt{3} - 2\xi_1/\sqrt{6} \quad , \tag{5-6}$$

where

$$\xi_1 = S\cos(H) \quad \textit{and}$$

$$\xi_2 = S\sin(H) \quad .$$

The main advantage of the IHS model is that hue describes the human perception of colour better than a red, green and blue combination. For example, an orange hue can be modified, making it brighter by changing intensity, or paler by changing the saturation, adjustments that are difficult to guess at intuitively in R,G and B space. Similarly, saturation and intensity can be held constant, and hue can be altered to a neighbouring value in the IHS hexone, a process that is counter-intuitive in RGB space. In image processing systems, the IHS model is often used for colour enhancements, followed by transformation to RGB for display. It can also be effectively used for displaying multiple datasets.

For example, Harris and Murray (1990) produced effective combinations of geological maps with radar. The geological map units were displayed as different hues, the intensity was proportional to radar reflectance, and the saturation was held constant. The SW quadrant of Figure 5-1 uses the same scheme. In Figure 5-8, a combination of a radar image and an airborne radioelement image is displayed using IHS space. The ability to combine information about the surface morphology, contained in the radar reflectance data, with information about the chemical composition of the rocks, as indicated by the radioelement measurements, greatly improves the interpretability of the scene. RGB displays of the same data require much more experimental "tweaking" to achieve the same degree of visual information.

Colour lookup tables

The display hardware uses a LUT to define the actual colours that are stored in the display memory. The simplest kind of colour LUT or palette is shown in Tables 5-1 and 5-2. Each colour number is linked to a trio of colour intensities, defining the red, green and blue. The appearance of the image can be changed instantaneously in many systems, by changing the display LUT. For the LUTs shown in Table 5-2, 2^8, or 8 bits, are required to store the colour number at each pixel. Alternatively, for graphics cards with 2^5 or 5 bits available at each pixel, the LUT is restricted to colour numbers ranging from 0 to 32.

In image display systems with 24 bits per pixel, capable of simultaneously displaying three 8-bit images together, three LUTs are employed, one for varying the intensity range of each of the three primary colours separately, as illustrated in Table 5-3.

Where cartographic images are displayed on a video monitor, a LUT is employed to control the colours assigned to various data types, whether they are pixels of a raster, or polygons, lines, points or annotations. Black and white "palettes" may comprise a series of grey values (equal red, green and blue intensities), often called greytones, or can refer to line patterns, such as hatch patterns and rock type symbols.

Table 5-1. Black and white lookup table (palette) showing variations in shades of grey.

#	Red	Green	Blue	Colour
0	0	0	0	Black
1	1	1	1	
2	2	2	2	
.	.	.	.	
49	49	49	49	Grey
.	.	.	.	
.	.	.	.	
254	254	254	254	
255	255	255	255	White

Table 5-2. Colour lookup table (palette) linking colour numbers to red, green and blue intensity values.

#	Red	Green	Blue	Colour
0	0	0	0	Black
1	255	0	0	Red
2	200	200	0	Yellow
.	.	.	.	.
49	240	100	80	Brown
.	.	.	.	.
.	.	.	.	.
254	0	0	255	Blue
255	255	255	255	White

Table 5-3. Three separate lookup tables for red, green, and blue colours, used for displaying three separate images as a colour composite image.

Image 1	Red	Image 2	Green	Image 3	Blue
0	0	0	255	0	160
1	25	1	220	1	170
2	30	2	215	2	185
.	.	.	.	.	
62	119	62	128	62	147
.	.	.	.	.	.
.	.	.	.	.	
254	227	254	15	254	38.
255	235	255	10	255	22

HARDCOPY DEVICES

Historically, the most common output devices for computer hardcopy of graphics were line printers and pen plotters. The former device produces pictures by using lines of text characters on a page, where each character represents one cell of a gridded image. Many of the earlier geological applications of computers made extensive use of line printers for creating contour maps and sections, see for example the books by Harbaugh and Merriam (1968), or Harbaugh and Bonham-Carter (1970). Greytone maps and images can also be produced with line printers by assigning heavy characters, like an X overprinted by a Z, to dark shades, and light characters, like a comma, to light shades. Dot matrix printers, which plot characters and graphics as a series of dots, offer a more flexible alternative (Romo, 1989). Dot matrix printers are inexpensive, but are being overtaken in popularity by laser printers, which have excellent resolution and can plot both characters and raster graphics. Pen plotters were, and still are, extensively used for vector graphics. Pen plotters, both drum and flatbed types, are suitable for plotting line work, symbols and labels. They are not well suited to plotting raster images, and are being largely superseded for GIS output by good quality raster devices.

Raster plotters that use inkjet, thermal wax and electrostatic principles are now widely used for hardcopy. Each pixel in the raster image becomes one dot, or a matrix of dots, on the output medium. Optical filmwriters that transfer a raster image on to photographic film are popular for production of high quality hardcopy. Although hardcopy technology changes rapidly, the general principles of making hardcopy of raster images are likely to remain the same. The most important principle for images reproduced with inks or dyes is that colours are determined using mixtures of the subtractive primaries, cyan, magenta and yellow. Because combinations of these are very limited in colour range, the appearance of mixing is simulated by a matrix of closely spaced dots made up of binary mixtures of the subtractive primaries. The eye then integrates and blurs the dots simulating particular colours.

For example, inkjet plotters create a raster of very fine dots by squirting fine drops of ink on to paper or transparent film. The inks are cyan, magenta and yellow. Whereas on a video monitor the additive primaries are added to black to produce a particular colour, on a hardcopy the subtractive primaries are "removed" from white. At each inkjet pixel, the inkjets are either ON or OFF, giving rise to eight possible binary combinations, Table 5-4. In practice, most plotters also use a separate black ink, because the black produced by mixing C,M and Y are often brown tinged, caused by the edges of the dots not being in perfect register.

As the dot size and ink colour cannot be changed, the range of output colours produced by direct mixing is very restricted. One of the methods used to produce a larger range of colours is known as **dithering**. Dithering works by using a matrix of dots to represent an image pixel. The dots in this **dither matrix** can be filled by any of the colours shown Table 5-4. The eye blurs the dots together, integrating the various colours to give the appearance of new colours. Dithering matrices are usually (2*2), (3*3) or (4*4). A typical dot size for an inexpensive table-top raster plotter is about 0.2 mm diameter. This means that each pixel occupies (0.8*0.8) mm, so that an image 250 pixels wide requires a page that is 20 cm wide.

A (4*4) dither matrix composed of sixteen dots that are either ON or OFF produces seventeen possible combinations for a given ink colour, see Figure 5-9B, and is used in an image processing system developed in Australia. The dot order shown in Figure 5-9A comes from Harrison and Jupp (1990). The ordering is critical, because inappropriate ordering can lead to distracting geometric patterns, such as herringbone or twill. The order in the matrix has been chosen to avoid horizontal, vertical and diagonal line structures. The ordering is such that no pair of adjacent cells contains sequential numbers, and all groups of 4-adjacent cells sum to 34, either within one dither or across adjacent dithers, as shown in Figure 5-9C. This type of structure is based on a "pandiagonal magic square" or "nasik" (Lippel and Kurland, 1971, as quoted by Harrison and Jupp, 1990).

Table 5-4. Combinations of binary mixtures of the three subtractive colours, cyan, magenta and yellow.

CYAN	MAGENTA	YELLOW	COLOUR
OFF	OFF	OFF	WHITE
OFF	OFF	ON	YELLOW
OFF	ON	OFF	MAGENTA
ON	OFF	OFF	CYAN
ON	ON	OFF	BLUE
ON	OFF	ON	GREEN
OFF	ON	ON	RED
ON	ON	ON	BLACK

To plot an image where the colours are defined by RGBs first requires a conversion to CMY values. For a (4*4) dither matrix, there are sixteen dot positions, so the intensities range from 0 (all OFF) to 16 (all ON). Thus, an 8-bit intensity range of 0-255 (or 0-100 if a percentage scale is employed) is converted linearly to the range 0-16. The red is plotted on an inverse cyan scale, the green on an inverse magenta scale and blue on an inverse yellow scale. For the sake of simplicity, assume that the RGB values have been compressed to the range 0-16, then an RGB of (16,0,0) is equivalent to CMY of (0,16,16), an RGB of (4,12,4) is equivalent to CMY of (12,4,12), RGB of (0,0,16) is CMY (16,16,0), and so on. Each of the subtractive primaries is dithered with the same dither matrix, as shown in the example of Figure 5-10. In practice, a straight conversion of RGB to CMY produces unsatisfactory results because additive colours are perceived rather differently than subtractive colours. In the additive system small changes in intensity are perceived more readily in the dark shades than in the light shades, whereas in the subtractive system the converse is true. A nonlinear stretch of the image histogram improves contrast in the darker colours before plotting.

Colour filmwriters do not use dithering to achieve colour differences. Each pixel is represented by a very small dot, produced by a beam of coloured light which exposes a photographic film. On the other hand, the reproduction of colour photographs with printers inks requires that **colour separation** be carried out, so that separate components of the photographic image are printed with different colours in a subtractive process. Thus although colour filmwriters can be used to make photographic slides (useful in their own right) or photographic colour prints, these prints must then go through a colour separation process before publication. More efficiently, the colour separation process can be carried out digitally, with the production of black and white transparencies for each of the subtractive primary colours and black, ready for publication.

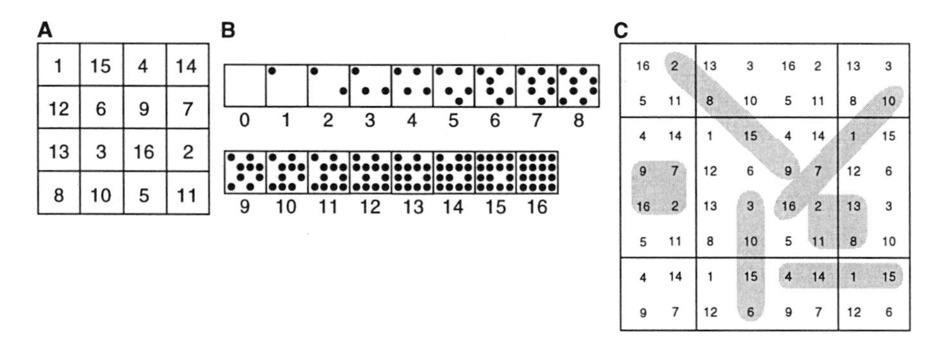

FIG. 5-9. A. A (4*4) array showing the numbers 1 to 16 in a sequence that defines the dot order in a dithering cell. **B.** (4*4) dithering cells labelled 0 to 16, indicating the dots that are turned ON for each of the 17 intensity levels. Note that the labelling is cumulative. Thus, for an intensity level of 4, the dot position for 4 (in row 1, column 3) is combined with the dot positions for 3, 2 and 1. **C.** A complete (4*4) dithering cell surrounded by eight neighbouring cells (shown partially), showing dot order, as in **A**. Note that the order numbers for any four-adjacent dots sum to 34. This property reduces the effects of undesirable repetitive patterns in the dithered image. Adapted from Harrison and Jupp (1990).

Where cartographic images of continuous field variables are plotted, the colour scale attempts to make the changes between adjacent colour numbers as smooth as possible, to give the perception of continuous variation. On the other hand, when a categorical scale variable, like soil units or metamorphic facies, are being plotted, colours are chosen to provide contrast between classes. When plotting units of small areal extent, like thin stratigraphic horizons or dykes, dark saturated colours should be used to make the units stand out.

VISUALIZATION OF SURFACES

In plan view, a map of a continuous field variable, such as topographic elevation, is traditionally displayed using contour lines. Although contours are effective for visualizing surfaces, colour-coded digital images, with colour LUTs carefully selected to give the illusion of gradually changing intensity or magnitude, convey greater realism than contours. Colour images with contour lines superimposed convey even more graphical information, combining the benefits of both colour and line.

Images of surfaces in plan view can be enhanced by the use of **hill shading**, a method which uses information about the illumination source. Regions from where the source are not visible are in shadow. Shading techniques have been used by traditional cartographers, particularly for topographic maps, but also for any (x,y,z) data, where (x,y) are geographic coordinates and z is a field variable such as magnetic intensity. The computations required for automatic hill shading are not particularly complex, see Chapter 7, and hill-shading algorithms are to be found in several GIS. Broome (1988) provides a computer program for hill-shading geophysical or other images, where the position of the sun can be interactively controlled with a computer mouse. Instead of shading being produced by topographic relief, a geophysical field variable is treated as elevation. Intrusive dykes, or other rock bodies with a magnetic susceptibility that contrasts strongly with the surrounding rocks, appear as

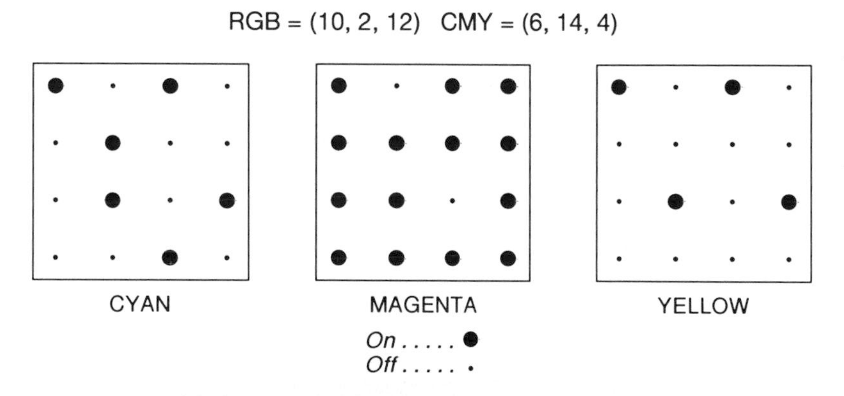

FIG. 5-10. Example of dither cells for the RGB colour (10,2,12), corresponding to the CMY colour cyan=6, magenta=14, and yellow=4. The dot order is the same as the one shown in Figure 5-9.

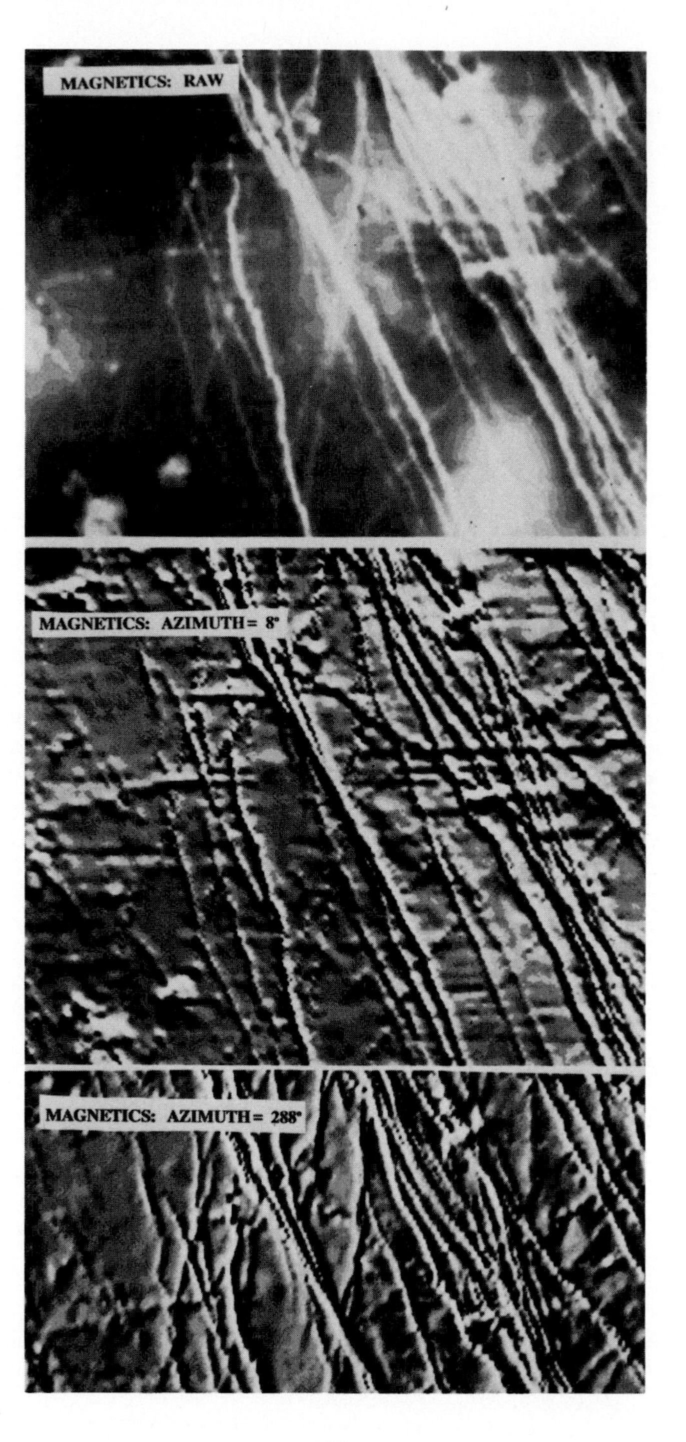
MAGNETICS: RAW
MAGNETICS: AZIMUTH= 8°
MAGNETICS: AZIMUTH= 288°

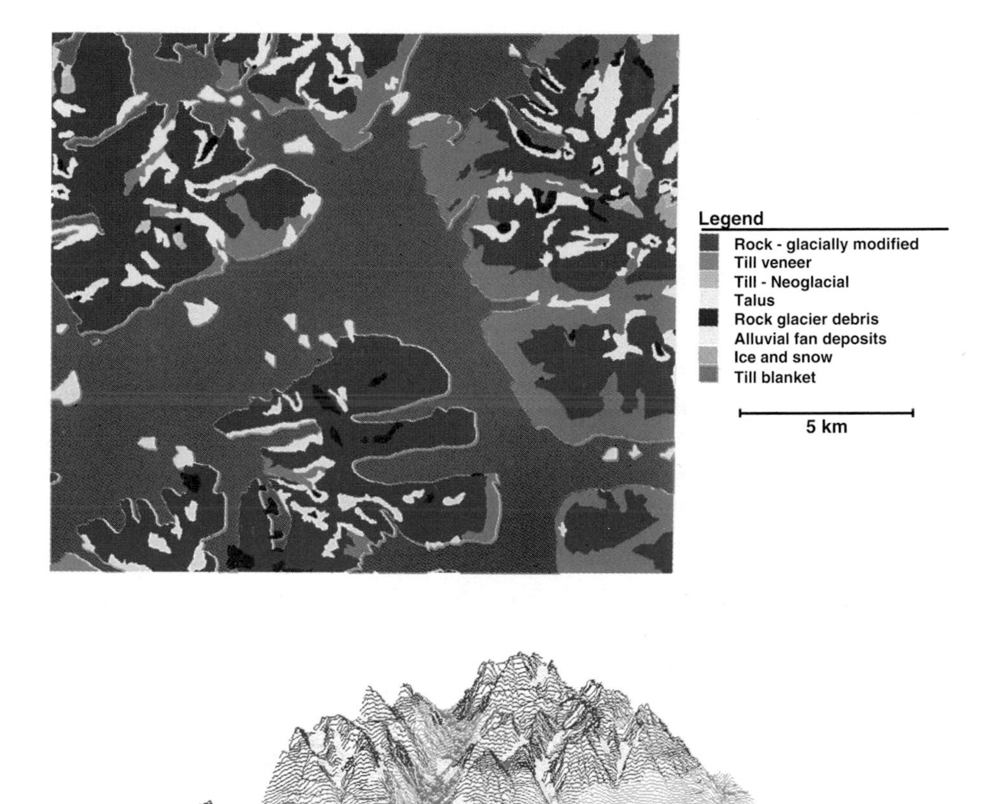

FIG. 5-12. Visual combination of a geological map with a digital elevation model (DEM) by means of a wire-frame diagram in perspective view. **Top.** Geological map of a small portion of the Richardson Lake area in Yukon Territory. **Bottom.** Wire frame showing digital elevation, viewed in perspective from the SE, with geological units "draped" over the surface, and with hidden lines removed. Data supplied by Valerie Sloan, University of Ottawa.

FIG. 5-11 (opposite). Hill-shading applied to an airborne magnetics image. **A.** Original "raw" image, before hill-shading has been applied. **B.** The same image illuminated from an artificial light source from the NNE, at an azimuth direction of 8°. Note that the dykes orientated NNW-SSE are clearly enhanced, because they are orientated at right angles to the light direction. **C.** The same image with the light source at azimuth 288°. Note that a set of dykes orientated at the NNE to SSW direction now become evident. Image supplied by Andy Rencz, Geological Survey of Canada.

topographic features on a shaded magnetics image, as illustrated in Figure 5-11. By changing the illumination direction, the appearance of features can be enhanced or suppressed, depending on their orientation to the light source.

Perspective displays are effective methods of portraying the shape and texture of surfaces. In a perspective view, the size of an object varies inversely with distance from the viewer. Graphic display systems often create perspective views with a "wireframe" model of a surface. The wire frame is a series of profiles parallel to the rows and or columns of the original grid (for a raster case), viewed by a perspective transformation. The effect is for parallel lines to converge with increasing distance, an important depth cue for human perception. Additional realism can be added by removing the edges and surfaces that would be hidden from the observer by the solid surface. Triangular meshes, as produced by Delaunay triangulation for example, can also be viewed as wire frames. Still further realism can be added by adding surface colour, reflectivity and texture, simulating an illumination source, smoothing the geometrical artifacts of the wire-frame geometry, and creating further depth cues such as the variation in haze due to atmospheric conditions, as summarized by Kraak (1989). Effective graphical overlays can be produced by "draping" one surface over the wire frame of another surface. For example, a geological surface can be draped over a topographic surface, as illustrated in Figure 5-12. Wireframe models can be rotated, tilted, and viewed from different positions. Interactive computation of large wireframe displays requires a considerable amount of computation, particularly to eliminate "hidden lines". Sophisticated computer graphics have been used extensively in the film industry, and GIS technology is reaping the benefits of applied research carried out in film and other disciplines.

Smith and Paradis (1989) show how 3-D geological solids can be modelled as blocks that can be sliced or cut by various surfaces to reveal forms and shapes in full 3-D. Van Driel (1989) illustrates several types of 3-D display, automatically generated from computers, from the traditional fence diagrams to sub-surface stereo wireframe views of earthquake aftershocks. The ability to visualize geological phenomena in 3-D is likely to make a major impact in many areas of the geosciences. Readers interested in 3-D display are referred to the books edited by Raper (1989), Turner (1992) and Pflug and Harbaugh (1992).

DYNAMICALLY-LINKED DATA VIEWS

The idea of dynamically-linked data views can be illustrated with the geological example shown in Figure 5-13. This illustration has been kindly supplied by John Haslett, who has published a series of studies illustrating geological and other applications of a software package developed by he and his colleagues for Macintosh computers (Haslett et al., 1990; Haslett et al., 1991). In Figure 5-13A, four data displays are shown together, as hardcopy output from a colour monitor. Each data view is a separate "window", representing different data spaces of a geochemical dataset from Wales. The geochemical samples were taken along streams. A number of metallic elements were determined from samples of sediment. The elements of interest are lead (Pb), copper (Cu), and zinc (Zn). The values have been transformed to logarithms, to reduce the positive skewness of the element distributions, and the diagrams are therefore labelled as LnPb, LnZn, and LnCu.

The four views shown in Figure 5-13A are: geographic (*x*,*y*) space, "frequency" space (as seen in a box-and-whisker diagram), (*x*,*z*) or (*y*,*z*) space (comprising element profiles along a stream), and geochemical space, (z_i, z_j).

1. The main view is a map (geographic space) showing streams in green with the locations of geochemical samples as red dots. The locations of mineral occurrences are shown as blue dots.

2. To the right of the map is a box-and-whisker plot (frequency space) of the logarithmically transformed Cu values. Frequency space could also be represented by histograms, cumulative frequency plots, and others.

3. In the lower left corner of the screen is a box containing three profiles, for the logarithms of Pb, Zn, and Cu. In this space, the horizontal axis is distance along a stream from an arbitrary origin, and the vertical axis is the geochemical element concentration.

4. In the lower right of the screen is a scatterplot of Pb versus Zn (geochemical space). In image processing systems, this space is sometimes called **feature space**, to distinguish it from geographic space.

The program is designed to be highly interactive, so that the user can select subsets of observations in any one of these four spaces, or in other spaces defined for the data. In the figure, a subset of stream samples has been chosen on the map view, as shown by the larger red dots. The subset identified in any one space is also identified, in this case by colour, on the other spaces. Thus, the box and whisker plot shows that the red points have significantly elevated Cu values compared with other samples. The profiles show the variations in element concentration along the stream. And in the geochemical space, the red dots comprise a group that mostly fills an area of the plot distinct from the main body of the points, suggesting that they are members of a separate statistical population.

In Figure 5-13B, a new screen display has been generated, showing the map and profile views only. The map has been "zoomed", so the scale is larger. Two individual points have been selected on the profiles, and the corresponding points are shown as large dots on the map.

The interactive viewing provides a marvellous graphical tool for understanding the spatial and chemical controls on the geochemistry of these stream sediments. Besides the four "spaces" viewed here, some other spaces that can be generated with Haslett's system include time series, and variogram clouds (variograms are discussed in Chapter 6). Selection of subsets of spatial objects can be carried on any one of the views.

The idea of dynamic graphics for exploratory data analysis in statistics (Cleveland and McGill, 1988) is now strongly rooted in statistical computing software such as S-Plus (Statistical Sciences, 1992). Linkages from GIS to statistical software that support dynamic graphics are beginning to be developed. Most systems already support the viewing of profiles, scatterplots and histograms, but without the benefit of multi-way dynamic linkages. However, the dynamic two-way linkage between geographic space and "table" space is found in some GIS, and is widely used for spatial query. Here, the term "table" space simply refers to a listing from an attribute table, shown in a window on the screen.

SPATIAL QUERY

As mentioned in the introduction, spatial query in a GIS involves either finding the spatial data objects (and their attributes) that are present at a specific location, or finding the locations at which spatial objects meet certain criteria. When these queries are carried out interactively, the results are often shown on a dynamically-linked view, similar to the linked views mentioned above. An example of the two types of spatial query are illustrated with reference to a small portion of a geological map, as illustrated in Figure 5-14.

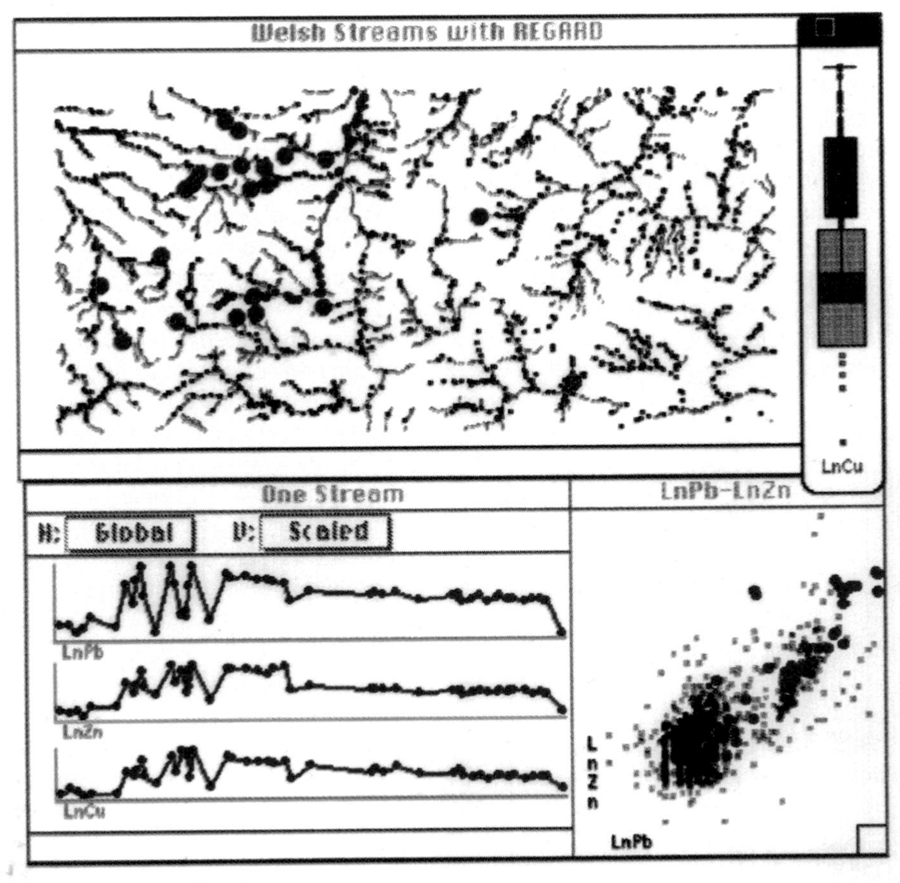

FIG. 5-13. A. Four dynamically-linked views of geochemical data from a stream survey in Wales. Points identified in one view are automatically highlighted in the other data views, as described in the text.

Query by Spatial Attributes

In Figure 5-14A, the polygon in red on the geological map is identified with the cursor. The question is: "What is at this location?". In this case, a "window" is displayed alongside the map containing part of the polygon attribute table, linked to the geological map. The record corresponding to the polygon object in question is highlighted, identifying the record number as 65, corresponding to class 19 of the geological map. In this case, two geometric polygon attributes are listed, indicating the area and perimeter of the polygon. A text field containing

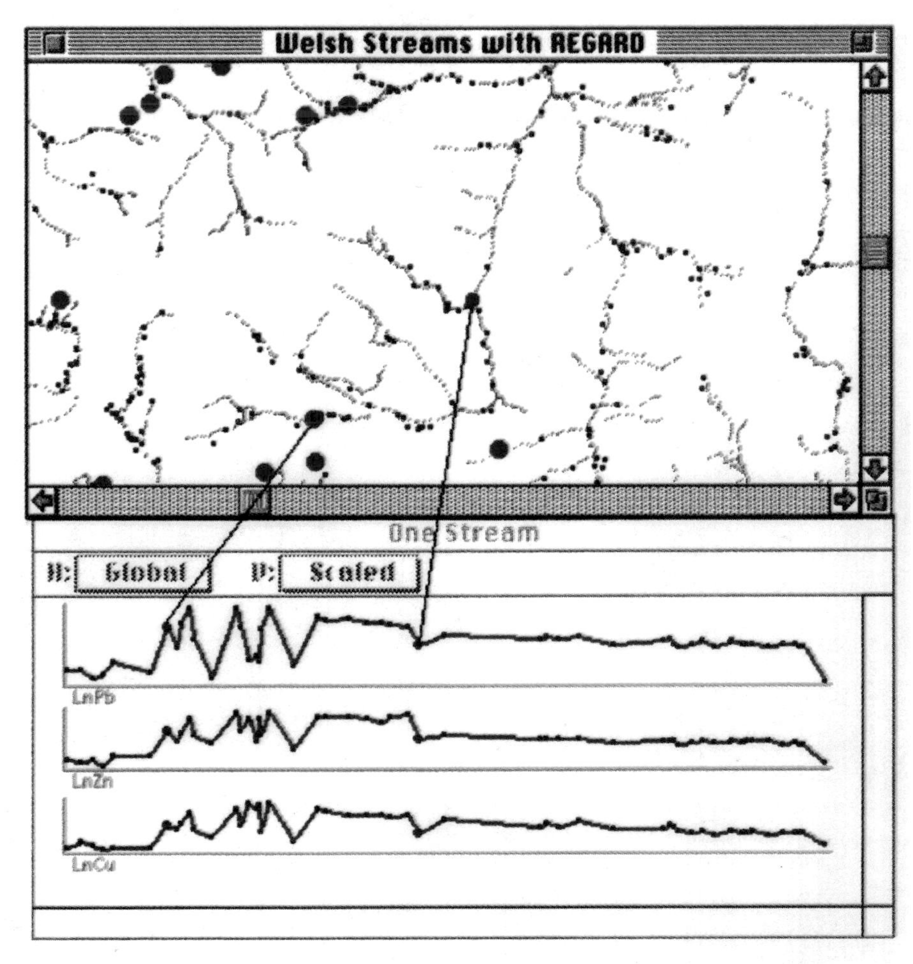

FIG. 5-13. B. A portion of the map in **A**, at an enlarged scale. Two points are highlighted to show their positions both on the map and on the geochemical profiles along a stream. Figures supplied by John Haslett, Department of Statistics, Trinity College, Dublin.

the name of the class records that the rock type is "migmatite". Other fields in the table can be selected for query, if applicable. By pointing with the cursor to different polygons on the map, the "table" view in the window is instantly updated.

Query by location can apply to any type of spatial object. Querying of points is of particular interest in many GIS studies. For example, given a map of point data, where the points are geochemical samples, the query might be to view selected attributes from the point attribute table corresponding to the point closest to the cursor. Alternatively, the request might be to find all the points within a circular, rectangular, or irregularly-shaped window, as defined interactively on the screen, to list their selected attributes, and summarize them statistically. Similarly the attributes of line objects can be identified interactively. The characteristics of individual (or groups of) pixels in a raster, to discover such properties as the "elevation above

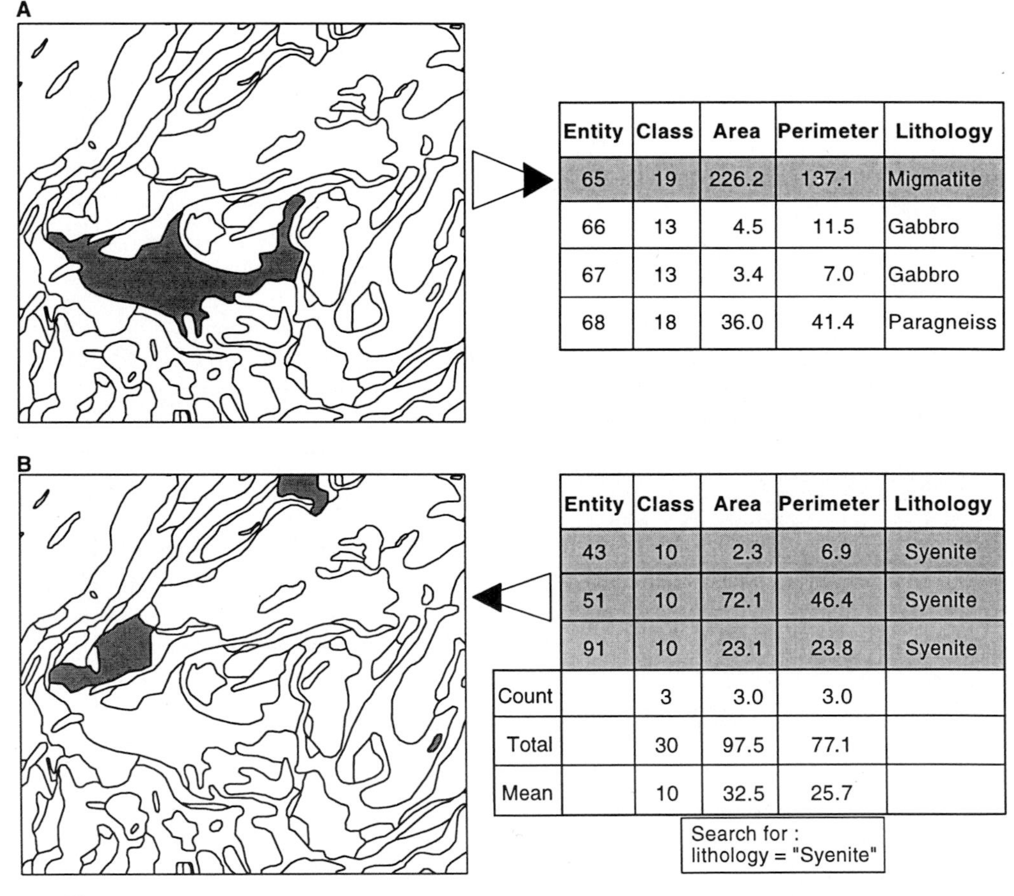

Entity	Class	Area	Perimeter	Lithology
65	19	226.2	137.1	Migmatite
66	13	4.5	11.5	Gabbro
67	13	3.4	7.0	Gabbro
68	18	36.0	41.4	Paragneiss

	Entity	Class	Area	Perimeter	Lithology
	43	10	2.3	6.9	Syenite
	51	10	72.1	46.4	Syenite
	91	10	23.1	23.8	Syenite
Count		3	3.0	3.0	
Total		30	97.5	77.1	
Mean		10	32.5	25.7	

FIG. 5-14. Interactive spatial query of a geological map. **A**. Identifying the attributes of a polygon *selected on a map view*, as indicated in an associated polygon attribute table. **B**. Identifying those polygons on the map that have the attribute called lithology equal to "syenite", as *selected from the polygon attribute table*.

sea level", the "Au content in till", the "rock formation", the "distance to the nearest fold axis", or other mapped variables in the database, can be simultaneously displayed in a window as the cursor is moved about.

Query by location thus employs a map view and a table view, with the selection of objects taking place on the map. The second type of query, finding the locations that satisfy certain criteria, works in reverse. Objects are selected on the basis of nonspatial attributes, from one or more tables. The results are shown on a map.

Query by Nonspatial Attributes

In Figure 5-14B, the search is for those polygons that contain the rock type "syenite". The search could be made by highlighting individual records in the table by hand with the cursor, with the corresponding polygons on the map being flagged. Alternatively, a search procedure can be implemented to look for all polygons that satisfy the criterion **lithology = "syenite"** resulting in the simultaneous flagging of the selected polygons on the map, and the highlighting of the corresponding records in the table view. Some systems may also carry out statistical summaries of the attributes of the selected attributes. The request could, of course, be much more complex, involving several attributes and possibly several types of objects, in a single operation. In most systems, complex queries are carried out with a language of the SQL type. SQL stands for Structured Query Language, and is widely used with relational databases, (Van der Lans, 1988). As in the case of query by spatial location, selection can apply to points, lines, polygons, or pixels in a raster. Query leads to the creation of a subset of objects. The outcome of the search may be to create a new file of spatial objects, or to generate a display with the selected objects highlighted or symbolized in some way.

The subject of manipulating spatial objects and their attribute tables with an algebraic query language is part of the more general topic of map modelling, to be introduced in later chapters. Map modelling also involves the use of an algebraic language for manipulating maps and attribute tables, but is not restricted to spatial query.

REFERENCES

Broome, J., 1988, An IBM-compatible microcomputer workstation for modelling and imaging potential field data: *Computers & Geosciences*, v. 14 (5), p. 659-666.

Cleveland, W.S. and McGill, M.E. (editors), 1988, *Dynamic Graphics for Statistics*: Brooks-Cole, Monterey, California, 455 p.

Harbaugh, J.W. and Merriam, D.F., 1968, *Computer Applications in Stratigraphic Analysis*: John Wiley & Sons, New York, 282 p.

Harbaugh, J.W. and Bonham-Carter, G.F., 1970, *Computer Simulation in Geology*: Wiley Interscience, New York-London, 575 p.

Harris, J.R. and Murray, R., 1990, IHS transformations for the integration of radar imagery with other remotely sensed data: *Photogrammetric Engineering and Remote Sensing*, v. 56, p. 1631-1641.

Harrison, B.A. and Jupp, D.L., 1990, *Introduction to Image Processing*: CSIRO-Division of Water Resources, Canberra, Australia, 255 p.

Haslett, J., Wills, G. and Unwin, A., 1990, SPIDER-an interactive statistical tool for the analysis of spatially distributed data: *International Journal of Geographical Information Systems*, v. 4 (3), p. 285-296.

Haslett, J., Bradley, R., Craig, P. and Unwin, A., 1991, Dynamic graphics for exploring spatial data with application to locating global and local anomalies: *The American Statistician*, v.45 (3), p. 234-242.

Kraak, M.J., 1989, Computer-assisted cartographical 3D imaging techniques: In *Three Dimensional Applications in Geographical Information Systems*, Edited by Raper, J., Taylor and Francis, London-New York, p. 99-113.

Lippel, B. and Kurland, M., 1971, The effect of dither on luminance quantisation of pictures: *IEEE Trans. on Comm. Tech.*, v. 19, p. 879-888.

Makower, J., Poff, C. and Berhgeim, L. (editors), 1990, *The Map Catalog*: 2nd Edition, Tilden Press, 364 p.

Mather, P.M., 1987, *The Computer Processing of Remotely Sensed Images*: John Wiley & Sons, New York-Chichester, 352 p.

Pflug, R. and Harbaugh, J.W. (editors), 1992, *Computer Graphics in Geology: Three-Dimensional Computer Graphics in Modeling Geologic Structures and Simulating Geologic Processes*: Lecture Notes in Earth Sciences., Springer-Verlag, Berlin, v. 12, 298 p.

Raper, J., (editor), 1989, *Three Dimensional Applications in Geographical Information Systems*: Taylor and Francis, London-New York, 190 p.

Rencz, A.N., Harris, J., Glynn, J., Labelle, G. and Baker, B., 1993, Presentation of integrated geoscientific map products: *Geological Survey of Canada, Open File*, 26 p.

Romo, J.M., 1989, Gray-scale maps with a personal computer: *Computers & Geosciences*, v. 15 (8), p. 1249-1263.

Smith, D.R. and Paradis, A.R., 1989, Three-dimensional GIS for the earth sciences: In *Three Dimensional Applications in Geographical Information Systems*, Edited by Raper, J., Taylor and Francis, London-New York, p. 149-154.

Statistical Sciences Inc., 1992, *S-PLUS Technical Overview Version 3.0*, Statsci Inc., Seattle, Washinton, 21 p.

Turner, A.K., (editor), 1992, *Three-Dimensional Modeling with Geographic Information Systems*: Kluwer Academic Publishers, Dordrecht, 443 p.

Van der Lans, R.T., 1988, *Introduction to SQL*: Addison-Wesley Publishing Co., Reading, Massachusetts, 348 p.

Van Driel, J.N., 1989, Three dimensional display of geologic data: In *Three Dimensional Applications in Geographical Information Systems*, Edited by Raper, J., Taylor and Francis, London-New York, p. 1-9.

CHAPTER 6

Spatial Data Transformations

INTRODUCTION

Many GIS operations can be regarded as transformations. In the two previous chapters, several types of transformations were discussed in connection with data input and visualization. For example, geometric transformations from one coordinate set to another are common both for registering input datasets to a working projection and for displaying output in perspective views. Another class of transformations consists of the changes between and within the principal types of spatial data objects, namely points, lines and areas, as shown in Table 6-1. Some of the GIS transformations in this table are used regularly, such as the conversion of points to areas. Others, such as the "skeletonizing" of areas to lines are used infrequently and are not universally available GIS functions. Not shown in the table are the transformations from data structures in the vector model to structures in the raster model, and

Table 6-1. Examples of elementary GIS transformations.

FROM/TO	POINTS	LINES	AREAS
POINTS	interpolation[1]	contouring[2]	Thiessen polygons[3]
LINES	line intersections[4]	line smoothing[5]	buffer zones[6]
AREAS	sample at points[7]	medial axis[8]	re-sampling at areas[9]

[1] Spatial interpolation from one set of points to another set of points, such as "gridding" prior to contouring.
[2] Joining points of equal value with iso-lines. Also "join-the-dot" operations in mathematical morphology.
[3] Generating Thiessen or Voronoi polygons from irregularly distributed points.
[4] Finding intersection points where lines, such as lineaments or faults, cross.
[5] Either by moving or "weeding" vertices to make a smoother line.
[6] Dilation of lines to create corridors at specified radii.
[7] Appending area attributes at points to a point attribute table. The points may be polygon centroids, or any other point data, such as positions of mineral deposits, or locations generated with a random number generator.
[8] Each polygon is reduced to a "skeleton", or medial axis line.
[9] Converting the attribute values of one set of polygons to another set of polygons. An example is the conversion of population density by county to population by geological province.

vice-versa. Raster-vector-raster conversions are widely used in GIS. Interconversion between data structures of the same data model are also common. An example is the transformation from a quadtree to a full raster, both structures belonging to the raster model.

The purposes of making transformations between data types are 1) to facilitate GIS operations that are more efficiently carried out on data in one form rather than in another, 2) for input or output, and 3) for operations that require that attributes from several sets of spatial objects be reduced to a single attribute table with a common set of objects. Consider the following scenarios where data transformations are involved.

1) A dataset of well records, where the fields are collar elevations, and depths to formation tops, is being employed to make an isopach map. The first step is usually to "grid" the irregularly-spaced points. This is a point-to-point transformation. An algorithm is then applied to "thread" isolines at pre-defined levels to create a vector structure containing labelled contour lines. Alternatively, a topological data structure might be generated, where the contour lines are now boundaries between polygons. These are point-to-line, or point-to-area transformations, respectively.

2) A geological map and a gravity map are to be compared. They are both in a raster format, A quick visual appraisal of the relationship between the two can be achieved by displaying the gravity data as a raster image in colour and superimposing the geological contacts as vector lines. This requires that the geological map be converted from a raster to a vector format. The polygon boundaries are either treated as "spaghetti", in which case the transformation is an area-to-line transformation. Alternatively, the arc-node topology is built for the geological polygons, allowing the polygons to be filled with cross-hatch or other patterns keyed to a geological legend, yet still leaving the coloured gravity image visible as a backdrop. This would then be an area-to-area transformation, changing from pixel objects to polygon objects.

3) Given a table of geocoded geochemical point data (soil samples) and a table of gold mine coordinate locations, the problem might be to evaluate the spatial relationship between the gold mines and arsenic levels in soil. There is no direct way of doing this, because the geochemical samples and the gold deposits are independent point sets, with different point locations. Either the arsenic levels at the gold mines must be estimated by interpolation from the geochemical point samples, or the distance to the nearest gold mine must be calculated for each geochemical sample location. The former approach involves point-to-point interpolation. The latter approach could be carried out directly, making distance calculations between the chemical points and the mine points using Pythagorean distance. Alternatively, the gold points could be buffered successively, generating a map showing distance to the nearest mine (point-to-area transformation). The distance to the nearest mine may then be appended as a new attribute to the geochemical attribute table (area-to-point transformation), allowing an examination of the relationship between arsenic level and distance to nearest mine to be evaluated with a scatterplot, or analyzed with a model.

4) Given a digital elevation raster model (DEM), and a digitized bedrock geology map in a topological vector format, the problem is to determine whether steeper surface slopes occur on dolomites than on other rock types. The first step is to transform the elevation raster to a slope raster, by using a neighbourhood derivative operator, see Chapter 7. The second step is to average the slope data over the geological units, a re-sampling by area (area-to-area) transformation. This results in the geological map polygons being reclassified by average

slope. The relationship between dolomite and surface slope can then be visualized by displaying the new slope map, or by inspection of a table, summarizing the slope characteristics of each geological unit. Such a table might be used to test whether the dolomite slopes are "significantly" steeper than slopes developed on other units. Alternatively, the geological map might be "draped" over the slope map in a perspective view, giving a visual impression of the relationship between slope and lithology.

This purpose of this chapter is to provide an understanding of the kinds of data transformations useful in GIS, some idea of how they work and where they can be applied. The discussion concentrates mainly on three topics of particular importance for geoscientific applications: converting point data to areas, dilation of line or area objects, and sampling area objects at points.

POINT-TO-AREA CONVERSIONS

Most geoscientific data are gathered at points. Geological maps are based largely on observations at outcrops, and usually these are regarded as points at the scale of the map. Geophysical measurements are taken at points, often along flight lines, down boreholes, on ground traverses or along a ship's tracks. Imaging sensors produce raster data structures that are really point observations on a regular lattice. Samples of bedrock and surface materials collected at points in the field are analyzed for compositional and textural attributes. The goal of analyzing and modelling spatial inter-relationships between one data source and another often requires that sample observations taken at points be interpolated to a surface representation using new point, polygon or pixel objects, because the sampling points for different surveys are not the same. For example, although multivariate analyses can be carried out on geochemical variables measured on a single set of point samples, comparison of data from one set of samples (such as a soil survey) with data from a different set (such as stream sediments) requires conversion to a common base for comparison.

A common approach is to convert attributes observed at points into the attributes of area objects. The result is to produce a map layer, either in raster or vector mode, that can then be visualized, analyzed and modelled as a semi-continuous or discontinuous piecewise surface, depending on the measurement scale of the attribute. Map layers produced from point data can then be combined with other map layers for analysis and modelling. Map layers of area objects can also be re-sampled at new point locations, such as the points on a regular grid, to build point attribute tables with multiple fields for multivariate and other kinds of data analysis. Alternatively, the re-sampling can occur over a new set of area objects, such as the cells of a grid, building a multi-field polygon attribute table for subsequent analysis. Thus, transformations may be bi-directional, points to areas and areas to points, depending on the requirements at hand.

Point-to-area conversion is carried out in numerous ways, depending mainly on the level of measurement of the point attribute being considered, and whether the points are natural spatial objects, or imposed objects for sampling. If the points are natural spatial objects, like sinkholes in karst topography, or lineament intersections, density mapping methods are appropriate. If the points are samples from a continuous or discontinuous field, then an attempt is made to reconstruct the field from the samples by making a digital model of the surface. If

A

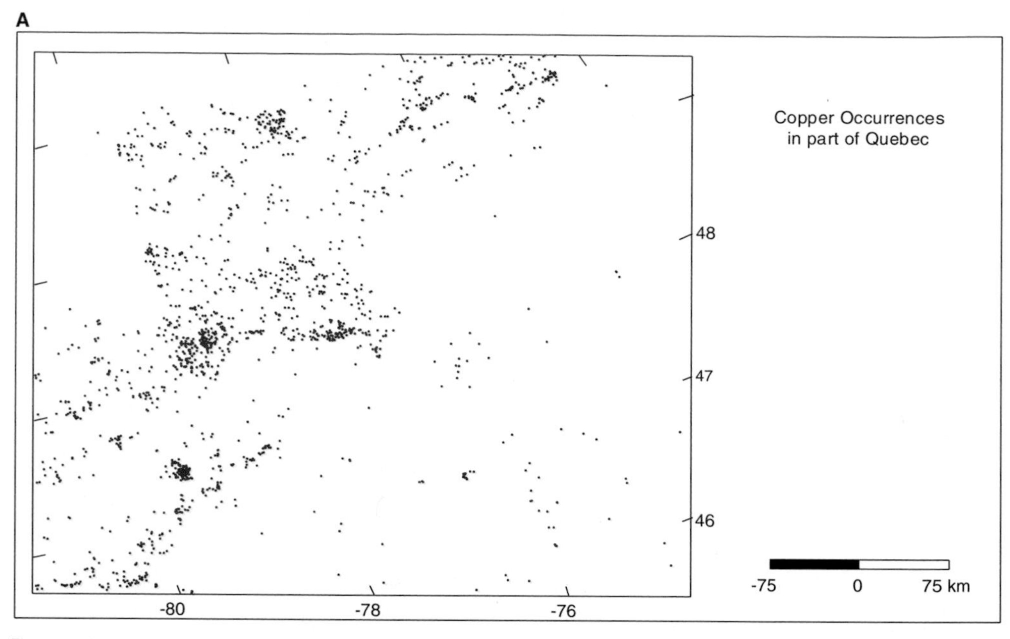
Copper Occurrences
in part of Quebec
48
47
46
-80
-78
-76
-75
0
75 km

B

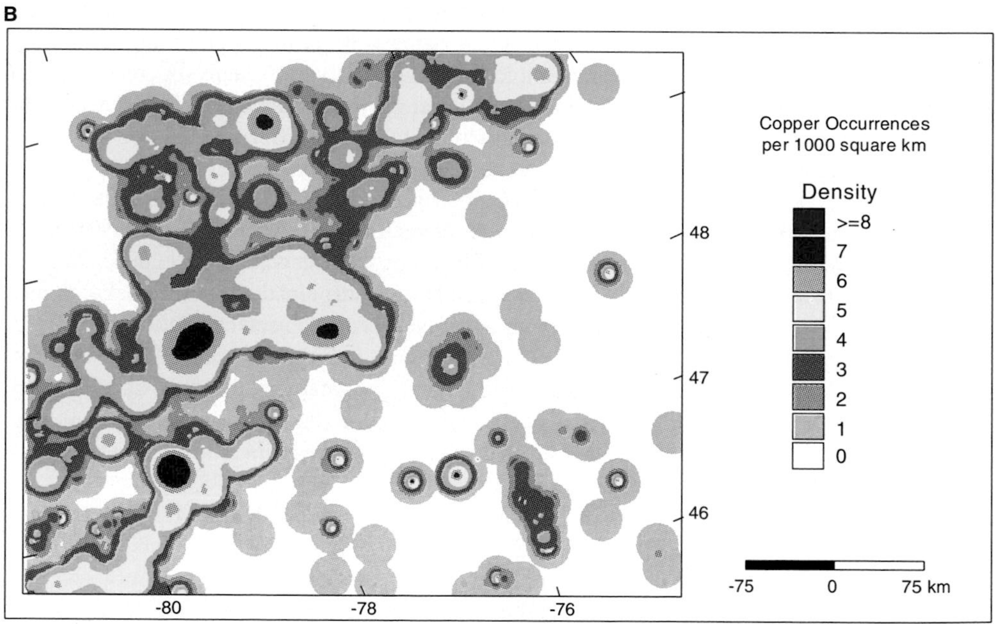
Copper Occurrences
per 1000 square km
Density
>=8
7
6
5
4
3
2
1
0
48
47
46
-80
-78
-76
-75
0
75 km

the attribute is one that is measured on a categorical scale, such as the numbers assigned arbitrarily to rock types on a geological map, the "surface" to be reconstructed consists of discontinuous pieces. Sample attributes measured on ordinal, interval and ratio scales, like geochemical elements and geophysical variables, are modelled as continuous or semi-continuous surfaces. Surface modelling of spatially continuous field variables involves interpolation from the irregularly-spaced samples to a grid of points that is normally treated as a raster. Each interpolated point is simply a cell over which the variable is constant. Alternatively, triangles are used to link the points using the TIN model and the surface is treated as a mosaic of triangular planar facets. Data can be stored in a TIN model, or the TIN used as an intermediate step in creating a grid.

Density mapping

In mapping point density, the goal is to create a digital model of a surface, but the "height" of the surface is not the value of a thematic point attribute, but is the point density, or number of points per unit area. This surface in practice is either modelled as a raster image, or as irregular polygons in a vector model. Consider the points shown on a map in Figure 6-1A. A simple method of generating a surface is by superimposing a grid and counting the number of points per cell. The resulting surface tends to be blocky unless the number of points per cell is large. A smoother surface is achieved using a moving window method (normally either a circular or rectangular window), counting the number of points in the window, and moving the window progressively over the map on a lattice to produce a raster. The density value for each lattice point is simply the count of the points occurring within the sampling window, divided by the area of the window, Figure 6-B. Clearly, the results depend strongly on the window area, and on boundary effects. As the window is made progressively smaller, the density surface is eventually reduced to a field of spikes, with a value of 1 at the points and zero elsewhere. As the area is increased, the surface becomes progressively more smooth, eventually reaching a value equal to the average point density for the whole region, although in practice this is achieved only if the region is the same shape as the window. Somewhere between these extremes is a size of window that produces reasonable results. Regions within a distance from the boundary equal to the sampling radius are biased downwards by the lack of information about points beyond the boundary. Boundary effects can be removed by including points outside the boundary in the calculations, or by creating a new boundary inside the original one, simply cutting off the affected zone.

For a discussion of more sophisticated statistical approaches to density mapping, such as kernel density estimation, readers are referred to Cressie (1991).

FIG. 6-1 (opposite). A. Copper occurrences in part of Quebec province. **B.** Density map showing number of copper occurrences per 1000 km^2. The calculations involved gridding the point density, by finding the number of points occurring within a moving circle, with a radius equal to $(1000/\pi)^{0.5}$, centred over each pixel.

The purpose of creating a density map is to create an area representation of point objects. This map layer can then be combined with other layers in a multi-layer GIS database for visualization and analysis. It may also be useful to determine whether the variation in point density is simply due to chance, or whether the shape of the reconstructed density surface is "real" rather than an artifact of sampling. One way of testing whether the density surface is meaningful is to test whether the configuration of points could have been produced as the outcome of a random process, like using a random number generator. Although the testing of point distributions does not come under the heading of being a data transformation, this is a convenient place to add a short digression on the topic.

If the points are random, or can be regarded as the product of an *independent random process*, then the expected density of points (such as the density that would occur if the number of points was very large) is constant for any location and the position of any point is independent of that of any other point. In practice, the actual density from a random process varies because the number of points on the map is too small. The expected density for small numbers of points can be modelled by a probability distribution, and the observed and expected distributions compared to see if the difference is significant.

Suppose that the region is subdivided into square cells, and the number of points in each cell is counted. Then the point density may be summarized as a table, or observed frequency distribution, showing the number of cells containing 0 points, 1 point, 2 points ..., n points, where n is the maximum number of points per cell. This observed frequency distribution is compared to a distribution that would be expected if the points were generated by a random process. The expected frequency distribution for an independent random process is given by the theoretical distribution known as a Poisson distribution. Independent random processes are also called Poisson processes. The formula for the Poisson distribution is

$$P(m;\lambda) = \frac{\lambda^m}{m!}\exp(-\lambda) \quad , \qquad (6\text{-}1)$$

where P is the probability that a cell contains m points, and λ is a density parameter equal to the average point density over the whole region. The observed frequency distribution is compared with that predicted by the Poisson formula, using the estimated density parameter, λ, and evaluating Equation 6-1 for values of m=0,1,2...n. The difference between the two distributions can be compared by a chi-square test or Kolmogorov-Smirnov test, as described in many statistical texts, for example Siegel (1956). Unwin (1981) provides a very clear explanation of the approach. The distribution of inter-point distances can also be used to test for spatial randomness of points, as summarized for example by Cressie (1991).

Methods for Point Samples

For point *samples*, methods of converting points to areas can be subdivided into two groups, depending whether or not spatial interpolation is involved. Non-interpolative methods are particularly appropriate when the point attribute is measured on a categorical measurement scale, but can also be useful in some cases for an attribute measured on an ordinal, interval or ratio scale.

Non-Interpolative Methods

Non-interpolative methods involve the assignment of one or more attributes of a point to a polygon. The polygon replaces the point and becomes the spatial object with which one or more attributes are associated. There must be a one-to-one mapping of points to polygons, although in some circumstances there is a many-to-one mapping. In the one-to-one case, the attribute table associated with the points is simply transferred to the polygons. Polygons can either be generated automatically by an algorithm, or drawn manually around a point based on subjective evidence. Several alternative methods are described and illustrated with the points shown in Figure 6-2A.

Method 1

In the simplest case, each point is associated with a cell of a grid, Figure 6-2B. Cells that contain no points are given **null** attributes, and the attributes of cells that contain more than one point are determined by aggregating attributes according to some rule, such as taking the mean, median, mode, maximum or minimum, depending on the scale of measurement. The appropriate rule for categorical data is normally the mode, or most frequent value. For ordinal scale attributes, the median can be used, and for interval and ratio scales the arithmetic mean is appropriate. In some cases, the geometric mean may be preferred, for example with geochemical data that has a positively-skewed frequency distribution. The advantage of this approach is its simplicity. This method has been used for small-scale geochemical mapping, using the median as an aggregation rule, for example by Garrett et al. (1990).

Method 2

Another simple method is to assign the points and their attributes to circular polygons, each circle centred on a point. Regions not covered by a circle are assigned null attributes. This method in satisfactory where points are not clustered, but in most situations, some circles overlap due to clustering, leading to an ambiguous decision about the size and average composition of the overlapping areas, Figure 6-2C. The advantage of the circular approach is that the radius of the circle can be chosen to reflect, at least subjectively, the size of the **zone of influence** of the point. This approach should be distinguished from the method of drawing symbols of various sizes to represent the attribute values at points. Variously-sized symbols are an effective method of graphical display, but the spatial objects they represent are still points, not areas.

Method 3

A method that overcomes the problem of polygons either having no points or more than one point is to generate Thiessen (also known as Voronoi) polygons. Thiessen polygons contain only a single point, and have the useful property that any location within a polygon

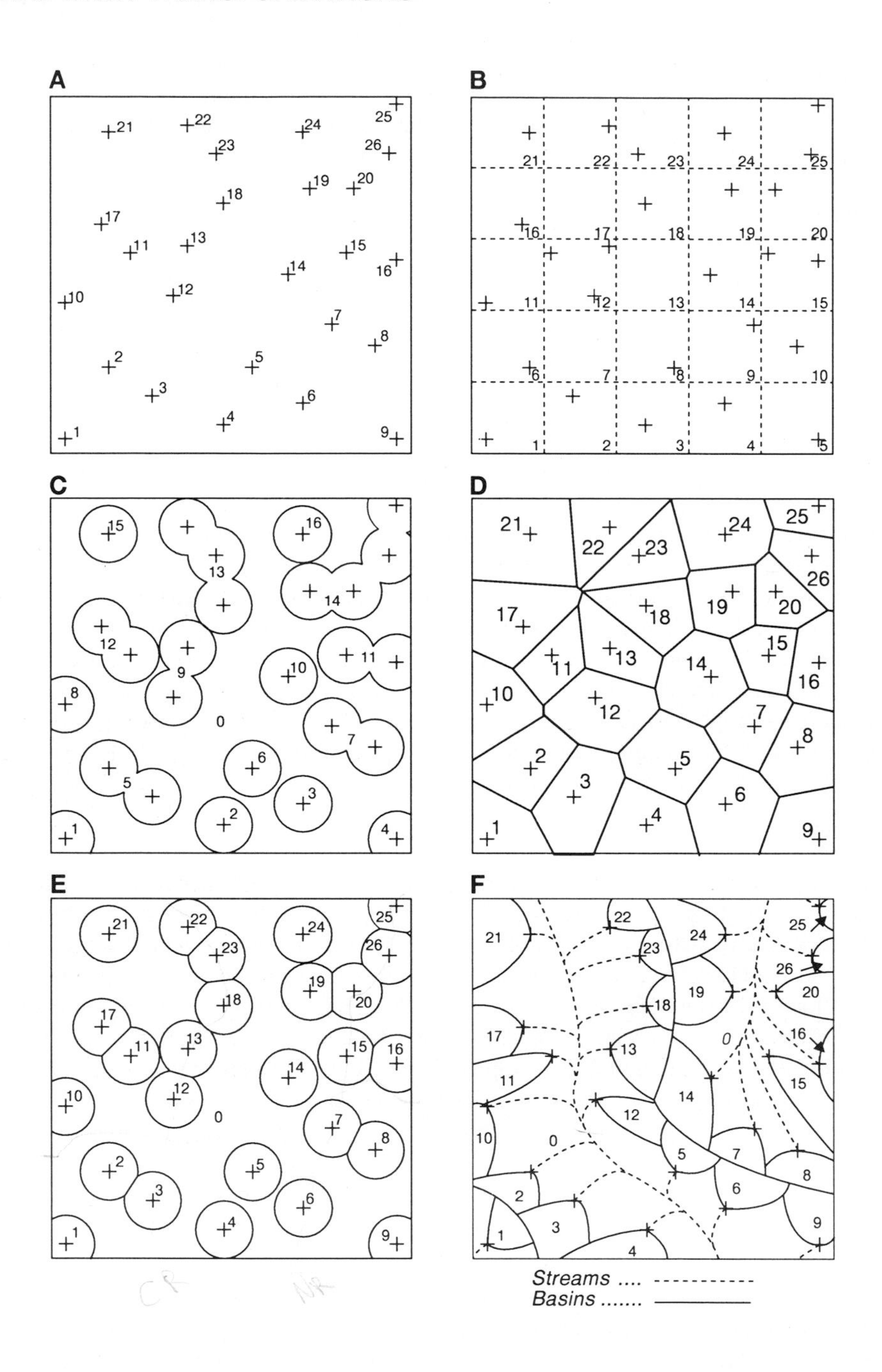
A
B
C
D
E
F
Streams ....
Basins ....

is closer (in Euclidean distance) to the associated point than to any of the neighbouring points. Thiessen polygons also have the property that the line linking a pair of neighbouring points is bisected at right angles by the intervening polygon boundary, Figure 6-2D. There is always a one-to-one relationship between points and polygons, except in the case of duplicate points at the identical location, in which case an aggregation rule must be applied. There are never any null regions, and polygons vary in size inversely with point density.

One of the uses of Thiessen polygons is for finding for a given point all the **natural** neighbours, i.e. all the points whose polygons are adjacent to the polygon of the given point (Watson, 1992). Thiessen polygons are also useful for generating grids from points regularly spaced on a square lattice, or for generating hexagonal rasters from points on a hexagonal lattice. The biggest disadvantage of using Thiessen polygons is that polygon size varies inversely with point density. This leads to situations where a region is assigned the attributes of a point that is far away, beyond any reasonable zone of influence. One way of circumventing this is to combine Thiessen with circles.

Figure 6-2E illustrates such a combination. Here the Voronoi polygons have been overlain by circular polygons. Any region that is outside a circular neighbourhood is thus eliminated and assigned null attributes. The radius of the circles can be set to reflect a reasonable estimate for the zone of influence. This approach is sometimes appropriate for samples taken along a set of traverses, where the distance between samples is small, but the distance between traverses is large.

Method 4

Sometimes it is more appropriate to assign points to irregularly-shaped polygons based on evidence that is independent of the point geometry. Consider the case of a large-scale geochemical map to be made where samples of water or sediment are taken along a stream. The zone of influence for one of these points is not circular, because the sample contains information about areas that are upstream and not downstream from the sampling site. The most likely zone of influence is either one that includes a narrow zone around the stream, upstream from the sample location (e.g. a "worm" map, Howarth and Sinding-Larsen, 1983), or is derived from the catchment basins of the stream. Figure 6-2F shows a polygon map generated by using the catchment basins associated with stream samples. This approach works well if the sampling has been carried out on small tributary streams, so that there is a one-to-one relationship between samples and catchment basins. Under these circumstances, the geochemical attribute table becomes a polygon (catchment basin) attribute table, and maps

FIG. 6-2 (opposite). A. Point distribution on a map. **B.** Points converted to a cellular grid. Cells with more than one point require an averaging rule for aggregation. Cells with no points are either "null", or can be assigned the average of the 4 closest neighbours. **C.** Points converted to circles. Because circles of closely adjacent polygons overlap, composite polygons are formed, requiring attribute aggregation. **D.** Points converted to Thiessen polygons. There is a one-for-one match of point to polygon, so that point attributes can be also used as polygon attributes (appropriate polygon pointers are required). **E.** Thiessen polygons confined to circular zones. Again, point attributes can be used as polygon attributes. **F.** Points converted to catchment basins. Like Thiessen polygons, this results in a single point per polygon, except where there are multiple points per basin.

are readily made by displaying the polygons with colour or pattern governed by the values of any selected field of the table, see for example Ellwood et al. (1986). Catchment basins can be determined from a topographic map, drainage lines and the point sample locations. Automatic methods (e.g. Jenson and Domingue, 1988) for calculating catchments from DEMs work satisfactorily in regions with mature drainage and accurate digital elevation data.

Method 5

The process of drawing boundaries on a geological map based on the attributes of outcrops (which can often be considered as points at the scale of mapping) also falls into this category of point-to-area conversion by a non-interpolative method. In this situation there is not a one-to-one relationship between points and polygons, and only a single attribute is normally considered for a particular set of polygons. The most common attribute to be used for mapping is the geological map unit, either a formal stratigraphic formation or an informal mapping unit. Boundaries of polygons corresponding to the contacts between map classes are manually digitized, based on the pattern of outcrops and their characteristics as well as independent evidence from structural and stratigraphic considerations, tempered by the interpretive judgments of the mapper, Figure 6-3. Transformations based on Thiessen polygons are unlikely to be satisfactory for geological maps, except possibly in the case of very dense point information.

Interpolative Methods

Where an attribute measured at sample points is a spatially continuous "field" variable, mappable as a single-valued surface, interpolation methods are appropriate for converting points to an area representation. Normally, only one attribute is interpolated at a time (although there is some literature on the interpolation of vector fields); the point attribute table cannot be simply transferred to become a polygon attribute table. The interpolation process involves estimating the value of the modelled variable at a succession of point locations, usually on a square lattice. This is the process of **gridding**. The gridded values are then treated as the pixels of a raster image. Alternatively, contour lines (isolines) are threaded through the

---

FIG. 6-3 (opposite). A point outcrop map converted to polygons by drawing geological contacts by hand. **A**. The polygons and lines are lakes and streams, used as a digital base for mapping in the field. The numbers are the map units identified at field localities (outcrops). Strike and dip and other structural measurements have been omitted here, but are present on the original. **B**. Boundaries between formations have been drawn subjectively. These reflect the geological interpretation, which considers the distribution of mapped units observed at the point locations, and the geological structure. In some cases, the contacts between units may also be directly observed, but not here. Note that the interpretation does not entirely satisfy the classification of the units at the stations. For example, the southwest boundary of the Prince Albert Group is gradational, with a number of stations within this unit (2) actually being classified as Tasijuaq Gabbro (3). Colour has been added to facilitate interpretation. Field data captured with FieldLog (see Chapter 4), and supplied by Mikkel Schau, Geological Survey of Canada.

A

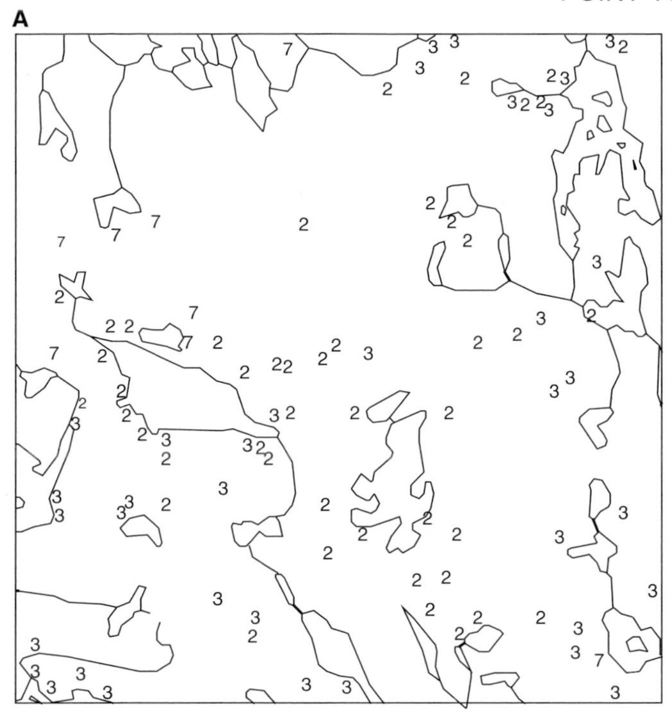

LEGEND

2 - Prince Albert Group

3 - Tasijuaq Gabbro

7 - Granite

B

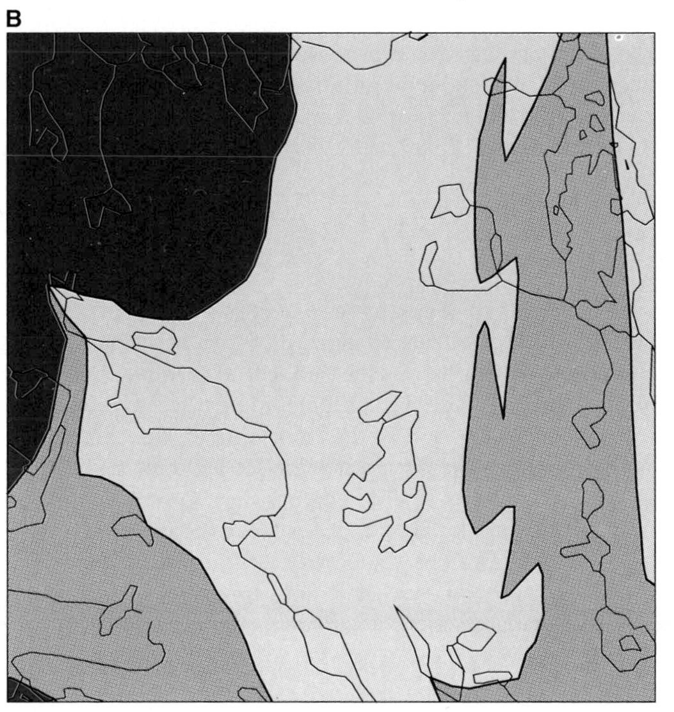

LEGEND

Prince Albert Group

Tasijuaq Gabbro

Granite

grid and the data is represented in a vector structure, either as labelled line objects, or as polygon objects whose boundaries are the contours. The whole process of converting point data to data structures that represent a continuous surface is called **contouring** or **surface modelling**. Besides gridding, many GIS allow the point data to be represented (and stored) as a triangulated irregular network (TIN).

Contouring by hand is the traditional method for representing surfaces on maps as isolines. Contour lines are only one method of symbolizing continuous surfaces. The variations in colour or pattern on a greytone or colour image are generally more effective than contour lines for visualization, sometimes further enhanced by combining both the contours and the image. Contour lines can be extracted from an image by raster-to-vector conversion. The reverse process of converting contour lines to a raster is also an important GIS function, by interpolating from the strings of vertices along each contour line to a regular lattice. Simply rasterizing a contour map as a "stepped" surface (with areas of constant elevation between contour lines) is unsatisfactory if the surface is to be used in applications where the surface slope is needed, such as hill-shading, or in analytical models of slope processes. Stepped surfaces are sometimes effective, however, in perspective displays.

There is an extensive literature on methods of contouring and surface modelling, see for example Watson (1992) and Lam (1983). The subject of **geostatistics** is largely concerned with the problem of characterizing and estimating **regionalized** variables; kriging in one of its many forms is the geostatistical approach to surface modelling, Isaaks and Srivastava (1989). The methods of geostatistics were originally developed for the estimation of ore grades in mining, but are now widely used in all branches of geoscience and other disciplines. Out of the many different approaches that have been proposed, three methods of surface modelling are briefly introduced here, namely **triangulation**, **distance weighting** and **kriging**. Readers interested in an up-to-date and comprehensive statistical treatment of this subject are referred to the books by Isaaks and Srivastava (1989), Cressie (1991) or Watson (1992).

Triangulation Method

In the triangulation method, sample points are joined by lines to form a mosaic of triangles. A **Delaunay** triangulation is used, which produces triangles that are as close to being equilateral as possible. Each triangle is treated as a plane surface. The equation for each planar-triangular facet is determined exactly from the attribute values (we will call them "heights", although the attribute need not be elevation) at the three vertices. Once the surface is defined as a mosaic of triangular planes in this way, the height of any new point on the surface can be calculated. Either a raster image or a set of contour lines in vector mode can be produced directly from the triangulated mosaic. The surface passes exactly through all the data points. The faceted nature of the surface can be smoothed by fitting local functions or surface patches to produce a less blocky appearance without deviating from the actual data points. Triangulation is a particularly popular method of surface modelling for topography, because it is readily able to accommodate surface discontinuities such as rivers, coastlines and cliffs. It is not widely used for geochemical and geophysical variables, but is commonly used for stratigraphic surfaces such as isopach maps and structural surfaces.

Using a cartesian coordinate system, with easting as x, northing as y, and z being the height of the surface, the equation for a triangular facet is

$$z = a + bx + cy \tag{6-2}$$

where a, b and c are coefficients. The three coefficients are solved for a particular triangle by applying the equation to each of the three vertices (also called nodes) of the triangle, the loci of data points where the values of (x,y,z) are known, leading to three simultaneous equations with three unknowns. From the data in Figure 6-4, the unknown point lies within the triangle whose vertices are given by the equations

$$\begin{aligned} 42 &= a + 3b + 5c \\ 32 &= a + 4b + 2c \\ 28 &= a + 2b + 3c \end{aligned} \tag{6-3}$$

whose solution is a=4.8, b=4.4 and c=4.8. By inserting these values into equation (6-2), the estimated height can then be calculated for any (x,y) point in the triangle from the relation

$$\hat{z} = 4.8 + 4.4x + 4.8y \quad . \tag{6-4}$$

Thus the elevation of the unknown point at location (3,4) is 37.2.

One of the advantages of using triangulation is that **breaklines** can be readily inserted into the surface where discontinuities occur. For example, cliffs and faults are two types of discontinuity modelled with breaklines. At each point on the breakline, the elevations on both sides of the discontinuity are entered as input. Other kinds of breaklines are shorelines, where the surface on one side is flat, and along rivers, where the breakline is "heighted" at each vertex along its length. Points along breaklines are included in the Delaunay triangulation. Special calculations are used at triangles adjacent to breaklines to ensure discontinuity and yield surface configurations that satisfy the continuity conditions elsewhere, McCullagh (1988).

Triangulation honours the data points, the surface passing exactly through each known data value. This is desirable in cases where the values at the data points are known to have relatively small errors, such as elevations determined by accurate surveying techniques. However, where the samples of a surface are associated with errors due to sampling and measurement that are relatively large as compared with the overall spatial variation, alternative interpolation methods that produce smooth surfaces, and do not necessarily honour the data points, are often used. In the triangulation method, interpolation is only affected by the heights at the three vertices. Thus the size of the zone of influence of a point is affected by the density of the surrounding points. In small triangles (dense points) the effective zone of influence of a single observation is correspondingly small, whereas in large triangles

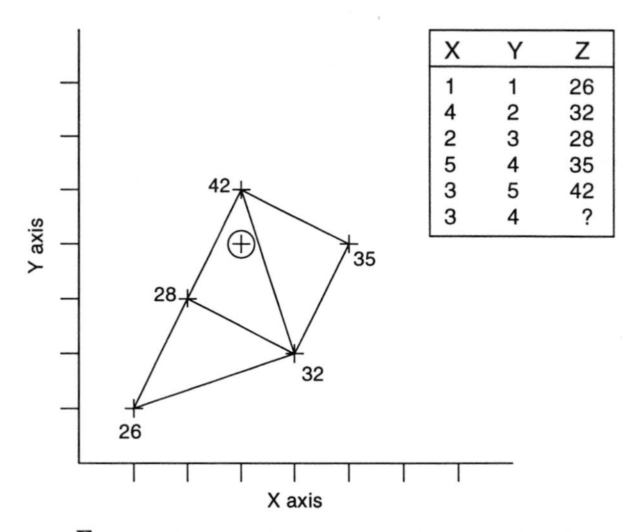

FIG. 6-4. Delaunay triangulation showing data points at vertices of triangles. The value of the attribute at the circled point can be estimated by assuming that the triangle in which it lies is a plane.

(sparse points) the zone of influence is large. For data that is both "noisy" and unevenly distributed, moving average methods that use circular or elliptical zones of influence are preferred. The inverse distance method is a simple method of this type; kriging is also a weighted moving average method, but is statistically based and requires more computation.

Inverse Distance Weighting

There are a variety of methods that use weighted moving averages of points within some zone of influence, usually circular. One of the simplest methods is based on weights that are inversely proportional to the square of the distance from the centre of the zone of influence. In general the formula for inverse distance weighting (IDW) is

$$\hat{z}_0 = \frac{\sum_{i=1}^{n} w_i z_i}{\sum_{i=1}^{n} w_i} \tag{6-5}$$

The hat over the z indicates an estimated value of surface height. The subscript 0 refers to the estimation point, and the subscript i refers to the sample points falling within the zone of influence, Figure 6-5A. The weights are related to distance by $w_i = 1/d_{i0}^2$, where d_{i0} is the distance from point i to point 0. This weighting relationship has the effect of giving data points close to the interpolation point relatively large weights, whereas those far away exert little influence on the estimated value. Instead of using inverse distances raised to the power 2, other exponents can be used, changing the rate of decay of the weighting function with increasing distance.

Another scheme in use for defining weights employs a three-parameter function shown graphically in Figure 6-5B. The three parameters are the radius of an inner circle, a, the radius of an outer circle, b, and a decay parameter, x/b, where x is the distance from the inner circle at which the weight is 0.5. This is analogous to the half-life of a radioactive isotope, except it is a "half distance", or distance at which the weight has "decayed" to 0.5. Within the inner circle, points are assigned a value of 1.0, and beyond the outer circle, points are assigned a weight of 0.0 (i.e. are not considered). The weights assigned to points lying at a distance between the inner and outer circles vary between 0 and 1 and depend both on distance and the decay parameter. For values of 0.5, the decay is linear; for values not equal to 0.5 the decay is non-linear as shown in the figure. If the decay rate is set to 1.0, all points within the outer radius are weighted equally, producing a "hat" function. At any estimation point, all weights are normalized so that the sum of the weights equals 1.0 , as in the inverse distance formula. The advantage of this approach is that points within the inner circle can be assigned an equal weight, which may be desirable in some circumstances. If the inner radius is set to 0, then the method is similar to inverse distance weighting, with the decay rate acting in a similar way to the distance exponent.

Weighting methods produce a surface that do not honour the data points, except in special cases. They have the advantage of being easy to use and understand. Their principal disadvantage arises from the arbitrary choice of parameters for the weighting function. By changing the distance exponent or the decay, very different surfaces can be produced, and it

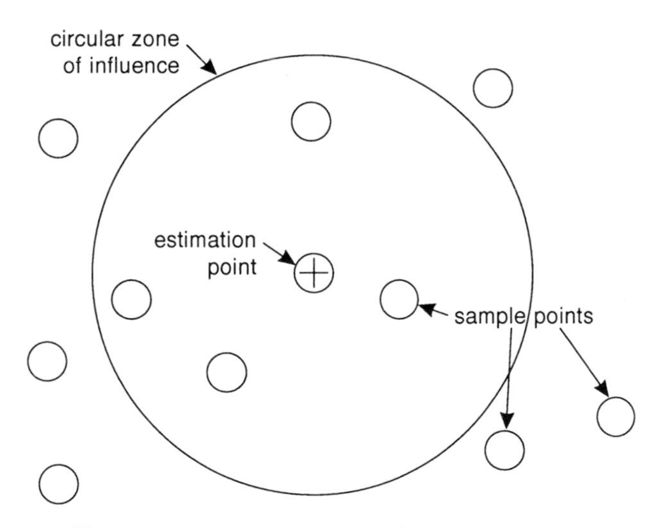

FIG. 6-5. A. Inverse distance weighting within a circular zone of influence.

is often unclear what principle should be followed in choosing parameter values. A more analytical approach is to use the method of **kriging**, where the weights are optimized at each interpolation point to produce a surface that satisfies statistical criteria.

Kriging

The following section is drawn, in part, on the treatment of this subject by Isaaks and Srivastava (1989). As with the inverse distance method, kriging involves taking the weighted sum of the points within a zone of influence, see Figure 6-5A, using the equation

$$\hat{z}_0 = \sum_{i=1}^{n} w_i z_i \tag{6-6}$$

but in this method the weights are constrained to sum to 1, avoiding the need to normalize by dividing by their sum. The weights are calculated from a set of (n+1) simultaneous linear equations, where n is the number of points used for the estimation at any one location, as before. In matrix form, the weight equations are

$$\mathbf{C} \cdot \mathbf{w} = \mathbf{d} \,,$$

which can be expanded to

$$\begin{bmatrix} C_{11} & C_{12} & \cdot & C_{1n} & 1 \\ \cdot & \cdot & \cdot & \cdot & \cdot \\ \cdot & \cdot & \cdot & \cdot & \cdot \\ C_{n1} & C_{n2} & \cdot & C_{nn} & 1 \\ 1 & 1 & . & 1 & 0 \end{bmatrix} \cdot \begin{bmatrix} w_1 \\ \cdot \\ \cdot \\ w_n \\ \mu \end{bmatrix} = \begin{bmatrix} C_{10} \\ \cdot \\ \cdot \\ C_{n0} \\ 1 \end{bmatrix} , \tag{6-7}$$

where the terms in **C** are spatial "covariance" values between the pairs of sample points and the terms in **d** are also spatial "covariances", but between the sample points and the point being estimated. The spatial covariance is a measure of spatial correlation. Points that are close together geometrically tend to be strongly correlated, whereas those spaced far apart tend to lack correlation. The spatial correlation of a variable with itself, commonly known as the autocorrelation, can be characterized using autocorrelation functions estimated from the data, as described further below. The term μ is a parameter known as the Lagrange multiplier, a dummy parameter that is used for forcing the weights to sum to one. Equation 6-7 is solved by pre-multiplying both sides by the inverse of **C**:

$$\mathbf{C}^{-1} \cdot \mathbf{C} \cdot \mathbf{w} = \mathbf{C}^{-1} \cdot \mathbf{d} \quad \text{or} \tag{6-8}$$

$$\mathbf{w} = \mathbf{C}^{-1} \cdot \mathbf{d} \,. \tag{6-9}$$

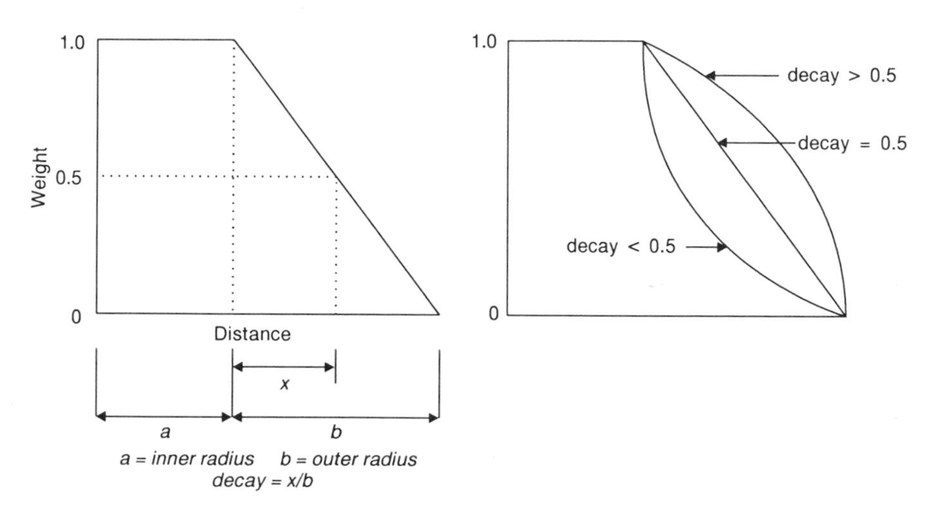

FIG. 6-5. B. Weighting parameters derived from distance-decay functions.

At every point to be estimated during the gridding operation, the covariance terms in **C** and **d** are determined, as will be described in a moment, then Equation (6-9) is solved for the weights, and the weights applied with Equation (6-6) to calculate the estimated value of the variable, the "height" of the modelled surface.

In some respects, the kriging equations can be compared to the simpler inverse distance equations. The **d** vector is analogous to the inverse distance, in that the spatial covariance is a type of statistical inverse "distance". For example the covariance, $C_{i\,0}$, between the point to be estimated, indicated by the subscript 0, and sample point i, is a value that varies inversely with the geometric distance $d_{i\,0}$. When the sample point is close to the estimation point, the covariance or degree of correlation is relatively large; when they are far apart the covariance is relatively small. In addition to these "distance" terms, there is also the effect of $\mathbf{C}^{-1}$. This also contains the statistical inverse "distances", where the distances are among the pairs of sample points, but the estimation point is not involved. The effect of this pre-multiplier is to adjust the weights for clustering of the points. Where two or more points are clustered together in the inverse distance method, the effect on the estimation point is weighted by the number of points in the cluster. This is undesirable, because it is merely an artifact caused by the irregular distribution of samples. But in kriging, the covariance between a pair of sample points that are close together is relatively large, because the points are geometrically close, having the effect of downweighting the affected points. The outcome of using the interpoint covariances, after inverting the matrix to produce $\mathbf{C}^{-1}$, is to adjust for the irregular spacing of the sample points.

To summarize, kriging considers three important factors in surface modelling -- distance, clustering and spatial autocorrelation -- as measured here by the spatial covariances. But what exactly are these covariances, and what do they mean?

Consider for a moment a scatterplot of two variables, x and y, (they need not be spatial variables). The covariance, C, between x and y is calculated by taking the average product of the deviations of each variable from their respective means:

$$C = \frac{1}{n} \sum_{i=1}^{n} (x_i - \overline{x}) \; (y_i - \overline{y}) \tag{6-10}$$

where there are n sample points. The covariance measures the degree to which x *co-varies* with y. If C is normalized by dividing by the product of the standard deviations of x and y, namely σ_x and σ_y, the covariance is converted to the correlation coefficient, r. The correlation coefficient ranges between a maximum of +1, indicating perfect correlation of the two variables, to -1, indicating perfect inverse correlation. A value of 0 indicates that no correlation exists, and that the two variables are independent. On the scatterplot, The degree of scatter of points from a 45° line is related to the strength of covariance or correlation. Another measure of the degree of spread about the line is the moment of inertia, defined as

$$\gamma = \frac{1}{2n} \sum_{i=1}^{n} (x_i - y_i)^2 \tag{6-11}$$

an expression that looks rather like the covariance, except that the squared differences between the variables are used instead of the products of deviations from the means.

Suppose that instead of x we substitute z_t and instead of y we substitute z_{t+h}, where z is a regionalized variable observed at location t, and at another location $t+h$, where h is a separation distance, also called a shift or lag. The spatial covariance of z with itself at separation distance h can also be measured by γ (or by C or r). By changing the separation distance (lag), a series of scatterplots can be generated, showing how the variable, z, is correlated with itself as a function of lag.The behaviour of this ***auto*correlation** is quantified as a function of h by $C(h)$, $r(h)$, or $\gamma(h)$. The plot of the covariance as a function of lag, $C(h)$, is called an autocovariance diagram, see Figure 6-6A. The comparable plot of the moment of inertia as function of lag, $\gamma(h)$, is called the **variogram**, see Figure 6-6B. The equivalent plot for the correlation coefficient is called an ***auto*correlogram.**

These functions, estimated from actual sample data values, characterize the spatial autocorrelation present in the data. The graphical scatterplot of the points on a variogram is called an experimental variogram, to distinguish it from a fitted line, or variogram model. One way of understanding the meaning of spatial autocorrelation is to consider a surface whose height is known at regular grid points. If the grid is duplicated, giving an identical copy of the original, and the two copies are now shifted in the x and y directions, any one of the autocorrelation functions can be measured at units of separation distance. At zero separation distance, the correlation or covariance values are at a maximum, because the data values are being compared with themselves; the variogram value is at a minimum, equal to 0. At a shift of 1 unit in the x direction, the covariance and correlation functions begin to decrease, whereas the variogram increases, as shown in Figure 6-6. With further separation, the covariance and correlation values continue to decrease (on average), and the variogram continues to increase. When the two surfaces have been moved apart to a distance known as the **range**, a, the covariance and correlation values reach zero and the variogram reaches a value called the **sill**. At distances beyond the range, the autocorrelation (whether measured by C, r or γ) values may vary, but essentially the two surfaces are independent. The range, and the shape of the autocorrelation function at separation distances less than the range, provide a quantitative measure of the nature of the spatial continuity of the surface.

Another important parameter of these functions is the **nugget** effect. At small separation distances, the covariance often decreases (the variogram increases) sharply. This is caused by variations due to sampling and measurement and is much more pronounced for some surfaces than others. For example, assay values in gold deposits change very rapidly over short distances, because the metal is not evenly disseminated but concentrated into nuggets. Thus, even at essentially zero separation distance, the autocorrelation is not 1. This is the origin of the term "nugget" in geostatistics. In addition, spatial autocorrelation functions are not necessarily isotropic, but often exhibit different behaviour with changes in orientation.

The practice of kriging requires that the covariance, correlation or variogram values be calculated for several orientations. Ignoring the anisotropic problem for this discussion, the next step is to fit a mathematical function (or model) to the measured values, so that the autocorrelation can be characterized with a small number of model parameters. The ultimate purpose of all this is to be able to obtain values of covariance quickly to set up the **C** and **d** matrices to solve the weight equations.

The most common models are the linear, spherical, exponential and Gaussian models. For example, the exponential model, shown graphically in Figure 6-6, has the following algebraic form for covariances:

$$C(h) = \begin{cases} C_0 + C_1 & \text{if } h = 0 \\ C_1 \exp(-3\frac{h}{a}) & \text{if } h > 0 \end{cases} \tag{6-12}$$

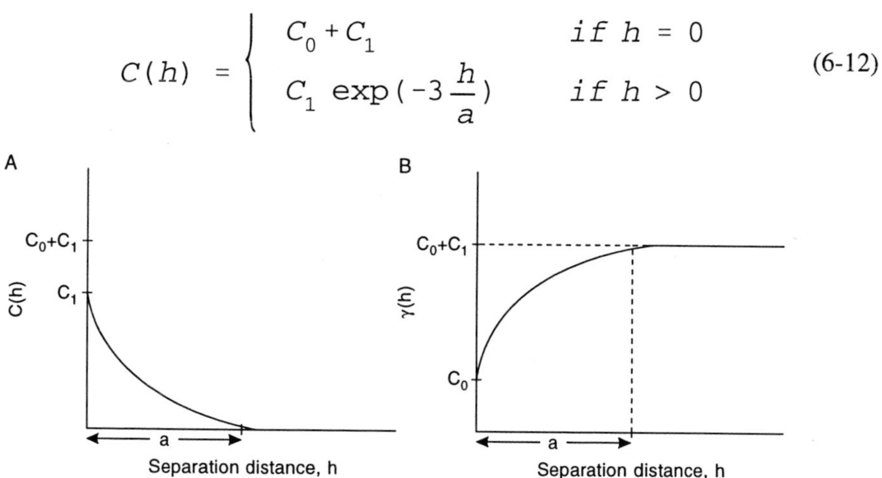

FIG. 6-6. Comparison of the covariance function and the variogram as models for characterizing spatial autocorrelation. An exponential model is illustrated. **A**. The covariance as a function of separation distance, or lag. The distance a is called the range. **B**. The variogram, showing that the shape is inverted from the covariance function. The value of the variogram at the distance equal to the range is called the sill. C_0 is the nugget effect.

and a similar form for the variogram is

$$\gamma(h) = \begin{cases} 0 & if\ h = 0 \\ C_0 + C_1(1 - \exp(-3\frac{h}{a})) & if\ h > 0 \end{cases} \qquad (6\text{-}13)$$

where C_0 is the nugget effect, providing the discontinuity at the origin, a is the range, the separation distance beyond which the variogram or covariance becomes more or less constant, and $(C_0 + C_1)$ is the sill. The sill is the variogram value for large separation distances, and is also the covariance value for zero separation distance. For a particular dataset, the values of C_0, C_1 and a are estimated so the model is a reasonable fit to the observed data values.

Now armed with numerical values of C_0, C_1, and a, assumed to be constant for the region of interest, the individual terms in **C** and **d** are calculated for any grid point. In practice, the sample points within a circular zone surrounding a grid point, or within an elliptical zone if the autocorrelation is anisotropic, are selected. Separation distances between sample points and distances from each sample point to the grid point are used to determine covariances from the model. Weights are then calculated from Equation (6-9) and applied to estimate the "height" of the surface at the grid point with Equation (6-6). This procedure is repeated for each grid point over the region to be mapped, and therefore involves a great deal of computation, considerably more than the IDW method. Figure 6-7 A and B show the results of kriging chromium values in stream sediments from samples taken in part of Vancouver Island.

At the same time as estimating the height of the surface, the uncertainty of the estimate can be calculated as

$$\sigma_R^2 = \sigma^2 - \sum_{i=1}^{n} w_i C_{i0} + \mu \qquad (6\text{-}14)$$

where σ^2_R is the estimation variance, σ^2 is the variance of the variable, calculated in the normal way as the sample variance, and the other terms are the same as those used in Equation (6-7). The estimation variance (also called the kriging variance) is often mapped at the same time as the surface itself. For example, the kriging standard deviation is shown as a surface in Figure 6-7C. Such a surface is a valuable source of information about the spatial variation of the uncertainty, due to the spatial interpolation process. In general, the uncertainty varies inversely with the density of the sample points, as might be expected, but it is also affected by the shape of the covariance function. The equations for the weights are formulated so that the estimation variance is minimized. In this sense, kriging is an optimal solution for gridding. It should be noted that the equations can be set up with the variogram or correlation functions as alternatives to the covariance functions.

One of the assumptions of ordinary kriging is that the mean of the variable is **stationary**, and does not show spatial trends. In practice, the variability of most surfaces can be decomposed into three components: a **trend** component (also known as the drift), a spatially autocorrelated component (also known as the **signal**), and a **noise** component--the unexplained residual variation. The trend component can be modelled separately by **trend surface analysis**. One way of dealing with spatial trends is to carry out a **trend surface**

analysis (by fitting a polynomial or Fourier series to the data), remove the drift component before kriging (Agterberg, 1974), and add the trend back afterwards. Alternatively, the method called **universal kriging** can be applied. Universal kriging combines trend surface analysis and ordinary kriging into a single operation, so that the trend and signal are estimated together.

Another kind of kriging, called **indicator kriging**, is used for "indicator" variables that have the values of 0 or 1. The estimates from indicator kriging are real numbers in the range (0,1), and can be treated as probability values. The kriged surface shows the probability that the indicator variable is 1. For example, an indicator approach can be applied to regional geochemical data. If a geochemical variable, such as copper, is transformed to an indicator variable by dividing the data values into anomalies (=1) and background (=0), indicator kriging of the resulting binary variable produces a surface showing the probability of the presence of anomalous copper. An accompanying map of the kriging variance shows the uncertainty of the probability surface due to spatial interpolation. There are several other kinds of kriging, such as disjunctive and factorial kriging, that are used in special circumstances. Also, the discussion has been limited to **punctual** interpolation, the estimation at point locations. In estimating grades and tonnages from mining assay data, **block** kriging is used to makes estimates over a volume of rock. More advanced treatments of geostatistics in mining are Journel and Huibregts (1978), David (1988), or Matheron and Armstrong (1987). For a discussion of geostatistics in petroleum geology, see Hohn (1988).

We now turn to another useful class of data conversions, the dilation of linear features. In fact the dilation process can be applied to points, lines and areas to produce maps showing proximity relationships. Like the point-to-area conversions, dilation operations are used in the preliminary stages of data treatment, prior to the principal stages of analysis and modelling.

DILATION OF SPATIAL OBJECTS

As the name suggests, the dilation of a spatial object leads to an expansion in size of the object. The term **buffering** is synonymous with dilation, and the term **spreading** has a similar meaning. Dilation can be applied to point, line and area objects. Dilation produces proximity "corridors" or buffer zones surrounding the point, line or area objects. Dilation can be carried out with either vector or raster data models, in a variety of data structures. Serra (1982) extends the notion of spatial dilation to greytone or continuous surfaces also.

The purpose of dilation is often to create a map which shows proximity to a selected feature. For example, distance to a transportation system, rail or road, may be very important in deciding where to site a waste disposal facility, or for affecting the economics of mine development. Maps showing proximity to roads or railways can be used as components of a multimap GIS analysis. In mineral potential mapping, proximity to particular geological contacts, to lineaments, or to faults, can be significant for modelling some kinds of mineral deposits. Again, maps showing proximity relationships are produced by dilation, to be used subsequently in modelling, where multiple layers are combined together to produce a mineral potential map. Special kinds of dilation operations are also used in **mathematical**

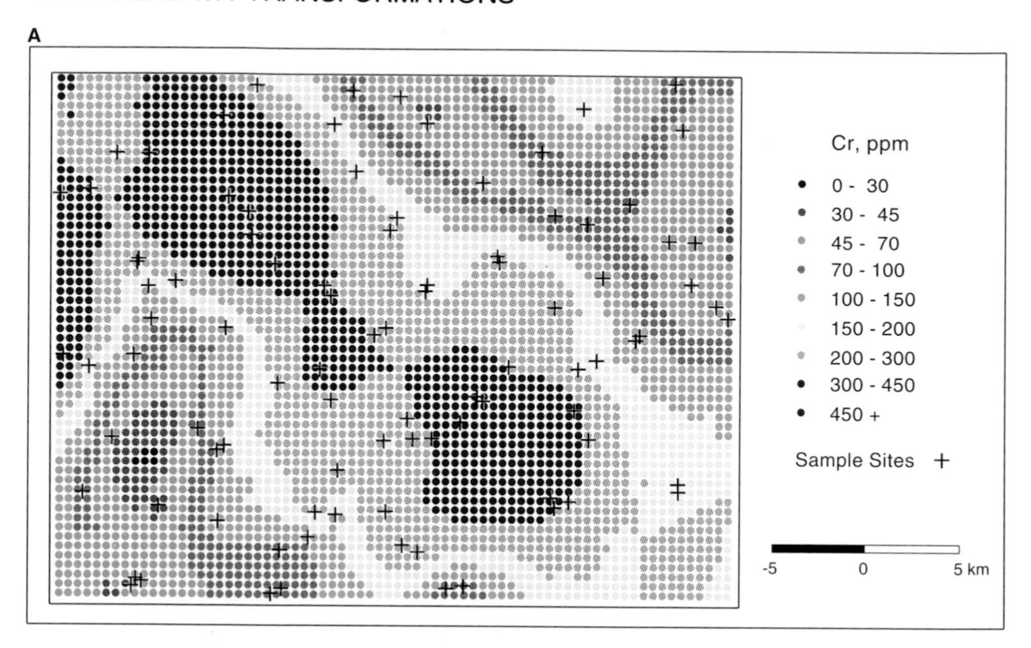

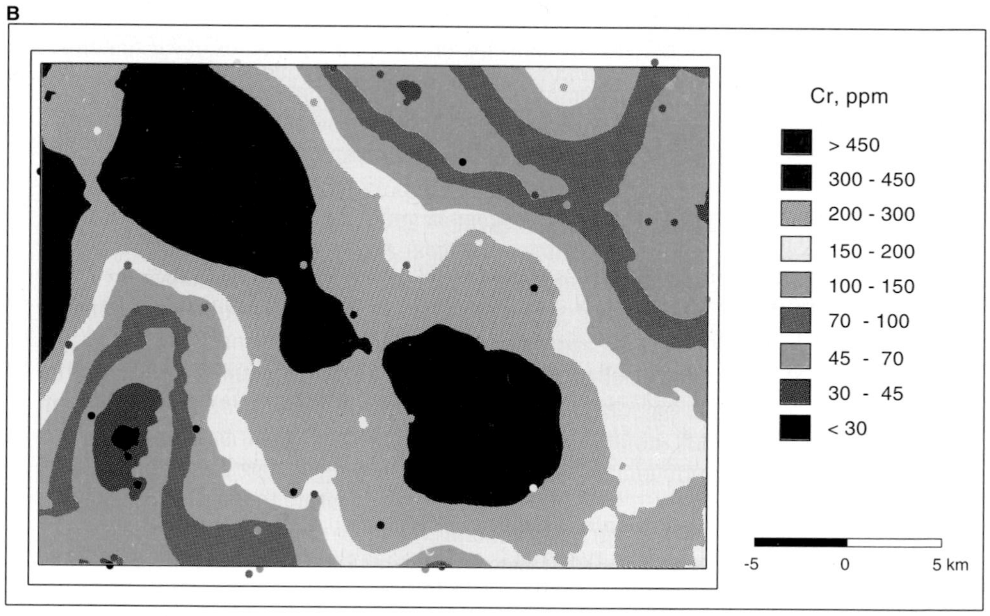

FIG. 6-7. Map of chromium in stream sediment, from part of Vancouver Island. **A**. Kriging estimates at regular points on a lattice, or grid. The points are displayed as filled circles, with colour according to the classification shown in the legend. **B**. Grid converted to a raster, with the original data points superimposed. The raster could be converted to vector to generate contour lines. Alternatively, a "threading" algorithm could be used to determine the positions of the contour lines directly.

C

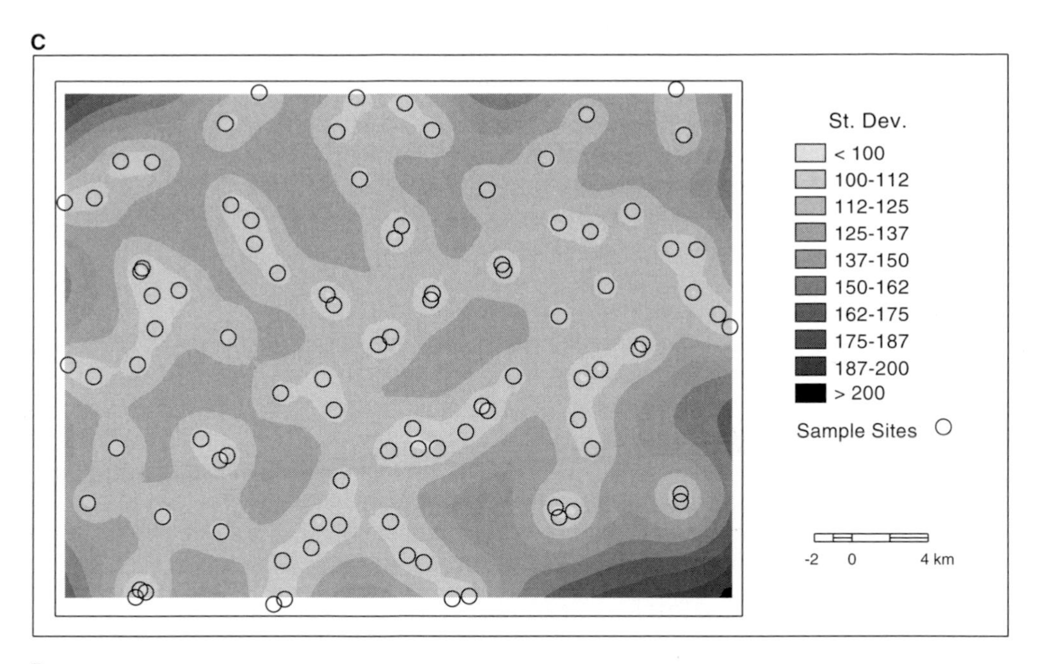

D

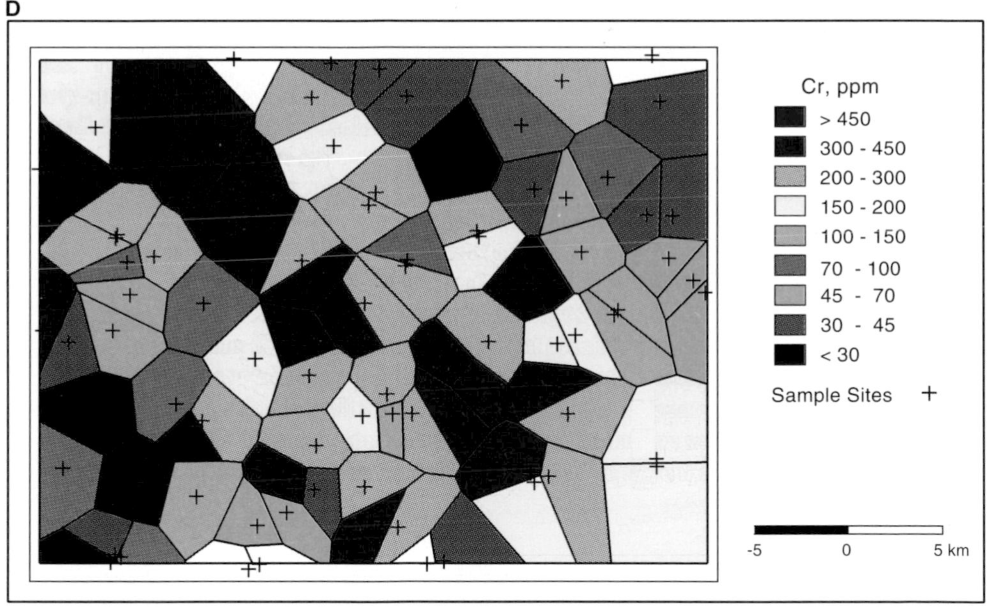

FIG. 6-7. C. Map of the kriging standard deviation. In general the kriging variance varies inversely with distance from the nearest sample point. **D**. Map of Thiessen (Voronoi) polygons of the same data, for comparison. The value in each polygon is constant, and equal to the value at the contained point.

morphology for the analysis of raster images. Mathematical morphology is a branch of spatial analysis dealing with the measurement, characterization and extraction of spatial features from raster images, Serra (1982). Dilation operations in mathematical morphology are part of a family of transformations that extend the notion of buffering to allow more complex changes to an image. For example, objects can be **eroded** (made smaller), **dilated**, or subjected to **opening** and **closing** transformations.

Dilating Linear Features

Figure 6-8A shows the locations of anticlines in part of the Meguma terrane of Nova Scotia. These lines are the surface traces of Devonian anticlinal fold axes. A number of small quartz-vein gold occurrences, some of them large enough to be mined, have been found in Lower Palaeozoic turbidites, many of them close to the hinges of these anticlinal folds. In order to quantify the spatial association of the gold occurrences to the fold axes, a distribution of distances of occurrences to the nearest fold axis can be calculated. Such a distribution can be obtained by first producing a buffer map by dilating the fold axes, followed by sampling the buffer map at the gold occurrence points, giving the distance of each occurrence to the nearest fold axis.

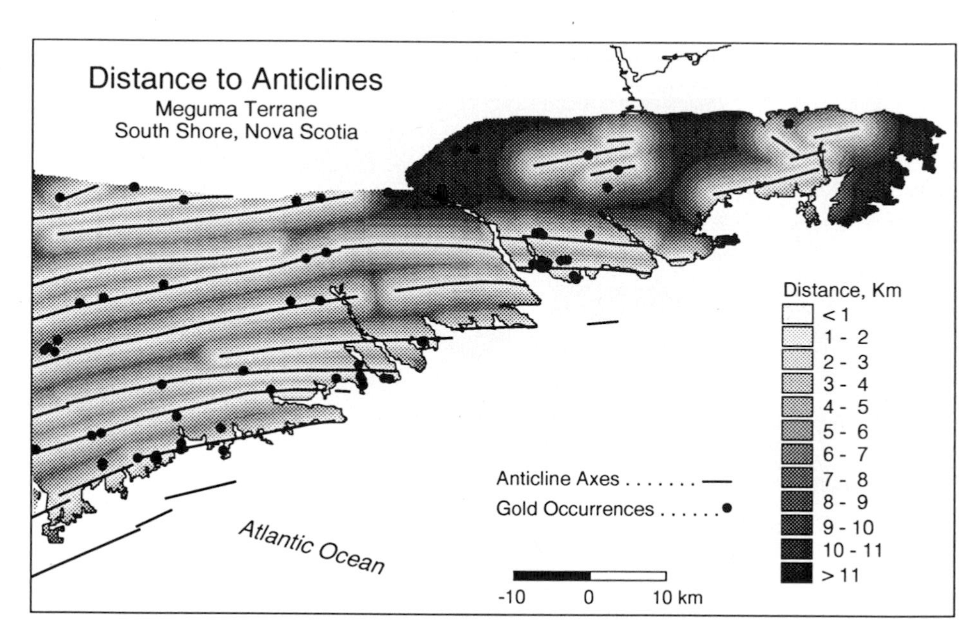

FIG. 6-8. A. Map to illustrate the dilation or buffering of linear features. Anticlinal fold axes in Meguma terrane, Nova Scotia have been successively dilated with corridors (250 m intervals) to produce a map showing proximity to the nearest fold axis. A classification has been applied so that the intervals on the map are 1 km. The points are locations of gold occurrences.

In this case, twenty buffer zones were created, at successive radii in intervals of 250 m. This produced a map (in this case a raster image) showing distance to the nearest fold axis, to the nearest 250 m. For display purposes, the map was reclassified into a smaller number of distance classes, with the interval being 1 km. During the buffering process, dilation of a spatial object proceeds until the predefined corridor radius is reached, or until another buffer zone from a neighbouring object, or a boundary, is encountered. Where two buffer zones that are "growing" from neighbouring spatial objects conflict, then the zone with the shortest radius takes priority, ensuring that the resulting map shows the *nearest* distance to the objects in question.

In Figure 6-8A, all fold axes have been dilated by successive buffer zones at 250 m intervals. Notice that the dilation operation has been constrained in places, where the buffer zone from one axis has simply merged with the buffer zone of the adjacent axis. Also notice that the buffer zones have been truncated by the coast, because a base map of land versus water was used to constrain the dilation.

The proximity maps can now be used to measure distance distributions of area and number of gold occurrences from the nearest fold axis. For example, Figure 6-8B shows a graph of cumulative corridor area versus distance (dashed line) calculated by adding together the cumulative area of each distance class on the map. It is simply a cumulative histogram of the

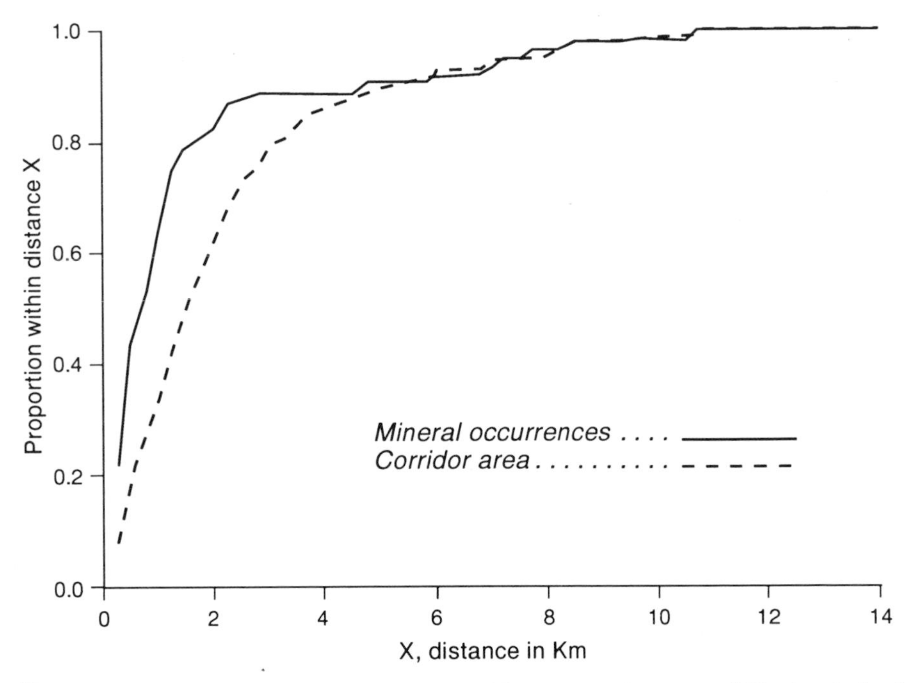

FIG. 6-8 B. Expected and observed distribution of gold occurrences from nearest fold axis, calculated from the buffer map as described in the text. Note that gold occurrences occur more frequently close to the fold axes than expected due to chance.

buffer map, shown as a continuous line. The figure also shows the cumulative number of mineral occurrences versus distance from the nearest fold axis. This is obtained by counting the cumulative number of occurrences in each distance class. The measurement of area is a standard tool of GIS, discussed further in Chapter 7. The counting of points is, in this case, a two-step operation. First the occurrence points are used to "sample" the buffer map, appending the distance class as an attribute to the point attribute table, as discussed below. A summation operation is then carried out on the attribute table to count the frequency of points in each distance interval.

The graph in Figure 6-8B allows the visual comparison of two cumulative frequency distributions. The area distribution (dashed line) is also an estimate of the **expected distribution** of an infinitely large number of points occurring over the map at random, independent of the fold axes. This line thus shows what the distance distribution would be like assuming an idealized situation (a model) of the gold points and the Devonian fold lines being generated by independent random processes. The graph shows that gold occurs more frequently close to fold axes than would be expected due to chance. We shall return to this topic in a later chapter.

Algorithms for buffering operations have been developed for a variety of data structures, both raster and vector. In raster mode, **mathematical morphology** operations extend the idea of buffering to include not only dilation, but also "erosion" of area objects. Erosion has the effect of shrinking an area object instead of growing it. Dilations and erosions can also be carried out anisotropically, so that objects can be expanded and shrunk in specific directions. There are many cases where this is desirable in geological applications. For example, one may want to express proximity to a granite batholith as a proximity map, but constrained only in one direction, not symmetrically in all directions. Fabbri (1984) discusses a number of examples of mathematical morphology applications, both to geological maps and to other kinds of images of geoscientific interest, such as thin sections of rocks. Serra (1982) provides a theoretical background, but also illustrates several applications to geology. The discussion here is limited to some very elementary operations pertinent to data transformations.

Mathematical Morphology Operations

In mathematical morphology, a binary raster image is treated as a set of points or pixels, forming a rectangular array or lattice. Figure 6-9(1) shows a simple binary raster image, ***A***, in which an area object, such as a granite outcrop on a map, is represented by points having a value 1 (black), with points elsewhere having a value 0 (white). ***A*** is thus a rectangular matrix of 1s or 0s. Suppose that another rectangular matrix, ***B***, containing 1s, and null values, is defined, in this case being as a (3*3) array, to be used as an **operator** on image ***A***. This operator in mathematical morphology is called a **structuring element**, see some (3*3) examples in Figure 6-9(9). For grids with square pixels, structuring elements are usually defined as square matrices with an odd number of rows and columns, so that there is symmetry round a central element, similar to the "kernel" matrices used in image filtering in the spatial domain. The structuring element is passed incrementally, in steps of one pixel, over the image. At each step, the value of the pixel in A, corresponding to the centre of the structuring element, is changed according to a set of rules. The change depends on the values of A in the neighbourhood of the centre pixel, and the configurations of 1s, and null values in the structuring element. The process is analogous to a **convolution** operation in image processing, but the rules are different.

A dilation operation is written as

$$C = A \oplus B \quad , \tag{6-15}$$

where C is a binary output image, consisting of another rectangular matrix with the same number of rows and columns as the input image A. The symbol $\oplus$ is called a Minkowski sum operator. As the structuring element B is centred over each pixel in A, the corresponding pixels in A and B are compared, and the value of the centre pixel determined for the output image C. The centre (output) pixel in C is assigned a value of 1 if the centre value in A equals 1, or if there is a match between any of the off-centre pixels of B with the corresponding pixels in A, otherwise the value equals 0. The effect is to leave the black pixels in A unchanged, but the white pixels are changed to black according to the configuration of the structuring element, and the proximity of the structuring element to black pixels in A. For image A in Figure 6-9(1), using the structuring element B_1 (Figure 6-9(9)), the effect of a dilation is to add a new rim of pixels all round the black object in A, as shown in Figure 6-9(2).

The erosion of an image is carried out in an analogous manner by a Minkowski difference, written as

$$F = A \ominus B \tag{6-16}$$

which causes a rim of black pixels to be turned white. Figure 6-9(4) shows the erosion of A by structuring element B_1, resulting in the shrinking of the black objects in A. Note that the large object is subdivided by this process.

By changing the configuration of the structuring element, different kinds of enlargement and shrinking of area objects are produced. In Figure 6-9(5) the structuring element B_2 causes a rim of pixels to be added only to the east side of the black object in A. In Figure 6-9(8) the effect of using B_3 as the structuring element is to grow the black objects only in the NW-SE directions.

Larger dilations and erosions can either be carried out by multiple passes with small structuring elements, or by a smaller number of passes with larger structuring elements. Thus a second dilation of A by B_1 produces the image in Figure 6-9(6).

Boundary effects become very noticeable in small images, because pixels adjacent to the boundary in the output image are affected by the unknown pixels surrounding the boundary of the input image. Depending how boundary effects are modelled, this effect can cause edge effects to migrate inwards from the boundary with successive operations.

Mathematical morphology operations are often carried out for images defined on a hexagonal rather than square lattice. This has the advantage that any pixel in a hexagonal grid has six adjacent pixels at the same distance, whereas in a square lattice the four diagonal neighbours are farther away by the factor $\sqrt{2}$ than the four square neighbours. It is easier to define a structuring element that approximates a circular disk in a hexagonal system, but with the obvious disadvantage that graphic display devices use square rasters. Thus if hexagonal logic is used, conversion to a square lattice is always needed for display.

Erosion and dilation operations can be combined to produce **opening** and **closing** operations. Openings and closings are useful for transforming images and for measuring the size distribution of spatial objects.

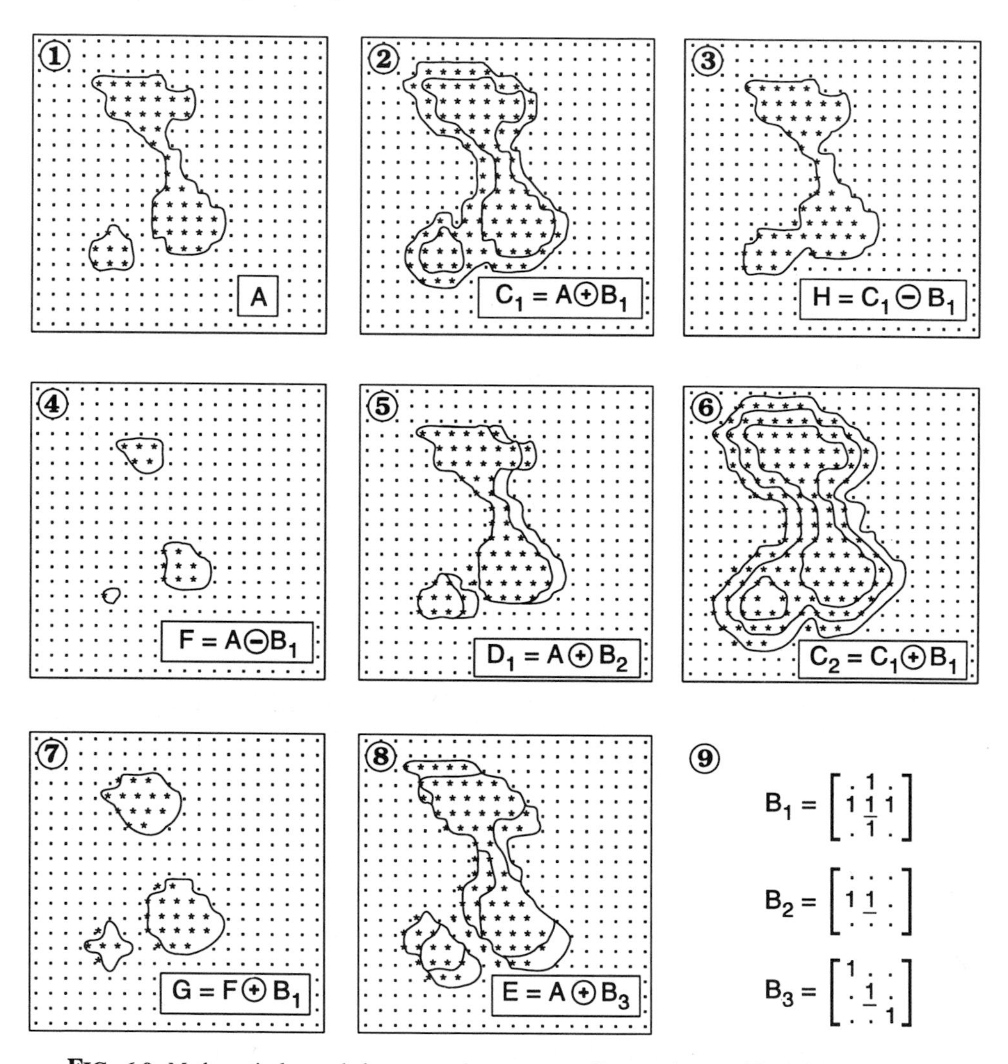

FIG. 6-9. Mathematical morphology operations on a small raster image. **(1)**. Original image **A**. **(2)**. Dilation of **A** using structuring element $\mathbf{B}_1$. **(3)**. Closing of image **A** by $\mathbf{B}_1$. **(4)**. Erosion of **A** by $\mathbf{B}_1$. **(5)**. Dilation of **A** by structuring element $\mathbf{B}_2$. **(6)**. Two successive dilations of **A** by $\mathbf{B}_1$. **(7)**. Opening of **A** by $\mathbf{B}_1$. **(8)**. Dilation of **A** by $\mathbf{B}_3$. **(9)**. Three (3*3) structuring elements.

A closing is defined as a dilation followed by an erosion. The effect, as shown in Figure 6-9(3), is to close the gap between the two separate black objects in *A*. Conversely, an opening is defined as an erosion followed by a dilation, causing large "blobby" objects to be subdivided and for small objects to disappear, as shown in Figure 6-9(7). Successive openings of a binary image by structuring elements that are disk-shaped results in progressive removal of "particles" according to their diameters. The operation is analogous to the sieving of sedimentary particles, where the holes in the sieve are the physical equivalent of the structuring element. By computing the area of the image, or number of black pixels, at each opening, the size distribution of spatial objects can be determined. The size distributions of "pores", i.e. objects composed of white pixels, can be determined by successive closings in a similar fashion.

These operations are used by "image analyzers", with software for processing and analyzing images obtained from light microscopes, electron microscopes and other sources. Mathematical morphology operations are also useful for analyzing spatial objects on maps and cartographic images. Fabbri (1984) illustrates a number of interesting geological examples.

SAMPLING TRANSFORMATIONS

We will now turn to data-type transformations that involve "sampling", or "resampling". The sampling of area objects at points is an important case. One application already discussed in Chapter 4 is the process of image resampling to rectify one set of raster pixels to another, for the purpose of geometric correction. Another application is digitizing, where regions are labelled by digitizing a point, sometimes called a "centroid", inside each polygon.The polygon attributes and the point attributes can be linked by sampling polygons at the centroids. This uses a "point-in-polygon" algorithm, and results in being able to tag polygons with the attributes of centroids. Another application is where a point attribute table is to be built by "sampling" at the point locations from multiple maps. The resulting attribute table can then be analyzed by statistical or other methods. We will also touch on the problem of "sampling" one set of area objects by another. For example, one may wish to use the spatial objects on a geological map to sample from geochemical and geophysical maps, to determine the average geochemical or geophysical response of each geological map unit. This process has received considerable attention in dealing with census, and other kinds of population data. Population statistics are reported by census tract, but then need to be resampled by county or by municipality. Some authors have referred to this as the modifiable areal unit problem (MAUP), Flowerdew (1991).

Besides resampling images for geometric correction, sampling of lines is also widely-used, mainly for the purpose of data compression. Lines digitized in a stream mode can be **weeded** to reduce the number of vertices. This process can be regarded as a type of resampling, and is carried out by a variety of methods, one being the Douglas-Peucker algorithm, as illustrated in Figure 6-10, another being a principal axis method, Cromley (1992); see also the references in the Glossary under the word "weeding".

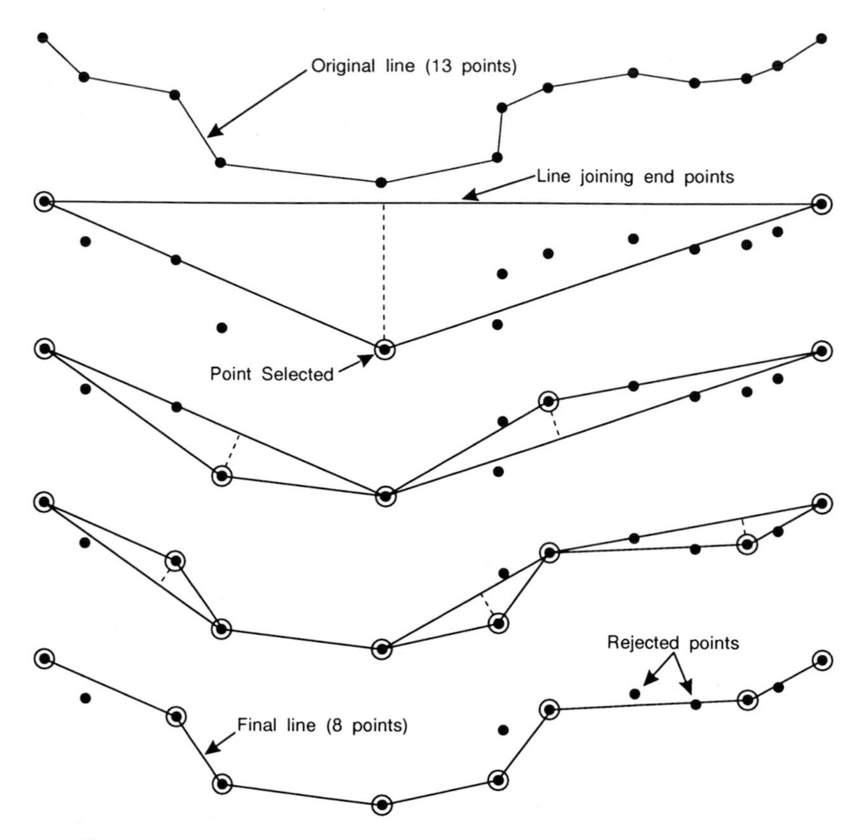

FIG. 6-10. Douglas-Peucker algorithm for line generalization. The original line contains 13 points, and the final line contains 8 points. The rejected points are said to be "weeded". A number of alternative algorithms for line weeding have been proposed.

Areas-to-Points

Area-to-point transformations involve the sampling of areal units at point locations. In some cases the points to be used for sampling are the locations at which field samples have been taken, in other cases the points are generated specifically for sampling. In Figure 6-11, the process of sampling a map layer with a pre-defined set of points is illustrated. We will discuss two aspects of this sampling process: 1) the method of finding the location of a point on a digitized map, and 2) the transfer of attributes from a polygon attribute table to a point attribute table.

In a raster data structure, the location of the pixel associated with a point location is straightforward, simply involving the conversion of the point coordinates to pixel coordinates by dividing the point coordinates by the pixel size, subtracting the coordinates of the origin, and rounding to the nearest integer. The attribute value for the pixel is then retrieved from the

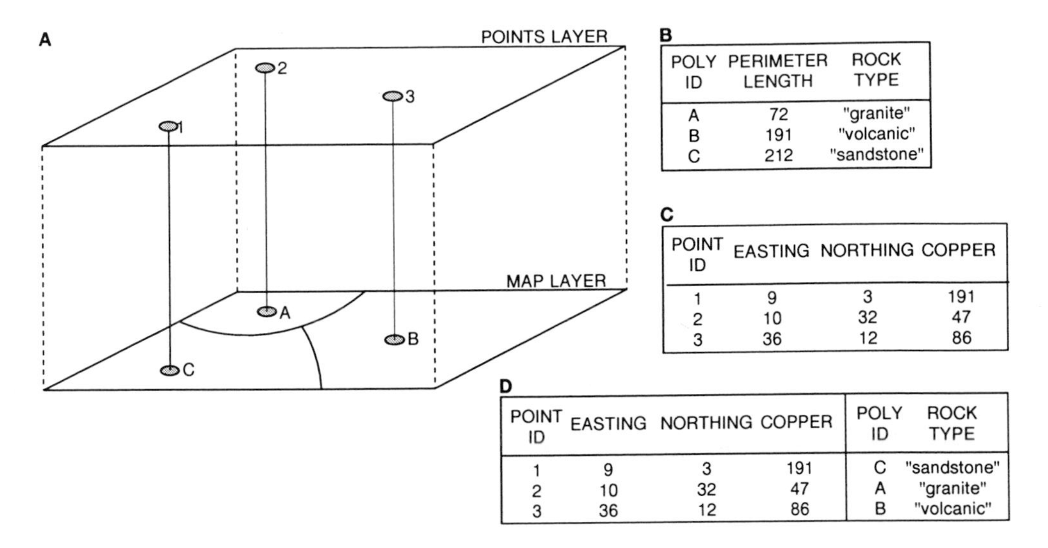

POLY ID	PERIMETER LENGTH	ROCK TYPE
A	72	"granite"
B	191	"volcanic"
C	212	"sandstone"

POINT ID	EASTING	NORTHING	COPPER
1	9	3	191
2	10	32	47
3	36	12	86

POINT ID	EASTING	NORTHING	COPPER	POLY ID	ROCK TYPE
1	9	3	191	C	"sandstone"
2	10	32	47	A	"granite"
3	36	12	86	B	"volcanic"

FIG. 6-11. Sampling maps at point locations, thereby generating new fields that are added to the point attribute table. **A**. Diagram to illustrate a layer of points, and a corresponding map layer consisting of 3 polygons. **B**. The polygon attribute table, with one record per polygon. **C**. The point attribute table, with one record per point. **D**. New point attribute table, with polygon attributes appended. In this simple example, the number of points happens to be the same as the number of polygons, but there can be any number of both points and polygons.

raster image file at the calculated pixel coordinates. In vector mode, the problem is more complicated, because each polygon is defined by a set of bounding arcs. One method of identifying the polygon in which a particular point occurs is by the "point-in-polygon" algorithm. The principle of this method can be illustrated graphically, as shown in Figure 6-12. The problem is to determine if the point, P, lies inside or outside the polygon. A line parallel to the y-axis is dropped from the point, and the number of intersections with the polygon boundary are determined. If the number is odd, the point must lie inside the polygon, no matter how convoluted the boundary. An even number of intersections means that the point must lie outside the polygon. Although this principle is simple, the search requires an extensive number of calculations, and a great deal of research has gone into finding efficient algorithms and data structures for this problem.

The transfer of attribute values from a polygon attribute table to a point attribute table is illustrated in Figure 6-11. Suppose that the polygons in A have the attributes of "polygon id", "perimeter length" and "rock type" held in a polygon attribute table, as shown in Figure 6-11B. The points to be used for sampling also have attributes in a point attribute table, see Figure 6-11C. They are the "point id", "easting", "northing" and "copper content". The sampling process (a point-in-polygon calculation if the polygons are in a vector structure) causes selected attributes from the polygon attribute table to be appended to the point attribute table. For example, the expanded point attribute table, see Figure 6-11D, includes the "polygon id" and the "rock type" for each point location. In this way, the characteristics of area objects can be "sampled" at points.

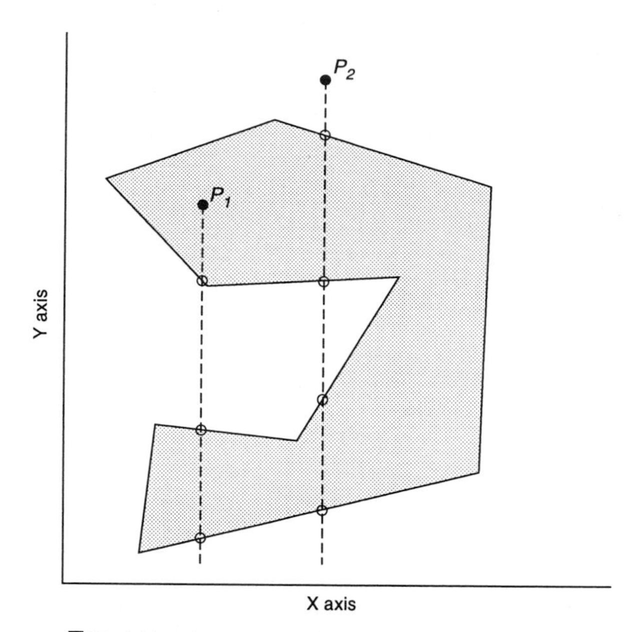

FIG. 6-12. Point-in-polygon algorithm. A line dropped from a point inside the polygon parallel to the *y*-axis will always have an odd number of intersections with the polygon boundary.

Consider a rather more complicated case, where a series of map layers are to be analyzed by point sampling. The goal is to generate a point attribute table, where the rows represent point locations, and the columns are attributes of the map layers. The resulting table is to be analyzed by statistical analysis to determine the relationships between the variables (i.e. between the map layers). In this case, suppose that the points to be used for sampling are on a regular grid, with a preset spacing. (Alternatively, the points might be generated using a random number generator, depending on the type of analysis). The sampling process can be visualized as pushing a set of "knitting needles", one per point, down through the stack of maps. The attribute of each map is picked up by the "knitting needle" and added to the point attribute table.

There are many applications of this type of sampling process. Probably the most common is in connection with the statistical analysis of multimap datasets. Most statistical software require the input dataset be organized as a flat file, with the rows (samples, cases, objects) being the records, and the attributes being the columns or fields of the file. Sampling a stack of maps at points and building a point attribute table, one field per sampled map, is one solution to the problem. The flat file represents a dataset suitable for univariate and multivariate statistical analysis, with the caveat that the points will very likely not conform to the requirements of an independent random sample, assumed for many statistical analyses.

Take another example, where the attributes of two sets of points are to be compared. The first might be soil samples, the second, vegetation samples. A table of geochemical attributes is associated with each. There are several ways to compare these two point datasets. One method, not necessarily the best, but having the possible advantage that no interpolation is involved, is to generate Thiessen polygons for the soil samples, transfer the soil point attributes to the soil polygons, then sample the soil polygons at the vegetation points, appending the soil attributes to each vegetation attribute record. This means that each vegetation sample can now be compared directly with the attributes of the closest soil sample. A disadvantage of this approach is that a different attribute table is produced by reversing the point datasets, generating the Thiessen polygons around the vegetation samples, and sampling with the soil points. Also there will be situations where the same soil sample is closest to several vegetation samples, or where one soil sample is omitted from the sampling, because in no case is it the closest to any vegetation sample. Nevertheless, these operations are straightforward in a GIS and provide a "quick-and-dirty" solution.

A alternative is to generate a surface model, by kriging for instance, for an attribute of the soil samples, and another surface model for the same attribute of the vegetation, and then to sample both surface models at a new set of points on a uniform grid. This new set of points could then be characterized by an attribute table, whose columns are the kriging estimates and kriging variances of the geochemical attributes from both sampling media.

Area-to-Area

The overlay of two or more maps is a most important GIS operation, and this topic is re-visited in Chapters 8 and 9. When the polygons of one map are superimposed on the polygons of another map, the result is to produce a new set of polygons common to both maps. The merged attribute tables produced by overlay operations can be used for modelling within the GIS or exported to other computer programs for specialized analyses. A polygon overlay operation can be regarded as a sampling process, where the new combined polygons are the "sampling" units.

Consider the two polygon maps in Figure 6-13. Map A consists of two polygons, map B of four polygons. The overlay produces an output map C, with six polygons, linked to an attribute table where the polygon identifiers for the two component maps are fields. Each polygon in C represents an area-weighted "sample" of maps A and B.

In a vector system, the generation of topological overlays is computationally demanding. The arcs from each map must be combined with the arcs of every other map in order to find intersection points. These intersections then become new nodes that split the intersected arcs. The partitioned arcs are reassembled into the new set of polygons that make up the geographic units common to all input maps. Finally each new polygon is labelled and linked to the polygon attribute table.

In a raster system, the overlay problem is more straightforward. The basic spatial objects are the pixels, and these are already common to all the input maps. Thus, during the overlay process, the computational geometry problems involved in the vector overlay are avoided. Attribute tables are often used, with records linked to polygons or "unique conditions" classes, as discussed in Chapter 8 and 9.

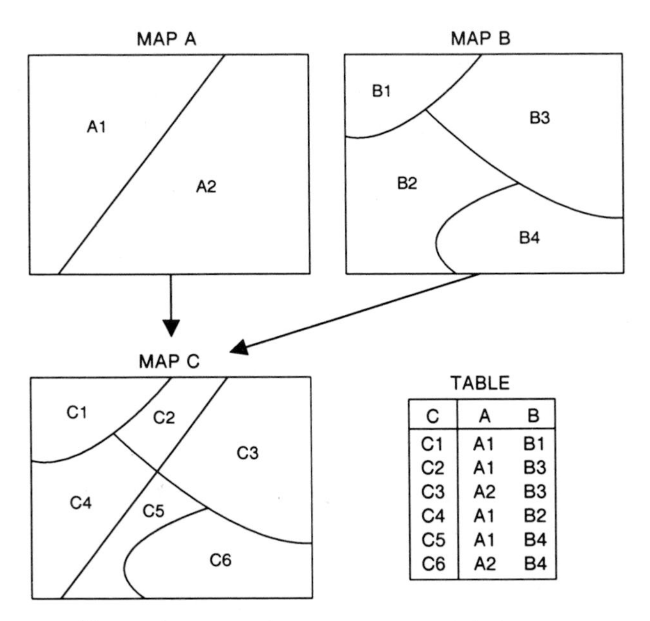

C	A	B
C1	A1	B1
C2	A1	B3
C3	A2	B3
C4	A1	B2
C5	A1	B4
C6	A2	B4

FIG. 6-13. Overlay of two polygon maps, producing a new set of polygons common to both maps. The polygons in C are now linked to the polygons of maps A and B in a polygon attribute table.

The attribute tables produced by an overlay operation can be aggregated in various ways, depending, amongst other considerations, on the measurement levels of the mapped quantities. Suppose that map A in Figure 6-14 is a geological map and that map B is a catchment basin map based on stream sediment sampling. Each polygon in map B is linked to a vector of geochemical element values, held in a catchment basin attribute table. The problem is to compute an estimated value of one or more of the geochemical elements, averaged over the geological map units.

If the attribute of interest is a field variable measured on an interval or ratio scale, and whose spatial distribution can reasonably be modelled as a smooth surface, the areal interpolation method suggested by Tobler (1979), called "pycnophylactic" interpolation can be used. Alternatively, if the value of the variable is spread evenly throughout the "source" zones, a simple area-weighted interpolation can be carried out, as described by Lam (1983).

Applying the latter approach to the case of the geochemical data values and the geological map units above, the average element values can be obtained by applying the formula:

$$\hat{z}_i = \frac{\sum_{j=1}^{n} a_{ij} z_j}{\sum_{j=1}^{n} a_{ij}} \tag{6-17}$$

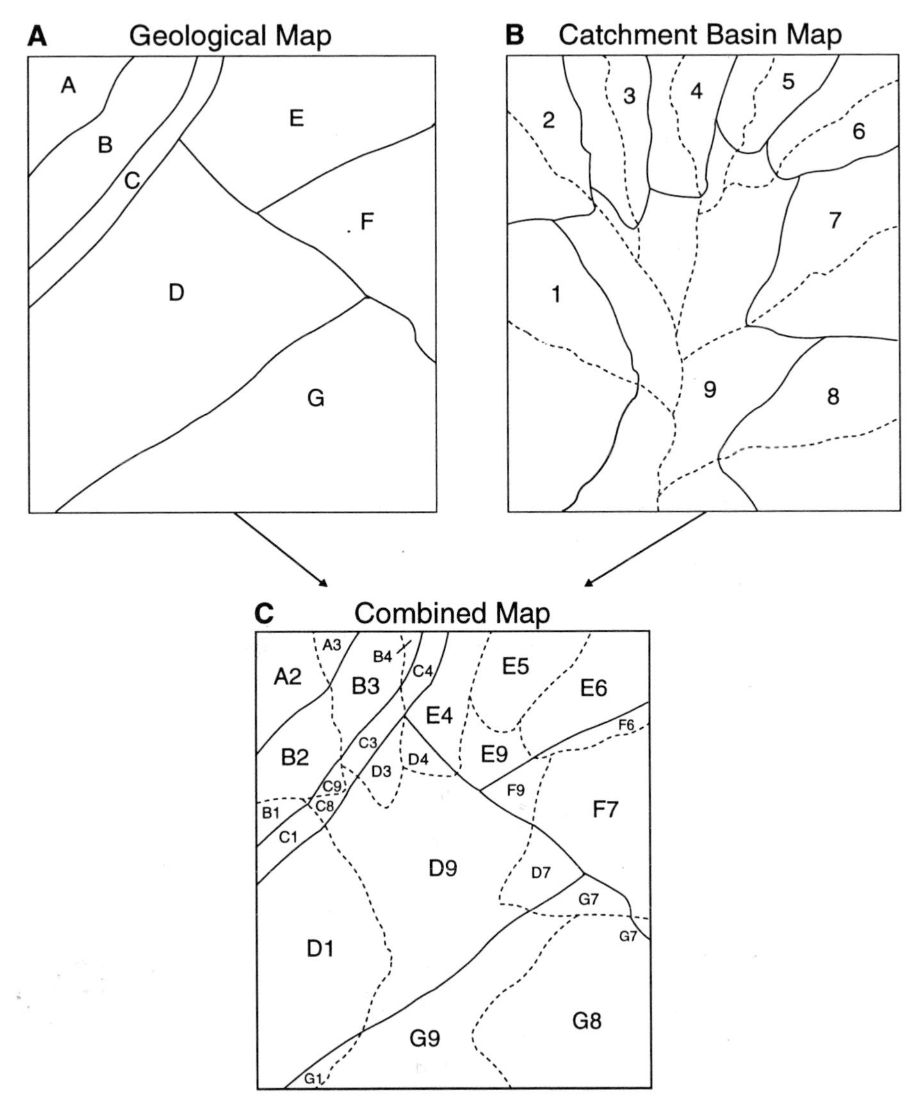

FIG. 6-14. Estimating average zinc content in stream sediment by "sampling" the catchment basin map with the polygons of the geological map. **A**. Geological map, with letters denoting map units. **B**. Catchment basins, denoted by the numerals 1-9. The dotted lines are the streams. The basin attribute table is linked to this map, and contains the geochemical data for the elements determined from stream sediment samples (1 sample per basin). **C**. Polygons produced by the overlap of the the two maps, each polygon labelled by a letter-numeral combination indicating the geology-catchment basin combination. The aggregation calculations for determining estimated element content of each geological map unit may be made with Equation (6-17), as shown in the text.

where the target zones (in this case the geological units) are denoted by the subscript i, the source zones (in this case the catchment basins) by the subscript j, z is the variable to be estimated in the target zones from the values in the source zones, and a_{ij} is the area of the overlapping target zone i and source zone j.

For example, in Figure 6-14, the application of equation (6-17) leads to the following expression for the zinc (Zn) content of geological map unit B:

```
Zn(B) = [area(B1)*Zn(B1) + area(B2)*Zn(B2) +
         area(B3)*Zn(B3) + area(B4)*Zn(B4)] / area(B)
```

where *Zn(B)* is the estimated zinc value. Similar expressions apply to any of the seven map units, or to any of the elements in the catchment basin attribute table. Note that Equation (6-17) must be modified if the variable being averaged consists of counts (like population numbers) rather than proportional measures (like parts per million).

Bonham-Carter et al. (1987) compared this approach for modelling the drainage geochemical characteristics of geological units with a mixing model approach. Using the geochemical element content as a response variable, area proportions of rock types in catchment basins were used as predictor variables in a regression model. A coefficient was determined for each rock type by least squares, minimizing the squared differences between measured and estimated element content, summed over all basins. The coefficients could be interpreted as mean element content values for each rock type ("target zone"), although the results tended to be unstable for rock types with small areal extent, as might be expected.

Care must be taken in these problems to distinguish between variables that are counts (like population numbers) as opposed to continuous measures like element content, or magnetic intensity. Counts must be expressed as densities (counts per unit area) for Equation (6-17) to apply.

Raster-to-Vector-to-Raster

In order to conclude this chapter, the transformation between the raster and vector data models are briefly mentioned. In systems that support both data models, this inter-conversion is regularly used. In some cases the boundary representation of the vector model is convenient and efficient, whereas in others the computations, like those for the multiple overlay of maps, are more readily made on the raster model.

Clarke (1990) and Foley and van Dam (1982) discuss algorithms for these transformations. A method for converting polygon data in vector mode to a run-length encoded raster is illustrated in Figure 6-15. A series of lines parallel to the x-axis are drawn across the map. The intersections between the horizontal lines and the polygon boundaries are calculated. The intersections partition each of the horizontal lines into strips whose length, after rounding to the pixel intervals form a run-length encoded transform of the vector data.

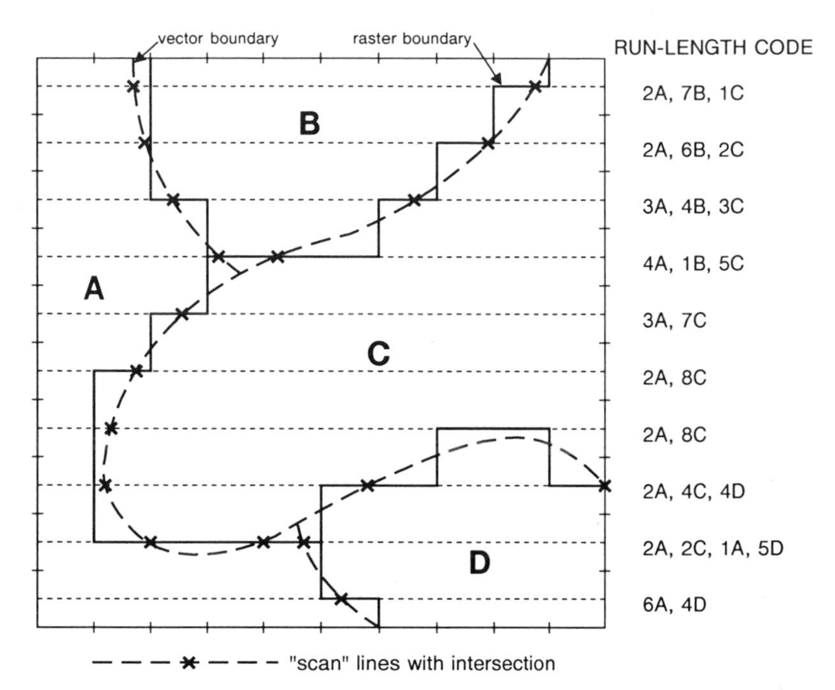

FIG. 6-15. Method of converting vector data in arc-node structure to a run-length-encoded raster structure. There are four polygons A, B, C, and D. The vector map is divided into horizontal strips, each one pixel wide. By moving from left-to-right within the length of each strip, the distance between the polygon boundaries can be measured in pixel units. The (run length, class) pairs are then linked together to create run-length code (see Chapter 3).

For the reverse transform, the problem is to find the boundaries of pixels where the pixel value changes from one class to another. Nodes where two or more boundaries intersect must then be extracted. Boundaries are then linked to form arcs, terminating in nodes. Smoothing may be carried out optionally to reduce the jaggedness of lines. Finally, polygons are labelled according to class value. Where the source of the raster data is classified remote sensing imagery, small polygons are often removed to reduce the "salt and pepper" appearance.

REFERENCES

Agterberg, F.P., 1974, *Geomathematics*: Elsevier, Amsterdam, 596 p.

Bonham-Carter, G.F., Rogers, P.J. and Ellwood, D.J., 1987, Catchment basin analysis applied to surficial geochemical data, Cobequid Highlands, Nova Scotia: *Journal Geochemical Exploration*, v. 29, p. 259-278.

Clarke, K.C., 1990, *Analytical and Computer Cartography*: Prentice Hall, Englewood Cliffs, New Jersey, 290 p.

Cressie, N.A.C., 1991, *Statistics for Spatial Data*: John Wiley & Sons, New York, 900 p.
Cromley, R.G., 1992, Principal line axis simplification: *Computers & Geosciences*, v. 18(8), p. 1003-1012.
David, M., 1988, *Handbook of Applied Geostatistical Ore Reserve Estimation*: Elsevier, Amsterdam, 232 p.
Ellwood, D.J., Bonham-Carter, G.F. and Goodfellow, W.D., 1986, An automated procedure for catchment basin analysis of stream geochemical data; Nahanni River map area, Yukon and Northwest Territories: *Geological Survey of Canada Paper 85-26*.(Annotated map).
Fabbri, A.G., 1984, *Image Processing of Geological Data*: Van Nostrand Reinhold, New York, 244 p.
Flowerdew, R., 1991, Spatial data integration: In *Geographical Information Systems*, v.1: Principles, Edited by Maguire, D.J., Goodchild, M.F. and Rhind, D.W., Longman Scientific and Technical, London, p. 375-387.
Foley, J.D. and van Dam, A., 1982, *Fundamentals of Interactive Computer Graphics*, Addison-Wesley, 664 p.
Garrett, R.G., Banville, R.M.P. and Adcock, S.W., 1990, Regional geochemical data compilation and map preparation, Labrador, Canada: *Journal of Geochemical Exploration*, v. 39, p. 91-116.
Hohn, M.E., 1988, *Geostatistics and Petroleum Geology*: Van Nostrand Reinhold, New York, 264 p.
Howarth, R.J. and Sinding-Larsen, R., 1983, Multivariate analysis: In *Statistics and Data Analysis in Geochemical Prospecting: Handbook of Exploration Geochemistry*, Edited by Howarth, R.J., Elsevier, Amsterdam, v.2, p. 207-289.
Isaaks, E.H. and Srivastava, R.M., 1989, *Applied Geostatistics*: Oxford University Press, New York-Oxford, 561 p.
Jenson, S.K. and Domingue, J.O., 1988, Extracting topographic structure from digital elevation data for geographic information system analysis: *Photogrammetric Engineering & Remote Sensing*, v. 54 (11), p. 1593-1600.
Journel, A.G. and Huibregts, C.J., 1978, *Mining Geostatistics*: Academic Press, New York, 600 p.
Lam, N.S., 1983, Spatial interpolation methods: a review: *The American Cartographer*, v.10(2), p. 128-149.
Matheron, G. and Armstrong, M., (editors), 1987, *Geostatistical Case Studies*: D.Reidel, Dordrecht-Boston, 248 p.
McCullagh, M.J., 1988, Terrain and surface modelling systems: theory and practice: *Photogrammetric Record*, v. 12(72), p. 747-779.
Serra, J., 1982, *Image Analysis and Mathematical Morphology*: Academic Press, London-New York, 610 p.
Siegel, S., 1956, *Nonparametric Statistics for the Behavioral Sciences*: McGraw-Hill, New York-Toronto, 312 p.
Tobler, W., 1979, A transformational view of cartography: *The American Cartographer*, v. 6(2), p. 101-106.
Unwin, D., 1981, *Introductory Spatial Analysis*: Methuen, London-New York, 212 p.
Watson, D.F., 1992, *Contouring: A Guide to the Analysis and Display of Spatial Data*: Pergamon Press, Oxford, 321 p.

CHAPTER 7

Tools for Map Analysis: Single Maps

INTRODUCTION

The analysis of spatial patterns on maps is the ultimate objective of many geological applications of GIS. Many of today's GIS packages excel in the collection, organization and visualization of spatial data, but are rather limited in their ability to be used for spatial analysis. This is partly due to the lack of a clearly developed theory of spatial analysis upon which GIS developers can draw. There is at present no real consensus in the GIS community about what computing tools should be included for spatial analysis or how they should be organized. Should a GIS contain a comprehensive toolbox of all the statistical, geostatistical and image processing functions that anyone is ever likely to need, or should a GIS simply act as a spatial data organizer, data viewer, and file server that interfaces with software that carries out specialized analyses? Many GIS already contain a modest toolbox of functions for analysis and modelling. It seems likely that this will expand and become more comprehensive in future developments of GIS. However, since the demands of the many fields that use GIS are diverse, it is unlikely that any single system will provide a complete range of all the functions that would ever be needed. Future GIS will probably meet the needs of 1) spatial research by improving the linkages with external software, and 2) accommodate the particular requirements of specialized applications by creating customized solutions. Thus in the former case, a new method of, say, multifactorial kriging, might be used with GIS by an efficient interchange of information (of data or subroutine calls) with a specialized external program. In the latter case, a standard procedure for, say, site evaluation for dam construction, could be packaged with a specialized procedure and user-friendly front-end designed solely for this one application.

One of the difficulties faced by GIS developers is due to the lack of consensus as to the methods needed for, and the definition of, spatial analysis. Is the analysis of data in a table linked to spatial objects in fact "spatial" if the operations can be carried out without regard to spatial variables? Many of the operations carried out in a GIS fall into this category, such as map reclassification, or the application of arithmetic or Boolean operators to the attributes of maps. On the other hand, operations that directly use spatial variables, either directly as in the fitting of mathematical trend surfaces, or indirectly as in the smoothing and filtering operations on raster images, are without question "spatial" in nature. However, although many GIS operations on maps are only spatial in the sense that the objects to which the operations are applied have geographic coordinates associated with them, such "non-spatial" operations can often lead to the discovery of significant spatial pattern. Take a simple example of processing a geophysical image with a contrast-stretch enhancement, thereby changing the area-class table or frequency distribution. This operation uses no spatial coordinates in the calculations, yet the result may reveal subtle patterns in the resulting image that, although present in the raw data, were simply not apparent before the enhancement was carried out. This is certainly a type of spatial analysis, because it is a process leading to the extraction of significant facts from spatial data.

For a description of spatial analysis operations applied to geological datasets, the book by Davis(1986) is an excellent introduction. There are many good image analysis textbooks, mainly centred on remote-sensing data, for example Mather (1987). The approach in this chapter (and chapters 9 and 10) is to discuss the operations commonly found in raster and vector GIS, with some limited treatment of methods not (yet) found in mainstream GIS. By way of introducing the subject, consider some examples of geological applications of operations on single maps and their associated attributes.

Examples of Applying Analytical Operations to Single Maps

1) A geological map has been digitized with the individual polygons labelled by map class, in this case the geological formation. The map attribute table links the map class to fields for a lithologic code, an age code and a formation name. The single map operation is to produce a new map **generalized** by combining formations according to age. Suppose the age codes are 1 for Mesozoic, 2 for Paleozoic and 3 for Phanerozoic, then the original map could be generalized into a new three-class age map by reclassifying or recoding the map class according to the geological age. The attribute table is thus being used directly as a lookup table. In a vector model, the boundaries of adjacent polygons belonging to the same age class now become redundant and are **dissolved,** an operation that is unnecessary in a raster model because polygon boundaries are not defined.

2) Given a stream-sediment geochemical data table, and a digitized map of catchment basins, the goal is to produce a series of geochemical maps, with classes based on percentiles of the element distributions. The keyfield of the attribute table is the basin number, and the other fields are element concentration values expressed as decimal numbers. A straight lookup between basin number and element value is not possible because a map class must be an integer number. The element field must also be linked to a second "lookup" or classification table, holding the breakpoints between percentile classes. As in case 1), the output map is a reclassified version of the input map. A variety of maps can be produced following the same procedure with derived attributes such as element ratios, or element combinations based on statistical models.

3) With the same data as in case 2), the goal might be to make a new map showing the spatial distribution of scores on the first principal component. Principal components analysis is a method to combine variables (in this case the geochemical elements) into a smaller number of new variables (the principal components) that are statistically independent of one another and "explain" as much of the co-variation between elements as possible. For example, elements that behave chemically in a similar fashion are strongly correlated and will have elevated "loadings" on the same principal component. Instead of making a separate map for each element, it may be

instructive to map a smaller number of principal component variables.This sounds like a multi-map problem, but with area objects (basins) linked to an attribute table, it can be treated as an operation on a single input map. The principal component variables are added as new fields to the attribute table. Principal component maps can then be made by reclassifying the basin map with the new fields. Although spatial coordinates are not used in conventional principal components analysis, there is a growing literature about spatial versions of this procedure, (Grunsky and Agterberg , 1988; Grunsky and Agterberg, 1991a; Grunsky and Agterberg, 1991b).

4) A total-field magnetics map, held as a raster image, is transformed to a slope map, by applying derivative operators in the x and y directions and calculating the magnitude of the resultant vector. Derivative maps are often useful for extracting important linear features not well-expressed in the original raster image. This operation is "spatial" in the sense that the derivative operator is applied to a pixel neighbourhood; the calculations cannot be carried out on each pixel independently.

5) A digitized soil map is summarized as a table or histogram, showing the area of each soil type. This output operation is simply a process of aggregation of the areas of the polygons on the maps by class type. It is nonspatial in the sense that it is independent of the spatial relationships in the data. In a raster, the pixels could be aggregated in any sequence, without affecting the results.

6) Spatial autocorrelation statistics are compared between two regions of the same geological map in order to characterize differences in map texture. This is a truly spatial operation, providing a statistical description of the autocorrelation characteristics.

Data Analysis and Modelling

One of the principal tools of GIS is what is often referred to as **map modelling**. A short digression is made here to discuss the idea of modelling as applied to the analysis of data on maps.

Data analysis can be defined as the extraction of significant facts embodied in a dataset; spatial data analysis therefore means the extraction of useful information from data that are distributed over space. Spatial data analysis is the process of seeking out patterns and associations on maps that help to characterize, understand and predict spatial phenomena. Consider the task of analyzing the relationship between the geochemical characteristics of soil in a region and the distribution of the underlying rock types. An understanding of such a relationship is essential for interpreting geochemical maps for exploration and environmental impact.

In the hypothetical example of Figure 7-1, a geological map consists of polygons of various rock types, and a geochemical dataset in point form comprises the sample locations and associated attributes in a table. An initial visual analysis of the zinc data shows a pattern of highs and lows that corresponds closely with the presence or absence of granite, as shown in Figure 7-1A. The Zn values appear to decrease with increasing distance from the granite contact, suggesting that the granite is Zn-rich compared with the surrounding rocks, and that surface processes have led to a dispersion of Zn away from the granite into nearby soils. To test the hypothesis of the correlation between Zn and proximity to the granite, the variation of Zn with distance from the granite contact is plotted in Figure 7-1B. This suggests a simple model of Zn decaying linearly with distance, as shown by the fitted line. The **model** appears to fit the data well, except that some relatively high values of Zn occur at intermediate-to-large distances from the granite, as indicated by some positive residuals. In fact by plotting the ratio [(observed Zn minus predicted Zn) / predicted Zn] in map view, as shown in Figure 7-1C, a *spatial* zone of high Zn values becomes apparent that simply is not obvious on the original map 1a. Furthermore, this zone turns out to correspond to a marshy region, where the Zn has been concentrated by the presence of organic material.

The purpose of this simple scenario is to bring out some of the ingredients of spatial data analysis in a GIS environment. First **visualization** leads to the recognition of a Zn **pattern** and its **spatial association** with granite. As a result, a **model** is proposed that describes the association. This is an **inductive** process, because conceptually a generalization is being made, based on a large number of particular instances, or data points on the map. The model in this case

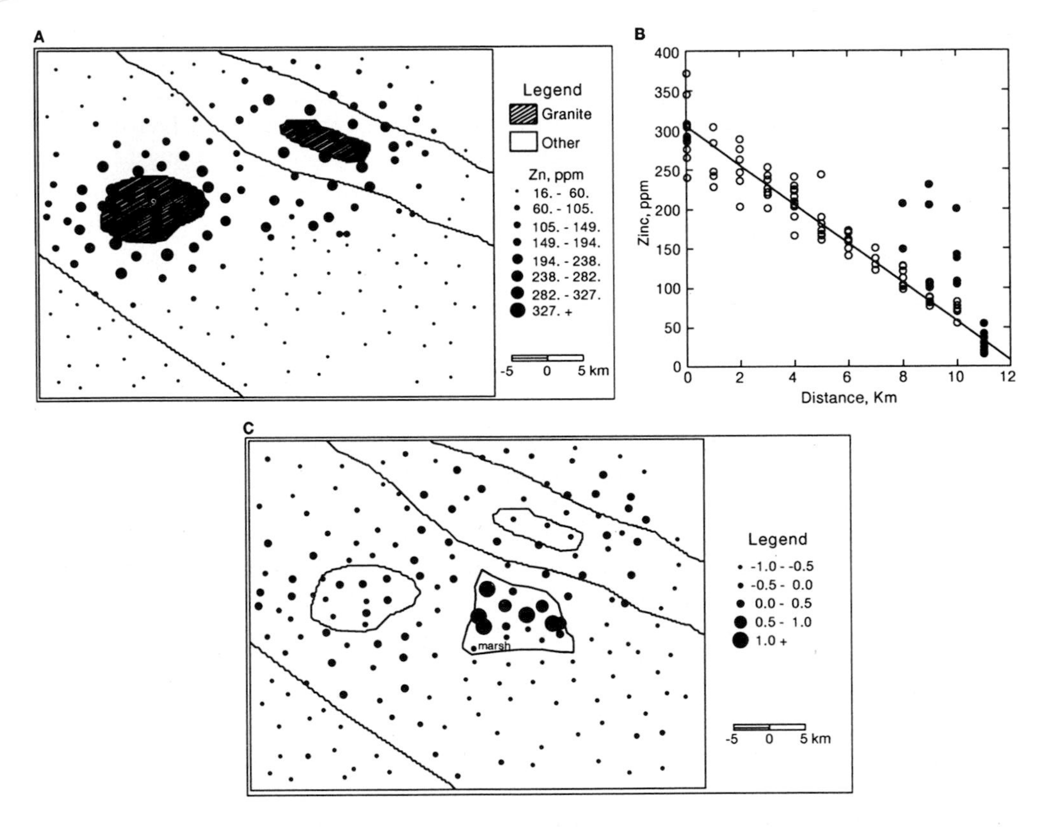

FIG. 7-1. A. Geological contacts superimposed on geochemical map of Zn in soil. Note that the Zn levels decrease with distance from the granite, as shown in the graph, **B.** A straight line model fitted to the points suggests positive residuals at some distance from the granite. The filled circles are samples from a marshy area, shown in C. The residuals are expressed as (observed-expected)/expected.

is a simple straight line fitted to the data, either by "eyeballing", or by least squares regression, where Zn is the dependent variable, y, and distance to contact, x, is the independent variable. The model is a functional relationship, that can be expressed algebraically, between the variables Zn and distance to granite. The advantage of applying the model is that the strength of association can be expressed quantitatively, with a correlation coefficient for example, and **predicted** and **residual** values of Zn can be calculated for each **observed** Zn value. Thus the application of the model leads to a **deductive** process, namely the evaluation of outcomes (predicted and residual values) for any number of data values. Deduction is therefore complementary to induction, moving from a generalization back to a set of particular instances. The model may not tell the whole story, but it provides an idealized framework for describing a relationship, for helping to provide insight and understanding to the geochemical data, and as a basis for predicting Zn values given only the geological map. A further analysis of the residuals might lead to the postulation of a modified model, with terms for other factors such as the presence of organic material in the soil.

Spatial data analysis encompasses a variety of activities that aid in the description, understanding and prediction of patterns and associations on maps. Modelling is a key ingredient of this activity, but measurement, statistical summary and visualization are also involved. Unfortunately, the word **model** is used to mean many things. In GIS, there are two common uses. One is the idea of a **data model**, an ideal schema for organizing data about the real world, as discussed in Chapter 2. In the present context, the word model means the symbolic representation of the relationships between spatial objects and their attributes.

In GIS, modelling is part of the analytical process of discovering, describing, and predicting spatial phenomena. Many GIS provide tools for defining models and carrying out modelling operations for prediction. Modelling operations may be carried out either directly on map layers themselves (using the class values) or indirectly on tables of attributes linked to spatial objects, or on combinations of both maps and their attributes. Tomlin (1990) describes a language for what is described as "cartographic modelling", and develops an algebra for spatial modelling. Cartographic or "map" modelling is carried out in GIS using a map algebra, that varies in syntax and implementation from one system to the next: there is no syntax or language structure that is used universally by GIS for modelling, at least at present. In this book, the approach is to employ a fairly basic and generic modelling language, as defined in the Appendix. This can be regarded as pseudo-code that can be translated according to the conventions of a particular GIS. A map model is a sequence of algebraic statements that result in the generation of a new output map from operations on one or more input maps.

Note that modelling in a GIS need not necessarily be carried out directly on a map: modelling is often a table operation. Thus, in the example of Figure 7-1, the linear model of Zn decreasing with distance from granite was fitted to two variables (the observed Zn field, and a field containing distance to the granite contact) in a point attribute table. The distance field was generated by appending proximity to granite values from a buffer map. By applying the fitted model, a new attribute field is then generated containing the predicted Zn levels, given the distance to granite. Thus we see, in this and later examples, that map modelling and table modelling are both essential operations.

Because modelling can often be highly specialized, with diverse requirements, GIS also provide links to external software that can carry out operations on shared data files. An attribute file is built in the GIS, exported to a program where a model is evaluated, resulting in one or more new attribute fields, then returned to the GIS for further analysis and display. External software may consist of statistical packages, spreadsheets, expert system shells, and customized software. It is fairly safe to assume that no single GIS will ever provide **all** the analytical tools needed for every eventuality, particularly for a research environment. Therefore the ability to couple the GIS to other computer programs is important, and emphasises the need to understand how the analytical tools that *are* available within a GIS can be used effectively.

Organization of GIS analysis tools

Analytical operations on maps can be divided into those that are truly spatial, because they involve spatial and/or topological attributes, and those operations that can be carried out independently of these attributes. Another way of organizing analytical operations in a GIS is on the basis of the type of output. Some operations produce map output, others produce tables or graphs. An example of map output is the calculation of a slope map from a map of a water table surface. Such an operation also explicitly uses information about a spatial neighbourhood. On the other hand, if a new field is calculated and added to a geochemical table showing the ratio of two elements, such as Zn/Pb, this results in table output: no spatial variables are involved, although the map display of Zn/Pb may well then lead to the recognition of spatial pattern. An example of generating an entirely new table is the calculation of the areas of each class on a map, or the measurement of the perimeter length of polygons. For organizational purposes one can therefore subdivide analytical operations according to whether or not spatial attributes are directly involved and whether the primary output is a map or a table, as summarized in Table 7-1.

This classification of analytical functions is with respect to maps and their attribute tables. The word map here implies a set of planar-enforced area objects, in either the raster or vector model. Attribute tables can be linked to area objects on the map (i.e. to polygon objects or to groups of polygons belonging to the same class), or to a set of point or line objects.

It should be noted that this classification of analytical operations is not the only one possible: operations can be organized on the basis of other criteria.

1. Map output, spatial variables not necessarily used

Map recoding, or reclassification, falls into this category. It simply means the reassignment of classes of the input map to new classes on the output map. One might argue that recoding is scarcely analytical, but the operation is so fundamental in a GIS and can often reveal new spatial pattern, that it is included here. Reclassification is a method used to generalize a map by combining polygons or combining classes. The input is a map, and the output of the operation is a new map.

Table 7.1. Examples of typical GIS operations on single map layers and associated attributes. Operations are grouped according to the type of output and whether or not spatial variables are involved.

Type of output	Spatial Attributes Used	
	Not necessarily	**Yes**
Map	Reclassification Map modelling	Spatial filtering Topological modelling Trend surface analysis Mathematical morphology
Table/ Other data views	Table modelling Aggregation Descriptive Statistics Box plots Scatterplots Time series	Spatial autocorrelation Variograms Power spectra

Reclassification of a map often involves the use of an attribute table. Also in this category are the operations carried out on one or more fields of an attribute table linked to a set of spatial objects. From a geochemical table linked to Thiessen polygons, for example, one might create a new map based on a model predicting Zn as a function of other geochemical attributes. This is either a one-step process using **map modelling**, or a two-step process, first employing **table modelling** to modify the attribute table, followed by reclassifying the polygon map with the new attribute field.

2. Table output, spatial variables not necessarily used

Operations that produce tables as output can generally be broken down into two types: **table modelling operations** where new fields are added as in category 1) above, and **aggregation operations** where summary statistics are calculated over a map or over some subset of spatial objects. These two cases are represented diagrammatically in Figure 7-2. In the first case, the addition of new columns or fields is the outcome of a modelling operation, where the rows, or spatial objects, remain unchanged. The new fields are simply new attributes derived from modelling operations on existing fields. In the second case, aggregation operations result in the reduction of the number of rows of the table, collapsing the number of spatial objects on the basis of subsets defined on one or more of the attributes. For example, suppose that aggregated values of two variables are to be computed for groups of spatial objects, as defined by the categorical attribute in the field labelled 3. The result is a new table with four rows (the number of categories of attribute 3) and two columns, the mean of attribute 1, and the maximum of attribute 2, ***by*** the classes defined by attribute 3. Instead of the mean and maximum, other statistical aggregation measurements might include the median, mode, variance, minimum, range, or the total number of observations. Aggregation can be carried out globally over the whole table (or equivalently the whole map), or over subsets defined by classes of one of the attributes. Such operations are to be found in any statistical package, controlled by statements like

Statistics (all) ***on*** **Zn to Cu,** ***by*** **Rocktype,**

implying that all the standard statistics are to be applied to the attributes between and including Zn and Cu in the table, subdivided into subsets based on the categories of Rocktype.

Attributes of spatial objects may be summarized and plotted in many ways other than map views. For example, an attribute table might contain annual records of rainfall or water level, taken at stations that refer to point objects (gauging stations) or area objects (lakes), in which case the output might be time series plots showing the variations in rainfall/water level over a multi-year interval and how they differ from one point station to the next.

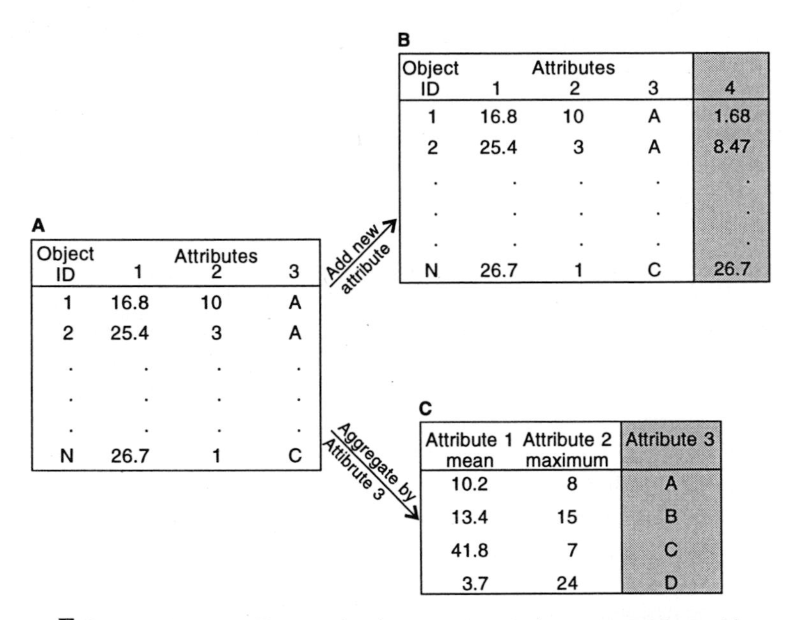

FIG. 7-2. Diagram to illustrate that the common operations on an attribute table (**A**) either add new fields in a modelling or measurement process (**B**), or condense the table by aggregation, summarizing one or more attributes based on subsets of spatial objects determined by a grouping attribute (**C**).

3. Map output, spatial variables used

Filtering operations on groups of neighbouring pixels in a raster model fall into this category, although spatial variables do not explicitly enter the calculations, but are implicit in the selection of pixels. Such operations produce a new raster output map or image. An attribute table is not generally involved, because the operation is carried out on a single attribute at a time. The purpose of filtering is to enhance and extract textural features from a raster map or image, and the calculations always involve groups of neighbouring pixels. The calculation of local measures of central tendency for smoothing purposes, the detection of edges, the conversion of surfaces into slope and aspect maps, and the production of illumination maps (also called insolation or hill-shading maps) are examples of neighbourhood operations.

Map modelling with either vector or raster data may also use spatial and topological information directly. For example, a geological map might be generated showing only those polygons north of latitude 45 and underlain by limestone and adjacent to igneous intrusions. Or, a binary map of lakes is converted into a new map showing lakes classified according to shape.

Trend surface analysis is a global method of fitting a polynomial function to continuous variables measured at points, where the independent variables are the spatial coordinates, Davis (1986). Trend surfaces can be helpful for recognizing and characterizing spatial trends and have been widely used in geology, particularly in the analysis of data from sedimentary basins.

Mathematical morphology operations, as briefly introduced in Chapter 6 under the topic of transformations, provide a battery of techniques for measuring the shape and size of "particles" (i.e.area or line objects) on raster images. The resulting geometric attributes can either be put in an attribute table, or used to classify spatial objects directly in an image. This technology has been widely exploited in "image analyzers" attached to optical and electron microscopes, but has not yet been widely adopted in GIS.

4. Table output, spatial variables used

Relationships between pairs of spatial objects separated by different distances can be very important in map analysis for characterizing texture and for prediction. By comparing each spatial object (point, line or area) with every other spatial object of its own type, useful information can be revealed about the spatial structure present in the data. For example, a plot of physical distance versus the squared difference in attribute value for a set of point samples produces the experimental variogram, discussed under point interpolation in Chapter 6. Spatial autocorrelation (the correlation of a map with a duplicate of itself under translation) can also be quantitatively characterized for any type of spatial object with attributes on any measurement scale, see Goodchild (1986). Indices of spatial autocorrelation calculated globally over a whole map are valuable for descriptive purposes, because they provide a measure of how similar objects are to their spatial neighbours. Autocorrelation indices can also be helpful for understanding patterns on maps and for prediction.

We now turn to a number of methods applicable to single maps for a more detailed discussion.

MAP RECLASSIFICATION

Classification is a method for simplifying complex phenomena, and is used in a GIS for generalization and enhancement. Geologists reduce the complexity of compositional and textural variation of rocks and soils by assigning them to discrete classes for mapping and description. Geochemists subdivide the range of a chemical element into discrete intervals for mapping purposes to enhance regions that have distinctive and unusual values. In a GIS, classification can either mean the process of assigning objects to classes on the basis of one or more attributes, or for a single attribute it can mean the list of breakpoint values that define class intervals (the cut-off values between one class and the next), or in a case of multi-attribute classification it can mean the process of defining multivariate classes. The latter function is known in image processing as "unsupervised classification", or in geology and biology as "cluster analysis". On the other hand "reclassification" is the process of changing the assignment of objects from one classification system to another.

Reclassification is carried out by converting an input map to an output map with a lookup table that recodes the old classes to new classes. The reclassification process depends on the measurement level of the attribute under consideration. In the case of categorical measurement levels, the recoding of classes is not constrained by the class order. With attributes that use interval, ordinal or ratio scales the reclassification usually respects the class sequence, but not necessarily.

Reclassification of a single map on the basis of several attributes in an attribute table can be carried out with statements in a modelling language. It can be regarded as a type of query operation, searching for objects that satisfy a single criterion or a set of multi-attribute criteria, but extending the process to produce a new map.

Consider the geological map in Figure 7-3A, part of Vancouver Island and the adjoining mainland. The geological map has 32 map units, as shown in the abbreviated legend. The map is fairly complex, and for some purposes reclassification is desirable for simplifying it and enhancing particular features. Several reclassifications are illustrated, the first one simply reduces the number of map units from 32 to 7, lumping together classes into major stratigraphic age groups. The second reclassification reduces the number of classes from 32 to 3, on the basis of sedimentary, igneous and metamorphic lithologies. The third reclassification reduces the number of classes from 32 to 2, but on the basis of a lithological *and* age criteria, making a map of Mesozoic diorites.

The actual mechanics of carrying out a reclassification depends on the GIS. The most common method is to use a look-up table that relates the class of the output map to the class of the input map. The geological map in the present case has an attribute table (Table 7-2), where the entities (rows) are the input map classes, and the keyfield is in column 1 with the class value. There are attribute fields with age codes (upper and lower age), and four lithology fields containing lithologic codes. The code for the most abundant lithology is in the field

"**LITH1**", the second-most common in "**LITH2**", and so on. The unit name is stored as a text variable called "**NAME**", and the final field called "**RECLASS**" contains the class values for a simplified geological age map. The first reclassification simply uses this last column. In map modelling pseudo-code, the reclassification process might be carried out with the following sequence of statements:

```
: at the current location, pick up the class value of the geological map,
: assign to temporary variable (INCLASS)
      INCLASS = CLASS('GEOLOGY')
: the value of the output class is found in the GEOLOGY attribute table, at
row=INCLASS, and column='RECLASS'
      OUTCLASS = TABLE('GEOLOGY', INCLASS, 'RECLASS')
: the result is mapped using the new class values
      RESULT(OUTCLASS)
```

The syntax for this pseudo-code is briefly described in Appendix I. When this sequence is invoked, the modelling operations defined in the statements are repeated systematically for every location on the map. The operations are thus repeated for every pixel in a raster, or every polygon in a vector model.

Taking an actual example, at a location where the input map class is 12, the corresponding row 12 of the **GEOLOGY** attribute table shows that the unit named "Kgr" (standing for Cretaceous granite) has the value 4 in the **RECLASS** field. Thus input class 12 is converted to output class 4. Similarly input class 27 is converted to output class 6, and so on. The result of applying this operation is to produce the output map shown in Figure 7-3B. The generalized units are helpful in providing a visual overview of the stratigraphic ages of rocks in the region.

In this first reclassification, the lithological and age codes have not been used, because the reclassification scheme occurs as a separate field in the table, allowing a direct lookup. Suppose, however, that the lithological codes used in the 4 fields "LITH1" to "LITH4" are pointers to a second table called **LITHOLOGY** (Table 7-3A), with another set of descriptive fields, and the age codes are pointers to a third table called **AGE** (Table 7-3B) with a field showing the absolute age in millions of years before present (MA) and the stratigraphic age as a text variable (in quotes). The reclassification process can take advantage of the *relational* structure of the tables to carry out a more sophisticated assignment of output classes. For example, the **LITHOLOGY** table has fields with binary indicator values that record whether the lithology in question is sedimentary, igneous or metamorphic (labelled "SED", "IGN" and "META" in Table 7-3A). Therefore the reclassification of the geology map on the basis of the most abundant lithology ("LITH1") to a map showing the distribution of sedimentary, igneous and sedimentary rocks might be carried out with the following pseudo-code, producing the output map shown in Figure 7-3C:

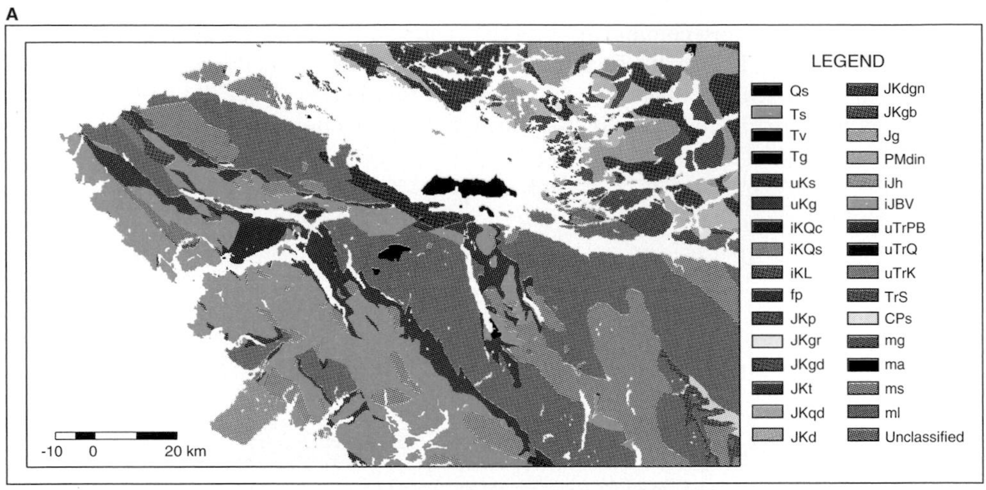

FIG. 7-3. Reclassification of a geological map. **A.** Original map with 32 classes. **B.** Generalized to 7 classes. **C.** Generalized further to 3 classes. **D.** Reclassification to 2 classes, to show only Mesozoic quartz diorite units.

```
INCLASS = CLASS('GEOLOGY')
LITH = TABLE('GEOLOGY', INCLASS, 'LITH1')
: use the lithology code as the keyfield of the LITHOLOGY table and lookup the
: indicator fields for sedimentary, igneous and metamorphic rocks
S = TABLE('LITHOLOGY', LITH, 'SED')
I = TABLE('LITHOLOGY', LITH, 'IGN')
M = TABLE('LITHOLOGY', LITH, 'META')
: let the output class be 1 for sedimentary, 2 for igneous or 3 for metamorphic
OUTCLASS = {1 IF S==1, 2 IF I==1, 3 IF M==1, 0}
RESULT(OUTCLASS)
```

Consider some examples of using this code. At a location where the geology class is 5, the LITH1 code is 46, and at record 46 in the LITHOLOGY table this corresponds to SILTSTONE and the SED indicator =1 for sedimentary. Similarly, for geology class 17 the lithology code is 59, which is a GNEISS and META=1 for metamorphic. The geology class 27 corresponds to LITH1=43, which has SED=1 and is a sedimentary greywacke, and so on. Note that in this case the sedimentary, igneous and metamorphic indicator attributes could be combined into a single three-state attribute, because the three states are mutually exclusive.

In a final example, suppose that the input map is to be reclassified into a binary map on the basis of both age and lithological criteria. This involves three linked relational tables: the first (**GEOLOGY**) is keyed to the map, and the second and third tables (**AGE** and **LITHOLOGY** respectively) are keyed to the **GEOLOGY** table. The criteria to be satisfied in making this new two-class map are:

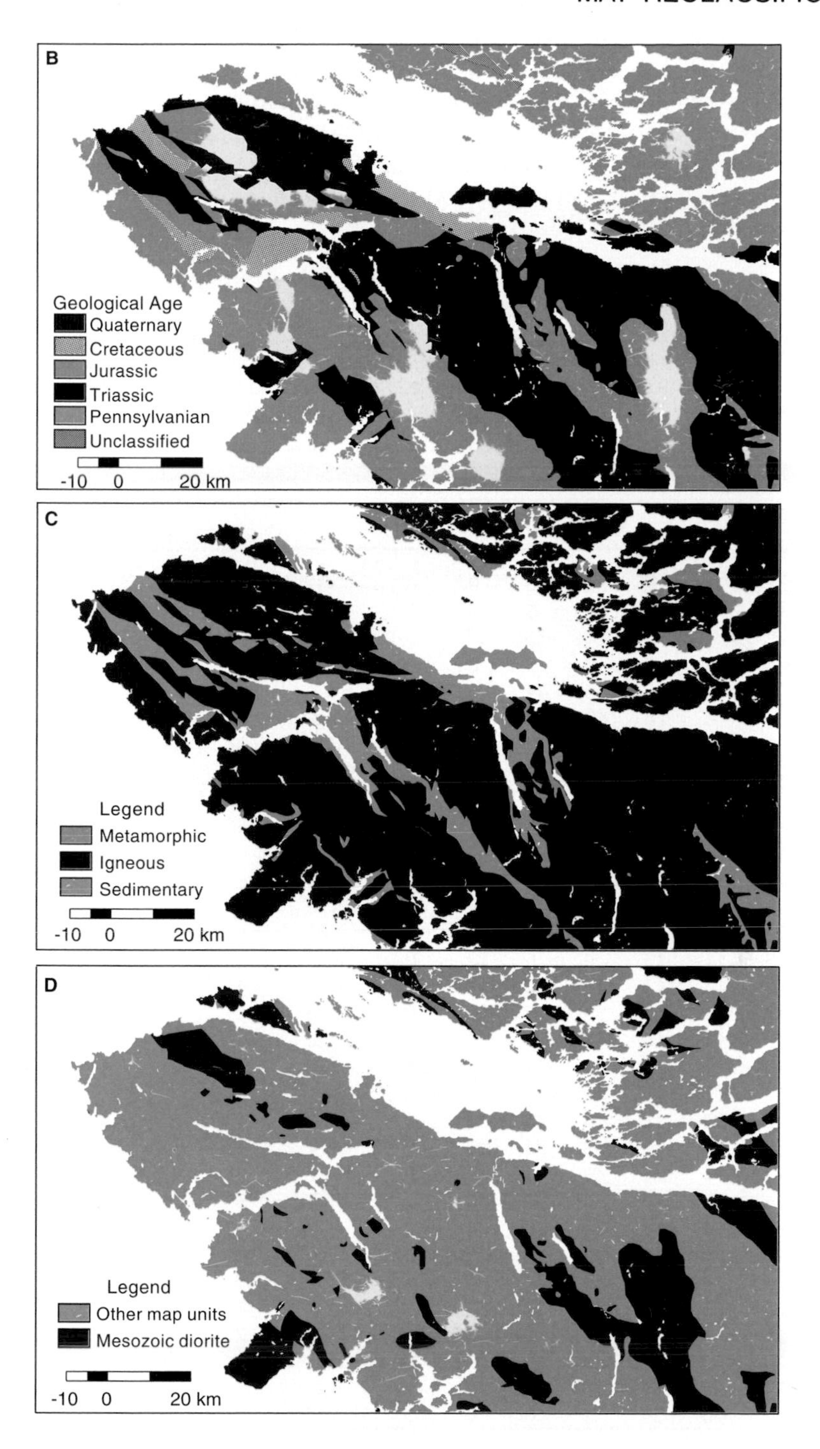
B
Geological Age
Quaternary
Cretaceous
Jurassic
Triassic
Pennsylvanian
Unclassified
-10 0 20 km
C
Legend
Metamorphic
Igneous
Sedimentary
-10 0 20 km
D
Legend
Other map units
Mesozoic diorite
-10 0 20 km

1) that the age of the rocks be Mesozoic (i.e. with an upper age ("UPAGE") greater than or equal to 65 million years before present (MA), and a lower age (LOWAGE) less than 235 MA), and

2) that the primary (most abundant) lithology (LITH1) is a quartz diorite.

The output map is to consist of two classes: the first class (=2) is where the above criteria are satisfied, and the second class (=1) is where they are not satisfied. Therefore the map modelling code might read as follows:

Table 7.2. Attribute table, with the name **GEOLOGY**, for the geological map in Figure 7-3A. Each row is linked by the keyfield (field 1) to the classes of the geological map. Field 9 is the field used directly for reclassification, containing the class values of the output map in Fig. 7-3B. Fields 2 and 3 (age codes) are pointers to an **AGE** attribute table (see Table 7-3B). Fields 4-7 are lithology codes, which point to a **LITHOLOGY** table, see Table 7-3A.

GEOLOGY TABLE

1	2	3	4	5	6	7	8	9	KEY TO FIELDS
1	1	2	55	0	0	0 '	QS'	1	1=CLASS
2	3	7	37	0	0	0 '	TS'	2	2=UPAGE
3	4	4	14	38	37	0 '	TV'	2	3=LOWAGE
4	6	6	9	0	0	0 '	TG'	2	4=LITH1
5	8	8	46	44	0	0 '	uKS'	3	5=LITH2
6	8	8	43	37	46	54 '	uKg'	3	6=LITH3
7	8	10	37	43	0	0 '	IKQc'	3	7=LITH4
8	8	10	46	49	43	0 '	IKQs'	3	8=NAME
9	10	10	43	37	46	0 '	IKL'	3	9=RECLASS
10	11	11	21	0	0	0 '	FP'	4	
11	8	13	43	46	37	0 '	JKP'	4	
12	11	13	3	7	0	0 '	Kgr'	4	
13	11	13	4	5	9	0 '	JKgd'	4	
14	11	13	5	9	4	0 '	JKt'	4	
15	11	13	9	5	13	0 '	JKqd'	4	
16	11	13	13	17	14	8 '	JKd'	4	
17	11	13	59	0	0	0 '	JKdg'	4	
18	11	13	14	0	0	0 '	JKgb'	4	
19	11	13	9	4	7	21 '	Jg'	4	
20	11	13	9	59	16	0 '	Pmdn'	4	
21	13	13	45	43	50	0 '	IJH'	4	
22	13	13	33	35	36	38 '	IJBV'	4	
23	14	14	49	44	50	43 '	uTPB'	5	
24	14	14	50	0	0	0 '	uTQ'	5	
25	14	14	34	36	24	50 '	uTK'	5	
26	14	15	48	15	45	0 '	Ts'	5	
27	20	20	43	45	50	0 '	CPs'	6	
28	50	50	24	16	52	45 '	mg'	7	
29	50	50	45	41	58	64 '	mG'	7	
30	50	50	58	0	0	0 '	ms'	7	
31	50	50	50	41	0	0 '	ml'	7	
32	50	50	1	0	0	0 '	UNCL'	7	

Table 7-3. Relational tables for lithology (A) and age (B), linked by keyfield to the codes in the GEOLOGY table.

A. LITHOLOGY TABLE

LITH	SED	IGN	META	NAME
1	0	0	0	'NOT IDENTIFIED'
2	0	1	0	'GRANITOID'
3	0	1	0	'GRANITE'
4	0	1	0	'GRANODIORITE'
5	0	1	6	'TONALITE'
6	0	1	0	'QUARTZ SYENITE'
7	0	1	0	'QUARTZ MONZONITE'
8	0	1	0	'QUARTZ MONZODIORITE'
9	0	1	0	'QUARTZ DIORITE'
10	0	1	0	'SYENITE'
11	0	1	0	'MONZONITE'
12	0	1	0	'MONZODIORITE'
13	0	1	0	'DIORITE'
14	0	1	0	'GABBRO'
15	0	1	0	'DIABASE'
16	0	1	0	'AMPHIBOLITE'
17	0	1	0	'ALASKITE'
18	0	1	0	'PYROXENITE'
19	0	1	0	'SYENOGRANITE'
20	0	1	0	'QUARTZ-EYE PORPHYRY'
21	0	1	0	'QUARTZ FELDSPAR PORPHYRY'
22	0	1	0	'ALKALI GRANITE'
23	0	1	0	'FELSIC VOLCANICS'
24	0	1	0	'INTERMEDIATE VOLCANICS'
25	0	1	0	'MAFIC VOLCANICS'
26	0	1	0	'FELSIC-INTER VOLCANICS'
27	0	1	0	'INTER-MAFIC VOLCANICS'
28	0	1	0	'FELSIC-MAFIC VOLCANICS'
29	0	1	0	'TRACHYTE'
30	0	1	0	'RHYOLITE'
31	0	1	0	'QUARTZ LATITE'
32	0	1	0	'DACITE'
33	0	1	0	'ANDESITE'
34	0	1	0	'BASALT'
35	0	1	0	'RHYODACITE'
36	0	1	0	'TUFF'
37	1	0	0	'CONGLOMERATE'
38	1	0	0	'BRECCIA'
39	1	0	0	'SANDSTONE'
40	1	0	0	'ARENITE'
41	1	0	0	'QUARTZITE'
42	1	0	0	'ARKOSE'
43	1	0	0	'GREYWACKE'
44	1	0	0	'SHALE'
45	1	0	0	'ARGILLITE'
46	1	0	0	'SILTSTONE'
47	1	0	0	'REDBEDS'
48	1	0	0	'UNDIFF SEDIMENTS'
49	1	0	0	'CALCAREOUS SANDSTONE'
50	1	0	0	'LIMESTONE'
51	1	0	0	'DOLOMITE'
52	1	0	0	'CHERT'
53	1	0	0	'IRON FORMATION'
54	1	0	0	'COAL'
55	1	0	0	'SURFICIAL'
56	0	0	1	'SLATE'
57	0	0	1	'PHYLLITE'
58	0	0	1	'SCHIST'
59	0	0	1	'GNEISS'
60	0	0	1	'PARAGNEISS'
61	0	0	1	'ORTHOGNEISS'
62	0	0	1	'MYLONITE'
63	0	0	1	'UNDIFF METASEDIMENTS'
64	0	0	1	'UNDIFF METAVOLCANICS'
65	0	0	1	'SKARN'

B. AGE TABLE

AGE	YEARS MA	NAME
1	0.0	'HOLOCENE'
2	0.0	'PLEISTOCENE'
3	2.0	'PLIOCENE'
4	6.0	'MIOCENE'
5	22.0	'OLIGOCENE'
6	36.0	'EOCENE'
7	58.0	'PALEOCENE'
8	65.0	'LATE CRETACEOUS'
9	88.0	'MID CRETACEOUS'
10	118.0	'EARLY CRETACEOUS'
11	145.0	'LATE JURASSIC'
12	160.0	'MID JURASSIC'
13	176.0	'EARLY JURASSIC'
14	195.0	'LATE TRIASSIC'
15	210.0	'MID TRIASSIC'
16	225.0	'EARLY TRIASSIC'
17	235.0	'LATE PERMIAN'
18	250.0	'MID PERMIAN'
19	260.0	'EARLY PERMIAN'
20	280.0	'PENNSYLVANIAN'
21	290.0	'MID CARBONIFEROUS'
22	315.0	'MISSISSIPPIAN'
23	345.0	'LATE DEVONIAN'
24	360.0	'MID DEVONIAN'
25	370.0	'EARLY DEVONIAN'
26	395.0	'LATE SILURIAN'
27	412.0	'MID SILURIAN'
28	422.0	'EARLY SILURIAN'
29	435.0	'LATE ORDOVICIAN'
30	460.0	'MID ORDOVICIAN'
31	472.0	'EARLY ORDOVICIAN'
32	500.0	'LATE CAMBRIAN'
33	515.0	'MID CAMBRIAN'
34	540.0	'EARLY CAMBRIAN'
35	570.0	'HADRYNIAN'
36	1000.0	'HELEKIAN'
37	1800.0	'APHEBIAN'
38	2600.0	'LATE ARCHEAN'
39	3000.0	'MID ARCHEAN'
40	3500.0	'EARLY ARCHEAN'
50	9999	'UNCLASSIFIED'

```
      INCLASS = CLASS('GEOLOGY')
: assign the upper age code to variable U
      U = TABLE('GEOLOGY', INCLASS, 'UPAGE')
: assign lower age code to variable L
      L = TABLE('GEOLOGY', INCLASS, 'LOWAGE')
      LITH = TABLE('GEOLOGY', INCLASS, 'LITH1')
: lookup the age in years MA for each age code in the AGE table
      UY = TABLE('AGE', U, 'YEARSMA')
      LY = TABLE('AGE', L, 'YEARSMA')
: look up the lithology name
      LITH$=TABLE ('LITHOLOGY', LITH, 'NAME')
: LITH$ is a character variable
: use conditional statement to assign OUTCLASS=2 if the conditions are satisfied,
: =1 otherwise
      OUTCLASS={2 IF UY>=65 AND LY<235 AND LITH$=='QUARTZ DIORITE',1}
      RESULT(OUTCLASS)
```

The map produced by this code is shown in Figure 7-3D. Three map units in the **GEOLOGY** table satisfy the criteria, namely classes 15, 19 and 20. Note that the same objective could have been achieved by manually adding a new reclassification field to the **GEOLOGY** table, identifying these three map classes. However, the modelling approach is considerably more flexible and powerful.

Classification tables using breakpoints

Instead of using a lookup table that matches an input class to an output class directly, the classification process can be controlled with a table of breakpoints. Classification tables of this type are appropriate for use with data measured on ordinal, interval or ratio scales, and are particularly useful for real number variables that can be negative as well as positive. Take, for example, the classification of geochemical maps for mineral exploration purposes, where the classification scheme is used to enhance the upper tail of the element frequency distribution. Geochemical elements usually have positively skewed frequency distributions, and therefore the goal is to create maps where the regions with anomalously-large element values are enhanced. A common scheme is to use a percentile scale, such as the 98th, 95th, 90th, 80th, 60th, and 25th percentiles, determined from a cumulative frequency distribution. Because geochemical element concentrations are recorded as decimal numbers, the cutoff values between classes are not necessarily integer numbers. In this case the assignment of class values is achieved by using a table of the breakpoints between classes. Breakpoint tables are widely used for classifying all kinds of continuous monotonic variables, whether they be applied to raw measurements, like geochemical or geophysical observations, or to derived

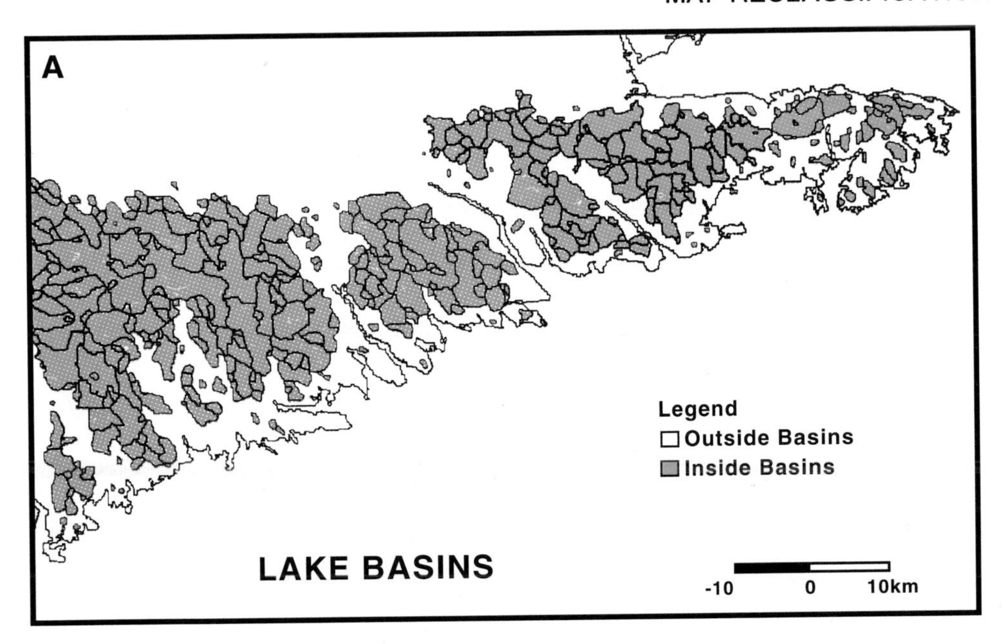

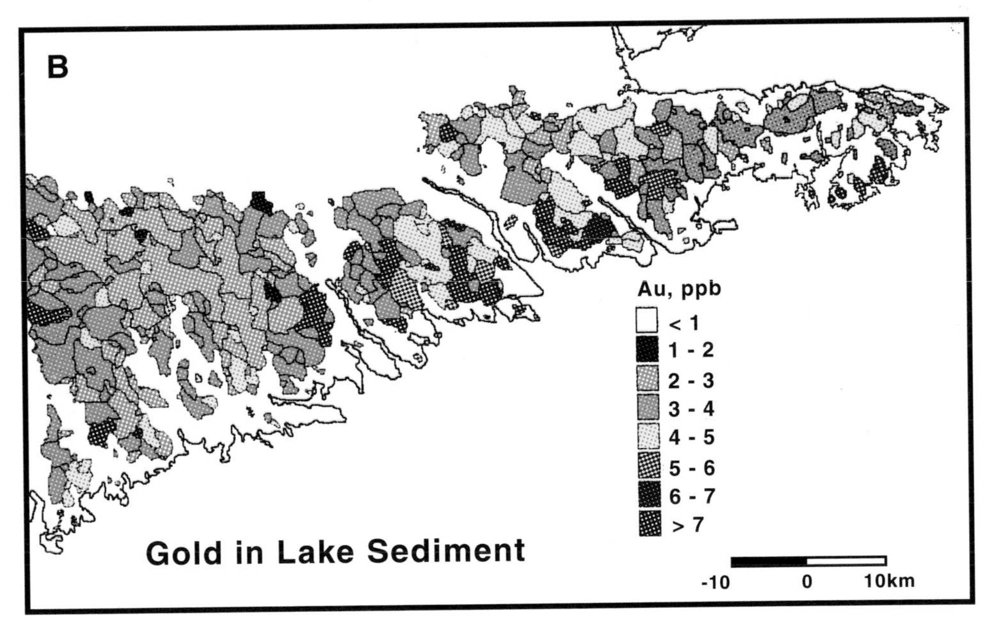

FIG. 7-4. Reclassification of a geochemical map. **A.** Lake catchment basins, with a separate class for each polygon. Each basin is linked to a record in an attribute table of geochemical variables. **B.** The basins have been reclassified on the basis of gold levels in lake sediment, using the attribute table and a classification table of breakpoints.

data obtained from modelling calculations.The following example illustrates the application of a breakpoint table for reclassifying a catchment basin map according to one or more geochemical attributes.

Consider a geochemical survey, where catchment basins are coded by basin number (Figure 7-4A), and the basin attribute table (Table 7-4) contains a number of fields for geochemical elements.The goal is to produce a series of new maps by reclassifying the catchment basins by using the element fields in a lookup operation. Suppose that a breakpoint table has been made for the element gold (Au). The classification table showing breakpoints is in Table 7-5, and the resulting Au map is in Figure 7-4B. The modelling statements to produce the reclassification of the basin map to the Au map are as follows:

```
: at the current location, get the class number of the lake basin
        INCLASS = CLASS('BASIN')
: use the basin attribute table to lookup the corresponding Au value
        AU = TABLE('BASIN', INCLASS, 'AU')
: classify using the breakpoint (classification) table called LSAU
        OUTCLASS = CLASSIFY(AU, 'LSAU')
: create new output map
        RESULT(OUTCLASS)
```

Notice that the modelling operation of the type

K = CLASSIFY(A, 'B')

causes a value of variable A to be converted to an integer K that is the required class value obtained from the classification table called 'B'.

Table 7-4. Basin attribute table, showing geochemical levels for 5 elements, in addition to geographic coordinates and sample number.

#	LAT	LONG	SAMPLE	Au	Sb	As	Th	W
1	44.8900	-62.4621	'771002'	4	0.6	60.1	5.3	1.1
2	44.8766	-62.4523	'771003'	9	0.3	18.8	3.4	1.9
3	44.8704	-62.4308	'771004'	7	0.2	8.4	2.9	2.4
4	44.8895	-62.4388	'771005'	5	0.2	10.1	3.1	1.0
5	44.9037	-62.4179	'771007'	5	0.8	22.8	5.0	1.0
6	44.9041	-62.4118	'771008'	5	0.8	10.2	4.2	1.2
7	44.9090	-62.4412	'771009'	5	0.3	5.5	4.3	1.0
8	44.9311	-62.4095	'771010'	5	0.2	7.5	2.4	0.8
9	44.9353	-62.4188	'771011'	5	0.2	5.7	3.7	1.0
10	44.9420	-62.4247	'771013'	4	0.2	9.6	6.4	0.8
.	.	.	. .					
.	.	.	. .					
434	45.2601	-61.3146	'771106'	5	0.3	11.5	1.7	1.3
435	45.2692	-61.2906	'771107'	6	0.3	10.5	3.6	1.2
436	45.2775	-61.2488	'771108'	3	0.1	10.3	2.5	1.5

Table 7-5. Elements used in constructing a classification table for Au in lake sediment, defining the intervals for making an Au map with breakpoints. Columns 1-4 show alternative methods of defining breakpoints. Breakpoints may be negative and/or decimal numbers. The actual table stored digitally does not require all these fields, and is abbreviated in a number of ways specific to particular GIS.

1	2	3	4	Ascending	Descending	Legend
3 (min)	-	3	-			
4	4	4	4	1	7	'3 - 3.99'
5	5	5	5	2	6	'4 - 4.99'
7	7	7	7	3	5	'5 - 6.99'
9	9	9	9	4	4	'7 - 8.99'
18	18	18	18	5	3	'9 - 17.99'
21	21	21	21	6	2	'18 - 20.99'
266(max)	266	-	-	7	1	'21 - 266'

Field 1 - Breakpoints defined for boundaries of each class
Field 2 - First class open ended, used for all Au values < 4
Field 3 - Last class open ended, used for all Au values >= 21
Field 4 - First and last classes open-ended
Field 5 -Ascending - output class values are in ascending order
Field 6 -Descending - output class values are in descending order
Field 7 -Legend- description

Take an actual reclassification of basin number 1. The input class is 1, corresponding to a Au value of 4 ppb in the first row of the **BASIN** table, which lies between breakpoints 4 and 5, so the output class is either 2 or 6, depending on the choice of ascending or descending output classes. On the other hand, basin 435 (Au=6 ppb) is classified as 3 (ascending classes) or 5 (descending classes). Classification tables can be saved for multiple use. The Au classification table could also be applied for defining the colour/size/type of symbol for a symbol plot of Au, or for defining the intervals on a contour map produced by an interpolation method. The number of breakpoints will depend on several factors such as the number of objects on the map, the number of possible colours that can be displayed simultaneously and the desired simplicity of the output maps. Often geochemical data are displayed on maps with between 6 and 20 classes. Geophysical data usually require a large range, and a full 8-bit range of 255 classes are often employed. By using a large number of intervals, subtle variations in the data are retained. As long as the original attribute table is kept, the data are not lost and further calculations can be based on the original data, not just on the class values. Note that nonlinear intervals between breakpoints result in the transformation of interval or ratio data to an ordinal scale in terms of the original measured quantities.

Sometimes a histogram view of the attribute to be classified can help in deciding where to make the breakpoints for the classification table. By interactively changing cutoff values on the histogram, (or cumulative frequency plot) with simultaneous display of the map view, the breakpoints can be optimized to maximize the spatial information. Classification lookup tables can be created in a variety of automatic ways. One common one is to use an "**equalization stretch**". This places the breakpoints to make the areas of classes on the map or image as close to equal as possible. A histogram of the resulting map classes will be approximately uniform, with the frequency of each class roughly equal. Another kind of automatic stretch is to make a linear transformation between the input values and the output values, preserving the original histogram shape. Alternatively, the histogram of the output can be made to approximate a Gaussian curve. There are a number of other methods not described here that produce enhancements of the data with lookup tables, as described in books on image processing.

OPERATIONS ON ATTRIBUTE TABLES

Two kinds of table operations are particularly important on GIS attribute tables. The first is the addition of "new" fields (columns) derived by operations involving a model on the "old" fields. The second is the statistical aggregation of selected fields over subsets of records or rows. An example of this last operation might be to find the average Zn value in catchment basins with less than 5 ppb Au, or to count the number of basins where Au + Ag is greater than 15 ppb. Although these operations can be carried out within a particular GIS using a modelling language, it is often convenient to export the tables to external packages (such as spreadsheets, or statistical programs), to take advantage of functionality that may be superior or easier to use, see Figure 7-5.

Adding new fields

The map modelling procedures discussed in the previous section produced maps showing "new" spatial objects by reclassification without saving the results in the field of an attribute table. However, the process can be broken down into 2 separate steps: the first step involves making the modelling calculations and saving the results as a new attribute field, and the second step requires the use of the new field as a lookup, possibly with the additional application of a classification table if the results are unsuitable as class values (such as being negative or decimal numbers). Thus map modelling and **table modelling** are closely related.

Arithmetic and logical operations carried out on one or more attribute columns to produce new attribute columns provide a powerful means of modelling with GIS data. The degree with which a GIS user is aware of the details of this operation varies depending on the user interface, but it is helpful to have some understanding of the mechanics of the operation because it gives a better idea of what may or may not be possible.

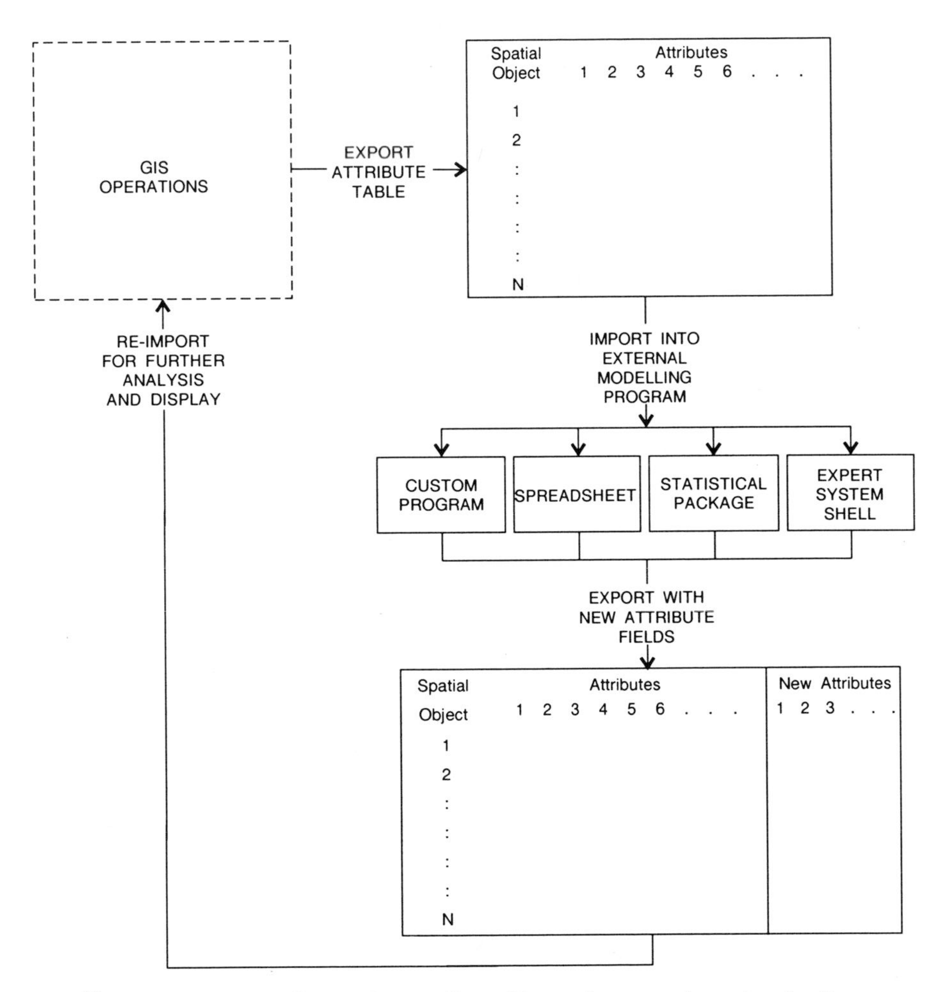

FIG. 7-5. Diagram to illustrate how attribute tables can be exported to external software packages for analysis and modelling, being returned to the GIS after adding new attributes, ready for further analysis and display.

Operations on data tables not only produce new fields, they can also be used for graphical display of histograms, scatterplots, boxplots, and for statistical summary calculations. Consider some simple examples of attribute table operations, using a table that contains geochemical point attributes, or equivalently, a polygon attribute table keyed to Voronoi polygons generated from the points.

1) Generate a new field based on logic operations. This is like an interactive spatial query, except that the result is assigned to a new attribute field. An example might be, generate and append a new field to the table, with the value 1 if As/Sb > 3 and Si > 30%, otherwise with the value 0. Display of these points or Voronoi polygons, symbolized according to the new field, shows where these conditions are satisfied in physical (geographic) space.

2) Plot the values of two attributes, Cd and Zn, on a scatterplot. Fit a line model to the data by least squares. Calculate and append to the table two new attributes based on the model: the predicted value of Zn, given the Cd value and the residual Zn value, equal to the observed value minus the predicted value. These expected and residual values can be mapped by reclassification of the Voronoi polygons (or simply the points), possibly revealing unexpected spatial pattern, as suggested in the example in Figure 7-1.

3) Identify anomalous objects on the scatterplot ("feature space") in 2). Code these objects in a new binary attribute field (anomaly present or absent). Reclassification of the polygons or points on the map according to this field enables the "feature-space" groupings to be evaluated in physical space.

4) Take the attribute table into a multivariate statistical package, opening the possibility of a wide variety of statistical modelling procedures. Some of the many operations that might be carried out that produce new attribute fields that can be appended to the table are: multiple linear and logistic regression analysis, discriminant analysis, principal components analysis, cluster analysis, decision tree analysis, neural network analysis, and analysis with expert system software.

As we shall see in Chapter 9, these same methods can be applied to attribute tables resulting from the overlay of multiple maps, the fields in the table being the classes of the input maps, and the spatial objects being the polygons produced by the overlay operation. In this way, multi-map combinations can be reduced to a single, but complex, combination map, for which the attribute table contains all the data from the individual map components, and the problem is reduced from a multiple map analysis to an operation on a single map and an associated attribute table.

SPATIAL, TOPOLOGICAL, AND GEOMETRICAL MODELLING

In the cases described above, spatial and topological information was not used for modelling. However, spatial (including geometric) and topological attributes can readily be added to the analysis, although the spatial objects must be treated as individual entities (unless they are pixels in a raster) rather than being grouped into classes. In a raster model, groups of pixels can belong to an irregular polygon, and can either be treated at a pixel level, or at a polygon level. For example, if the easting and northing of the geometric centroids of polygon objects are appended to the polygon attribute table, they can be used in models such as polynomial trend surface analysis. Similarly, topological information such as the attributes of adjacent polygons, or geometric information, such as the shape or size of individual polygons, can be added as fields to the polygon attribute table. The attribute table to be used for modelling in such cases must refer to individual spatial objects, such as polygons, not to classes of polygons, because spatial variables are not sufficiently specific if they refer to general map classes. For example, the geometric centroid of a class called "granite", comprising many non-adjacent polygons, would probably have little physical meaning. On the other hand, the centroid coordinates (or perimeter length, or whether the granite is in contact with shale, etc.) could be extremely important attributes of individual granite **polygons**, but are not relevant to the collection of polygons that make up the **class** called granite. Similarly, a measure such as the average squared difference of a variable between a polygon and its adjacent neighbours, weighted by the perimeter length of the adjacent boundary arcs, only makes sense when applied to individual polygon entities. Measures of this type can be used to express autocorrelation characteristics, and have been used as independent variables in logistic models of forest fire prediction (Chou et al.,1990).

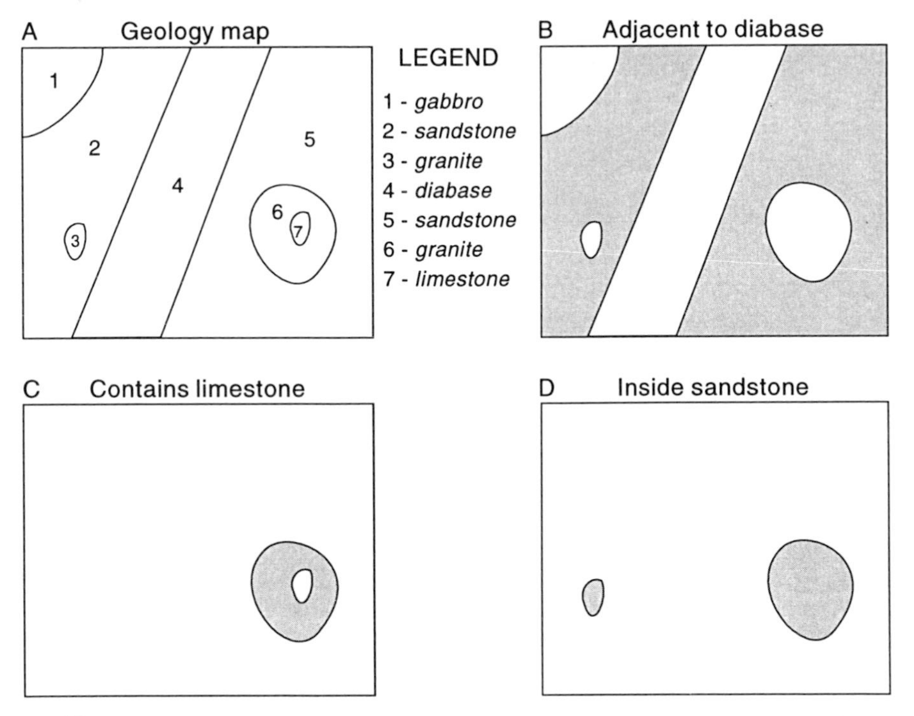

FIG. 7-6. Reclassification of a geological map using topological operations. **A.** Map with individually labelled polygons. Each polygon is a separate class. **B.** Polygons **adjacent to** diabase polygons. **C.** Polygons **adjacent to and containing** limestone polygons. **D.** Polygons **contained by** sandstone polygons.

An example of modelling using topological attributes of polygons is illustrated in Figure 7-6. This is a reclassification of a geological map using individual polygon entities (Figure 7-6A) whose topological characteristics are known. In the first case, Figure 7-6B, polygons are assigned to class 1 if they are **adjacent to** polygons belonging to the class "diabase", else they are assigned to class 0. In the second case, Figure 7-6C, the selection is based upon a containment criterion, whether the polygon **contains *and* is adjacent to** polygons classified as limestone. In the final case, Figure 7-6D, a "contained by" criterion is used, in which polygons that are **inside** polygons classified as sandstone are selected.

Topological modelling can also be applied to line objects. For example, topological attributes are needed to select all the contacts between felsic intrusions and map units that contain more than 50% carbonate sediments. Such an operation may be precursor to the production of a proximity map showing distance to the selected contact. Attributes of individual line objects are also needed for selecting lines that have particular geometric characteristics, such as orientation, length, or shape. In modelling mineral potential, for example, it is often necessary to select lineaments and other lines that have particular orientations, in order to generate maps that can be used as evidence of mineralization.

An example of applying geometrical attributes of polygon entities for modelling is illustrated in Figure 7-7. Here, a map of lakes in part of western Quebec (Figure 7-7A) has been classified to illustrate variations in lake shape. The area in km^2, A, and perimeter length in km, P, were calculated for each of the 38 lake polygons, as shown in Table 7-6. In addition, a new field was calculated and added to the table (by table modelling) to express the "thinness ratio", defined as $4\pi A/P^2$, one of the many shape factors that have been used in geology (see Davis, 1986). The map of lakes was then reclassified by the thinness ratio in the polygon attribute table, using a table of classification breakpoints, producing the map shown in Figure 7-7B. A wide variety of shape measurements are possible, either using polygon boundaries in vector mode, or applying mathematical morphology operators in raster mode, for this kind of geometrical modelling.

Methods that assume that individual data records are an independent random sample from some population run into difficulties where statistical inference is concerned, because spatial objects are often spatially autocorrelated to some degree. Objects close together geographically tend to be more similar to one another than those far apart, so that neighbouring spatial objects are *not* independent, yet the idea of independent random samples is fundamental to most statistical hypothesis testing and modelling. A common approach is to apply these methods ignoring the autocorrelation problem, treating the results in a descriptive and exploratory sense rather than using them for statistical inference. There are also methods of modifying statistical models to account explicitly for spatial autocorrelation effects, as described for example by Griffith (1988), and Monte Carlo methods that avoid the usual assumptions of statistical models, but they are beyond the scope of this book.

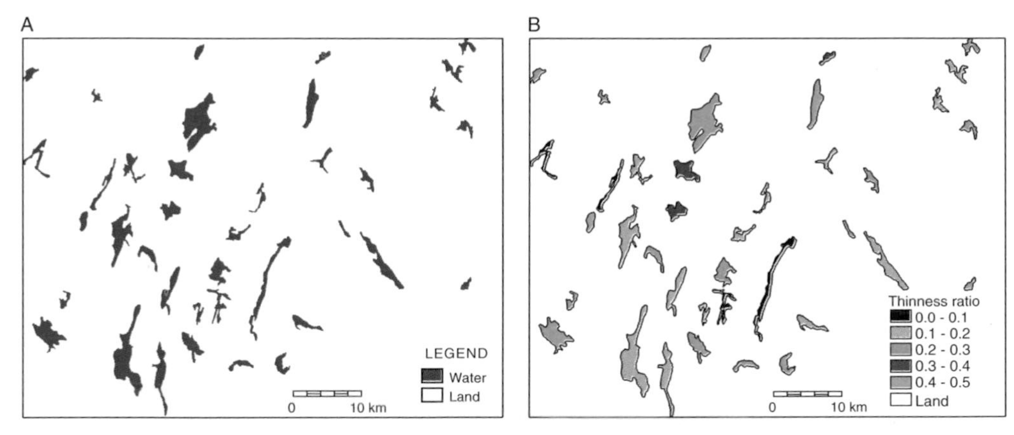

FIG. 7-7. Reclassification of map of lakes on the basis of lake shape. **A.** Original map of lakes from an area in west Quebec. **B.** Lakes classified according to the thinness ratio. Each lake polygon is treated as a separate spatial object with unique geometrical attributes, such as area, perimeter, and shape.

Table 7-6. Attribute table for lake polygons. The keyfield is the polygon number, unique for each lake. The attributes are the area, perimeter, and thinness ratio (TRATIO), used as a shape parameter. The thinness ratio is calculated from the area and perimeter fields.

#	AREA	PERIM	TRATIO
1	10.5	26.0	0.19
2	5.9	19.0	0.21
3	2.1	11.8	0.19
4	2.7	9.0	0.41
5	20.4	43.5	0.13
6	6.4	23.0	0.15
7	2.2	14.4	0.13
8	4.6	30.0	0.06
.	.	.	.
33	2.0	9.0	0.31
34	3.0	16.4	0.14
35	1.5	7.2	0.36
36	2.7	12.8	0.21
37	3.8	18.9	0.13
38	2.7	13.9	0.18

AGGREGATION OPERATIONS ON TABLES

Consider the attribute table in Table 7-7A, describing the depth to bedrock, the bedrock type and the surficial geology class over 10 polygons. The question might be: "What is the average depth to bedrock for each bedrock type?". If the differences between bedrock types are large, it may suggest that the thickness of overburden is in some way related to the bedrock geology. It might also be useful to take the average depth to bedrock and use it as an attribute of a bedrock geology map, so that the geological units could be classified and displayed according to this derived variable.

Table 7-7A. Attribute table for surficial geology map

Polygon id	Depth to bedrock	Bedrock type	Surficial class
1	10	A	I
2	50	B	I
3	30	B	II
4	30	B	I
5	18	C	I
6	26	C	II
7	22	C	II
8	10	A	II
9	11	A	III
10	3	A	III

B. Aggregation by bedrock type. The surficial class is the mode.

Depth to bedrock							Bed-rock type	Surf-icial class
Count	Total	Mean	S.Dev	Min	Max	Range		
4	34	8.5	3.7	3	11	8	A	III
3	110	36.7	11.5	30	50	20	B	I
3	66	22.0	4.0	18	26	8	C	II

Aggregation operations can be used to calculate a variety of summary statistics. Here, in Table 7-7B, we have calculated not only the mean depth by bedrock type, but also the number of polygons, the total depth, the standard deviation, the minimum, maximum and range in depth for each of the three bedrock types. In addition the modal surficial class has been determined for each bedrock type. Note that if the polygon map were to be reclassified by bedrock type, the new aggregated table could be used as its attribute table, using the bedrock type as the keyfield.

The extreme case of aggregation is where the operation is carried out over the whole table or map, without subsets, to give global statistics. Where the table to be aggregated contains a field with polygon area, the summation operation produces area statistics. This is an "**area analysis**" operation and is usually a specific task that can be carried out on any polygon attribute table, or directly on a map by map class. Such information might be generated for a geological map, for example, providing a valuable descriptive summary of the relative frequency of each map unit, as shown in Table 7-8.

Aggregation operations are standard tasks for statistical and spreadsheet packages, so that if the GIS does not provide this function internally, it can be readily carried out externally on exported attribute tables.

Table 7-8. Table of areas for each class of a reclassified geological map of part of Vancouver Island and adjoining mainland. The original map is shown in Figure 7-3A, but this reclassification differs from those illustrated in Figure 7-3B to D.

Class[1]	Legend[2]	Area, km^2	Area, %	Cumulative Area, %
2	Mafic igneous	3945.9	34.34	34.34
3	Intermediate igneous	5673.3	49.37	83.70
4	Felsic igneous	431.4	3.75	87.46
6	Argillaceous seds.	824.3	7.17	94.63
9	Limestone	406.7	3.54	98.17
11	Conglomerate	108.5	0.94	99.11
13	Unclassified	101.8	0.89	100.00
TOTAL		11491.9		100.00

[1] **Note that map class is the keyfield. It need only contain values of map classes with nonzero areas.**
[2] **This text field has been added for explanatory purposes.**

We now turn to a group of operations that specifically use information about spatial neighbourhoods. Some of these operations are similar to aggregation operations, where summary statistics are calculated over small spatial subsets (moving windows) of the map.

OPERATIONS ON SPATIAL NEIGHBOURHOODS

Using the raster model there are a class of operations that specifically deal with pixel neighbourhoods. In a raster, the data structure ensures that a zone of spatially adjacent pixels are close together in the raster file, and the addressing system by row and column makes their access straightforward. Neighbourhood operations on raster images are generally called filtering operations. Filtering of an image generates a new image in which some feature or group of features is enhanced and characterized. Filters can be used to smooth data, removing local "high frequency" variation, to enhance edges or directional features, such as lineaments, to characterize texture by the scale of local roughness, to derive shape information such as slopes, to calculate the illumination over a surface given the zenith and azimuth of a light source, and to carry out other operations.

Filtering operations were originally developed for analysis of time series data. A short digression to examine filtering of time series helps in the understanding of filtering of maps. Readers are referred to Davis (1986) for a clear introduction to the subject.Take the case of a time series of water depth in a well, recorded daily over tens of years, producing a curve of depth versus elapsed time. By calculating a moving average, over a "window" of three (or more) successive points, the curve can be smoothed and generalized; the local variation is subdued, enhancing the longer term trends. The smoothing operation filters out the high frequencies, leaving the low frequencies, and is called a "low-pass" filter.If the smooth curve is subtracted from the original curve, the residual is the high frequency information, and a filter that produces this in one step from the original is a "high-pass" filter.There are also "band-pass" filters that will isolate particular frequency intervals.These filters, which are a series of weights applied to the values within the moving window, operate in what is known as the "time domain". By changing the size of the window and the weighting of points within the window, different components of the time series can be enhanced or removed. However, instead of operating within the time domain, the curve can be decomposed by analysis in the "frequency domain". In fact, a time series can be broken down into a series of sinusoidal waveforms of varying wavelength and amplitude, that when added together again approximate the original.These waveforms are expressed mathematically as a harmonic series, consisting of cosine and sine terms as follows:

$$f(t) = a_0 + \sum_{n=1}^{k} a_n \cos n\omega t + \sum_{n=1}^{k} b_n \sin n\omega t \quad , \qquad (7\text{-}1)$$

where $f(t)$ is the water depth as a function of time, t, ω is the frequency equal to $2\pi/T$, and T is the length of the time series. The term a_0 represents the mean water depth and the summation terms are contributions of a set of sine and cosine terms of increasing frequency. The coefficients a_n and b_n are estimated by a fitting procedure, usually a Fast Fourier Transform (FFT) algorithm, maximizing the fit of the series to the observed data. The amplitude of the nth term, or harmonic, is equal to

$$A_n = \sqrt{a_n^2 + b_n^2} \quad , \tag{7-2}$$

and the phase angle or displacement of the first crest of the sinusoid for the nth harmonic is

$$\Theta = \tan^{-1}\frac{b_n}{a_n} \quad . \tag{7-3}$$

A time series can be decomposed, or filtered, in the "frequency domain", producing the amplitude (or "power") spectrum, which summarizes the relative contributions of the different frequencies in the data. The harmonic model, with estimated coefficients, may be used to produce a synthetic or predicted time series curve, either including all frequencies, or restricting the frequencies to a particular interval. The synthetic curve is, therefore, a filtered version of the original, produced by processing in the frequency domain.

We now turn to the application of filters to spatial data. A profile across a continuous surface is exactly equivalent to a time series, except that distance is substituted for time. In extending the model to maps, the single spatial dimension is replaced by two spatial dimensions. Filtering can either be carried out in the frequency domain as before, or in the **spatial** domain instead of the time domain. Image data are often filtered in the frequency domain, using a FFT with two spatial variables, the result being a two-dimensional power spectrum. Discussion of filtering in the frequency domain is beyond the scope of this book, and readers are referred to Fuller (1966), Davis (1986) or Robinson (1982) for geological treatments of the subject.

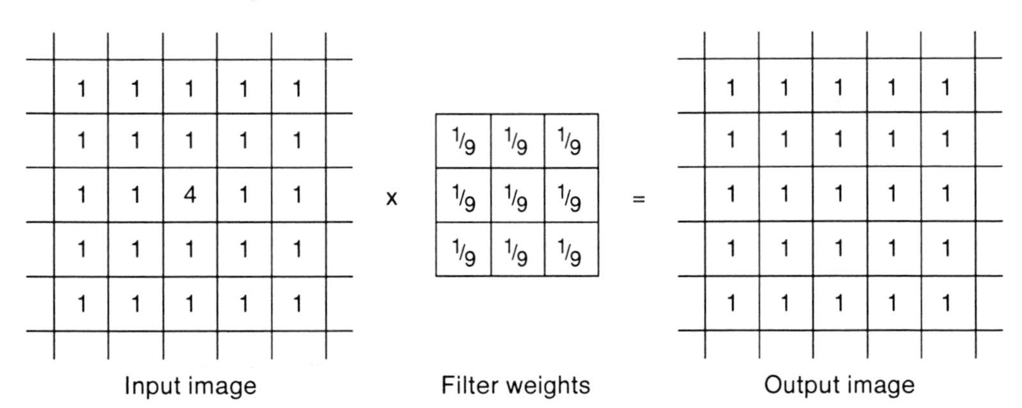

FIG. 7-8. Filtering of a raster image by convolving with a square kernel. In this case the filter is "low-pass" because the high frequency information is removed by smoothing. The values in the output image are actually 12/9=1.33, but are rounded down to the nearest integer by truncation. Thus in this case, smoothing entirely removes the centre spike.

On the other hand, filtering of images in the *spatial domain* with moving windows is a basic function found in image analysis systems and raster GIS. The weights in a rectangular moving window (also called a "kernel", "template" or simply "filter") are applied in a summation operation to each pixel and its neighbours, transforming the input image to a filtered output image. In the simplest case the kernel is square, with an odd number of pixels on each side to ensure symmetry about the central pixel, as shown in Figure 7-8. The kernel is applied to the image by a process known as "convolution", an operation in which the central pixel in the output image is determined as a sum of the product of kernel values multiplied by the corresponding input image values. At a single pixel at location (x,y), the convolution process is

$$\sum_{i=-m}^{m} \sum_{j=-n}^{n} f(x+i, y+j)\, g(i,j) \quad , \tag{7-4}$$

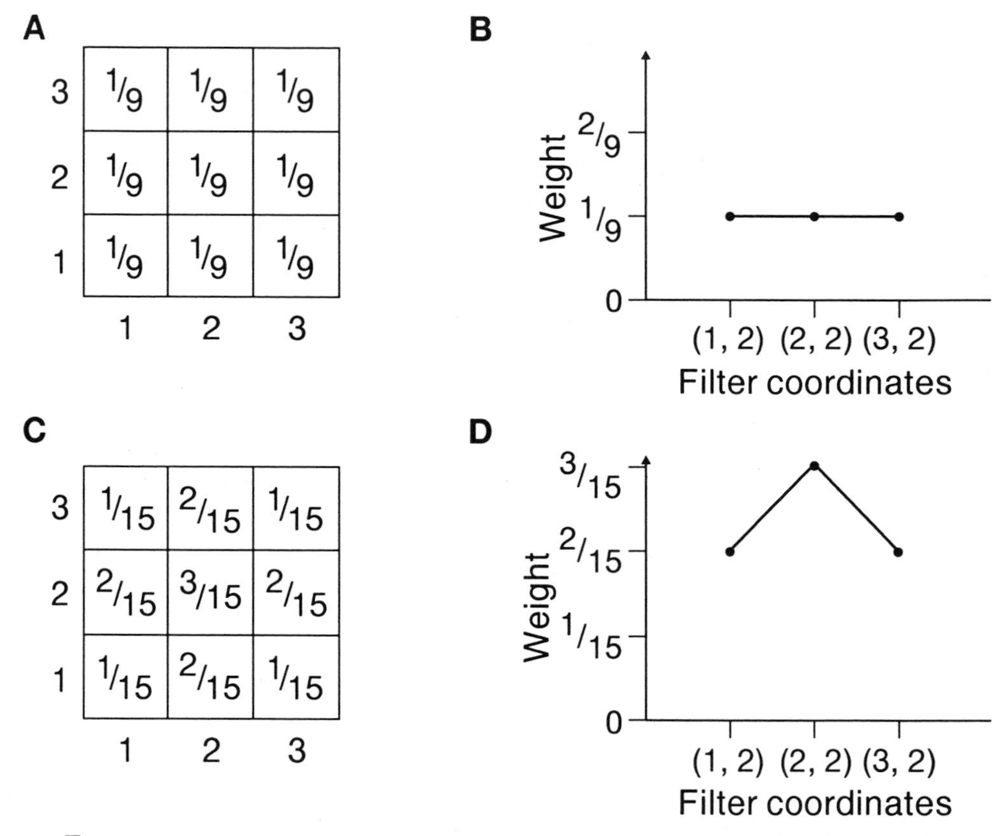

FIG. 7-9. A. A (3x3) kernel for smoothing. **B.** Profile through the weight kernel. **C.** Another (3x3) smoothing kernel and its profile (**D**). The kernel in C produces a "softer" response than the kernel in A.

where there are $2m+1$ pixels across the kernel, $2n+1$ pixels down the kernel, f is the value of the input image and g is the kernel weight.

Smoothing or low-pass filters

One of the simplest filters takes the mean of the values over the spatial neighbourhood of the kernel, see Figures 7-8 and 7-9. Here the input values are integers and the output values are also integers, with the decimal values lost by truncation. Alternatively, the output values can be converted with a classification lookup table, re-scaling the output to integers. This is a "low-pass" filter, because it smooths the data, removing the high-frequency variation and allowing the low-frequency information to "pass" through. If one row of the kernel values are plotted, as shown in Figure 7-9B, the shape of the weighting function is seen to be rectangular. A "better" smoothing kernel has a weighting function shown in Figure 7-9C and D. The output is no longer the mean value over a neighbourhood, but a weighted mean value, and produces a "softer" response in the output. Notice that the weights are normalized to sum to 1, otherwise the mean value of the output image differs from the mean value of the input image.

Because the convolution process involves arithmetic operations, the mean filter is not suitable for ordinal and nominal scale data. However, a **median** filter can be applied to ordinal (as well as interval and ratio) data, in which the output pixel value is simply the middle value, or 50th percentile, of the pixels within the neighbourhood. Similarly, a **modal** filter can be used with categorical data, where the output is the most frequent value over the neighbourhood. Modal filters are sometimes used for cleaning up noisy images, where isolated pixels are different from neighbouring regions, as often is the case with LANDSAT images that have been classified. Note that median and modal filters are not convolution operations, because Equation (7-4) does not apply.

Smoothing filters can also be constrained so that the central pixel is only replaced by the filtered value if the difference between the original and filtered value is greater than some threshold. This allows extreme values to be removed without changing intermediate values.

High-pass filters for edge detection

High-pass filters use kernels that allow high frequency information to pass through, removing the low-frequency information. They thus have the effect of enhancing local variation and accentuating areas of rapid change, or edges. One way of carrying out this operation is to subtract a low-pass (smoothed) image from the original. An "edge enhanced" image is the original image to which the high-pass image has been added. As the high-pass image is defined as the original minus the low-pass image, the edge enhanced image is thus the original image multiplied by 2, minus the low-pass image. This operation can be very useful with Landsat imagery for accentuating features related to rapid change such as geological contacts or lake boundaries. A common edge enhancement filter is the "Laplacian", which for a (3x3) kernel has the weights.

0	1	0
1	-4	1
0	1	0

After convolving the image with this kernel, the effect is to accentuate any differences between the central pixel and the average of its four adjacent neighbours.

Directional filters

Some of the high-frequency information on images is directional, like lineaments on satellite data. Directional kernels can be used to enhance local image differences (high-frequency). They can also be used to calculate the local **slope** magnitude and direction of maximum slope, also called **aspect**. Slope and aspect are equivalent to the stratigraphic angle and direction of dip of bedded strata. Two simple (3x3) kernels for computing slope and aspect are

-1	0	1
-2	0	2
-1	0	1

and

-1	-2	-1
0	0	0
1	2	1

producing two output images, the first being the slope component in the E-W (x) direction, the second the slope component in the N-S (*y*) direction. In other words, these are the local first derivatives of the surface, $\partial z/\partial x$, and $\partial z/\partial y$, respectively. The slope components can be combined to produce slope and aspect:

$$slope = \sqrt{\left(\frac{\partial z}{\partial x}\right)^2 + \left(\frac{\partial z}{\partial y}\right)^2} \quad and \qquad (7\text{-}5)$$

$$aspect = \mathrm{arc}^{-1}\left(\frac{\partial z/\partial y}{\partial z/\partial x}\right) \quad . \qquad (7\text{-}6)$$

Usually the derivative filters are applied after an initial smoothing of the data, otherwise high frequency noise is accentuated.

Hill shading

The slope and aspect of a surface can be applied for calculating the illumination of a surface by an infinitely distant light source (like the sun). The differences in illumination produce a shading effect that is often very effective for enhancing textural, shape and directional features, as illustrated in Chapter 4. Given the azimuth angle, α, and zenith angle, β, of the light source (angle from the vertical), the illumination is the cosine of the incidence angle, γ, as illustrated in Figure 7-10, and can be calculated from the formula

$$\cos(\gamma) = \cos(slope) * \cos(\alpha) + \sin(slope) * \sin(\alpha) * \cos(\alpha - aspect) \quad . \tag{7-7}$$

Figure 7-11A shows an image of topographic elevation (DEM) for the state of Nevada. In Figure 7-11B the same elevation information has been enhanced by an illumination operation, using a light source at an azimuth of 280° and zenith angle of 45°. The image reveals information about geological structures that are not readily detected on the original.

A variation on the illumination method has been applied by Jupp et al. (1985), as reported in Harrison and Jupp (1990) for calculating "exposure" of underwater reefs to the prevailing wind direction, but it could also be used to produce similar results with any digital elevation surface. Essentially their method calculates exposure, E, from the first derivatives of the surface using the relation

$$E = -\left(\frac{\partial z}{\partial x}\sin\theta + \frac{\partial z}{\partial y}\cos\theta\right) \quad , \tag{7-8}$$

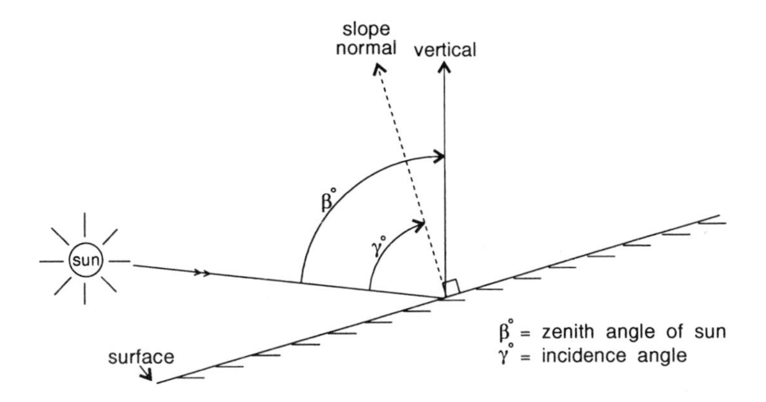

FIG. 7-10. Diagram to show the relationship of surface slope angle, the normal to the slope, the zenith angle of the sun and the incidence angle. Because the sun is distant, the suns rays are parallel. The sun azimuth is not shown, because the view is a profile.

where θ is now the azimuth of the wind direction. The effect is similar to illumination with a light source on the horizon, i.e. with a zenith angle of 90°.

Harrison and Jupp also report on work by Smith (1983), who applied second derivative filters to a surface, and used the resulting curvature characteristics to identify and extract various surface morphology categories that could be described in terms of basins, hills, saddles, gullies and ridges. At each pixel, they calculated four second partial derivatives using appropriate kernels. These four partial derivatives form at each pixel a (2x2) matrix as follows:

$$\begin{bmatrix} \frac{\partial^2 z}{\partial x^2} & \frac{\partial^2 z}{\partial x \partial y} \\ \frac{\partial^2 z}{\partial y \partial x} & \frac{\partial^2 z}{\partial y^2} \end{bmatrix} \tag{7-9}$$

from which two eigenvalues, λ_1 and λ_2, were calculated by matrix methods. These eigenvalues, plus two further quantities derived from them called the Laplacian, $(\lambda_1 + \lambda_2)/2$ (not to be confused with the Laplacian edge kernel), and the discriminant, $(\lambda_1 - \lambda_2)/2$, were used to discriminate between curvature types as shown in Table 7-9.

Texture filters

Many neighbourhood operators have been proposed for characterizing the texture of a surface, e.g. Haralick (1979). Textural features can be important for distinguishing surface types based on roughness. Often images of remotely-sensed or airborne geophysical images

Table 7-9. Surface curvature parameters used by Smith (1983) to distinguish between surface morphological features, as reported by Jupp and Harrison (1990).

	λ_1	λ_2	**Laplacian**	**Discriminant**
Flat ground	~0	~0	~0	small
Basin	>>0	>>0	>>0	small
Hill	<<0	<<0	<<0	small
Saddle	>>0	<<0	~0	large
Gully[1]	>>0	~0	>>0	large
Ridge	~0	<<0	<<0	large

[1]Channel

have textural characteristics that vary with the underlying geology, whereas the intensity values themselves may show little contrast between geological lithologies or domains. One of the simplest textural measures is the local variance, equal to

$$Variance = \sum_{j=1}^{m} \sum_{i=1}^{n} \frac{(z_{ij} - \bar{z})^2}{nm} \tag{7-10}$$

When the local variance is zero, the roughness is also zero; as the variance increases, so does the roughness.

For a measure of texture that can be used on any type of measurement scale, the local entropy, or information statistic, defined as:

$$Entropy = - \sum_{i=1}^{k} (p_i \log p_i) \tag{7-11}$$

Table 7- 10. Calculation of entropy as a measure of texture on a (3x3) window. Notice that entropy is minus the sum.

A

1	7	2
3	4	2
1	2	4

i	p_i	$p_i \log p_i$
1	2/9	-0.1452
2	3/9	-0.1590
3	1/9	-0.1060
4	2/9	-0.6532
7	1/9	-0.1060
Sum	1	-0.6614

B

1	2	1
1	1	1
2	1	1

i	p_i	$p_i \log p_i$
1	7/9	-0.0849
2	2/9	-0.1452
Sum	1	-0.2301

can be used, where p_i is the proportion of the cells in the neighbourhood with value i, and k is the number of pixels in the neighbourhood. Table 7-10A shows a (3x3) neighbourhood, with the associated values of i and p_i, leading to an entropy value of about 0.66. In Table 7-10B, a similar calculation is shown where the entropy is 0.23. As the entropy increases, the degree of roughness also increases.

An even simpler measure that can also be used with data on any measurement scale is the diversity, which is simply the number of different pixel values, or class values, within the neighbourhood, either as an integer count or as a proportion of the total number possible. Diversity values for the data in Table 7-10 are 5 and 2, for A and B respectively. An alternative is to count the number of boundaries between adjacent pixels that have different values within the neighbourhood.The numbers of boundaries for the same data are 11 and 5, respectively.

JOIN-COUNT STATISTICS

One of the important operations on single maps that generates table output rather than a new map is the calculation of autocorrelation statistics. A number of measures of spatial autocorrelation are used with area objects on maps. In Chapter 6, autocorrelation statistics for

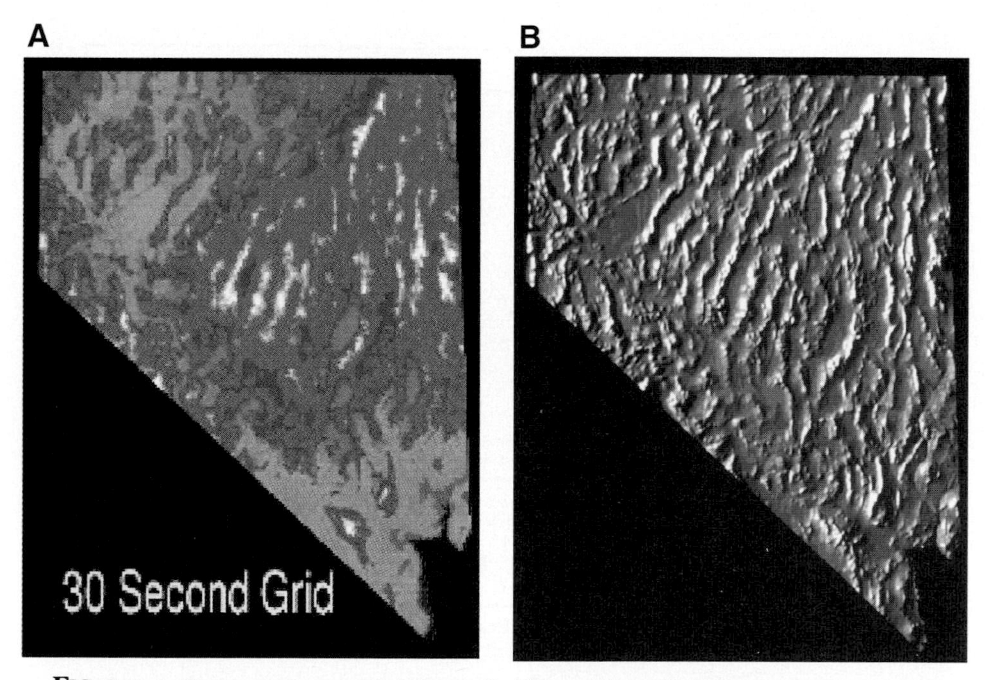

FIG. 7-11. A. Digital elevation model of Nevada. **B.** After applying artificial illumination from the NW at a zenith angle of 45°. Diagram supplied by Mark Mihalasky, University of Ottawa.

point sample data were introduced. It is also sometimes useful to characterize spatial autocorrelation on polygon maps. Recognition of the existence and strength of autocorrelation is useful descriptively, for helping to understand spatial pattern, and for prediction.

In raster mode, the data on maps can be treated as points on a lattice, and the geostatistical methods of variograms, autocovariance and autocorrelation functions can be applied. For example, Fabbri (1984) and Agterberg and Fabbri (1978) demonstrate how autocorrelation and autocovariance functions can be calculated using mathematical morphology operators on geological map patterns. In vector mode, some other measures have been used to summarize the autocorrelation behaviour of maps, although these methods have not been widely applied in geology. Two approaches, one involving **join-count statistics**, suitable for nominal scale data, the other involving **Geary** and **Moran** indices are applied to interval and ratio data. Much of the following treatment is based on the discussion in Goodchild (1986), see also Griffith and Amrhein (1991).

Join-count statistics, as the name implies, measure the number and type of joins between adjacent polygons on a map. The frequency with which the various join types occur yields information about the clustering and spatial autocorrelation present on a polygon map with binary attributes. Suppose that two map patterns are compared, Figure 7-12. In pattern A the shaded regions are clustered together spatially, whereas in pattern B the shaded regions are not spatially adjacent, although the non-shaded regions are in some cases adjacent to one another. Call the

Table 7-11. Contiguity or weight matrix, **W**, for the polygon maps in Figure 7-12. The value of w_{ij} records the adjacency of polygon *i* with polygon *j* as 1, 0 otherwise. Zero values are put in the principal diagonal.

		Polygon												
		1	2	3	4	5	6	7	8	9	10	11	12	13
Polygon	1	0	1	0	0	1	1	0	0	0	0	0	0	0
	2	1	0	1	0	0	1	1	0	0	0	0	0	0
	3	0	1	0	1	0	0	1	1	1	0	0	0	0
	4	0	0	1	0	0	0	0	0	1	0	0	0	0
	5	1	0	0	0	0	1	0	0	0	1	1	0	0
	6	1	1	0	0	1	0	1	0	0	0	1	0	0
	7	0	1	1	0	0	1	0	1	0	0	1	1	0
	8	0	0	1	0	0	0	1	0	1	0	0	1	0
	9	0	0	1	1	0	0	0	1	0	0	0	1	1
	10	0	0	0	0	1	0	0	0	0	0	1	0	0
	11	0	0	0	0	1	1	1	0	0	1	0	1	1
	12	0	0	0	0	0	0	1	1	1	0	1	0	1
	13	0	0	0	0	0	0	0	0	1	0	1	1	0

shaded polygons black, B, and the non-shaded polygons white, W. Then the contacts between polygons fall into 3 groups: BB or black-black joins, WW or white-white joins and BW or black-white joins. The frequency of observed joins of each type can be counted from each

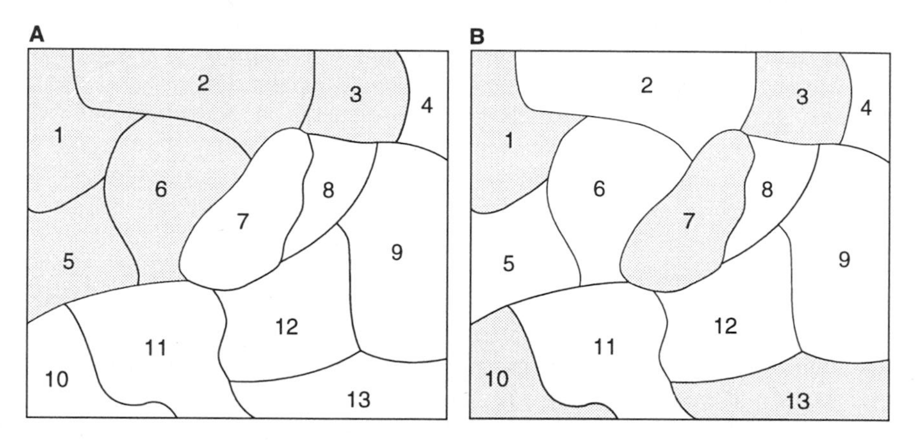

FIG. 7-12. Map composed of 13 polygons to illustrate join-count statistics where the attributes are binary. Pattern A is positively autocorrelated, B is negatively autocorrelated.

Table 7-12. Join-count statistics for the patterns A and B in Figure 7-12. The expected join counts are the same for both patterns. Note that A has fewer black-white (BW) joins than expected, B has more than expected, indicating positive autocorrelation for A, negative for B.

	A, Observed	B, Observed	Expected[1] (either)
BB	6	1	3.46
BW	9	17	13.85
WW	12	9	9.69
J	27	27	27.00
n_B	5	5	
n_W	8	8	
n	13	13	

[1] From Equations 7-12, 7-13, and 7-14.

pattern and compared. In addition, the expected frequencies can also be calculated, allowing the observed frequencies to be compared with those expected from a polygon map with the same frequency of black and white polygons, organized at random. The expected values are derived by considering all possible permutations of the polygons, Griffith and Amrhein (1991). The expected values are:

$$E(BB) = \frac{J\, n_B\, (n_B - 1)}{n(n-1)} \quad , \tag{7-12}$$

$$E(WW) = \frac{J\, n_W\, (n_W - 1)}{n(n-1)} \quad , \quad and \tag{7-13}$$

$$E(BW) = \frac{2J\, n_W\, n_B}{n(n-1)} \quad , \tag{7-14}$$

where n_B is the number of black polygons, n_W is the number of white polygons and $n = n_B + n_W$. J is the total number of joins, equal to (BB + WW + BW). The total number of joins also equals one half of the 1s in the **contiguity** or **weight** matrix, **W**, as shown in Table 7-12. The elements of **W** are 1 for adjacent polygons, 0 otherwise. This matrix can be based on other measures of contiguity, such as boundary length, or on measures of distance, using polygon centroids. The weights in the principal diagonal are zero, i.e. $w_{ii} = 0$ for all i. This weight matrix is required in full for the Geary and Moran indices.

For the patterns in Figure 7-12, the observed and expected join-counts are summarized in Table 7-12. Notice that because the black polygons in A are clustered together, and also the white polygons, the number of BW joins are relatively small. In fact the expected number of BW joins is just under 14, for a random distribution of the 5 black polygons and 8 white polygons. On the other hand, pattern B has only 1 BB join, 9 WW joins and 17 BW joins, indicating that the polygons are negatively autocorrelated, with more BW joins than expected due to chance. Notice that the sum of BB + WW + BW is 27 for the expected as well as the observed joins, and this is exactly half the number of 1s in **W**, Table 7-12. In this case the expected number of joins of each type lies somewhere between the values for positively autocorrelated pattern A and negatively autocorrelated pattern B. Join-count statistics can be calculated for any binary map in vector or raster. Where the data are multistate instead of binary, each state can be treated separately as black, with all other states as white.

Geary and Moran coefficients are used with data that are measured on interval or ratio scales.The contiguity matrix is employed as a substitute for physical distance between area objects, and the measures of autocorrelation have some similarity to those used for point data.The Geary index, G, is defined as

$$G = \frac{\sum_{i=1}^{n} \sum_{j=1}^{n} w_{ij}\, c_{ij}^{*}}{2 \sum_{i=1}^{n} \sum_{j=1}^{n} w_{ij} \sum_{i=1}^{n} (z_i - \bar{z})^2 \,/\, (n-1)} \quad , \tag{7-15}$$

where $c_{ij}^* = (z_i - z_j)^2$, and z_i is the value of the attribute at polygon i. Note that the $(z_i - z_j)^2$ term is similar to the moment of inertia used for the variogram.The expected value of G for a random situation with no autocorrelation is 1; $0 < G < 1$ for positively autocorrelated and $G > 1$ for negatively autocorrelated patterns.

The Moran index, I, on the other hand has a scale that increases with increasing autocorrelation, like a correlation coefficient, and counter to the Geary index. The formula is

$$I = \frac{\sum_{i=1}^{n} \sum_{j=1}^{n} w_{ij} \, c_{ij}}{\sum_{i=1}^{n} \sum_{j=1}^{n} w_{ij} \sum_{i=1}^{n} (z_i - \bar{z})^2 / n} \quad , \tag{7-16}$$

where $c_{ij} = (z_i - \bar{z})(z_j - \bar{z})$. Here the c_{ij} term is like a covariance, and the Moran index is similar to the normal correlation coefficient in that the denominator contains an expression for the attribute variance.The Moran I is basically negative for negative autocorrelation and positive for positive autocorrelation, except that the expected value for no autocorrelation is slightly negative. Under a null hypothesis of no autocorrelation, the expected values of G and I are:

$$E(I) = -1/(n-1) \quad , \tag{7-17}$$

$$E(G) = 1 \quad . \tag{7-18}$$

In order to test whether a particular measured value of G or I differs significantly from their expected values, and assuming that each of the n attribute values is drawn independently from some population of normally distributed values, the variances of the indices can be calculated from:

$$var(I) = \frac{n^2 S_1 - n S_2 + 3 S_0^2}{S_0^2 \, (n^2 - 1)} - E(I)^2 \quad , \text{ and} \tag{7-19}$$

$$var(G) = \frac{(2 S_1 + S_2)(n-1) - 4 S_0^2}{2(n+1) \, S_0^2} \quad , \tag{7-20}$$

where the quantities S_0, S_1 and S_2 are defined as follows:

$$S_0 = \sum_{i=1}^{n} \sum_{j=1}^{n} w_{ij} \quad ,$$

$$S_1 = \sum_{i=1}^{n} \sum_{j=1}^{n} (w_{ij} + w_{ji})^2 / 2 \quad \text{and}$$

$$S_2 = \sum_{i=1}^{n} (w_{i.} + w_{.i})^2 \quad .$$

Table 7-13. Two ratio scale attributes of the polygons in Figure 7-12, used to illustrate the calculation of Geary and Moran indices of autocorrelation. These are number of landslides over a fixed time interval and number of landfills in present use. The attributes could also be used as counts per unit area, by dividing by polygon area.

Polygon	Number of Landslides	Number of Landfills
1	5	5
2	8	0
3	7	8
4	1	1
5	9	7
6	7	3
7	3	9
8	4	1
9	1	7
10	3	3
11	3	1
12	1	4
13	0	3

The subscripted $w_{i.}$ and $w_{.i}$ are the row and column totals of the weight matrix, respectively. A z-score can be used as a test statistic and is calculated as

$$z_I^* = \frac{(I - E(I))}{var(I)} \tag{7-21}$$

for the Moran index, and

$$z_G^* = \frac{(G - E(G))}{var(G)} \tag{7-22}$$

Table 7-14. Autocorrelation statistics for the landslide and landfill patterns.

	Landslides	Landfills
Geary, G	0.63	1.41
E(G)	1.00	1.00
var(G)	0.03	0.03
z_G*	-2.04	2.23
α(G)	0.021	0.013
Moran, I	0.27	-0.37
E(I)	-0.08	-0.08
var(I)	0.02	0.02
z_I*	2.37	-1.92
α(I)	0.009	0.027

for the Geary index. Since z_I* and z_G* are expected to be normally distributed with zero mean and unit standard deviation under the null hypothesis of no autocorrelation, the significance level, for either index, can be found from a table of areas under a normal curve corresponding to the calculated z-score. The smaller the value of α, the less likely the null hypothesis, and the greater confidence in its rejection.

In order to illustrate the calculation of these indices, suppose that the same polygon maps in Figure 7-12 are used, but instead of binary attributes, the two variables – number of landslides and number of landfills – are employed (Table 7-13). The results shown in Table 7-14 indicate that the landslide data are positively autocorrelated, because $0 < G < 1$ and $I > 0$. The z_G* score is -2.04, corresponding to a tabled value of the probability level, α, of 0.021. Thus a value of G this low would only be expected in 21 cases out of 1000 due to chance, indicating a positive autocorrelation much stronger than for a random pattern. Similarly, for the same data, the Moran I as large as 0.27 would be expected due to chance only in 9 cases out of 1000, confirming the positive autocorrelation of the spatial pattern of landslides.This might then lead to speculation about the cause of the pattern, and to a search of other patterns over the region that may be correlated with landslides, such as soil type, slope magnitude, agricultural practice, and other attributes.

In the case of the data on landfills, we notice that $G > 1$ and $I < 0$, indicating negative autocorrelation. Again these results can be shown to be significant, with α levels of 0.013 and 0.027 respectively, allowing the null hypothesis to be rejected. A possible "explanation" for the negative autocorrelation of landfill polygons might be that landfills are close to population centres, and population centres are negatively autocorrelated. Such a hypothesis could be further tested with population data.

The Geary and Moran indices are summary statistics, giving a parameter that characterizes the degree of autocorrelation over a whole polygon map. They can also be calculated over a moving window in raster data to examine the regional variation in autocorrelation (Prosser and Aitken, 1990), producing map output. Chou et al. (1990) applied Moran *I* statistics to the distribution of wildfires in California, but used weighting matrices based not only on contiguity, but also distance between centroids, areas of adjacent polygons, and lengths of the boundary common to adjacent polygons. They found that the contiguity values themselves performed satisfactorily, yielding Moran *I* values that were highly significant. On the basis of this result they derived a new mapped variable for their polygon data that expressed the degree of similarity of a polygon to its neighbours, using contiguity and attribute value. They called this a spatial weighting function, *SWF*, defined as

$$SWF_i = \frac{\sum_{j=1}^{n} w_{ij} z_j}{\sum_{j=1}^{n} w_{ij}} \tag{7-23}$$

for the *i*-th polygon. This has the effect of providing a local autocorrelation measure that they found to be highly significant in a logistic regression that predicted the occurrence of fires as a function of regional maps such as slope and aspect, annual precipitation, vegetation type and others. SWF was calculated for each polygon, where the z value was a binary variable, 1 for fires present in a polygon, 0 otherwise. The SWF values were used along with the other independent variables and was found to the dominant factor in explaining wildfire distribution. This illustrates the importance of spatial autocorrelation effects for prediction, as well as for map description.

REFERENCES

Agterberg, F.P. and Fabbri, A.G., 1978, Spatial correlation of stratigraphic units quantified from geological maps, *Computers & Geosciences*: v. 4, p. 285-294.

Chou, Y., Minnich, R.A., Salazar, L.A., Power, J.D. and Dezzani, R.J., 1990, Spatial autocorrelation of wildfire distribution in the Idyllwild Quadrangle, San Jacinto Mountain, California: *Photogrammetric Engineering & Remote Sensing*, v. 56 (11), p. 1507-1513.

Davis, J.C., 1986, *Statistics and Data Analysis in Geology*, Second Edition: John Wiley and Sons, New York, 646 p.

Fabbri, A.G., 1984, *Image Processing of Geological Data*: Van Nostrand Reinhold, New York, 244 p.

Fuller, B.D., 1966, Two-dimensional frequency analysis and design of grid operators: *Society of Exploration Geophysicists Mining Geophysics*, Vol. II. Theory, Editor, Hanson, D.A., p. 658-708.

Goodchild, M.F., 1986, *Spatial Autocorrelation*: Geo Books, Norwich, England, v. 47, 55 p.
Griffith, D.A., 1988, *Advanced Spatial Statistics – Special Topics in the Exploration of Quantitative Spatial Data Series*: Kluwer Academic Publishers, Dordrecht-Boston, 273 p.
Griffith, D.A. and Amrhein, C.G., 1991, *Statistical Analysis for Geographers*: Prentice Hall, Englewood Cliffs, New Jersey, 478 p.
Grunsky, E.C. and Agterberg, F.P., 1988, Spatial and multivariate analysis of geochemical data from metavolcanic rocks in the Ben Nevis area, Ontario: *Mathematical Geology*, v. 20 (7), p. 825-861.
Grunsky, E.C. and Agterberg, 1991a, FUNCORR: a FORTRAN-77 program for computing multivariate spatial autocorrelation: *Computers & Geosciences*, v. 17 (1), p. 115-131.
Grunsky, E.C. and Agterberg, F.P., 1991b, SPFAC: a FORTRAN-77 program for spatial factor analysis of multivariate data: *Computers & Geosciences*, v. 17 (1), p. 133-160.
Haralick, R.M., 1979, Statistical and structural approaches to texture: *Proceedings IEEE*, v. 67, p. 768-804.
Harrison, B.A. and Jupp, D.L., 1990, *Introduction to Image Processing*: CSIRO-Division of Water Resources, Canberra, Australia, 255 p.
Jupp, D.L.B., Mayo, K.K., Kuchler, D.A., Van R. Classen, D., Kenchington, R.A., and Guerin, P.R., 1985, Remote sensing for planning and managing the Great Barrier Reef of Australia: *Photogrammetria*, v. 40, p. 21-42.
Mather, P.M., 1987, *Computer Processing of Remotely-Sensed Images*: John Wiley and Sons, Chichester-New York, 352 p.
Prosser, R. and Aitken, S.C., 1990, Analyzing geographic information using grid-based spatial autocorrelation statistics and directional autoregression models: *GIS/LIS '90 Proceedings*, v. 2, p. 497-505.
Robinson, J.E., 1982, *Computer Applications in Petroleum Geology*: Hutchinson Ross, Stroudsburg, PA, 164 p.
Smith, G.B., 1983, Shape from shading: an assessment: *Proceedings NASA Symposium on Mathematical Pattern Recognition and Image Analysis*, June 1-3, 1983, Johnson Space Center, Houston, Texas, p. 543-576.
Tomlin, C.D., 1990, *Geographic Information Systems and Cartographic Modeling*: Prentice Hall, Englewood Cliffs, New Jersey, 249 p.

CHAPTER 8

Tools for Map Analysis: Map Pairs

INTRODUCTION

Probably the single most significant aspect of GIS is the ability to combine spatial data from different sources together. The purpose of making such combinations is to identify and describe spatial associations present in the data, and to use models for analysis and prediction of spatial phenomena. Mineral potential mapping is a good geological example; spatial associations between known zones of mineralization with geological, geophysical and geochemical maps can be explored with the help of GIS tools, and models used to combine maps and predict undiscovered deposits. Before looking at multi-map combination (Chapter 9), this chapter examines the methods of comparison and combination of maps in pairs.

As in Chapter 7, the discussion is centred on maps as sets of area objects, in either the raster or vector data models. Sets of point and line objects are also important for spatial analysis: often point and line data are converted to areas (as planar-enforced maps), using the transformation methods of Chapter 6. The emphasis in this book is thus placed on vector or raster overlays of polygons or pixels, at the expense of topics like network analysis and point pattern analysis. Network analysis is particularly important in GIS applications dealing with transportation and communication, Lupian et al. (1987). Networks may also be relevant to some fields of geoscience, like drainage networks in hydrology. Similarly, point patterns find application

Table 1. Categories of two-map operations based on type of output and the importance of spatial attributes

	Spatial attributes used	
	Not necessarily	**Yes**
Map output	Map overlay Map modelling	Neighbourhood analysis Topological modelling
Table output	Map overlay Table modelling Area analysis Correlation	Spatial cross-correlation

in the spatial distribution of mineral deposits, seismic epicentres and other point events. The book by Diggle (1983) is a statistical treatment of point pattern analysis. Openshaw et al. (1987) discuss the analysis of point patterns with what they call a Geographical Analysis Machine (GAM), applied to the identification of clusters in the incidence of disease; Openshaw et al. (1989) have also applied similar methods for evaluating sites for nuclear waste disposal.

As in Chapter 7, operations on map pairs can be subdivided on the basis of whether the output is a new map or a table (or both), and whether spatial attributes play an essential role, see Table 8-1. Of course, spatial location is involved in all these operations, in the sense that maps are compared, or matched, at points of overlap defined in physical space. However, many operations can be carried out on matched data items without regard to locational, topological or geometrical attributes, and in this sense are nonspatial operations. Typical examples of operations according to this four-fold division are as follows:

1) *Map output, spatial attributes not necessarily used.* This category involves the derivation of new maps from pairs of input maps, sometimes using attribute tables for one or both input maps, according to a model. The process is often called map **overlay**, and in the vector model results in the generation of a new set of polygons that is common to both input maps. A simple geological example is the overlay of a slope map with a soil map. Visual analysis of this combination may lead to a description and understanding of the spatial association between slope and soil type; the combination may also be used with a model to generate new maps that predict slope stability and landslide potential.

2) *Table and other non-map output, spatial attributes not necessarily used.* Map overlay can also result in producing table output, summarizing the attributes of polygon objects that have the characteristics of both input maps. These new combined polygon objects comprise a single overlay map, thereby reducing two (or more) input maps to a single output map. Any of the methods of the previous chapter can now be applied to the overlay map and its associated table. For example, table modelling may be carried out on overlay tables, as an alternative to direct map modelling, and the results used to reclassify the overlay map. Analysis of map pairs can also lead to summary statistics, often output in the form of tables (or in the form of data views that complement the map view), and to statistical indices, such as correlation measures, for quantifying spatial associations. For example, an area cross-tabulation of soil type and slope magnitude can lead to a quantitative characterization of the association between these two maps. Alternatively, a box-and-whisker plot might be used to show differences in slope between soil classes, using a data view in "frequency" space, aggregating objects by a categorical attribute (soil class).

3) *Map output, spatial attributes used.* This category includes operations in raster mode that involve a pixel neighbourhood of two maps simultaneously, or operations in vector mode that use topological attributes such as adjacency and containment with two maps. For example, in raster mode one might produce a map of soil diversity within a particular radius, but only if the slope is uniform over the same radius. Or in vector mode, the goal might be to produce a map showing slope polygons adjacent to soil polygons with particular attributes. There are a number of advanced methods in geostatistics that fall into this category that are beyond the scope of this book. For example, co-kriging is a geostatistical method for spatial interpolation similar to ordinary kriging, but is applied to pairs of map variables.

4) *Table and other non-map output, spatial attributes used.* The measurement of spatial associations between two maps can be examined using spatial cross-correlation analysis. Here the goal is not simply to look at the association between the two maps in register, but to examine how the correlation behaves with systematic shifts in register in different directions. This kind of analysis produces tabular or graphical output summarizing cross-correlation (or co-variogram) as a function of shift or lag, like an autocorrelation function (or variogram), and helps to assess the significance of particular cross-correlation values. The cross-spectral analysis of maps in the frequency domain also fits in this category. Vector overlaps of binary maps can be examined using join-count statistics, that help evaluate whether particular joins occur more than would be expected due to chance.

The discussion in this chapter is mainly limited to categories 1) and 2). The first part of the chapter discusses overlay and modelling of map pairs; the second part describes methods of map correlation which vary according to the measurement level of the data.

TWO-MAP OVERLAYS AND MAP MODELLING

Two-map overlays with some basic arithmetic and Boolean operators are powerful tools for examining the spatial patterns caused by interactions of one map with another. In generating a two-map overlay, the goal is to combine the inputs according to a set of rules (the map model) that determines for each location the class of the output map from the classes of the input maps. In each of the following examples, the two input maps and the resulting

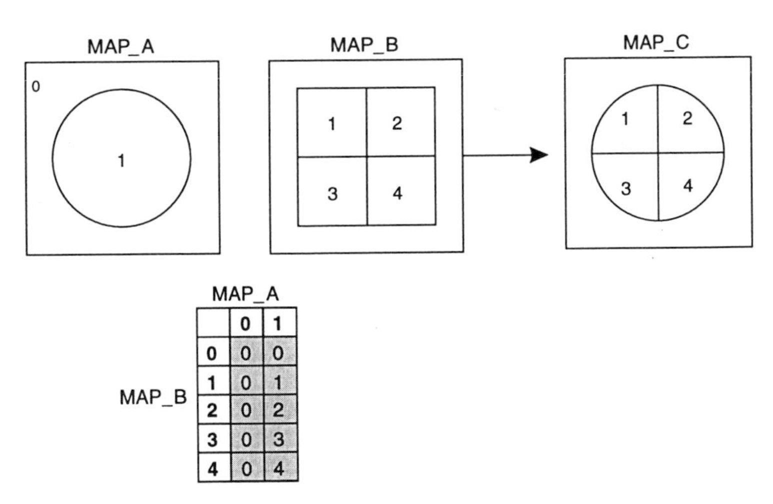

MAP_B \ MAP_A	0	1
0	0	0
1	0	1
2	0	2
3	0	3
4	0	4

FIG. 8-1. Two-map overlay in which binary MAP_A is **imposed** on multi-class MAP_B to produce MAP_C, as defined by the map modelling statements in the text. The truth table shows the outcome of applying the model, giving the results for all combinations of the input map classes.

output map are shown. In addition, a "truth table" shows the output map class for each combination of the input map classes, and a series of statements in the map modelling pseudo-code (see Appendix) shows the algebraic form of the model.

For example, Figure 8-1 shows the result of "**imposing**" a binary map on to a multi-class map. MAP_A has two classes, 1 and 0, MAP_B has 5 classes, 0 to 4, and MAP_C, the output, also has 5 classes, as shown in the truth table. The impose operation can be expressed with the following map modelling statements:

> ***:lines starting with a colon are comments***
> ***:get the class values of MAP_A and MAP_B at the current location***
> **:MAP_A must be binary.**
> ***:The output will be nonzero only at locations where MAP_A is equal to 1***
> **A = class('MAP_A')**

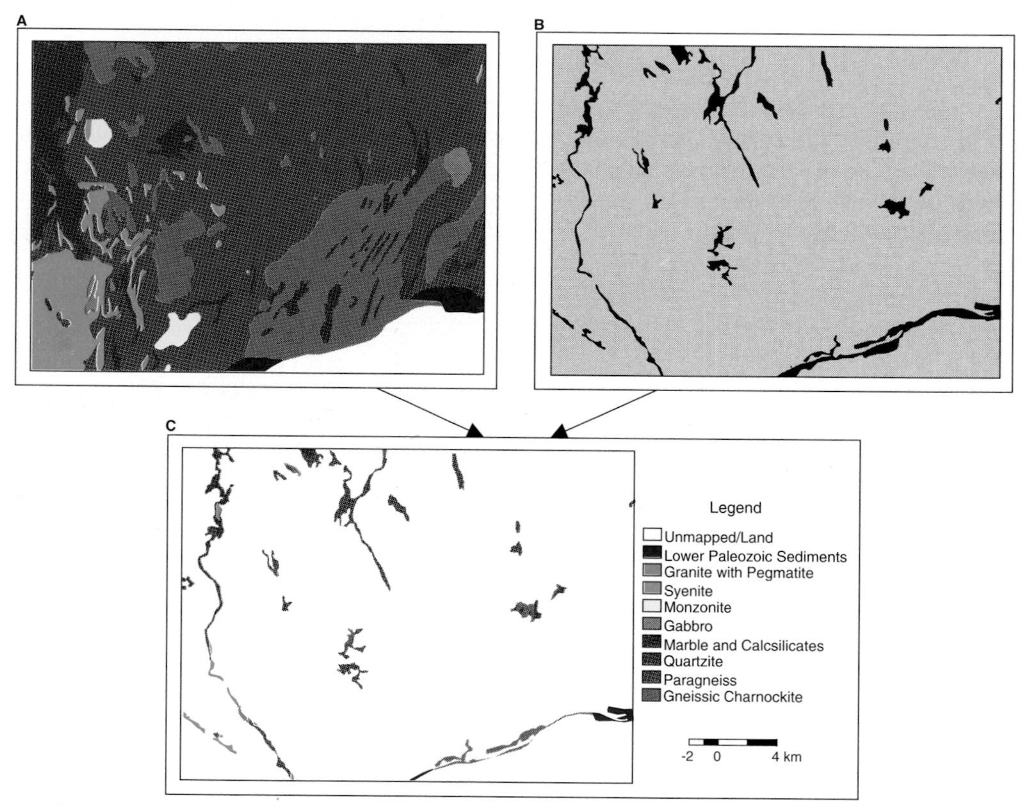

FIG. 8-2. A. Geological map. **B.** Map showing locations of rivers and lakes. The water is class 1 (black) and the land is class 0 (grey). **C.** Map produced by imposing A on B with the result that the geology is only shown where the water occurs. Where the land occurs, the output class is set to 0.

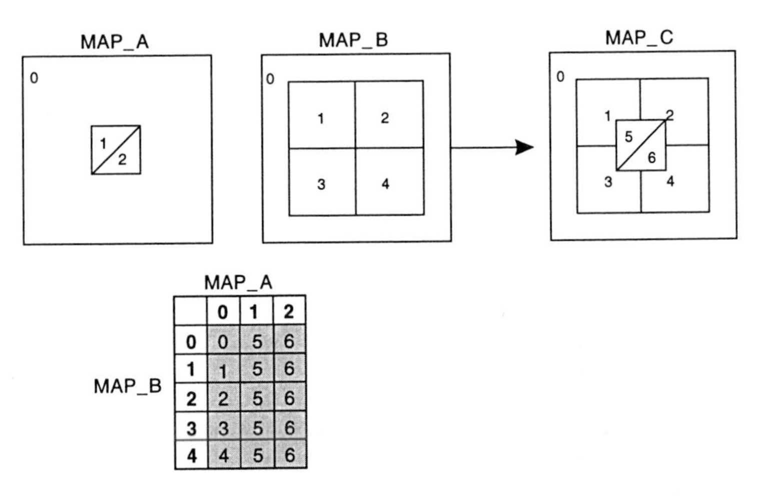

MAP_B \ MAP_A	0	1	2
0	0	5	6
1	1	5	6
2	2	5	6
3	3	5	6
4	4	5	6

FIG. 8-3. Two-map overlay using the **stamp** operation, as defined in the map modelling statements in the text.

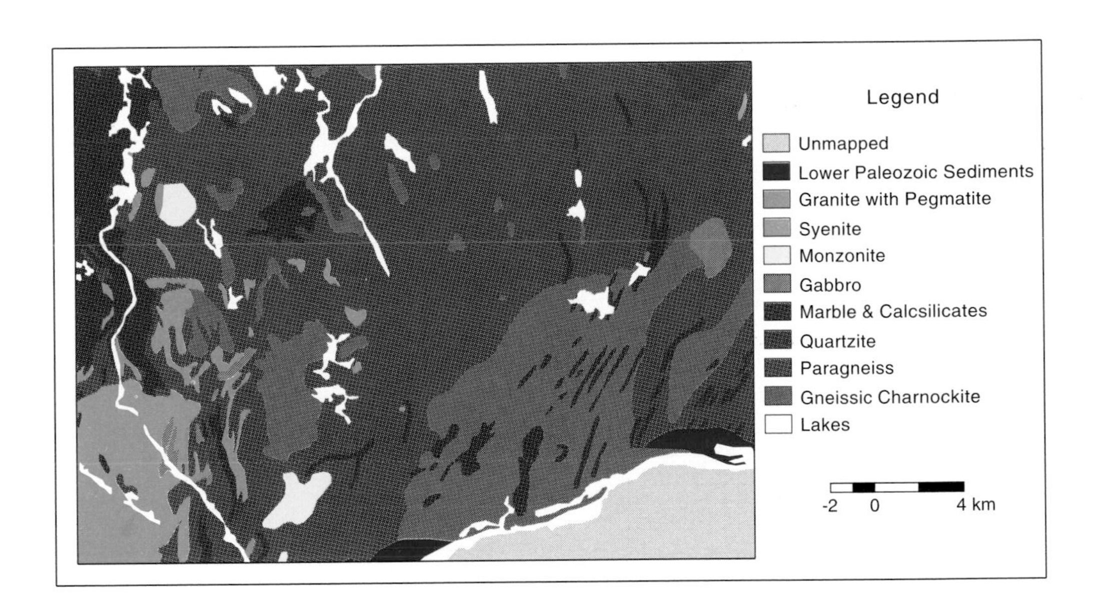

FIG. 8-4. Two-map overlay in which the water/land map (same as in Figure 8-2B) is **stamped** on the geology map (Figure 8-2A). The result is to cause areas where water occurs to be assigned to one larger than the maximum class value of the geology map. In this case the lake class is coloured white.

```
:lines starting with a colon are comments
:get the class values of MAP_A and MAP_B at the current location
:MAP_A must be binary.
:The output will be nonzero only at locations where MAP_A is equal to 1
        A = class('MAP_A')
        B = class('MAP_B')
:multiply the class values together
        C = A * B
:output the result as a new map
        RESULT(C)
```

The impose operation is clearly useful for restricting the areas on an output map according to a binary map that acts as a mask. Often the binary map is used as a basemap to define the extent of the study region. Figure 8-2 shows the result of imposing a binary map of lakes (lakes=1, land=0) on a geological map. The new map is a set of polygons with the land excised as if with a "cookie cutter", because where land is present the class of the output map is set to 0. The resulting map shows the spatial relationship of lakes to geology as well as providing an aid to identifying location by reference to the surface hydrology.

Suppose, however that MAP_A is "**stamped**" on MAP_B, as shown diagrammatically in Figure 8-3. The map modelling statements for this operation are:

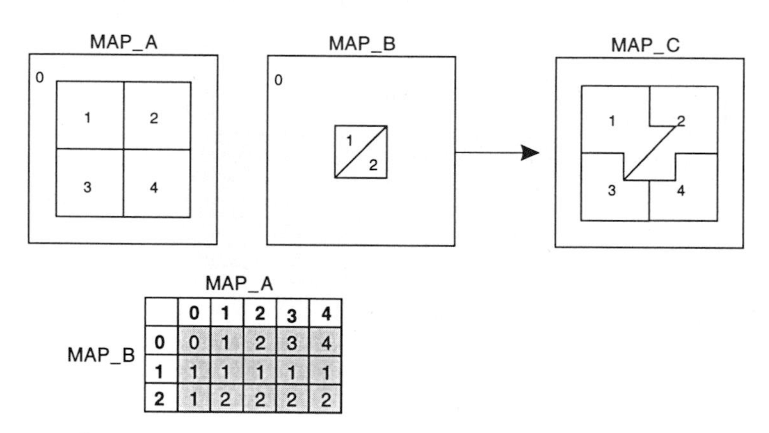

MAP_B \ MAP_A	0	1	2	3	4
0	0	1	2	3	4
1	1	1	1	1	1
2	1	2	2	2	2

FIG. 8-5. Two-map overlay illustrating the **join** operation, in which MAP_B takes priority over MAP_A.

```
:get the input map class values as before
        A = class('MAP_A')
        B = class('MAP_B')
:use function MAXCLASS() to find max class of MAP_B
        MAXB = MAXCLASS('MAP_B')
:use conditional statement
        C = {B if A==0, A + MAXB}
:this causes C to be set to B if A equals 0,
:otherwise C is set to the sum of A plus MAXB
        RESULT(C)
```

This results in MAP_A (classes greater than 0) being **stamped** over the top of, and obliterating the classes of, MAP_B. This can be useful in examining spatial relationships between two maps, particularly if MAP_A is binary. For example, in Figure 8-4 notice how the lake pattern of MAP_A is stamped on the geological map, so that only the geology beneath land is visible. In this example, a similar result could also be produced by recoding MAP_A with water=0 and land=1, followed by an impose operation, except that instead of the water in C having a class number equal to the maximum class of the geology map plus 1, it would have class 0.

In Figure 8-5, two maps have been "**joined**". In this operation, MAP_B takes precedence over MAP_A, as defined in the following statements:

```
:get the input map classes
        A = class('MAP_A')
        B = class('MAP_B')
:C is set to B if B is greater than zero, else C is set to A
        C = {B if B > 0, A}
        RESULT(C)
```

This logic can also be used for **joining** more than two overlapping map sheets together in a single operation, as shown in Figure 8-6. In this case, the conditional statement contains more than two expressions, and execution continues from left to right until an expression is satisfied. Thus if four maps B1, B2, B3 and B4 are combined, the conditional statement might read

```
C = {B1 if B1 > 0, B2 if B2 > 0, B3 if B3 > 0, B4}
```

which would cause B1 to take precedence over B2, B2 over B3, and B3 over B4 in the join operation.

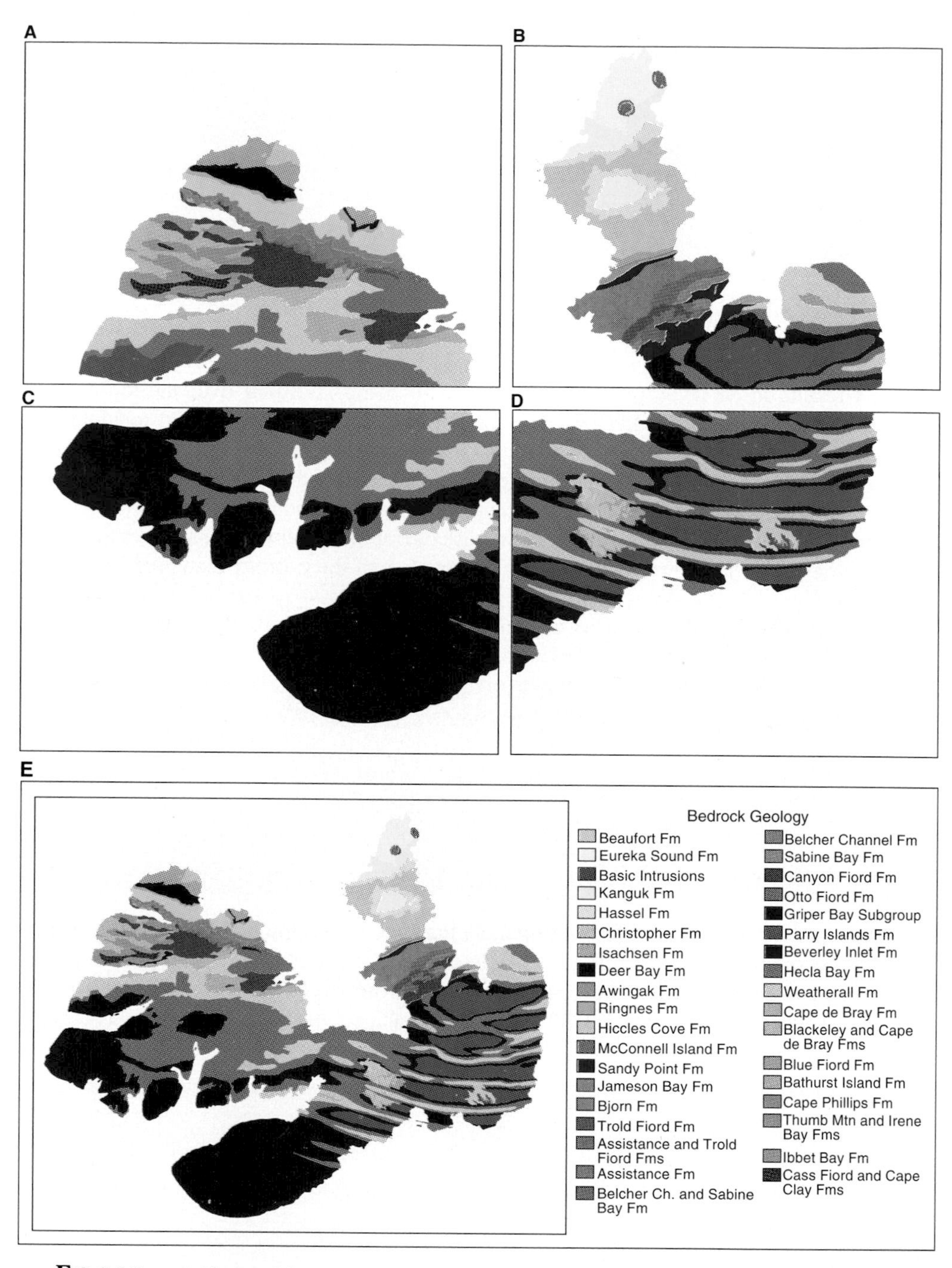

FIG. 8-6. Example of using a join operation to mosaic four map sheets together, Melville Island, Canadian Arctic.

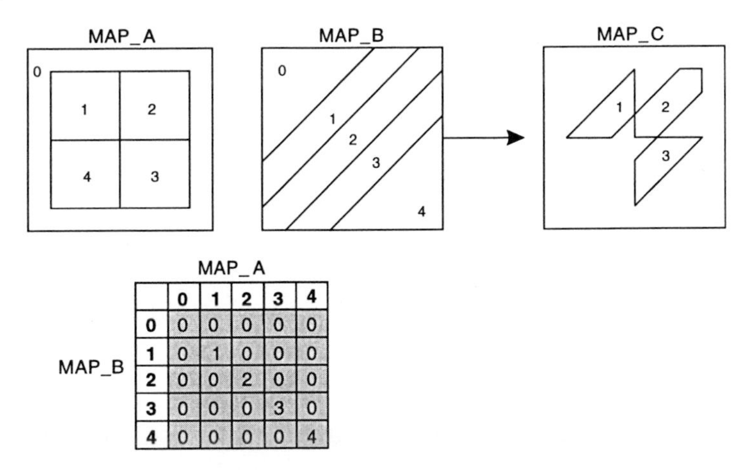

	0	1	2	3	4
0	0	0	0	0	0
1	0	1	0	0	0
2	0	0	2	0	0
3	0	0	0	3	0
4	0	0	0	0	4

FIG. 8-7. Two-map overlay to illustrate a **compare** operation. Where the two input maps disagree, the output map is set to 0, but where they have the same class value, the class value is unchanged.

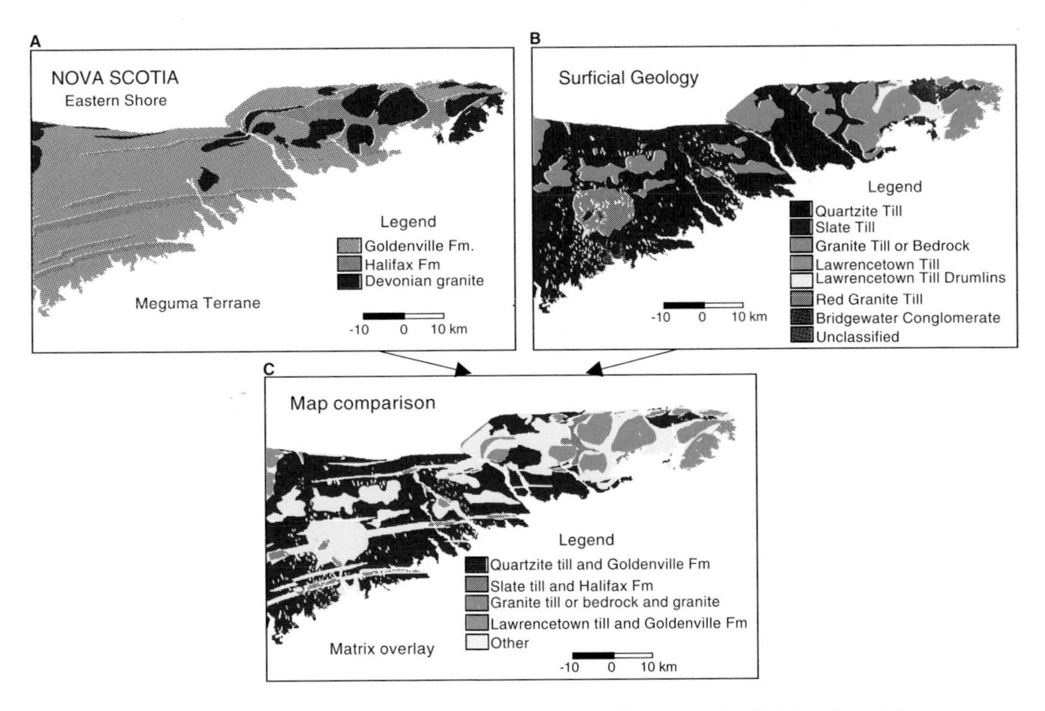

FIG. 8-8. Example of comparing a map of bedrock geology with a map of surficial geology, Meguma Terrane, Nova Scotia. The new map classes were defined using a matrix overlay, allowing more flexibility than the compare model of Figure 8-7.

Many GIS supply special editing software for correcting mistakes along seams between adjoining sheets, because almost inevitably adjacent map sheets fail to link up perfectly. Where such operations are carried out in vector mode, edge-matching can be accelerated by semi-automatic methods.

A simple method for **comparing** two maps that have the same number of matched classes is shown in Figure 8-7, controlled by the following modelling statements:

```
:get input map classes as before
        A = class('MAP_A')
        B = class('MAP_B')
:the result is A if A equals B
        C = A * (A == B)
:the expression (A == B) has the value 1 if the equality is true,
:otherwise it has the value 0
        RESULT(C)
```

This immediately reveals where the two maps are the same, where they differ, and which classes show the greatest correlation. In Figure 8-8, a bedrock geology map is compared to a surficial geology map, using a variation on the method illustrated in Figure 8-7 to accommodate the fact that there are a different number of classes on the two input maps. The correlation between the Quartzite Till and the Goldenville Fm is clearly demonstrated. Numerical measures of correlation between units on the two maps are discussed in a later section.

Sometimes two maps can be usefully combined by weighting and adding the individual class values. Suppose that MAP_A and MAP_B in Figure 8-9 are geochemical anomalies, but that MAP_B is more important than MAP_A, then the coincidence of the two sets of anomalies might be modelled with the following statements:

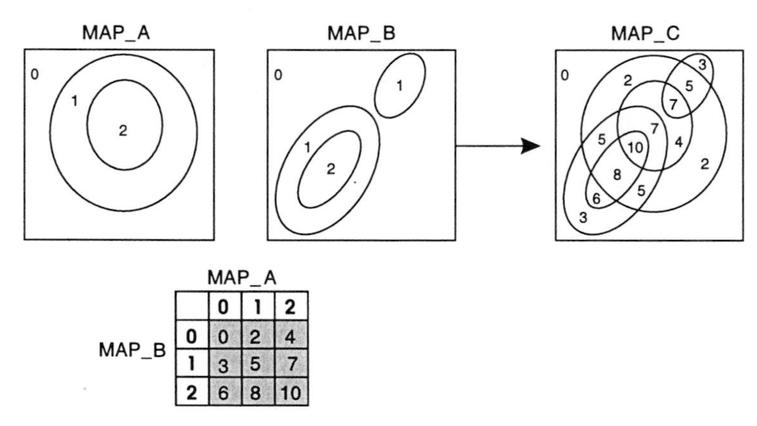

FIG. 8-9. Two-map overlay illustrating the weighted sum of two input maps.

```
:get input map classes as before
        A = class('MAP_A')
        B = class('MAP_B')
:A is weighted by a factor of 2, and B by a factor of 3
        C = 2*A + 3*B
        RESULT(C)
```

Figure 8-10 shows a geochemical example of this operation applied to two geochemical anomaly maps.

Finally, Figure 8-11 illustrates a rather more involved combination, using both Boolean and arithmetic operators. This simply illustrates the flexibility of using map modelling to combine maps together. The modelling statements are:

```
:get input map classes
        A = class('MAP_A')
        B = class('MAP_B')
:use a conditional statement that reads, C is set to
:A if B equals 2, else to B minus 1 if B is greater than 0, else to zero
        C = {A if B == 2, B-1 if B > 0, 0}
        RESULT(C)
```

A method that is called a **matrix overlay** in some systems allows complete flexibility of assigning output classes in terms of the input class combinations. Each combination of classes from the two input maps is assigned an output class in a two-dimensional array, exactly comparable to the truth tables shown for the preceding combination methods. This array may be built, or modified from a template, with a text editor. In the case of the arbitrary assignment of output class values from the combination of two categorical scale input maps, the matrix method is more efficient than enumerating all the possible map combinations in the modelling language. On the other hand, if the result can be stated as a function of the input maps through algebraic statements, the map modelling is more powerful since it can readily handle input maps with a large number of classes.

Given these map modelling tools, the number of possible ways of combining pairs of maps becomes almost infinite, constrained only by the imagination of the user. Although these modelling operations do not explicitly use spatial attributes, unexpected spatial pattern is often revealed in the output. We now turn to cases where each of the input maps is associated with an attribute table. The attribute tables contain thematic attributes, and in some cases, spatial, geometric and topological attributes.

Map Modelling with Attribute Tables

Consider a case where both input maps have attribute tables, and the modelling language is employed to combine two input maps. The modelling language can utilize not only map classes, but also items in attribute tables, at records that are linked to spatial objects on the map by a keyfield.

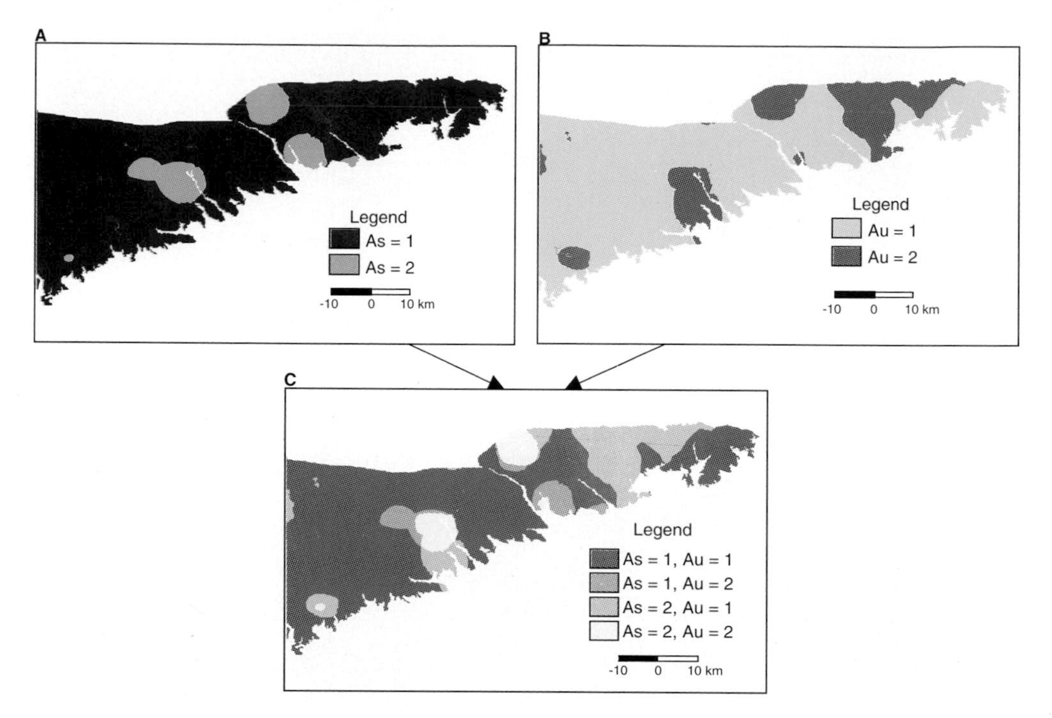

FIG. 8-10. Example of the weighted sum of two geochemical anomaly maps, from the Meguma Terrane of Nova Scotia.**A.** Arsenic anomaly map. **B.** Gold anomaly map. **C.** Combination of A and B. The actual class numbers are 5 (AS=1, Au=1), 7 (As=1, Au=2), 8 (As=2, Au=1) and 10 (As=2, Au=2).

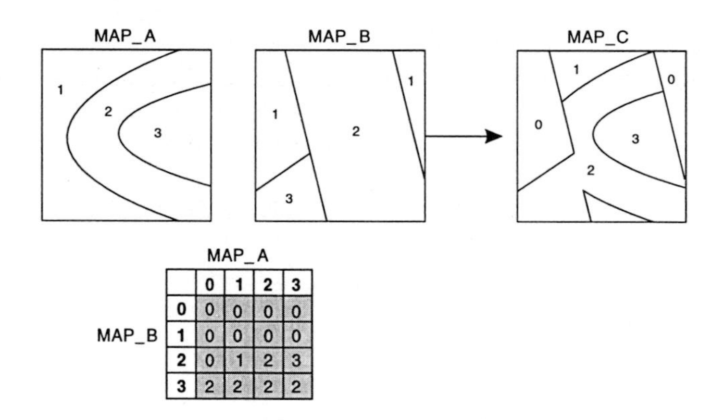

	0	1	2	3
0	0	0	0	0
1	0	0	0	0
2	0	1	2	3
3	2	2	2	2

FIG. 8-11. Two-map operation to illustrate a model for combining maps with both logical and arithmetic operators.

In Figure 8-12, a geological map is to be combined with a geochemical map (the polygons could be catchment basins). Each map is linked to an attribute table by map class. There are three fields in the **GEOLOGY** table, the first being the **class number** (keyfield providing the link to the map also called **GEOLOGY**), the second being a **lithology** code, and the third being an **age** code. The table called **GEOCHEMISTRY** is a polygon attribute table with four fields: a keyfield (providing a link to the map also called **GEOCHEMISTRY**) and three fields for geochemical elements. The goal is to produce a new map classified into 2 classes: the first class

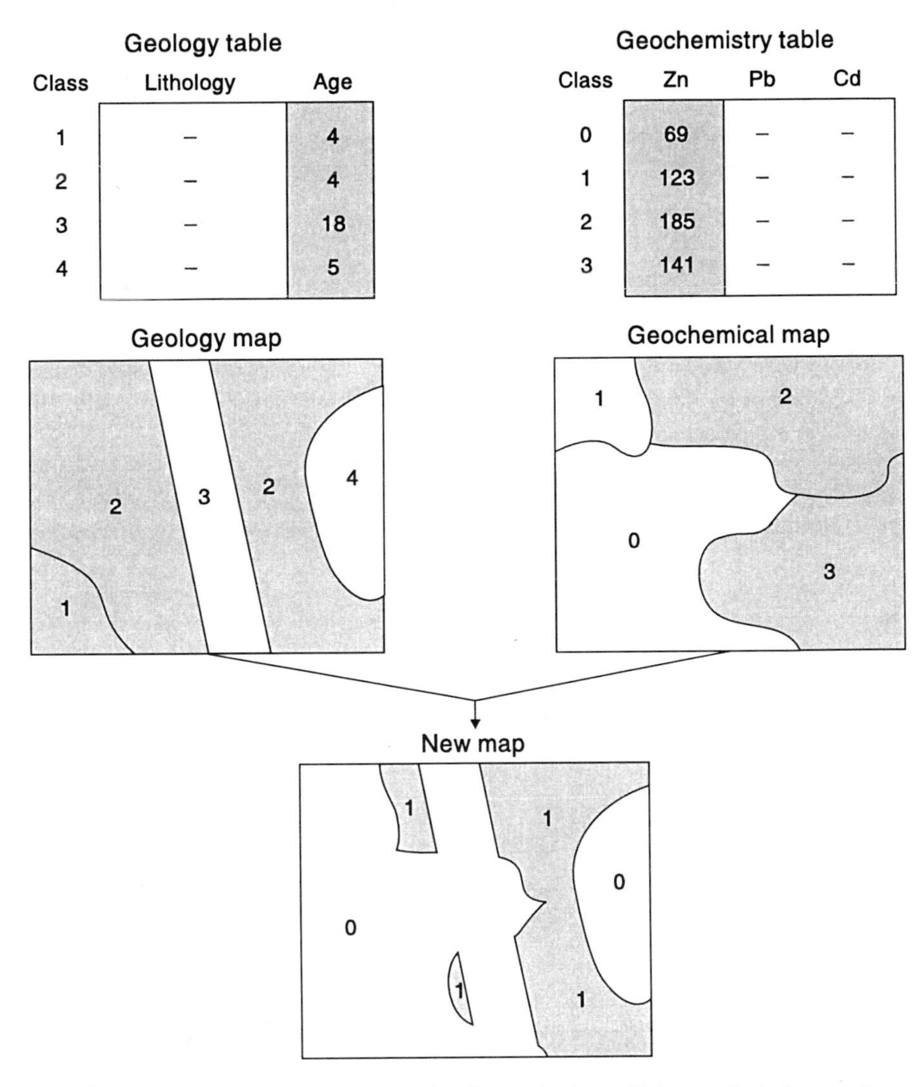

FIG. 8-12. Diagram to illustrate the use of attribute tables for combining a geological map and a geochemical map using a map modelling statements.

consists of those areas that have an age code of 4 and a zinc value of greater than 140 ppm. The second class contains all the remaining areas. Map modelling pseudo-code to carry out this operation is as follows:

```
:get the class values for the geology and geochemistry maps, at the
:        current location
         GEOL = class('GEOLOGY')
         GEOCHEM = class('GEOCHEMISTRY')
:lookup the age code in the geology table for this map class
:the expression "table('A', B, 'C')" means
:        fetch the value in the table called 'A' at record B and field 'C'
         AGECODE = table('GEOLOGY', GEOL, 'AGE')
:lookup the zinc value in the geochemistry table for this polygon
         ZINC = table('GEOCHEMISTRY', GEOCHEM, 'Zn')
:set output to 1 if conditions are satisfied, else set to 0
         OUT = {1 if AGECODE == 4 AND ZINC > 140, 0}
:OUT is the class value of the new map
         RESULT(OUT)
```

The output is a binary map showing the regions where these conditions are satisfied. In this case, the output is a non-negative integer, so that a classification table of breakpoints is not required and the result can be directly mapped.

In another example, suppose that a catchment basin map of arsenic (As) in lake sediment, X, is to be compared with point samples of As values in soil, Y, transformed to a map by using Thiessen polygons. The catchment basin map and the Thiessen polygon map are both linked to large polygon attribute tables, of which As is but one of many fields. Because the mean value and range of As in the two sampling media are very different, making a direct comparison of the two maps potentially misleading, it is decided to transform both variables to the range 0-1 by the equation

$$X_i^* = \frac{X_i - X_{min}}{X_{max} - X_{min}} , \tag{8-1}$$

where X_i and X_i* are the i-th values of the original and transformed variable, respectively. The variable Y is similarly transformed to Y^*. The output from the model is then the difference X^*-Y^*, which is much more readily interpreted (see for example Herzfeld and Merriam, 1990). The map model for this operation might read as follows:

```
: get the class of the catchment basin at the current location
        INCLASS1 = CLASS('BASIN')
: lookup the As value in the associated table called LAKESED
        X = TABLE('LAKESED', INCLASS1, 'AS')
: transform to a value in the range (0-1). The minimum and maximum values
are assumed to be known, previously calculated with a table aggregation
function
        XSTAR = (X - XMIN) / (XMAX - XMIN)
: get the class of the soil map at the current location
        INCLASS2 = CLASS('SOIL')
: lookup the As value in the associated SOIL table
        Y = TABLE('SOIL', INCLASS2, 'AS')
: transform to the range (0-1)
        YSTAR = (Y - YMIN) / (YMAX - YMIN)
: compare the two values by taking the difference
        Z = XSTAR - YSTAR
: classify the result using a classification table of breakpoints called 'DIFF'
        OUT = CLASSIFY(Z, 'DIFF')
        RESULT(OUT)
```

Here, the SOIL map is linked to an attribute table also named SOIL, but the catchment basin map (BASIN) is linked to a table named LAKESED (i.e. the map and associated attribute table need not have the same name, as long as the class or polygon numbers on the map match index numbers in the keyfield of the table).

The minimum and maximum values could either be entered manually or calculated with a table function. Although the maps and associated tables have both been called by the same name, this is not necessary. The model is evaluated for each location in turn, directly producing a new map with classes defined by the classification table called DIFF. Note that the breakpoints in this table must allow for negative as well as positive numbers.

Instead of applying a uniform transformation to the interval (0-1), an alternative approach is to convert each variable to a standard normal deviate by subtracting the mean and dividing by the standard deviation, before making the comparison. Because geochemical elements usually have positively skewed distributions, a better method for geochemical data is either to transform to logarithms before converting to a standard normal deviate, or to standardize by transforming to ranks or percentiles. Transformation to ranks is discussed further below under rank correlation.

Readers may be wondering how the index numbers in a keyfield of a polygon attribute table are assigned the correct polygon or class numbers of a map, to ensure that the map and table are correctly linked. There are several ways of accomplishing this; three are mentioned here, although the mechanics of building attribute tables are system-dependent. The first method is to build an index field for the table manually with a text-editor, given the polygon or class numbers from a hardcopy source, or by interactive spatial query. Clearly this is possible only for uncomplicated maps with few spatial objects or classes. The second method, already discussed in Chapter 6, is to generate centroids for the polygons, then to build a point attribute table for these points and

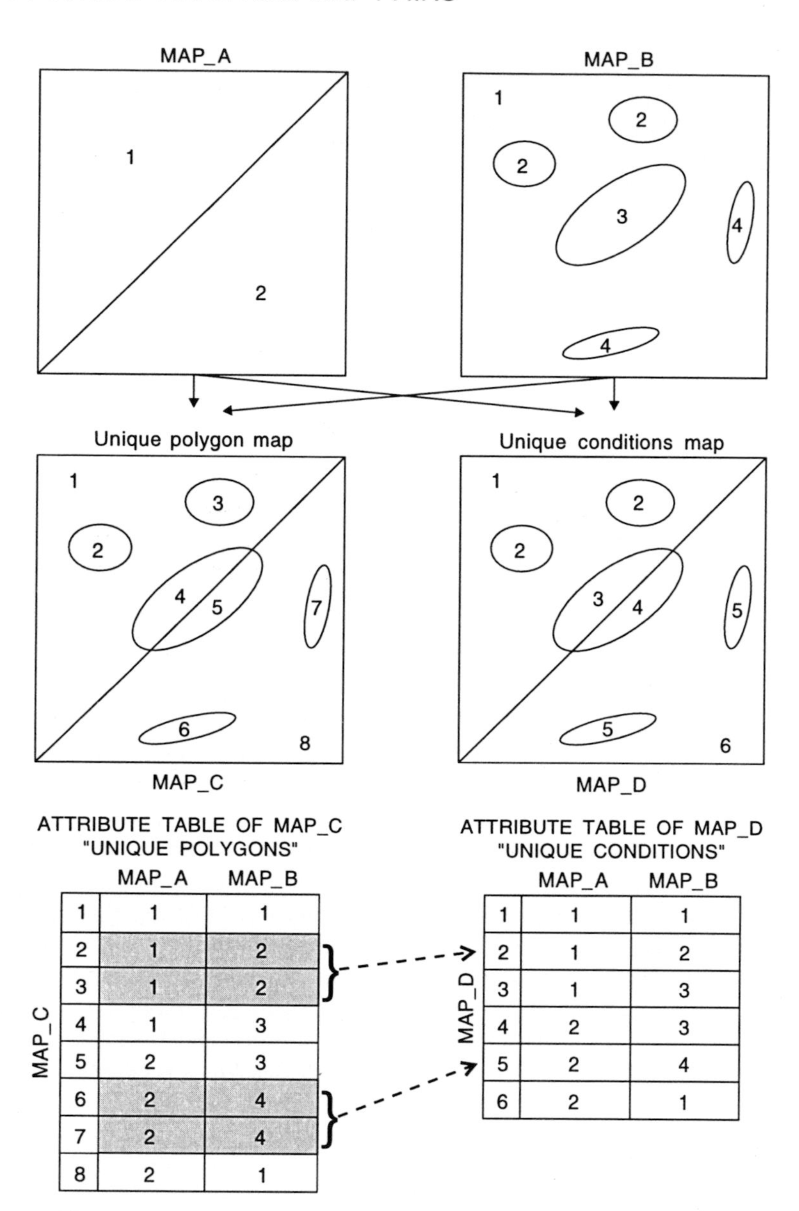

MAP_C	MAP_A	MAP_B
1	1	1
2	1	2
3	1	2
4	1	3
5	2	3
6	2	4
7	2	4
8	2	1

MAP_D	MAP_A	MAP_B
1	1	1
2	1	2
3	1	3
4	2	3
5	2	4
6	2	1

FIG. 8-13. Diagram to illustrate the generation of an overlay map and an associated attribute table. The polygons on the overlay map are formed by the combination of polygons from the two input maps. There are two kind of overlay, one with a unique class for each polygon ("unique polygon" overlay), the other producing "unique conditions" classes, where polygons with the same input map characteristics are combined.

append the polygon or class number as a new field (to become the keyfield) to the table. This is the normal method used in digitizing a map (Chapter 4). The third method is to create an attribute table during an overlay operation, as discussed in the next section.

Map Overlays and Overlay Tables

The map modelling operations described above have directly produced map output from input maps and their associated tables. However, instead of directly producing the final map as the only output, an alternative is to produce an *overlay map* that is linked to an *overlay table*. Modelling operations are then carried out on the table, not directly on the maps, and the results are assigned to a new field or column in the overlay table. The overlay map can then reclassified according to the new field, with or without a classification table of breakpoints. Overlay operations of this type are basically of two varieties: those that assign a unique class value, or entity number, to each individual polygon created by the overlay process, and those that allow two or more polygons to be assigned to the same class if they have the same characteristics or conditions of the input maps. These two varieties can be called unique polygon maps, or entity maps, and **unique conditions maps**, respectively, although a variety of names are used by commercial GIS.

The distinction between these two kinds of overlay map are illustrated in Figure 8-13, which shows the overlay of MAP_A and MAP_B producing *either* MAP_C and its attribute table (unique polygons), *or* MAP_D and its attribute table (unique conditions). Each polygon of MAP_C is assigned a unique class value, so that the number of map classes equals the number of polygons. MAP_D is similar to MAP_C, except that each row in the attribute table is unique, having the same vector of input map values, giving rise to the name "unique conditions". MAP_D and its table can easily be derived from MAP_C and its table by aggregating all those polygons that have the same characteristics of the input maps. The unique conditions map will always have the same or fewer classes than the unique polygon map. The unique conditions map is preferred for normal modelling operations, because the number of rows of the attribute table are minimal, therefore speeding up modelling computations. For example, suppose two binary maps are being combined in an overlay operation. The unique conditions map consists of only 4 classes or unique conditions (A and B present, A and B absent, A present B absent, and A absent B present), so the attribute table has only 4 rows. This means that the modelling operation on the attribute table (table modelling) need only be carried out 4 times, which is obviously very efficient. On the other hand, the unique polygon map for this case might have thousands of polygons, each requiring a separate entry in the table and separate modelling calculations. In some cases the saving in computation time between these two methods is significant.

The unique polygon map is essential for modelling operations that require additional spatial and topological attributes. Suppose that the size, shape, location or topological characteristics of individual polygon entities are being used in a model, then a unique conditions overlay is not appropriate. For example, suppose all polygons whose centroids fall within a rectangular window, or all polygons larger than 10 square km, or all polygons adjacent to a particular granite, are to be used in a modelling operation, then the unique polygon method must be applied, because each polygon has a unique set of spatial attributes.

Table modelling with an attribute table, produced by overlay, results in the addition of one or more fields containing model results. It is then a simple matter to reclassify the overlay map with selected fields of this attribute table, using classification breakpoint tables where necessary. The principles of overlay modelling are the same for multiple maps as for map pairs, except that the number of unique conditions can become very large. Modelling with an attribute table is impractical where the number of records becomes too large (limits vary depending on the system). In such cases, map modelling must be used directly, without building an intervening table. This is usually the case in image processing systems, where the number of unique conditions for combinations of two or more 8-bit images (255 classes per image) can become prohibitively large. Image processing systems sometimes use a data structure called a "hash table" (see Mather (1987) for example) for alleviating the problem, but this is outside the scope of the present discussion.

Although modelling operations are often carried out within the GIS with a modelling or macro language, the attribute tables associated with unique conditions or unique polygon overlays can be exported to carry out specialized modelling outside the GIS in other software packages. Because the overlay process produces a single new map from two or more original maps, the treatment of an overlay map and associated attribute tables is directly comparable to the analysis and modelling of single maps and their attributes, as described in Chapter 7. We now turn to the question of identifying and quantifying correlations between pairs of maps.

CORRELATION BETWEEN TWO MAPS

When two maps are combined with a model, the result is a new map that can be visually inspected for spatial associations. It is also useful to quantify spatial associations between maps with summary statistics, either on a global basis, or on the basis of subsets defined by spatial or thematic criteria.This permits a comparison of the strength of a spatial association between one map pair and another map pair.The identification and ranking of spatial correlations on maps are important in many areas of earth science. Exploration geologists need to identify map patterns that are indicative of mineral deposits. Environmental geologists need to be able to correlate geochemical patterns with possible causes of pollution, or with geological patterns that provide a "natural" explanation for chemical distributions. Geomorphologists need to measure the associations between landslide regions and maps of soil type, vegetation type, or other factors. In medical geology, the distribution of disease may be spatially correlated with geochemical patterns in soil or groundwater. Insurance companies need to correlate patterns of earthquake damage with geological features to establish zones according to risk.

Usually it is desirable to know not only that pattern X is spatially associated with pattern Y, but that the association is significantly greater than might occur due to chance. Unfortunately, although the association of one map with another can be measured and described quantitatively, the significance of a particular correlation value in a probability sense is much more difficult to evaluate with spatial data than with nonspatial data.This is because spatial data usually do not satisfy the assumptions of classical statistical models, particularly in regard to the independence of samples. Remember that in classical hypothesis testing, one of the basic assumptions is that an "independent random sample" is used, and that the samples are "independent and identically distributed". Suppose that two binary maps are to be compared at a series of randomly selected point locations. If the average distance between the sample locations is large, then the assumption of independence of samples may be reasonable, but as the number of random samples is increased and the separation between points becomes small, then samples are no longer independent in a statistical sense. In general we find that the greater the number of samples from a spatial dataset, the closer together they are, and the more they are spatially autocorrelated, which means that they are not independent. Consequently, summary measures of correlation between maps are generally useful in a descriptive sense only, and the statistical significance of tests about null hypotheses are in error because the estimates of variance of the correlation statistics are too small. Instead of using the term "significant", which is loaded with probability connotations, it is safer to talk about "unusual" or "interesting" correlations, words that draw attention to an association, rather than implying that a null a hypothesis of zero correlation can be rejected at some level of probability, see Openshaw (1991). Thus inter-map correlation measures are generally useful for *exploring* relationships rather than *confirming* them.

The second aspect of map correlation that is most important in a GIS environment is measurement scale. The product moment correlation coefficient, the usual Pearson's r, is not suitable for attributes measured on nominal or ordinal scales, and these levels of measurement are widely used in GIS. Often, the raw data on which spatial associations are to be measured can be cast as an area cross-tabulation, with the classes of one map being the rows, and the classes of the second map being the columns. This form of the data can be visually inspected, graphically displayed, and employed to derive measures of correlation. If the map classes are nominal only (implying that rows or columns of the area table can be moved around without affecting the information content) then a variety of correlation measures, or "association" measures, can be employed. Several of these are based on chi-squared statistics, and assume that the area cross-tabulation can be treated as a contingency table. Entropy measures, also called information statistics, are also useful. For pairs of maps with the same number of classes, where the classes are directly comparable and matched between the two maps, the kappa coefficient of agreement, often employed for evaluating classification accuracy of remote sensing images, is an appropriate correlation measure. In situations involving binary maps, "matching coefficients" (as used in cluster analysis), and measures based on the notions of probability and odds are helpful. Ordinal data can either be treated as nominal data, or rank correlation measures can be applied. Of course, if both maps are measured using continuous interval or ratio scale variables, then Pearson's r is appropriate. Some of these measures, and examples of hów to calculate them, are discussed below.

In general, the global correlation of mapped variables in geology over a whole study area is seldom very great. Too many processes are operating, generating a naturally noisy dataset, and global correlations are hardly to be expected. More often, correlations occur locally, under particular sets of conditions. It is therefore a mistake to calculate inter-map correlations without actually looking at the spatial patterns with a map view, because the "baby can get thrown out with the bathwater". Local correlations are often apparent visually that are simply lost in an overall statistical summary. Initial display of map combinations can lead to the definition of local domains over which particular correlations *are* unusual and potentially

Table 8-2. Unique conditions table resulting from the overlay of a bedrock geology map (Figure 8-8A) and a surficial geology map (Figure 8-8B). There are 24 possible unique conditions, because one map has 3 classes, the other 8 classes, but two conditions do not occur reducing the actual number from 24 to 22. Each polygon on the overlay map is keyed to the unique condition number. Thus the unique conditions map could be reclassified, with this table as a lookup, by bedrock class or surficial class (thereby reconstituting the original maps), or by area.

Unique condition #	Bedrock class	Surficial class	Area, km^2
1	1	1	1438.63
2	3	3	344.89
3	1	3	292.86
4	2	1	176.80
5	3	1	103.68
6	2	3	89.70
7	1	5	87.69
8	1	6	85.31
9	2	2	81.39
10	1	8	62.94
11	2	5	42.23
12	2	6	40.41
13	1	2	34.40
14	3	5	15.39
15	1	4	14.17
16	3	2	11.13
17	2	8	10.23
18	3	8	9.95
19	3	4	1.19
20	2	7	1.01
21	1	7	0.79
22	2	4	0.44

Table 8-3. Area cross-tabulation for the same maps summarized in the unique conditions overlay table. The same information is simply presented in a different form. The units are km^2. Each cell of this table corresponds to a unique condition class. Notice that two cells of the table contain zero area, as already noted in the caption to the unique conditions table (Table 8-2).

	Bedrock geology map			
Surficial geology map	**1.Goldenville Fm**	**2.Halifax Fm**	**3.Granite**	**Total**
1.Quartzite Till	1438.63	176.80	103.68	1719.11
2.Slate Till	34.40	81.39	11.13	126.92
3.Bedrock / Granite Till	292.86	89.70	344.89	727.45
4.Lawrencetown Till A	14.17	0.44	1.19	15.80
5.Lawrencetown Till B	87.69	42.23	15.39	145.30
6.Red granite Till	85.31	40.41	0.00	125.73
7.Bridgwater Conglom.	0.79	1.01	0.00	1.80
8.Unclassified	62.94	10.23	9.95	83.13
Total	2016.80	442.22	486.23	2945.24

interesting.The initial appraisal of inter-map correlations can often be effectively carried out by superimposing polygon outlines from one map on to a colour display of the other map, or by using the RGB or IHS colour assignments to make colour composite images, as discussed in Chapter 4.

Another factor that makes the assessment of spatial associations difficult to interpret is the effect of edges or boundaries, and the choice of a region for making measurements. This can easily be seen by examining the sensitivity of measures of spatial association to changes in the size or shape of a study area. Clearly, if the correlation measure changes drastically with minor shifts in the boundary of the study region, the results are not robust and interpretations based on the results are suspect.

Area Cross-Tabulations

Consider the spatial correlation of bedrock and surficial geology in the Meguma terrane of Nova Scotia. An area cross-tabulation is simply a two-dimensional table summarizing the areal overlap of all the possible combinations of the two input maps. It contains the same information as a unique conditions overlay table, but is organized differently, as shown in Tables 8-2 and 8-3 . Notice that the relationship between the surficial geology (Figure 8-8A) and bedrock geology (Figure 8-8B) is apparent both on the map and in the cross-tabulation. We see that class 1 of the bedrock map (Goldenville Fm) and class 1 of the surficial map (Quartzite till) are spatially associated. Given the presence of one of these units, there is a strong probability that the other unit is also present. The Slate Till is predominantly occurring on the Halifax Fm, although the Halifax is also overlain by a variety of other surficial units. Thus although the presence of Slate Till strongly suggests the presence of Halifax, Halifax is not such a good predictor of Slate Till. The Granite mapped on the bedrock map predominantly is associated with the surficial unit that is mapped as either Bedrock or Granite Till (shown here as a single map unit), but again the Bedrock/Granite Till unit is also found associated with the Goldenville Fm. These associations are reinforced by making a map showing where these associations are present, Figure 8-8C, created with a matrix overlay. The area cross-tabulation, however, is not very easy to interpret, because the frequency of each map unit differs between classes, and it is difficult to know whether a particular combination occurs more frequently than might be expected if the two maps were generated by independent processes. Various measures can be used to quantify the degree of association between maps, and in each case the calculations are based on the numbers from the area cross-tabulation.

Note that an overall measure of whether one map is correlated with another is not always very illuminating. Correlations of particular units on one map with particular units on another map are usually more enlightening, and such correlations may also be restricted geographically or thematically, due the complexity of real-world processes. The asymmetry of correlations, as indicated here, is also an important factor; the presence of one unit may be a good predictor of the presence of another unit, but the converse is not necessarily true.

An issue that bears on correlation measures applied to areas is the question of measurement units. Clearly the size of an area has a magnitude that is constant, but the number used to describe the size varies with the measurement unit. Thus 1 km^2 is the same area as 10^6 m^2, yet the number is a million times larger. Correlation measures that are affected by the choice of units are less desirable than those that are independent of units. Furthermore, there is a tendency to think of the areal unit of measurement as representing a sampling unit, or count. Thus by picking small units, an apparent but misleading impression of sample size is created. In fact, the smaller the sampling unit (pixel size in a raster for example), the greater is the effect of spatial autocorrelation and consequent dependency of neighbouring spatial objects. The tendency to think of the count of area units as the sample size, n, must therefore be avoided, as the "true" n is likely to be much smaller.

Although the areas recorded in area tables are sensitive to measurement units, area proportions are not. Area proportions are the areas divided by total area. Proportions are the same for any areal units, whether the measurement is a count of pixels, a measurement of a polygon from the bounding vertices, or a count of sample points (on a lattice or random). Even with proportions, however, the problem of spatial dependency due to autocorrelation is still present, invalidating the traditional statistical tests of significance. As we shall see, chi-square statistics are affected by the choice of units, but entropy and some other correlation measures are unchanged.

Nominal Scale Data

Many different measures of association for nominal scale data have been proposed. In the most general case of two maps each with multiple classes, probably the most common measures are based either on chi-squared statistics or on entropy statistics. For the special case of two maps with the same number of matched classes, the kappa coefficient of agreement is often quoted. Where the two maps are binary, a host of similarity or "matching coefficients" have been used, as developed mainly for use in cluster analysis, see Romesburg (1990) for a review. A variety of measures based on conditional probability, such as the odds ratio, Yule's coefficient of association, and the "contrast" can be applied to binary maps. They are introduced here, partly because they are useful correlation measures in their own right, and partly because of their relationship with "weights of evidence", a method for combining multiple binary maps together for predictive modelling, discussed further in Chapter 9.

In using the widely known chi-square statistic, the area cross tabulation is used like a contingency table. Let the area table between map A and map B be called matrix ***T***, with elements T_{ij}, where there are i=1,2.. n classes of map B (rows of the table) and j=1,2.. m classes of map A (columns of the table). The marginal totals of ***T*** are defined as $T_{i.}$ for the sum of the i-th row, $T_{.j}$ for the sum of the j-th column, and $T_{..}$ for the grand total summed over rows and columns. If the two maps are independent of one another, with no correlation between them, then the expected area in each overlap category is given by the product of the marginal totals, divided by grand total. Thus the expected area T_{ij}* for the i-th row and j-th column is

$$T_{ij}^{*} = \frac{T_{i.}\,T_{.j}}{T_{..}} \quad . \tag{8-2}$$

Then the **chi-square** statistic is defined as

$$\chi^2 = \sum_{i=1}^{n}\sum_{j=1}^{m} \frac{(T_{ij} - T_{ij}^{*})^2}{T_{ij}^{*}} \quad , \tag{8-3}$$

the familiar *(observed-expected)*^{2}*/expected* expression, which has a lower limit of 0 when the observed areas exactly equal the expected areas and the two maps are completely independent. As the observed areas become increasingly different from the expected areas, chi-square increases in magnitude and has a variable upper limit. Two commonly quoted coefficients of association based on chi-square values are the Cramers coefficient, V, and the contingency coefficient C, The former is defined as

$$V = \sqrt{\frac{\chi^2}{T_{..}\;M}} \quad , \tag{8-4}$$

and the latter as

$$C = \sqrt{\frac{\chi^2}{T_{..} + \chi^2}} \tag{8-5}$$

Table 8-4. Chi-square values, reported to 1 decimal place, for the area cross-tabulation between bedrock and surficial geology maps. A large value of chi-square can result from the observed area being greater or less than the expected area. Although chi-square values are affected by the units of measurement, but they can be useful within a single table in a relative sense. For example, note that one of the largest values (203.9) is for the overlap of the Slate Till on the Halifax Fm., a bedrock unit that also contains slate.

	Bedrock geology map		
Surficial geology map	**1.Goldenville Fm**	**2.Halifax Fm**	**3.Granite**
1.Quartzite Till	58.1	25.6	114.3
2.Slate Till	31.7	203.9	4.6
3.Bedrock/Granite till	84.6	3.5	420.8
4.Lawrencetown Till A	1.0	1.6	0.8
5.Lawrencetown Till B	1.4	19.1	3.1
6.Red Granite Till	.0	24.6	20.8
7.Bridgwater Conglom.	.2	2.0	.3
8.Unclassified	.7	.4	1.0

where M is the minimum of (n-1, m-1). Chi-square varies in magnitude depending on the degree with which the expected areas equal the observed areas. Chi-square is also strongly dependent on the units of measurement, being proportional to the size of the areal unit. Thus if the lineal units change by a factor of 100 from metres to centimetres, then areal units increase by a factor of 10,000, and chi-square likewise. The magnitude of the contingency coefficient, C, is independent of measurement units, and varies between 0 (indicating no correlation between the maps) to a maximum value less than 1, depending on χ^2 and the total area. Cramers V also has a value of 0 for independence, and a maximum value that is less than 1, depending on χ^2, total area and the area table dimensions. Where these measures are applied to tables of counts based on independent random sampling, the statistical significance of associations can be tested, but with areal measurements this is clearly inappropriate. Nevertheless, chi-square values can provide an *exploratory* and *descriptive* measure of spatial correlation between maps if these limitations are appreciated. The table of chi-square values calculated from Table 8-3 is shown in Table 8-4. The chi-square value for the whole table is approximately 1023, which is not very helpful because there is no basis for judging how significant this value is, being dependent on the units of measurement. On the other hand, the relative variation in chi-square values within the table is quite instructive. The largest value is for the Granite with Bedrock/Granite Till association, the second-largest for the Halifax with Slate Till association, and so on. Some of the other large χ^2 values in the table correspond to cases where the observed area is less than the expected area, which is one of the disadvantages of using this measure, although it is simple to distinguish between positive and negative associations if the expected values are output as part of the results. For this case, $V = 0.4169$ and $C = 0.5079$. Although it is not possible to assess the significance of these coefficients statistically, they are both substantially greater than 0.

Entropy measures, also based on the area cross-tabulation matrix ***T***, can also be used for measuring associations. Suppose that the T_{ij} values are transformed to area proportions, *p*, by dividing each area element by the grand total $T_{..}$. Thus $p_{ij}=T_{ij}/T_{..}$, and the marginal proportions are defined as $p_{i.}=T_{i.}/T_{..}$ and $p_{.j}$ as $T_{.j}/T_{..}$. Then entropy measures, also known as **information statistics** can be defined using the area proportions as estimates of probabilities. Proportions are dimensionless, so entropy measures have the advantage over chi-squared measures of being unaffected by measurement units.

Assuming that an area proportions matrix for map A and map B has been determined from ***T***, then the entropy of A and B are defined as:

$$H(A) = -\sum_{j=1}^{m} p_{.j} \ln p_{.j} \quad \text{and} \qquad (8\text{-}6)$$

$$H(B) = -\sum_{i=1}^{n} p_{i.} \ln p_{i.} \quad ,$$

where ln is the natural logarithm.The **joint entropy** of the combination , *H(A,B)*, is simply

$$H(A,B) = -\sum_{i=1}^{n} \sum_{j=1}^{m} p_{ij} \ln p_{ij} \quad . \qquad (8\text{-}7)$$

Then a "**joint information uncertainty**", *U(A,B),* can be used as a measure of association, and is defined as

$$U(A,B) = 2\left[\frac{H(A)+H(B)-H(A,B)}{H(A)+H(B)}\right] \quad , \qquad (8\text{-}8)$$

which varies between 0 and 1. When the two maps are completely independent, then $H(A,B) = H(A) + H(B)$ and $U(A,B)$ is 0, and when the two maps are completely dependent, $H(A) = H(B) = H(A,B) = 1$, and $U(A,B)$ is 1. This measure can be interpreted as a symmetrical combination of two uncertainty measures, one being the uncertainty with which B predicts A, the other being the uncertainty with which A predicts B. For a discussion of these quantities and their calculation, see Press et al. (1986). Note that although the significance tests can be applied to entropy statistics based on frequency counts and independent random sampling, such tests are inappropriate for area cross-tabulations, as before.

Of particular interest for pairs of nominal-scale maps containing the *same number of classes* and where the classes are *matched*, is the **coefficient of agreement, kappa**. Originally proposed by Cohen (1960) for use with psychology data, kappa measures the *amount of agreement* between attributes, and corrects for the *expected* amount of agreement. Kappa has been quite widely used for measuring classification accuracy of LANDSAT images, where the true classification is sometimes available as "ground truth". Consider the data in Table 8-5, derived from a paper by Rosenfield and Fitzpatrick-Lins (1986). There are five classes on each map, and the order of the two sets of classes is arranged to be the same, so that the classes are matched. The results of the classification can be expressed as the usual cross-tabulation, T, often referred to in remote sensing literature as the "confusion matrix". The values in the principal diagonal reflect the amount of agreement, and the off-diagonal elements show the amount of "confusion", where the classes of the maps are mismatched. The kappa coefficient is defined here in terms of expected area proportions with the formula

Table 8-5. Calculation of kappa coefficient of agreement from a confusion matrix. **A.** The confusion matrix, T (same as area cross-tabulation), from Rosenfield and Fitzpatrick-Lins (1986). **B.** Matrix of observed proportions, with elements p_{ij}. Shaded values in principal diagonal represent areas of agreement. **C.** Expected proportions matrix, q_{ij}, with principal diagonal shaded. These are the proportions expected in the null situation of no correlation, or independence. **D.** Conditional kappa and area proportion of agreement for each class. **E.** Table of chi-squared values for comparison.

A. CONFUSION MATRIX

Rows: Classified Landsat image; columns: Geological Map

Class	1	2	3	4	5	Total
1	148	0	1	1	1	151
2	1	50	6	0	0	57
3	8	15	39	6	0	68
4	2	3	7	25	1	38
5	0	0	1	1	6	8
Total	159	68	54	33	8	322

B. OBSERVED PROPORTIONS MATRIX

Rows: Classified Landsat image; columns: Geological Map

Class	1	2	3	4	5	Total
1	0.460	0.	0.003	0.003	0.003	0.469
2	0.003	0.156	0.019	0.	0.	0.177
3	0.025	0.047	0.121	0.019	0.	0.212
4	0.006	0.009	0.021	0.078	0.003	0.117
5	0.	0.	0.003	0.003	0.019	0.025
Total	0.494	0.211	0.167	0.103	0.025	1.000

C. EXPECTED PROPORTIONS MATRIX

Rows: Classified Landsat image; columns: Geological Map

Class	1	2	3	4	5	Total
1	0.232	0.099	0.078	0.048	0.012	0.469
2	0.088	0.037	0.030	0.018	0.004	0.177
3	0.104	0.045	0.035	0.022	0.005	0.212
4	0.058	0.025	0.020	0.012	0.003	0.117
5	0.012	0.005	0.004	0.003	0.001	0.025
Total	0.494	0.211	0.167	0.103	0.025	1.000

D. CONDITIONAL KAPPA VALUES

Class, i	K_i	Proportion correct
1	0.961	0.980
2	0.844	0.877
3	0.488	0.574
4	0.619	0.358
5	0.744	0.750

E. CHI-SQUARE MATRIX

Classified Landsat image / Geological Map

Class	1	2	3	4	5	Total
1	72.33	31.89	23.36	13.54	2.02	143.1
2	26.18	119.73	1.33	5.84	1.42	154.5
3	19.48	0.03	66.78	0.13	1.69	88.1
4	14.98	3.15	0.06	114.38	0.00	132.6
5	3.95	1.68	0.09	0.04	169.32	175.1
Total	136.92	156.48	91.62	133.93	174.45	693.4

$$K = \frac{\sum_{i=1}^{n} p_{ii} - \sum_{i=1}^{n} q_{ii}}{1 - \sum_{i=1}^{n} q_{ii}} \tag{8-9}$$

where p_{ij} is the observed area proportion defined above for entropy, and q_{ij} is the *expected* area proportion for the *i*-th row and *j*-th column under the assumption of no association between maps, obtained from the product of the marginal totals $p_{i.}$ and $p_{.j}$. *n* is the number of matched classes. In contrast to chi-square, which compares the observed and expected frequencies for each overlap category, kappa looks only at the categories in the principal diagonal of the proportions matrix. The term Σp_{ii} is the total observed agreement, whereas Σq_{ii} is the total expected agreement due to chance. Kappa varies between -1 (perfect disagreement) to +1 (perfect agreement), with a value of 0 indicating that the agreement is no better than that expected due to chance. In addition, besides the overall kappa value, a *conditional* kappa value, can be calculated for each class, showing the breakdown of agreement by class. Conditional kappa for the *i*-th class, is defined as

$$K_i = \frac{p_{ii} - q_{ii}}{p_{i.} - q_{ii}} \quad . \tag{8-10}$$

The calculation of kappa is illustrated in Table 8-5. After determining the observed and expected proportions (Table 8-5 B and C), the proportion of observed agreement is (0.460+0.155+0.121+0.078+0.019)=0.832, whereas the proportion of expected agreement is (0.232+0.037+0.035+0.012+0.001)=0.317, so kappa is (0.810-0.317)/(1-0.317) or 0.7544. Without adjusting for the expected agreement, the observed agreement is 83.2% (the proportion of the total area that is correctly classified), but after removing the expected agreement due to chance, the value of kappa is 75.44%. The conditional kappa for class 1 is (0.460-0.232)/(0.469-0.232)=0.962, slightly different from the value shown in Table 8-5 D due to rounding error. The conditional kappa values show that individual classes vary in agreement from 96% to 49%, and that a simple estimate of the correct proportion overstates the degree of correlation in each class.

It is also instructive to compare chi-square and kappa values for these data. Table 8-5 E gives the individual chi-square terms and the totals. As with kappa, chi-square corrects for the expected overlap area, but each *individual* term is *divided* by the expected value. Further, chi-square is summed over all the terms in the table, not just over the principal diagonal, so that large deviations from expected that occur in off-diagonal locations can make major contributions to the total. This illustrates that chi-square is really a goodness-of-fit measure rather than a goodness-of-agreement measure. Kappa is easier to interpret than chi-square, but is limited to the situation of comparing maps with the same number of matched classes, whereas chi-square (and entropy) can be applied to any area cross-tabulation. Inkley et al. (1984) describe a method, called the Linear Assignment Method, to find the best assignment of classes of one categorical map with the classes or a second map, also categorical. The number of classes must be the same on each map, or nearly so. This method could be used as a precursor to calculating kappa, in cases where the matching of classes is not obvious.

The expression for the variance of kappa is given in Bishop et al. (1975), and is reported in Hudson and Ramm (1987). Where kappa is applied to a random sample of pixels, as is usually the case for checking the accuracy of LANDSAT classifications, and assuming that kappa is normally distributed, the null hypothesis that kappa is zero can be tested. However, if the confusion matrix is the total area cross-tabulation, such a significance test is not justified, as discussed before.

Binary maps

The comparison of binary maps is a special case of comparing nominal scale maps, and any of the above methods apply. The need to compare binary maps occurs regularly in geology, as for example in comparing two different kinds of binary anomaly patterns, or matching the presence of a particular lithology to a distinctive geophysical or geochemical indicator pattern. Indicator patterns (simply binary maps with a value of 1 or 0) are often used for estimating mineral potential by combining anomalies of different kinds, so that pairwise associations are important to understand and characterize. Furthermore, binary patterns are simpler to deal with than multi-class maps, where the number of possible overlap conditions is large and unwieldy.

Besides the chi-square, entropy and kappa measures described above, any of the so-called similarity or "matching" coefficients used in cluster analysis (Romesburg, 1990) can be employed for binary map comparison. In addition, methods based on "odds ratios" are briefly introduced, partly because of their relevance to the topic of "weights of evidence", used for combining map patterns in the next chapter, but also because they are useful correlation measures in their own right.

To illustrate the calculations with a geological example, consider the bedrock and surficial geology maps used for the chi-square calculations. Two binary map patterns have been selected from these multi-class maps by reclassification, and the overlap is shown in Figure 8-14. One binary pattern is the Granite unit (class 1 on the geology map). This is called map A, and the pattern is either present or absent. Similarly, map B is the binary pattern produced by the presence or absence of the Bedrock/Granite Till (referred to as Granite Till hereafter) unit from the surficial map. The overlap relationships of these two binary maps are summarized as the area cross-tabs $\boldsymbol{T}$ matrix and as a Venn diagram in Figure 8-15. The goal is to examine various possible measures of correlation between the two binary map patterns. How well can the presence of Granite be predicted from the presence of the Granite Till, for example?

Table 8-6. Relationship between probability, *P*, odds, *O*, and logits, ln (*O*), the natural logarithms of odds. Odds are often, but not necessarily, expressed as a ratio. Note that a probability of one half corresponds to odds of 1 (the same as 50/50 or "evens") and a logit of zero.

P	*O*	ln*O*
.0	0	$-\infty$
.1	1/9	-2.20
.2	1/4	-1.39
.4	2/3	-0.41
.5	1/1	0.00
.6	3/2	0.41
.8	4/1	1.39
.9	9/1	2.20
1.0	∞	∞

A commonly-used **similarity or matching coefficient** used for cluster analysis, where samples are described by a number of binary attributes, is Jaccard's coefficient, C_J. In terms of the area cross-tabulation matrix, $\boldsymbol{T}$, this coefficient is defined as

$$C_J = \frac{T_{11}}{T_{12} + T_{21} + T_{11}} \quad , \tag{8-11}$$

where T_{11} is the area of "positive match", with both patterns present, and T_{12} and T_{21} are the areas of "mismatch". Notice that the area of "negative match", T_{22}, is ignored, so the similarity of patterns as measured by Jaccard's coefficient is not affected by regions where both patterns are absent. Thus, for the data in Figure 8-15, C_J is 345 / (382 + 141 + 345) = 0.3975. Jaccard's coefficient ranges between 0 (complete dissimilarity) to 1 (complete similarity). No correction is made for the *expected* degree of matching, as in the case of chi-square or kappa.

A similar similarity measure is the so-called **simple matching coefficient,** C_A, which is defined as the total area where the patterns match (positive and negative) divided by the total area. With areal data, this coefficient is known in the geographic literature as the coefficient of areal association, e.g.Taylor (1977). The coefficient ranges between 0 and 1, as before. For comparison with Jaccard's coefficient, the areal coefficient of association for the same data is (345 + 2077) / (345 + 2077 + 382 + 141) = 0.8244. Notice that the value is appreciably greater, due to the inclusion of negative matches in the expression. In this example, the area of negative match is 2077 km^2, by far the largest of the 4 overlap regions. Notice that C_J and C_A are dimensionless and therefore independent of the units of areal measurement.

We now turn to the idea of the probability and odds, as applied to the relationship between two binary map patterns. Probability can be expressed as odds, or vice-versa, using the relation O = P / (1 - P), see Table 8-6. Odds values less than 1 correspond to probabilities less than 0.5, and very small probabilities are nearly the same as odds. Logits are the natural logarithms of odds. The logit scale is therefore centred about 0, corresponding to a probability of 0.5, with negative values for odds less than 1/1 and positive values for odds greater than 1/1. In the following description, the background to a measure called the *odds ratio* is explained. Logits are used in logistic regression models and for Bayesian weights of evidence modelling, as described later.

First, consider the idea of probabilities in relation to the Granite and Granite Till binary patterns from Figures 8-14 and 8-15. The four possible overlap conditions are summarized graphically and correspond to the four elements of the area matrix, $\boldsymbol{T}$. The area proportions for each overlap condition are $p_{ij}=T_{ij}/T_{..}$ as before, and the proportions are treated as estimates of probabilities. Thus the probability of pattern A occurring in the study region is P{A}, estimated by $p_{.1}$, the probability of B occurring is P{B}, estimated by $p_{1.}$, the probability of A and B occurring together is P{A∩B}, estimated by p_{11}, the probability of A and not B occurring is P{A∩$\overline{\text{B}}$}=p_{21}, and similarly P{$\overline{\text{A}}$∩B}=p_{12}, and P{$\overline{\text{A}}$∩$\overline{\text{B}}$}=p_{22}. The bar over the letter A or B means *not*, and ∩ is the symbol for a logical intersection or Boolean AND operation.

The conditional probability of B occurring given the presence of A is written as P{B | A}. From probability theory, conditional probability is defined as P{B | A} = P{B∩A} / P{A}. Thus the probability of pattern B occurring, given the presence of A can be expressed as a probability ratio, or equivalently as a ratio of area proportions, or directly in terms of areas, satisfying the relationships

$$P\{B|A\} = \frac{P\{B\cap A\}}{P\{A\}} = \frac{p_{11}}{p_{.1}} = \frac{T_{11}}{T_{.1}} \quad , \tag{8-12}$$

which follows from the basic definition of conditional probability, followed by the substitution of area proportions as estimates of probabilities, and finally as a ratio of areas. Given the data for the Granite (A) and the Granite Till (B), the unconditional probability of the Granite Till occurring, P{B}, is 727/2945 = 0.2469, whereas the conditional probability of the Granite Till given the presence of the Granite, P{B|A}, is 345/486=0.7098. Clearly this indicates a strong association between the Granite and the Granite Till, because where the Granite occurs the probability of the Granite Till also occurring increases sharply from .2469 to .7098. If the conditional probability is now converted to conditional *odds*, O{B|A} by the relation between odds and probability mentioned above, then

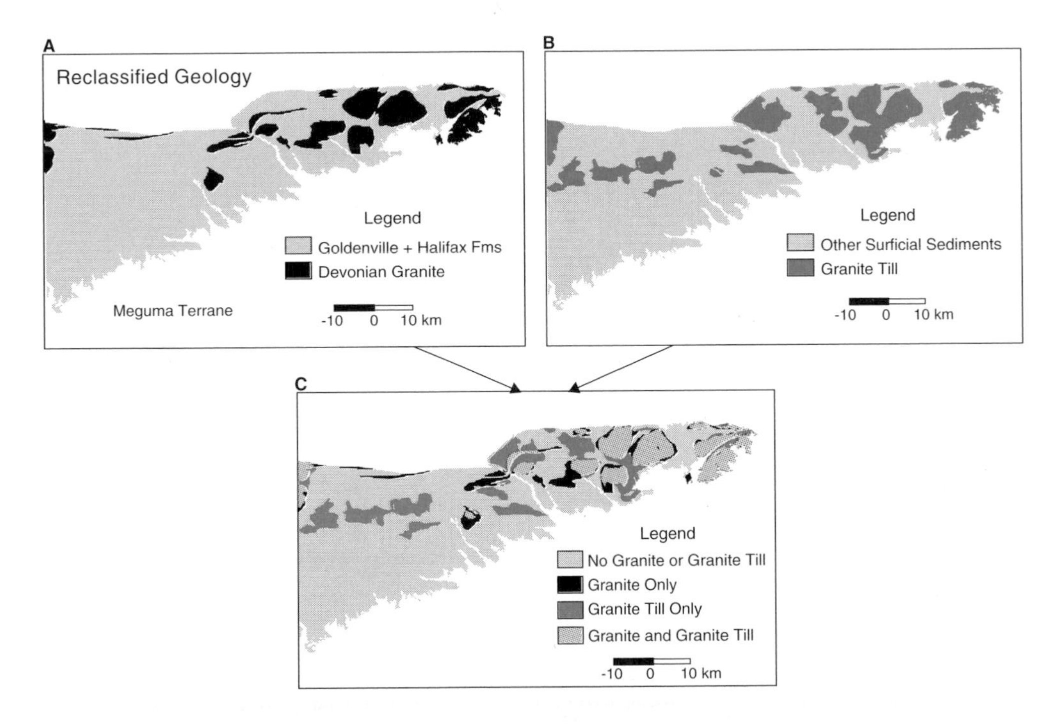

FIG. 8-14. Comparison of two binary maps, made by reclassifying the multi-class bedrock and surficial maps shown in Figure 8-8. Map A is the Granite from the bedrock geology map, map B is the Granite Till from the map of surficial units (this unit also includes exposed bedrock with little to no surficial cover).

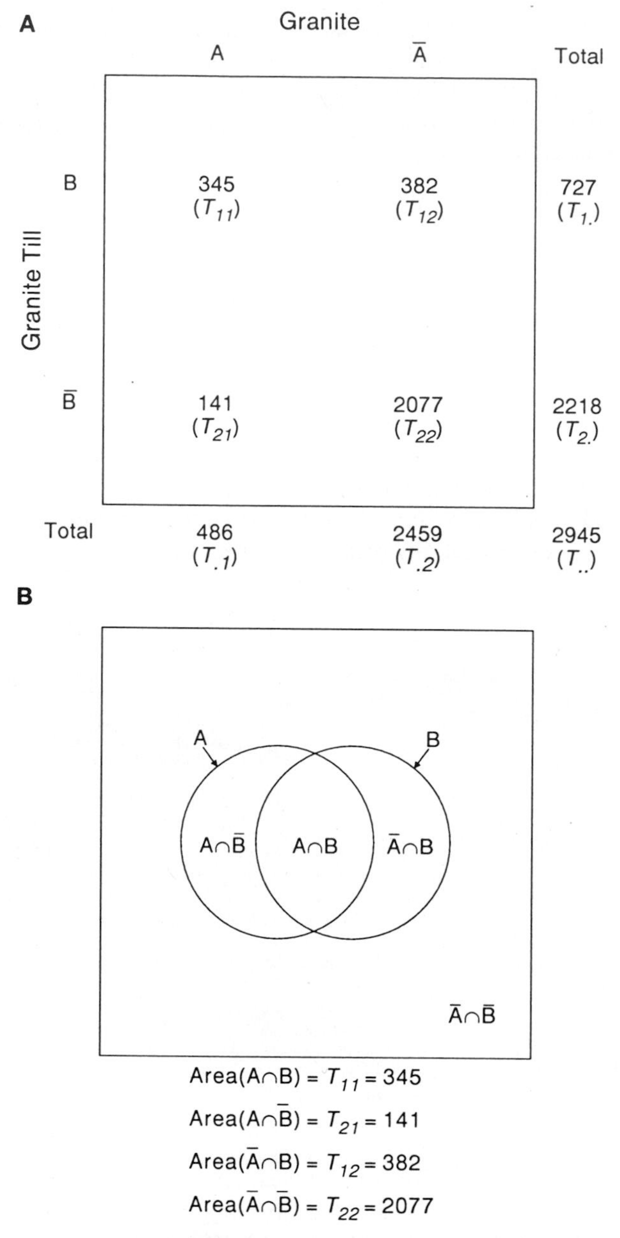

FIG. 8-15. A. Area cross-tabulation showing the overlap values for the Granite and Granite Till. **B.** Venn diagram summarizing the overlap relationships between the two binary maps in graphical form (areas not drawn to scale). Numbers are areas measured in km^2.

$$O\{B|A\} = \frac{P\{B|A\}}{1 - P\{B|A\}} = \frac{P\{B|A\}}{P\{\bar{B}|A\}} , \tag{8-13}$$

which follows because 1 - P{B|A}=P{$\bar{B}$ | A}. The probability of Granite Till being absent, given the presence of granite, P{$\bar{B}$ | A}, is $p_{21}/p_{.1}$, (using the same logic as Equation 8-12) so O(B | A) can now be defined as

$$O\{B|A\} = \frac{p_{11}/p_{.1}}{p_{21}/p_{.1}} = \frac{p_{11}}{p_{21}} = \frac{T_{11}}{T_{21}} , \tag{8-14}$$

because the ratio of the areas is the same as the ratio of the area proportions. In terms of our example data, this now means that the odds of Granite Till occurring given the presence of Granite are (345/141)=2.45. In other words if Granite is present then the odds of Granite Till also being present are nearly 25 to 10, as compared to the unconditional odds of Granite Till being present if the bedrock unit is unknown, which is 727/2218=0.3278, or approximately 3 to 10. Knowing that Granite is present therefore increases the odds that Granite Till is also present by about 8 times.

Instead of having to chose which map is the dependent one in the relationship, it would be preferable to have a measure that was symmetrical in the sense that neither map is the causative agent that precedes the other. Accordingly, a similar argument to the above is used to derive an expression for the conditional odds of B given the *absence* of A, O{B | $\bar{A}$}:

$$O\{B|\bar{A}\} = \frac{p_{12}}{p_{22}} = \frac{T_{12}}{T_{22}} , \tag{8-15}$$

and applying this formula to the example data produces (382/2077)=0.1839. Thus the knowledge that the Granite is absent reduces the odds of the presence of Granite Till from more than 3 in 10 to less than 2 in 10. Notice that the absence of Granite carries much less information than the presence of Granite in predicting the presence of Granite Till, which is not surprising from a geological point of view.

We can now combine the two conditional odds expressions (Equations 8-14 and 8-15) together in various ways to provide a symmetrical measure of association between the two binary patterns. The simplest measure is the **odds ratio** O_R, defined as

$$O_R = \frac{O\{B|A\}}{O\{B|\bar{A}\}} = \frac{T_{11}T_{22}}{T_{12}T_{21}} , \tag{8-16}$$

which is straightforward to calculate from the area cross-tabulation, being the product of the terms on the principal diagonal, divided by the product of other two terms. For the example data, O_R= (345*2077) / (382*141) = 13.3. The numerator is a measure of agreement between the patterns, and the denominator a measure of disagreement. The odds ratio is always positive, being greater than 1 for patterns that are positively associated, 1 if the two patterns are independent and less than 1 if they are negatively associated. Transforming to a logit scale, by taking the natural logarithm of the odds ratio, produces a closely-related index of association called the **contrast**, C_W. In this case, the contrast therefore has a value of ln(13.3) = 2.59. The contrast is 0 when the patterns overlap only by the expected amount due to chance, is positive for positive associations and negative for negative associations.

The difference of the square roots of the conditional odds, divided by the sum of their square roots provides another convenient association measure, referred to here by the Greek letter α. This measure was originally proposed by **Yule**, as quoted by Fleiss (1991), and can be expressed in terms of the overlap areas by

$$\alpha = \frac{\sqrt{O\{B|A\}} - \sqrt{O\{B|\bar{A}\}}}{\sqrt{O\{B|A\}} + \sqrt{O\{B|\bar{A}\}}} = \frac{\sqrt{T_{11}/T_{21}} - \sqrt{T_{12}/T_{22}}}{\sqrt{T_{11}/T_{21}} + \sqrt{T_{12}/T_{22}}} . \quad (8\text{-}17)$$

α ranges in value between -1 and +1 like a correlation coefficient, with 0 implying independence of the two patterns. For the present example, α=0.5699 indicating a strong positive association.

Any one of these measures of associations between binary patterns can be applied to the comparison of multi-class categorical maps by treating each combination of map classes as a binary case, lumping together other classes in the same manner as was done to isolate the granite and granite till patterns from the bedrock and surficial geology maps, respectively. By this means, a matrix of binary coefficients can be produced, one for each element of the area cross-tabulation or ***T*** matrix. Inspection of these coefficients usually gives more insight

Table 8-7. Odds ratio values for the bedrock and surficial geology maps, treating each overlap combination as a binary case, as discussed in the text. Notice that the Granite with Granite Till relationship stands out as the largest value, and the Halifax Fm. with Slate Till value close behind. Odds ratios >1 indicate a positive association, values between 0 and 1 indicate a negative association.

	Bedrock geology map		
Surficial geology map	**1.Goldenville Fm**	**2.Halifax Fm**	**3.Granite**
1.Quartzite Till	5.75	.42	.14
2.Slate Till	.16	12.17	.47
3.Bedrock/Granite Till	.19	.74	13.25
4.Lawrencetown Till A	4.01	.16	.41
5.Lawrencetown Till B	.69	2.46	.59
6.Red Granite Till	.97	2.85	.00
7.Bridgwater Conglom.	.36	7.22	.00
8.Unclassified	1.45	.79	.68

into the overlap relationships of note than either the area values themselves or the chi-square or entropy values that give an overall correlation number. For example, Table 8-7 shows odds ratios calculated for all combinations of classes on the bedrock and surficial geology maps.

The "**contrast**", C_W, is also interesting because of its relationship to the weights of evidence, used for quantifying spatial associations between binary map patterns and for predictive modelling, as will be described in Chapter 9. Note here that the contrast can be expressed as the difference in the natural logarithms of the conditional odds by the relation

$$C_W = \ln O\{B|A\} - \ln O\{B|\bar{A}\} \quad . \tag{8-18}$$

However, the contrast is normally expressed as the difference between the weights in "weights of evidence". This topic is briefly introduced here because the weights can be very nicely derived in terms of logits, and related to the terms of the (2x2) area table.

The idea behind **weights of evidence** is that several binary patterns can be combined together to predict another binary pattern. We might, for example, want to predict the presence of a buried granite, given a series of binary patterns for geochemical and geophysical map patterns as indicators. A pair of weights W^+ and W^-, are determined for each predictor pattern, depending on the measured spatial association between the pattern and the granite. The weights may then be combined from each pattern to make a predictive map for granite. Taking a single predictor pattern, B, the positive weight, W^+, can be expressed as the difference between the unconditional or *prior* logit of B, and the conditional or *posterior* logit of B.

$$W^+ = \ln O\{B|A\} - \ln O\{B\} = \ln \left[\frac{O\{B|A\}}{O\{B\}}\right] \quad . \tag{8-19}$$

Then by substituting the expression for conditional odds from Equation 8-14, and an expression for unconditional odds, W^+ can be expressed in terms of the overlap areas in the ***T*** matrix:

$$W^+ = \ln\left[\frac{T_{11}/T_{21}}{T_{1.}/T_{2.}}\right] = \ln\left[\frac{T_{11}T_{2.}}{T_{21}T_{1.}}\right] \quad . \tag{8-20}$$

Similarly, W^- is the difference between the *prior* logit of B and the *posterior* logit of B, given the *absence* of pattern A. Substitution of Equation 8-15 leads to an expression for the negative weight, W^-:

$$W^- = \ln O\{B|\bar{A}\} - \ln O\{B\} = \ln\left[\frac{T_{12}T_{2.}}{T_{22}T_{1.}}\right] \quad . \tag{8-21}$$

Having computed the weights , they are now applied to predict the response pattern. The *posterior* logit (or probability after a simple transformation) is generated as a map using the model

$$\ln O\{B|A\} = \ln O\{B\} + W^+ \qquad and \tag{8-22}$$

Table 8-8. Sensitivity of measures of association to changes in the area cross-tabulation. Cases I to VI are slightly different versions of the same **T** matrix, simulating various effects. The table shows the effects on selected association measures. **Case I.** Original *T* matrix. **Case II.** Enlarge boundary. **Case III.** Close agreement. **Case IV.** Agreement of absence. **Case V.** Strong disagreement. **Case VI.** Almost independent.

I

610	49
41	108

II

610	149
141	108

III

610	1
1	108

IV

108	49
41	610

V

41	610
108	49

VI

657	170
152	35

Case	χ^2	C	U	C_A	κ	O_R	α	C_W
I	328	0.54	0.35	0.89	0.64	32.8	0.70	3.49
II	56	0.23	0.05	0.71	0.23	3.1	0.28	1.14
III	704	0.70	0.96	0.99	0.99	65880	0.99	11.10
IV	328	0.54	0.35	0.89	0.64	32.8	0.70	3.49
V	328	0.54	0.35	0.11	-0.28	0.03	-0.70	-3.49
VI	0.3	0.02	0.00	0.68	-0.18	0.89	-0.03	-0.12

$$\ln O\{B|\bar{A}\} = \ln O\{B\} + W^{-} \quad , \tag{8-23}$$

where the *prior* logit, ln *O{B}*, is assumed to be constant for the whole map area. Equation (8-22) is used where A is present, and (8-23) where A is not present. This is the basis of the weights of evidence method, but instead of 2 binary maps (giving only a trivial result), there can be many maps, and a pair of weights is calculated for each map pattern to be used for prediction, as discussed in Chapter 9.

Before moving on from the subject of associations between binary maps, the various measures of association are reviewed and compared with respect to the data summarized in Table 8-8. Six cases are presented, each represented by a (2x2) area matrix, *T*. Case I is the starting point, to be used as a basis for comparison. Notice that out of the total area of 808 units, 610 units are occupied by the presence of both A and B, 108 units by A and B, with only a small area where the patterns are mismatched. Case II is the same as Case I, but the off-diagonal elements in the area matrix have both been increased by 100. This simulates the effect of increasing the size of the study area with "padding" that contains neither pattern A or B. Case III is to show the effect of reducing the off-diagonal elements to almost 0, creating a situation where patterns A and B are almost completely matched, or in perfect agreement. In Case IV, the values of T_{11} and T_{22} are interchanged, so that there are more areas of negative matching than positive matching. Case V is complementary to Case I in that the matching areas in the principal diagonal are switched with the mismatch areas, to examine the effect of disagreement between patterns as compared with agreement. Finally, Case VI was generated by using values close to the expected areas for Case I under an assumption of independence or no association between map patterns.

First, notice that the measures calculated for Case II are greatly reduced as compared with those of Case I, illustrating the sensitivity of any of these measures to the choice of the study region. A modest change in the boundaries of the region, where the new area contains the presence of neither pattern, can radically alter the results. The implication of this is that it is all too easy to either overlook interesting associations, or conversely, to be impressed by apparently large associations, simply due to the choice of the region for measurement. In short, choosing appropriate spatial domains for measuring associations is important.

In Case III, the extensive overlap between the two patterns results in elevated values of χ^2, O_R, and C_W. Notice that U, C_A, κ and α are all close to 1 (the maximum), but C is only 0.7 because the maximum is less than 1. Case IV (agreement of absence, or large number of negative matches) is similar to Case I in that all the association measures fail to distinguish between the two types of agreement. On the other hand, Jaccard's coefficient, C_J, not shown in the table, *does* distinguish between positive and negative matches, being 0.87 for Case I and 0.55 for Case II. The results for Case V illustrate that χ^2, C and U yield the same results as Cases I and IV, even though the patterns are strongly mismatched. Yule's α and C_W are strongly negative, κ somewhat less strongly so, whereas O_R and C_A are small positive values. Finally, in Case VI (patterns independent), all association measures are small except C_A, which has a rather large value of 0.68 due to the failure of this coefficient to adjust for the effect of associations due to chance.

To conclude, α and κ both yield similar results and have the nice property of ranging between -1 and +1 like a regular correlation coefficient. χ^2, C and U should probably be avoided for binary cases, because they do not distinguish between large interactions due to agreement and large interactions due to disagreement. κ is useful for both binary and multi-class maps where the classes are matched. In general, care should be used in interpreting C_A because chance associations are not compensated. O_R and C_W are interesting because of their link to weights of evidence modelling, and because their computing formulas are relatively easy to remember! Finally, if matching areas of positive agreement are more important than matching areas of negative agreement, then C_J might be considered as a suitable candidate, although expected matching is not compensated.

Although these association measures can be instructive for map comparison, care must be exercised in their interpretation because they are sensitive to choice of the measurement domain and because their statistical significance is normally not possible to estimate by conventional means. Computer-intensive methods of significance testing are required, using Monte Carlo simulation procedures (e.g. Knudsen, 1987; Openshaw et al., 1987, 1990), a topic beyond the scope of the present discussion. Notice that all these measures, and the measures for ordinal, interval and ratio scale data to be discussed next, are based on object-to-object comparison and ignore the association between an object and its near neighbours.

Interval and Ratio Scale Maps

Continuous variables, like topographic elevation, Bouguer gravity or intensity values on a remote sensing image, can be compared with the product moment correlation coefficient, *r*:

$$r = \frac{\sum_{i=1}^{n} (x_i - \bar{x})(y_i - \bar{y})}{\sqrt{\sum_{i=1}^{n} (x_i - \bar{x})^2 \sum_{i=1}^{n} (y - \bar{y})^2}} , \tag{8-24}$$

where *x* and *y* are the values of the two maps, *x* and *y* are their respective means, and there are $i=1,2... n$ spatial objects. *r* varies between =1 (perfect correlation) through 0 (no correlation or independence) to -1 (perfect negative correlation). The spatial objects could be sample points, pixels or irregular polygons. For correlation based on polygons, it is desirable to use an area-weighted form of the above formula, so that large polygons will have more influence on the result than small polygons. The formula is then

$$r_W = \frac{\sum_{i=1}^{n} T_i (x_i - \bar{x})(y_i - \bar{y})}{\sqrt{\sum_{i=1}^{n} T_i (x - \bar{x})^2 \sum_{i=1}^{n} T_i (y_i - \bar{y})^2}} , \tag{8-25}$$

where T_i is the area of the *i*-th polygon, and $\bar{x}$ and $\bar{y}$ are now *area weighted* means. This formula is useful for calculating correlation coefficients from attribute tables associated with unique polygon or unique conditions overlays.

Ordinal scale data

Very commonly in GIS, the data is ordinal rather than interval or ratio in scale. This either arises because the original measurements are themselves based on rankings, or because interval or ratio scale data have been transformed to ranks in producing a map. For example, a common transformation of geochemical data is to produce classes based on unequal percentiles, emphasising the upper tail of the frequency distribution in order to enhance anomalous values.

The appropriate correlation measures for such data are rank correlation coefficients. For example, Spearman's rank correlation coefficient, r_S, is defined as

$$r_S = \frac{\sum_{i=1}^{n} (R_x - \bar{R}_x)(R_y - \bar{R}_y)}{\sqrt{\sum_{i=1}^{n} (R_x - \bar{R}_x)^2 \sum_{i=1}^{n} (R_y - \bar{R}_y)^2}}, \quad (8\text{-}26)$$

where R_x and R_y are the ranks of x and y respectively, and the bar indicates mean value as before. As can be seen, Equation (8-26) is the same as (8-24), except that ranks are substituted for the original values. Ranks are determined by simply sorting the data into ascending order, forming the integer sequence i=1,2,..n, then the rank R=i. The mean rank is then the middle value, an integer if n is odd, a half integer if n is even. The sum of the ranks of x or y is the same as the sum of the integers 1 to n, i.e. $0.5n(n+1)$. This leads to a simplified computing formula for r_S, as discussed in most elementary statistics books, that must be adjusted if there are tied ranks. Where rank correlation is to be computed for areal data, the number of tied ranks often becomes very great, so in general it is easier to transform the raw data to ranks and then to use Equation (8-26). Having transformed to ranks, the ordinary product moment correlation formula can be used, and tied ranks require no special correction. In calculating ranks R_x from x, if some of the x's have identical values it is conventional to assign all these ties the mean of the ranks they would have had if their values had been slightly different. Tied values will either be assigned an integer value or half integer, depending on the number of values with a particular x.

As before, if the areal data comes from irregular polygons (or unique conditions classes), a weighted version of (8-26) can be used which is the same as Equation (8-25) but substituting ranks. In this case, the ranks should also reflect area weighting, determined from a cumulative area distribution.

Consider the data in Table 8-9, where an overlay of Map A (variable x) and Map B (variable y) has resulted in 10 polygons. Both maps have only 5 classes, and the classes are ordered reflecting a scaled sequence. Notice that several ties exist, which is inevitable where there are more polygons than different values of x or y. In order to determine the area-weighted values of x, the values of x are treated in an ascending sequence. There are two polygons with x=1, the first having an area of 80 units, the second with an area of 37 units, a total of 117. Thus R_x for x=1 is 117/2=58.5. There is only one polygon with x=2, having an area of 71 units. The area-weighted rank for x=2 is 117+(71/2)=152.5. Similarly R_x for x=3 is 117+71+((17+18)/2)=205.5. Where x=4, the rank is 117+71+35+((10+3+22)/2)=240.5, and for x=5, R_x is 117+71+35+35+((30+51)/2)=298.5. The ranks of y are calculated in similar fashion. Then using the area weighted calculation of r_s gives a value of 0.822. This indicates a strong positive correlation between the two maps, as is clear from the table.

Alternatively, if the data have already been accumulated into an area cross tabulation between the two maps, as shown Table 8-9B, the calculation of weighted ranks is shown and the non-zero entries of the table can be used to carry out the calculation as before, giving the same result.

Table 8-9. Calculation of area-weighted rank correlation coefficient between two maps with ordinal scale data. **A.** Attribute data from a unique polygon (or unique conditions) overlay, with area weighted ranks appended. **B.** The same data organized into an area cross-tabulation.

A. Overlay attribute table

Polygon	Map A (x)	Map B (y)	Area	R_x	R_y
1	5	4	30	298.5	242
2	4	3	10	240.5	211
3	3	5	17	205.5	253.5
4	4	5	3	240.5	253.5
5	4	4	22	240.5	242
6	1	1	80	58.5	49
7	1	2	37	58.5	152
8	2	2	71	152.5	152
9	5	5	51	298.5	253.5
10	3	1	18	205.5	49

B. Area cross-tabulation.

Map B (y)	Map A (x) Class	1	2	3	4	5	Sum	Cum.	R_y
	1	80	0	18	0	0	98	98	49
	2	37	71	0	0	0	108	206	152
	3	0	0	0	10	0	10	216	211
	4	0	0	0	22	30	52	268	242
	5	0	0	17	3	51	71	339	253.5
	Sum	117	71	35	35	81			
	Cum.	117	188	223	258	339			
	R_x	58.5	152.5	205.5	240.5	298.5			

Table 8-10. Measures of map correlation based on combinations of measurement scales.

	Nominal	**Ordinal**	**Interval/Ratio**
Nominal	χ^2, O_R, α, C_W, etc.	Median by nominal class	Mean by nominal class
Ordinal		Rank correlation coefficient	Rank correlation coefficient
Interval/ ratio			Covariance, Correlation coefficient

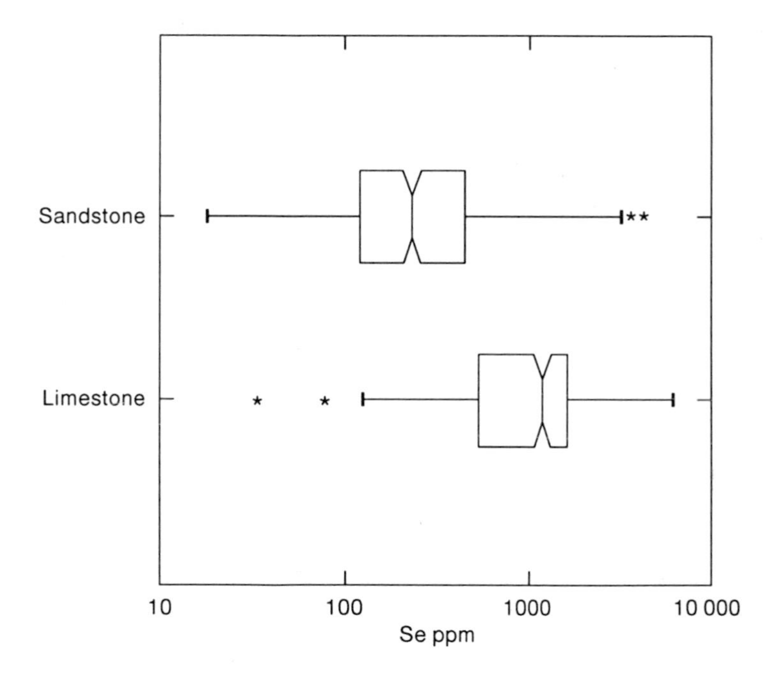

FIG. 8-16. Boxplot comparing a geological map (categorical measurement level) consisting of two classes (sandstone and limestone) with a geochemical map (ratio scale of measurement) of selenium (Se). The values used for making the plot could either be measured at the original sample points, or they could be points or pixels on a regular lattice used to sample a selenium map.

Mixed measurement scales

In practice, the collection of spatial datasets available for any particular GIS project comprises attributes from mixed scales of measurement, and the coefficients for quantifying spatial associations described thus far have assumed that both members of a map pair have the same level of measurement. Table 8-10 shows the six possible combinations based on mixtures of measurement scales, of which the three in the principal diagonal are the cases already discussed.

In the case of an ordinal scale map variable and a categorical map, an effective method of comparison is to use a boxplot to summarize the ordinal variable for each nominal map class. The boxplot, Figure 8-16, summarizes the cumulative frequency distribution, showing the 95th and 5th percentiles with the free ends of the "whiskers", and a central rectangular box that spans from the 25th to 75th percentiles. The 50th percentile, or median value, is usually shown in a notch whose width represents the 95% confidence interval on the median, (Tukey, 1977). The boxplot does not give a number to express correlation, but does show graphically whether the ordinal variable is partitioned into distinctive groups by the nominal scale classes. The notch on the median cannot be reliably estimated for areal data by conventional means, and should either not be used, or used only in a descriptive sense, and not for formal hypothesis testing.

For a map of interval or ratio scale data, compared with a categorical map, again the box plot is to be recommended, with the addition of the mean as a valid measure of central tendency. Multiple F-tests can be used to test the equality of means, between classes, but the usual caution about statistical significance applies.

For the comparison of an ordinal scale map to an interval/ratio scale map, it is probably best to treat the interval/ratio data as ranks and use a rank correlation coefficient. This might occur, for example, in comparing a geochemical map with a map of distance from a point or linear feature, expressed by a map with buffer zones that may vary in width.

The option of recoding the variables to binary form, then using binary measures of association, is often a practical solution.

OTHER TOPICS

As the reader will appreciate, making quantitative comparisons between one map and another is quite complex, and the treatment devoted to it in this chapter is a brief introduction. We have not dealt with methods, for example, that consider the spatial neighbourhood of objects, and comparison has been limited to an "entity-to-entity" approach. Agterberg and Fabbri (1978) applied the idea of geometric covariance measurements by measuring the amount of overlap between two binary images as a function of shifting the images with respect to one another in various directions. Such operations produce a two-dimensional covariance function that gives insight as to how the spatial correlation changes with separation between images, whether the correlation is spatially isotropic (it often is not due to trends) and a measure of the expected correlation at large separation distances. Serra (1982) is an important source of information on this subject. In analyzing pairs of geophysical images, the cross-correlation relationships can be examined in the frequency domain with cross-spectra,

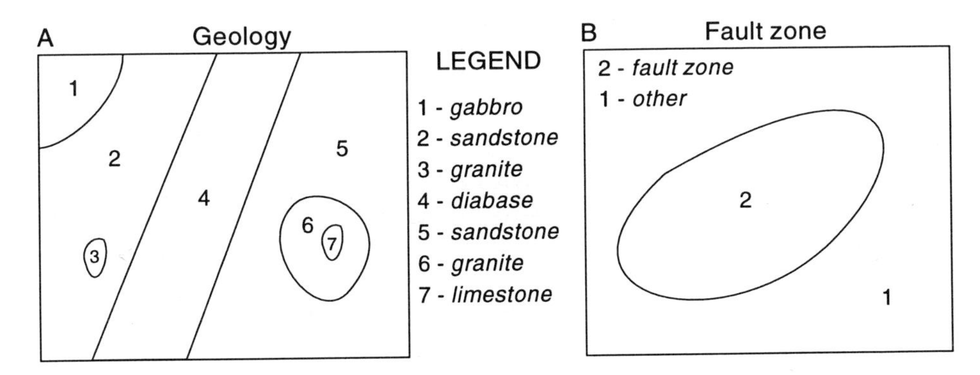

C Unique polygon table

	Geology	Fault zone
1	1	1
2	2	1
3	2	2
4	3	2
5	4	1
6	4	2
7	4	1
8	5	1
9	5	2
10	6	2
11	7	1
12	6	1

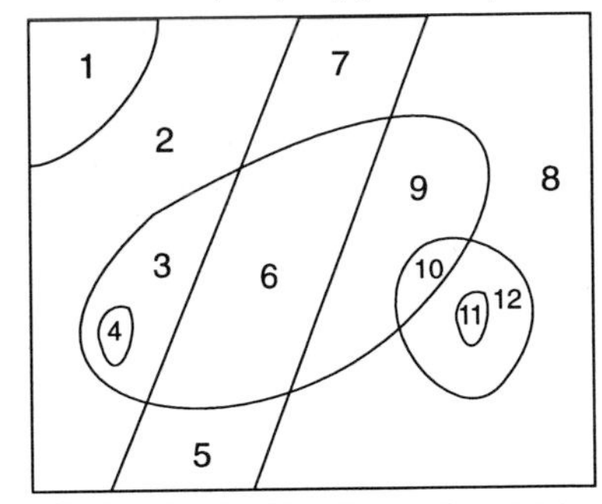

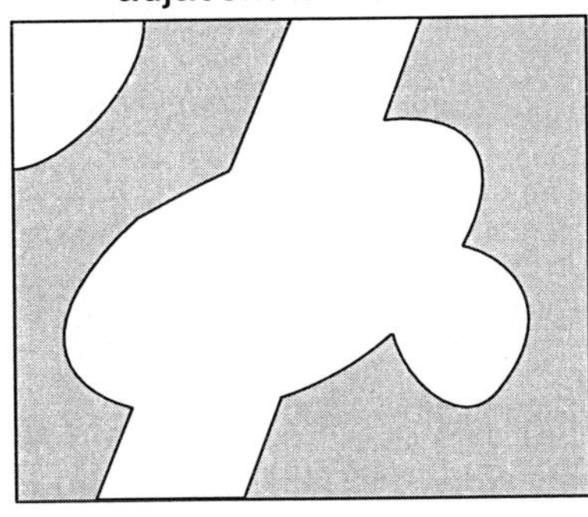

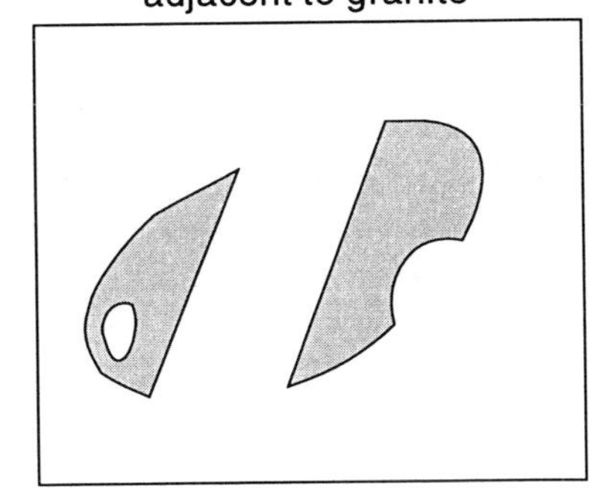

FIG. 8-17. Illustration of topological modelling based on the overlay of **A.** a geological map, and **B.** a map showing a fault neighbourhood. **C.** An attribute table for the unique polygon map, shown in **D**. Note that each polygon in **D** is linked to a record in the attribute table. **E.** Output map showing areas that are **adjacent to** the fault zone **and adjacent to** the diabase. **F.** Ouput map showing areas that are **inside** the fault zone **and adjacent to** the granite.

that reveal the spatial frequencies and phase shifts at which the data show associations. There are also a number of developments in spatial statistics and their application to geological problems that have relevance to GIS, but they are outside the scope of this book.

One subject is worth a brief mention, because it does use functionality available in current GIS. That is the application of map models with topological attributes extracted from two maps combined in a polygon overlay. Figure 8-17 shows a simple geological map, overlain by a map showing a zone round a fault. Given the unique polygon map, the goal is to use map modelling to produce new maps showing : 1) regions adjacent to the fault zone and adjacent to the diabase, and 2) regions inside the fault zone that are also adjacent to the granite. This type of topological modelling utilizes adjacency and containment attributes that can be extracted from topological data structures. It should also be noted that spatial information about proximity to contacts and other linear features, once buffered to produce an areal map, can then be treated like any nonspatial attribute.

SUMMARY

The combination and display of map layers in pairs is an important aspect of GIS, because it allows the examination of spatial relationships between spatial phenomena. Although the ultimate goal of most GIS studies involves multiple data layers, the relationship between map pairs is often an exploratory first step, and may determine how features of one or both maps are to be enhanced or extracted for subsequent analyses.

Modelling operations on pairs of maps to produce new integrated map layers can be carried out with Boolean and arithmetic operators on both the map class values and attributes held in attribute tables. In some cases topological attributes can be directly used in the modelling process.

Methods of quantifying the degree of correlation between one map and another utilize data from an area cross-tabulation. Where categorical maps are being compared, measures using chi-square suffer from being dependent on the units of area measurement, whereas entropy measures are based on area proportions and are not effected by areal units. The kappa coefficient of agreement is suitable for comparing maps with matched classes. For binary maps, the odds ratio and coefficients derived from it are simple to calculate and relatively straightforward to interpret. Rank correlation coefficients are appropriate for comparing ordinal scale maps, and with polygon data, area weighting should be used to account for polygons (and unique conditions classes) of varying size.

REFERENCES

Agterberg, F.P. and Fabbri, A.G., 1978, Spatial correlation of stratigraphic units quantified from geological maps: *Computers & Geosciences*, v. 4, p. 285-294.

Bishop, M.M., Fienberg, S.E. and Holland, P.W., 1975, *Discrete Multivariate Analysis: Theory and Practice*: MIT Press, Cambridge Massachusetts, 587 p.

Cohen, J., 1960, A coefficient of agreement for nominal scales: *Educational and Psychological Measurement*, v. 20 (1), p. 37-46.

Diggle, P.H., 1983, *Statistical Analysis of Spatial Point Patterns*: Academic Press, London, 148 p.

Fleiss, J.L., 1991, *Statistical Methods for Rates and Proportions* – Second Edition: John Wiley and Sons, New York-Chichester, 321 p.

Herzfeld, U.C. and Merriam, D.F., 1990, A map-comparison technique utilizing weighted input parameters: In: *Computer Applications in Resource Estimation – Prediction and Assessment for Metals and Petroleum*, Editors Gaál, G. and Merriam, D.F., Pergamon Press, Oxford-New York, p. 43-52.

Inkley, D.B., Anderson, S.H., Mankowski, S.G. and McDonal, L.L, 1984, An objective method for comparing resource distributions: *Geographical Analysis*, v. 16(4), p. 369-372.

Hudson, W.D. and Ramm, C.W., 1987, Correct formulation of the Kappa coefficient of agreement: *Photogrammetric Engineering & Remote Sensing*, v. 53 (4), p. 421-422.

Knudsen, D.C., 1987, Computer-intensive significance-testing procedures: *Professional Geographer*, v. 39 (2), p. 208-214.

Lupian, A.E., Moreland, W.H. and Dangermond, J., 1987, Network analysis in geographic information systems: *Photogrammetric Engineering and Remote Sensing*, v. 53(10), p. 1417-1421.

Mather, P.M., 1987, *Computer Processing of Remotely-Sensed Images*: John Wiley and Sons, Chichester-New York, 352 p.

Openshaw, S., 1991, Developing appropriate spatial analysis methods for GIS: In: *Geographical Information Systems. Volume 1. Principles*, Eds. Maguire, D.J., Goodchild, M.F. and Rhind, D. W., Longmans Scientific & Technical, London, p. 389-402.

Openshaw, S., Charlton, M., Wymer, C. and Craft, A., 1987, A Mark 1 geographical analysis machine for the automated analysis of point data sets: *International Journal of Geographical Information Systems*, v. 1 (4), p. 335-358.

Openshaw, S., Carver, S. and Fernie, J., 1989, *Britain's Nuclear Waste: Siting and Safety*: Belhaven Press, London-New York, p. 121-176.

Press, W.H., Flannery, B.P., Teukolsky, S.A. and Vetterling, W.T, 1986, *Numerical Recipes: The Art of Scientific Computing*: Cambridge University Press, Cambridge, 818 p.

Romesburg, H.C., 1990, *Cluster Analysis for Researchers*: Robert E. Krieger Publishing Company, Malabar, Florida, 334 p.

Rosenfield, G.H. and Fitzpatrick-Lins, K., 1986, A coefficient of agreement as a measure of thematic classification accuracy: *Photogrammetric Engineering & Remote Sensing*, v. 52 (2), p. 223-227.

Serra, J., 1982, *Image Analysis and Mathematical Morphology*: Academic Press, London-New York, 610 p.

Taylor, P.J., 1977, *Quantitative Methods in Geography*: Houghton Mifflin Company, Boston-Atlanta, 386 p.

Tukey, J.W., 1977, *Exploratory Data Analysis*: Addison-Wesley, Reading, Massachusetts, 686 p.

Serra, J., 1982, *Image Analysis and Mathematical Morphology*: Academic Press, London-New York, 610 p.
Taylor, P.J., 1977, *Quantitative Methods in Geography*: Houghton Mifflin Company, Boston-Atlanta, 386 p.
Tukey, J.W., 1977, *Exploratory Data Analysis*: Addison-Wesley, Reading, Massachusetts, 686 p.

CHAPTER 9

Tools for Map Analysis: Multiple Maps

INTRODUCTION

The ultimate purpose of most GIS projects is to combine spatial data from diverse sources together, in order to describe and analyze interactions, to make predictions with models, and to provide support for decision-makers. The purpose of this chapter is to present some models of interest to geoscientists, to show how they can be implemented in a GIS environment, and to illustrate the models with reference to two applications: site selection for a landfill, and mineral potential mapping. Using GIS for site selection (whether the site be for a landfill, for nuclear or other toxic waste, for a dam, for a pipeline, for residential development, etc.), involves finding locations or zones that satisfy a set of criteria. If the criteria are defined as a set of deterministic rules, the model consists of applying Boolean operators to a set of input maps, and the output is a binary map, because each location is either satisfactory or not. Alternatively, each location may be evaluated according to weighted criteria, resulting in a ranking on a suitability scale. The subsequent selection process then benefits from the ability to assess suitability *rankings*, rather than simply presence/absence, and from the knowledge of spatial patterns of suitability. GIS modelling of mineral potential also involves calculating a suitability, or mineral favourability, from geoscientific maps. The potential is calculated by weighting and combining multiple sources of evidence. The assignment of weights to maps is carried out either by analyzing the importance of evidence relative to known mineral deposits, or by using the subjective judgement of mineral deposits geologists. The principles employed in modelling the site selection and potential mapping processes are, as will be seen, similar to each other in many respects.

Models used for the site selection process are usually **prescriptive**. That is, they involve the application of a set of criteria that are set out as good engineering practice, and may result from a blend of scientific, economic and social factors. On the other hand, mineral potential mapping involves the use of **predictive** models, because the ultimate purpose of determining mineral potential is to discover new deposits. Potential maps are either used by mineral exploration companies, for increasing their chances of finding a new mine, or by government agencies to assess the mineral resources of a region as an aid in land-use planning.

Types of Models

Although it is difficult to classify GIS models, it is useful to distinguish between three types based on the kinds of relationships that the models represent, as shown in Table 9-1. In a very general sense, a GIS model can be thought of as the process of combining a set of input maps with a function to produce an output map.

output map = *f* (2 or more input maps)

The function, *f*, takes many different forms, but the relationships expressed by the function are either based on a theoretical understanding of physical and chemical principles, or they are empirical, based on observations of data, or some blend of theory and empiricism.

In the theoretical category, finite-difference or finite-element models of fluid flow based on equations of motion derived from mechanical principles are widely used in fields such as groundwater flow and oceanic and atmospheric circulation. In order to model lake circulation, for example, maps of water depth, bottom slope, inflow and outflow and wind stress (orientation and direction) are the input, and a velocity field is output, e.g. Simons (1971), Bonham-Carter and Thomas (1973). Such models predict the flow field from the application of theoretical principles (applications of the Navier-Stokes equations in this case), and are checked by comparing the output with measured data. This kind of specialized simulation modelling is generally not part of a general-purpose GIS, but simulation software can be coupled with a GIS for defining input data and for visualizing results. However, some simple models that are based on physical principles can be applied within a general-purpose GIS. An example is the correction of gravity data for the effects of topography, as in the computation of Bouguer anomaly maps.

Semi-empirical relationships are used in many geoscience fields, for example in models of slope processes, and in sedimentation simulation models, Tetzlaf and Harbaugh (1989). In applying GIS to slope stability and landslides, Wadge (1988) discusses the application of several physically-based and empirical models. van Westen (1993), in a comprehensive study of GIS applied to slope stability zonation in a district of Columbia, South America, discusses the problems of applying the "infinite slope" model to the estimation and mapping of a slope safety factor. Although based on theoretical mechanical principles, several terms in the infinite slope equation must be estimated by empirical means, and the approach is best described as semi-empirical. Using this model, slope stability (and slope safety) can be

Table 9-1. A classification of models used in geology. See also Harbaugh and Bonham-Carter (1970, Chapter 1). These are symbolic models, as opposed to physical scale models.

		Example	
Class	**Type of Relationship**	**Field of Application**	**Model**
Theoretical	Physical/chemical principles	Groundwater flow	Equations of motion
Hybrid	Semi-empirical	Sediment transport	Transport equations
Empirical	Heuristic or statistical	Mineral prediction	Statistical regression

mapped as output, using a variety of input maps, such as slope angle, soil depth, soil strength, and depth to the groundwater table.

Models for predicting mineral potential, based on statistical relationships or on heuristic relationships, are examples of empirical models. The physical and chemical principles governing the formation of mineral deposits are for the most part too complex for direct prediction from mathematically-expressed theory. The prediction of mineral deposits must rely mostly on empirical relationships, with the aid of descriptive "deposit models". Study of mineral deposits has led to the recognition of a large number of classes of deposit, or deposit types. Each deposit type consists of a number of actual deposits, regarded as being sufficiently similar in terms of their characteristics, to be treated as a descriptive "model" that guides the search for new deposits of the same type. The description of a deposit model does include, as far as possible, an evaluation of the chemical and physical processes that control deposit formation, but the model cannot be expressed in purely mathematical terms. In applying GIS to the mapping of mineral potential, deposit models play a role both in the selection and derivation of maps that are likely to be good predictors of the deposit-type under consideration, and for the assignment of weights to the various predictor maps. The assignment of weights can either be carried out using statistical criteria, using an actual study region to estimate the spatial relationships between predictor maps and the response map (estimated with known mineral deposits, or sometimes with anomalous geochemical zones), or the weights can be estimated on the basis of expert opinion. These two types are sometimes called "data-driven" and "knowledge-driven" models, as shown in Table 9-2. In data-driven modelling, the various input maps are combined using models such as logistic regression, weights of evidence or neural network analysis. Knowledge-driven models include the use of fuzzy logic, Bayesian probability and Dempster-Shafer belief theory, Chung and Moon (1991), Chung and Fabbri (1993), sometimes applied with inference networks in an expert system, An et al. (1992).

Mineral potential models can be directly implemented in a GIS, or in more complex cases, such as expert systems, be modelled with independent computer programs that are coupled with the GIS for input and output.

Table 9-2. Mineral potential models, divided into data-driven and knowledge-driven types.

Type	Model parameters	Example
Data-driven	Calculated from training data	Logistic regression Weights of evidence Neural network
Knowledge-driven	Estimated by an expert	Fuzzy logic Dempster-Shafer belief theory

Modelling with GIS

In practice, GIS provide only a basic modelling toolkit, often insufficient for specialized modelling requirements. General-purpose GIS are not employed to model fluid flow problems, for example, because it is simpler to write a specialized computer program tailored for such an application, unfettered by the overhead of a general-purpose system that satisfies the needs of a whole variety of applications. Similarly, it is unlikely that GIS will ever provide the full range of tools needed for statistical analysis, for expert system modelling, for neural network analysis, and so on. The specialized needs of particular GIS applications are too diverse to encompass all types of model within a single system.

As discussed in Chapters 7 and 8, the basic modelling language that allows operations on maps and their associated attribute tables is adequate for some of the simpler models. Where specialized modelling programs are to be used outside the GIS, tools are provided for creating data files in a form suitable for the analysis. Output from modelling can be returned to the GIS for further analysis and display. Alternatively, dynamic linkages are used, either with subroutine calls from the GIS modelling language, or employing dynamic data exchange (DDE) capabilities.

Interchange of Map Data Files with External Software

The form in which data can be exported to external software for multi-map modelling varies from one system to another. In a raster format, such as BIL (band interleaved by pixel, see Chapter 3), data can be exported as a stack of raster images, each raster having the same number of rows and columns. On the other hand, in vector format, a polygon overlay can be used to create a map formed by the intersection of all the maps that are to be combined. On such a map, each polygon is a unique spatial object, and is linked by a keyfield to a record in an attribute table. This type of unique polygon table was described in Chapter 8 for the two-map case, and can be extended for multiple maps. Alternatively, if geometric attributes of individual polygons are not needed for modelling, a unique conditions map, and associated table, can be generated. Polygons having the same vector of attributes are combined into unique conditions classes, as discussed in Chapter 8. In either case, the unique conditions, or the unique polygons, table can be exported, preferably in an ASCII format, ready to be processed by external software. New attribute tables are generated with the modelling results, containing the unique polygon number or unique conditions class as the keyfield, ready to be re-imported into the GIS for display, and further analysis.

With a unique polygon or unique conditions map, modelling operations are carried out on a *single* map and its associated attributes. All the modelling operations discussed in Chapter 7, for single maps and their attributes, now apply. In this way the multi-map problem is reduced to a "single" map problem. Sometimes, the creation of a unique conditions map can greatly reduce the number of computations needed for modelling.

For example, suppose that 10 binary maps are combined by unique conditions overlay. There are only $2^{10}=1024$ unique combinations possible at any single location. Thus the unique conditions table will have no more than 1024 records (it will usually have less, because almost invariably some conditions do not occur in practice), and the modelling calculations need be carried out only once per record. The result of the model is added as a new field to the unique conditions attribute table, and the unique conditions map is then reclassified by this field. If the modelling is carried out pixel-by-pixel in a raster, the modelling operation must be repeated once per pixel, possibly several million times for a single study area. Thus although the creation of a unique conditions overlay requires computation expense, the subsequent number of repeated modelling calculations may be reduced several thousand times. Further, the same unique conditions files can be used for any number of cycles through the modelling process, changing parameters or altering the modelling equations.

Unique conditions do not always provide such dramatic savings, however. If the input maps are 8-bit images, then the number of possible unique conditions is equal to 256^n, where n is the number of maps. Thus even 3 images would require up to 256^3= 16,777,216 unique conditions, so that unless the image size was more than 4,000 by 4,000 pixels, it would be more economical to carry out the modelling pixel-by-pixel.

Modelling Applications

In this chapter a number of models for combining maps together, suitable for implementation in a GIS, are discussed. Applications used as examples deal mostly with mineral resources, but some of the initial models are illustrated with reference to a hypothetical landfill site selection problem. Probably the simplest and best-known type of GIS model is based on **Boolean operations**, illustrated first. Next, the method sometimes known as an **index overlay** is introduced, again a popular method that is customized in many GIS, and is applicable where maps are to be added together in a weighted combination. This is followed by a discussion of the methods of **fuzzy logic**, which allow for more flexible combinations of weighted maps, and can be readily implemented with a GIS modelling language. All methods to this point are based on subjective empirical models, with the rules, weights or fuzzy membership values being assigned subjectively, using a knowledge of the process involved to estimate the relative importance of the input maps. The next section puts the modelling process on to a probabilistic framework, so that the weighting of individual map layers is based on a **Bayesian probability** model. In particular, the **weights of evidence model** is presented in a map context, with examples showing application to mineral potential mapping. The relationship of weights of evidence to the methods used in the **Prospector expert system** are explained, and a very simple example of a Prospector-style inference network is presented. Some other important models are briefly mentioned, particularly the logistic regression model and the Dempster-Shafer model, although discussion of many other multivariate statistical models that can be applied to GIS data is omitted.

BOOLEAN LOGIC MODELS

An excellent introduction to the use of Boolean operations for reasoning with geological maps is to be found in Varnes (1974) and in Robinove (1989).

Landfill site selection

Suppose that a municipality is seeking potential sites for a new landfill. The guidelines established for evaluating the suitability of an area for waste disposal have been set out (perhaps in a consultant's report or government publication) as a set of Boolean rules. The ones used here are fictitious, but serve to illustrate the principles involved.

The site must be located so that all of the following conditions are satisfied. The location must:

1. **Be in an area where unconsolidated surficial material is more than a minimum thickness, AND**
2. **Be in material that has a low permeability, preventing rapid leaching of soluble matter into the groundwater, AND**
3. **Be where the average surface slope is not steep (slope < some limit), AND**
4. **Be where the underlying bedrock is not a fractured limestone, AND**
5. **Be where river flooding is unlikely (outside the limits of the 100-year flood, for example), AND**
6. **Be where the present zoning is agricultural, not municipal or industrial, AND**
7. **Be where the land suitability is not classed as prime arable farmland, AND**
8. **Be within a certain distance of the present municipal limits, AND**
9. **Be within a certain distance of a major road, AND**
10. **Not be designated as environmentally sensitive.**

There are certainly other factors, such as visibility, exposure to the prevailing wind direction, and cost of the land for example, but this list serves to illustrate the principles of applying a **prescriptive** Boolean model. A final important constraint is that the selected land parcel must be at least some minimum size, such as 0.5 km^2, because although a smaller parcel might be suitable in every other way, it would be unusable for a landfill simply by virtue of its size. Note that this constraint involves a geometric, rather than a purely thematic, attribute. The first step is to construct a series of maps for the area under consideration, to which these rules can be systematically applied. The maps and their sources are summarized in Table 9-3 and illustrated in Figure 9-1, A to J.

Typically, map modelling statements for combining the 10 maps together to produce a single binary output map where class 1 indicates areas that satisfy all the above conditions, and class 0 indicates all remaining areas, might look like this:

Table 9-3. Maps used in landfill site selection problem, see Figure 9-1, A to J. The data are fictitious.

Map Name	Description	Source
OVERTHIK	Overburden thickness	Mapped from well records
PERMEAB	Overburden permeability	Mapped from well records
SLOPE	Slope of ground	Calculated from DEM
GEOLOGY	Geological map	Geological survey
FLOOD	Extent of flooding	Hydrological map
ZONING	Zoning map	Municipal government
SUITAB	Suitability for agriculture	Land use map
MUNIBUF	Distance from city limits	Buffer round city limits
ROADBUF	Distance to major road	Buffer round roads
ECOLOG	Sensitive ecology	Ecological survey

```
: At current location, determine if condition for each input map is satisfied
: The conditions, C1 to C10, are either TRUE (=1) or FALSE (=0)
: See Table 9-5 for a summary of the map classes
        C1 = class('OVERTHIK') > 4
        C2 = class('PERMEAB') < 2
        C3 = class('SLOPE') < 2
        C4 = class('GEOLOGY') <> 4
        C5 = class('FLOOD') <> 1
        C6 = class('ZONING') > 1
        C7 = class('SUITAB') > 1
        C8 = class('MUNIBUF') < 4
        C9 = class('ROADBUF') < 6
        C10= class('ECOLOG') == 1
: Combine conditions with Boolean "AND" operators
: The variable OUTPUT is either TRUE (=1) or FALSE (=0)
        OUTPUT = C1 AND C2 AND C3 AND C4 AND C5 AND C6 AND
        C7 AND C8 AND C9 AND C10
: Map result as a binary 2-class map
        RESULT(OUTPUT)
```

The initial statements in this procedure convert information about each of the input maps into a binary (true, false) or (1,0) form. The first line can be read as "make variable C1 (for condition 1) equal to 1 (TRUE) if the class of the overburden thickness map is greater than

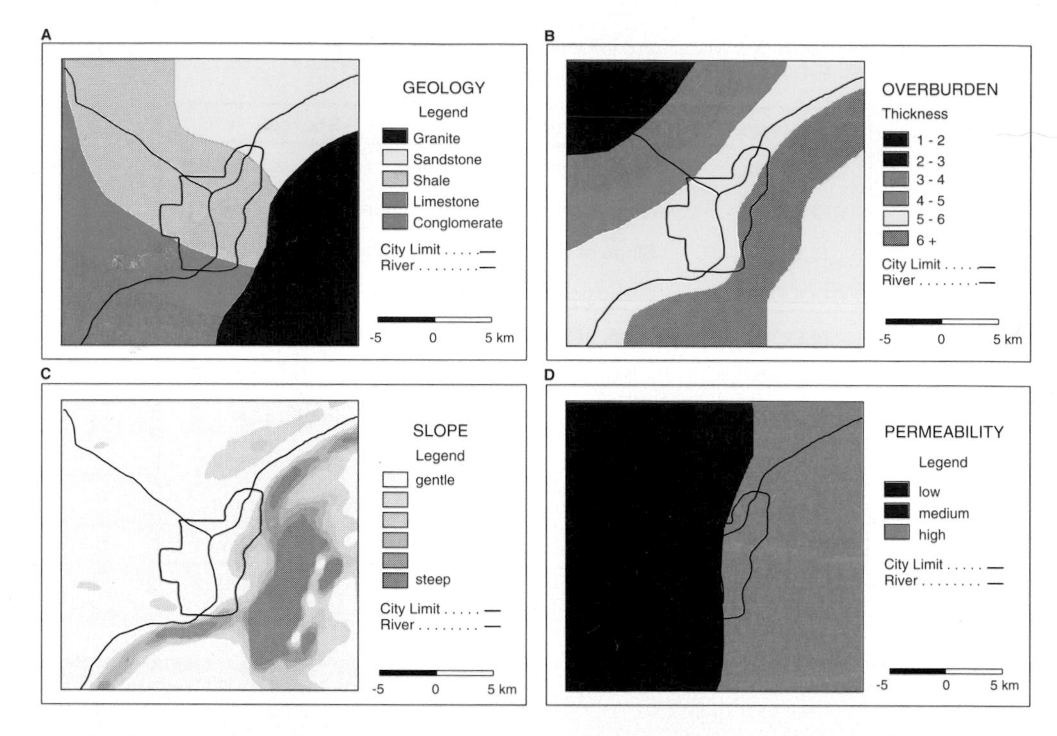

FIG. 9-1. Input maps used in landfill study. **A.** Geological map, with location of river and municipal boundary. **B.** Map of overburden thickness in metres, interpolated from point data. **C.** Slope map, derived from a DEM by applying derivative operators. **D.** Permeability of overburden materials.

4, else make variable C1 equal to 0 (FALSE)". Note that a conditional statement could also be used for this purpose. Thus the lines

C1 = {1 if class('OVERTHIK') > 4, 0} *and*
C1 = class('OVERTHIK') > 4

are equivalent. The actual assignment of class values for each of the input maps must, of course, be known from their legends, see Table 9-5 (later in the chapter). In the first 10 statements of the procedure, 10 temporary variables, C1 to C10, are created, whose values are either 1 or 0 depending whether the respective condition is true or false. The final statement applies the Boolean AND operator to all the temporary variables, so that the variable

---

FIG. 9-1 (opposite). E. Distance from municipal boundary, using 1 km buffer zones. **F.** Simplified zoning map, showing 4 classes of zoning.

E

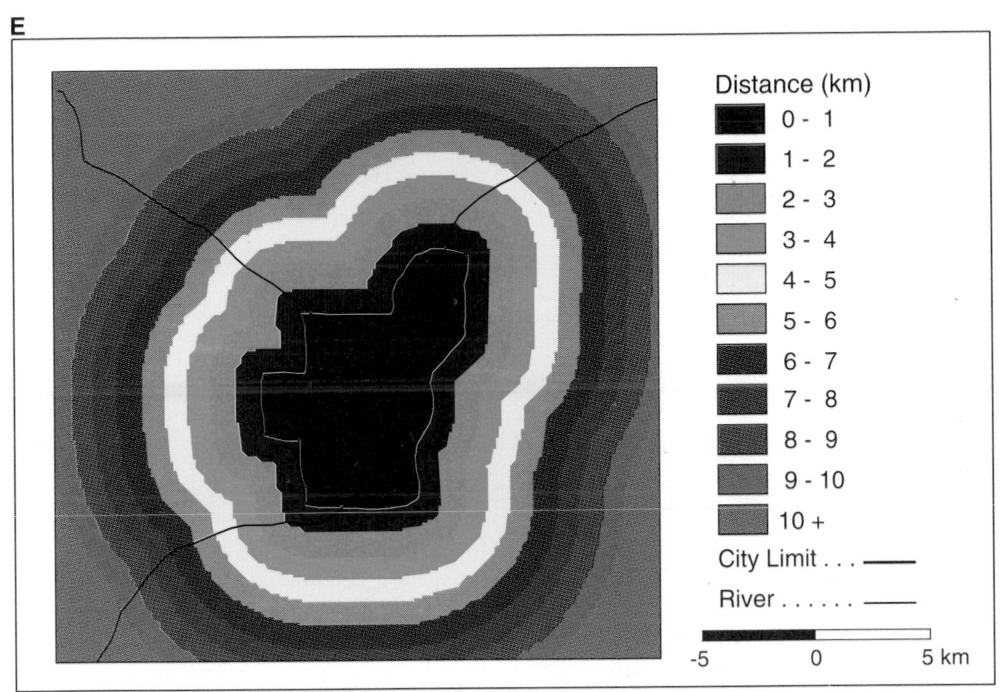
Distance (km)
0 - 1
1 - 2
2 - 3
3 - 4
4 - 5
5 - 6
6 - 7
7 - 8
8 - 9
9 - 10
10 +
City Limit . . .
River . . . . . .
-5
0
5 km

F

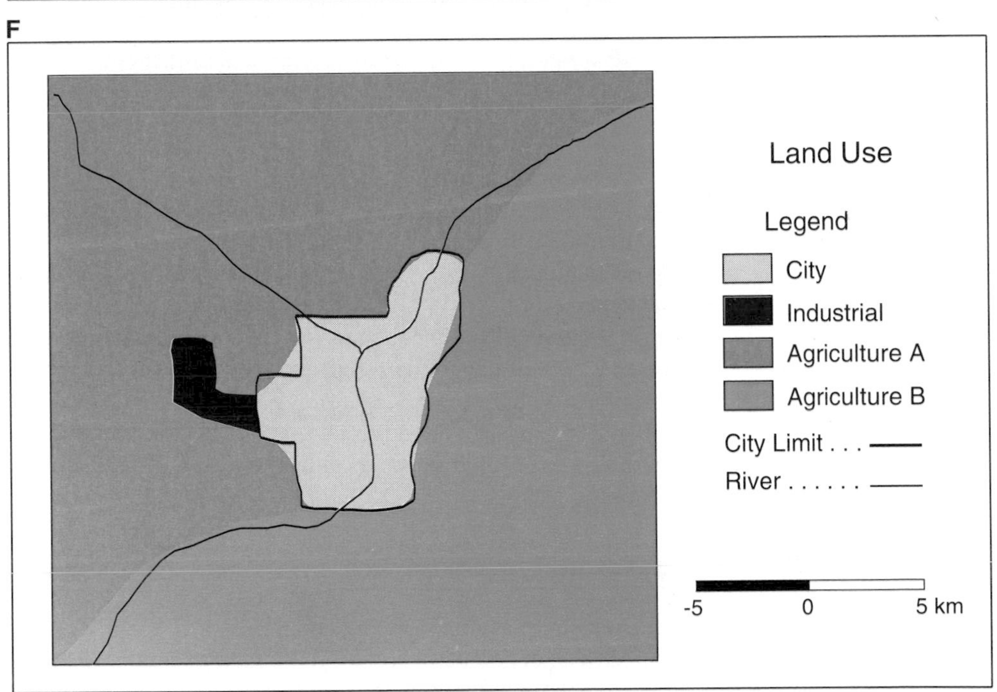
Land Use
Legend
City
Industrial
Agriculture A
Agriculture B
City Limit . . .
River . . . . . .
-5
0
5 km

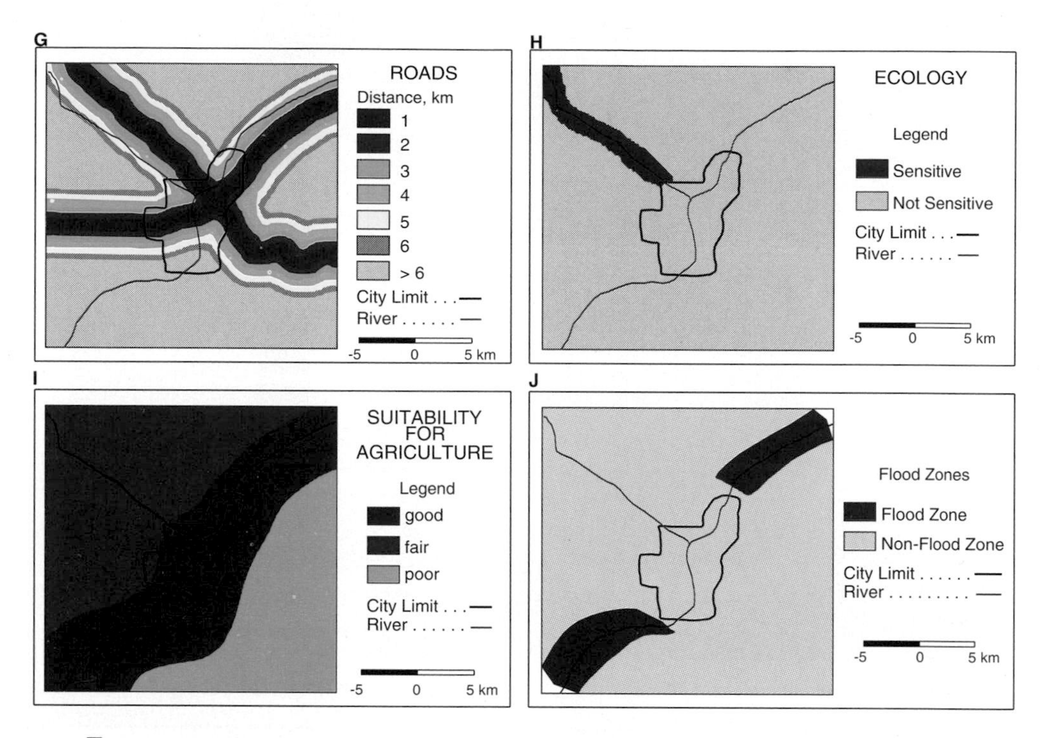

FIG. 9.1 G. Distance from roads, made by buffering. **H.** Map showing location of an ecologically sensitive area, surrounding one of the rivers. **I.** Suitability for agriculture, based on soil factors. **J.** Flood zones, based on past flood records.

OUTPUT is 1 only if all the 10 conditions are true, otherwise OUTPUT is 0. The procedure is applied repeatedly to each location over the region of interest (each pixel in a raster, block in a quadtree, unique polygon or unique class from a polygon overlay), so that OUTPUT becomes a new map, with two classes, 1 or 0.

The output map is shown in Figure 9-2A. It contains just two polygons where all conditions are satisfied. An area analysis of the polygon entities shows that one polygon is 0.4 km^2, and the other one is 1.4 km^2. Therefore the larger polygon is selected, because it exceeds the minimum threshold of 0.5 km^2. We now turn to a second application, making a mineral evaluation map from a variety of input maps, to illustrate the application of the Boolean model for potential mapping.

Mineral Potential Evaluation

The datasets used to illustrate mineral potential mapping come from a GIS study of gold potential in part of the Meguma terrane, of southeast Nova Scotia. A large number of small gold deposits and minor gold occurrences are found in the Goldenville Fm and Halifax Fm,

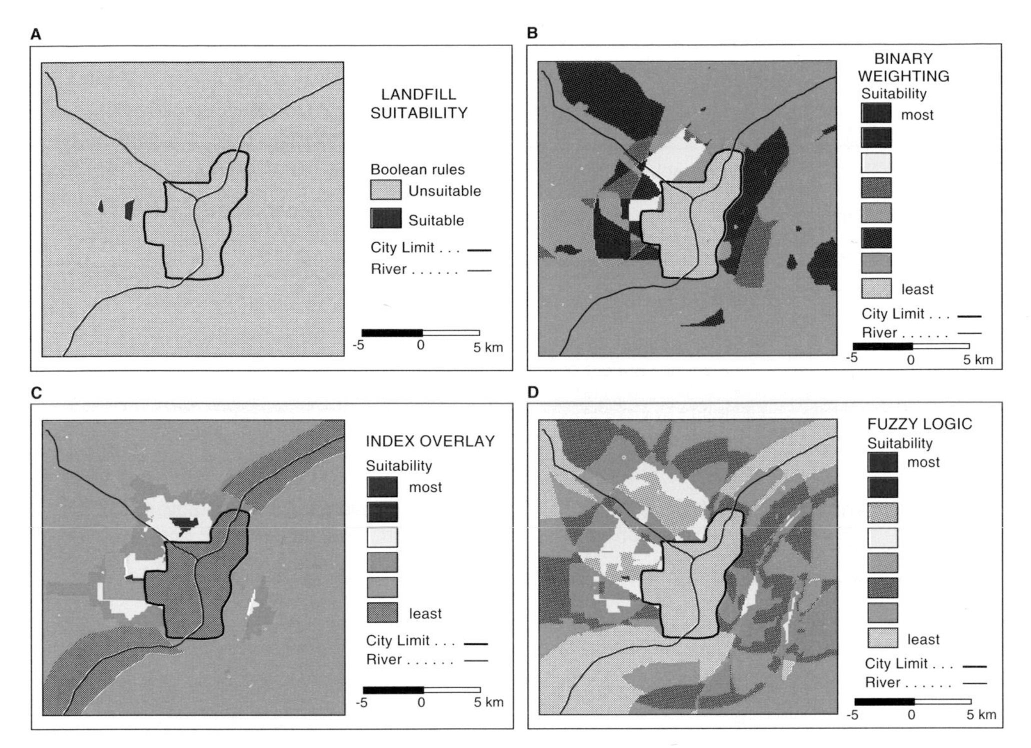

FIG. 9-2. Landfill site selection output maps. **A**. Suitability map made using deterministic Boolean rules. Only two polygons satisfy all the constraints. Maps B, C and D are mentioned later in the chapter. **B**. Output from binary weighting procedure. The areas are ranked, giving greater flexibility for decision-making. **C**. Output from an index overlay. Again, the areas are ranked according to suitability, but the particular choice of scoring has resulted in a rather different solution to **A** and **B**. **D**. Suitability using fuzzy logic. The ouput is ranked according to membership of a fuzzy set, the set comprising areas that satisfy the proposition "*This location is suitable for a landfill*". The degree with which the proposition is satisfied is scored on a scale from 0 to 1, then classified to make a map.

Lower Paleozoic turbidite formations that were intruded by granites and folded during the Devonian period. The gold occurs as native metal in quartz veins. Many of the occurrences were found during the latter half of the 19th century. The deposits are similar in type to the gold deposits in the Ballarat district of Victoria, Australia. Bonham-Carter et al. (1988) and Wright (1988) discuss the source of the data and its manipulation in a GIS.

The principal factors of the deposit model that are relevant to mineral mapping potential in a GIS may be summarized as follows. The deposits often occur at or close to the crests of folds. The minerals associated with gold in the deposits include arsenopyite and stibnite. Thus As and Sb are important geochemical elements that can be used as pathfinders for gold in drainage sediments and vegetation. In the region under study, the deposits occur mostly in the Goldenville Fm (a sand-silt unit); in other parts of the Meguma terrane, gold-bearing veins are also found in the

Table 9-4. Maps used for mineral potential example, Meguma terrane, Nova Scotia.

Map Name	Description	Source
GEOL	Bedrock geology	Digitized from published map
BASIN	Lake catchment basins*	Digitized from hand-drawn basins
BIOAS	Arsenic (As) in vegetation	Interpolated from point attribute data
BIOAU	Gold (Au) in vegetation	Interpolated from point attribute data
ANTI	Distance to axial traces of Devonian anticlines	Traces digitized from map and buffered
GOLDHAL	Distance to Goldenville-Halifax contact (at surface)	Contacts extracted as vectors from GEOLOGY map, buffered
NWLINS	Distance to NW lineaments	Digitized lineaments, labelled by orientation, selected if NW-SE and buffered

*Each basin is linked to an individual record in an attribute table called 'BASIN' containing geochemical and other fields.

Halifax Fm (a slate unit); this deposit type does not occur in the Devonian granite intrusions. Some geologists have suggested that proximity to the Goldenville-Halifax contact may be an important factor in some of the deposits. Others have suggested the possible importance of the granite as a source of gold-bearing fluids. Another hypothesis is that structures oriented in a NW-SE direction are associated with some of the deposits, possibly due to the reactivation of faults in the basement, thereby controlling the migration of gold-bearing fluids.

Given a map of the geology, geochemical data from samples of lake sediments and vegetation, regional geophysical survey data, lineaments interpreted from satellite images, and other data sources, the goal is to create a series of maps to be used for predicting gold, and to combine them with map modelling, as described for the base-metal deposits in Chapter 1. In its simplest form, the deposit model can be cast as a series of conditional statements, like the landfill model. The following criteria can be applied as a prescriptive model for mapping zones with gold potential. Each selected location must:

1. Be underlain by either the Goldenville Formation or the Halifax Formation, AND
2. Be within a sampled lake catchment basin that is geochemically anomalous, having
arsenic (As) concentrations in sediment greater than 50 ppm, OR
antimony (Sb) concentrations in sediment greater than 0.8 ppm, OR
gold (Au) concentrations in sediment greater than 9 ppb, OR
tungsten (W) concentrations in sediment greater than or equal to 3 ppm, OR
3. Be within biogeochemically anomalous zones, having

arsenic (As) concentrations in Balsam Fir twigs greater than 3.2 ppm, OR gold (Au) concentrations in Balsam Fir twigs greater than 12 ppb, AND

4. **Be within 1.25 km from the nearest anticline axis, AND**
5. **Be within 1 km from the surface trace of the Goldenville-Halifax contact, AND**
6. **Be within 1 km of the nearest NW-SE lineament.**

As in the landfill case, the steps in producing a new map showing areas that satisfy these criteria must start by generating a map for each factor, as summarized in Table 9-4, and illustrated in Figures 9-3, A-J. Note that Boolean OR (logical union) operations are used to combine the geochemical maps. This is because the presence of any one of these pathfinder geochemical elements at anomalous levels (as opposed to joint occurrence) is judged to be significant in the location of gold deposits. The geochemical data are then combined with other factors with Boolean AND (logical intersection) operations.

The first condition specifies that only the Goldenville Formation or the Halifax Formation, as defined on the geological map, be considered as possible areas for gold deposits. This excludes the Devonian granites, which never host this deposit type. The second condition uses lake catchment basins for making the geochemical maps, as shown in Figure 9-3. The attribute table containing geochemical values of lake sediment samples is linked by keyfield to polygons of the map called BASIN. The biogeochemical data, derived from a vegetation survey, was interpolated to create a separate map for each of the relevant geochemical attributes, because the zone of influence around each sample location is not identified with a specific catchment, as is the case for drainage data. Thus kriging was used to interpolate from the original sample points, to create the BIOAS arsenic map and the BIOAU gold map, shown in Figure 9-3. Buffering was used to generate the "distance to" maps that make up the last three criteria. They are called ANTI for the proximity to the traces of anticlinal fold axes, GOLDHAL for proximity to the Goldenville-Halifax contact, and NWLINS for proximity to the NW lineaments.

The fold axes were simply digitized from the geological map and buffered. The Goldenville-Halifax contact was derived from a topological vector file. A search of the arc topology table resulted in the selection of all arcs with Goldenville on one side and Halifax on the other. The selected arcs were then buffered to produce the proximity map. The NW lineaments were derived from a combination of lineaments recognized on a vertical gradient magnetics map and a LANDSAT image. The lineaments were analyzed to determine their average orientation, by using the azimuth of the line joining first and last vertices, but other methods could have been employed. Those lines orientated between 315^0+22.5^0 and 315^0-22.5^0, i.e. lying in a 45^0 interval centred about the northwest direction, were selected and buffered.

The map modelling procedure is listed as follows:

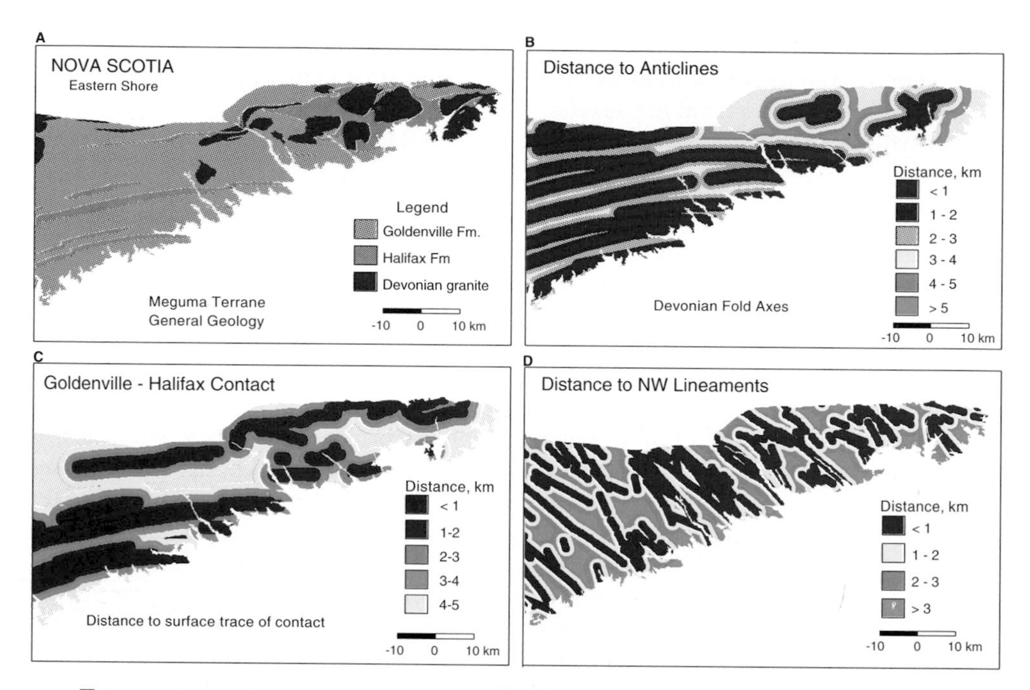

Figure 9-3. Input maps for gold potential study, Meguma terrane, Nova Scotia. **A**. Geological map, showing Lower Paleozoic sedimentary rocks (Goldenville and Halifax Fms) intruded by Devonian granites. **B**. Map showing proximity to the surface traces of Devonian anticline axes, made by buffering. **C**. Proximity to the Goldenville-Halifax contact, made by extracting the contact lines from a topological file, followed by buffering. **D**. Proximity to NW lineaments. Lineaments were digitized from a LANDSAT image, sorted and tagged by orientation, selected by orientation class, and buffered.

```
: At current location, determine whether or not map conditions are satisfied
        M1 = class('GEOL') == 1 OR class('GEOL') == 2
: Get the current class number for the catchment basin map
        ROW = class('BASIN')
: Lookup lake sediment geochemical values from table called 'BASIN'
: For each element determine whether the conditions are TRUE or FALSE
        M2 = table('BASIN', ROW, 'AS') > 30
        M3 = table('BASIN', ROW, 'SB') > 0.8
        M4 = table('BASIN', ROW, 'AU') > 9
        M5 = table('BASIN', ROW, 'W') >= 3
```

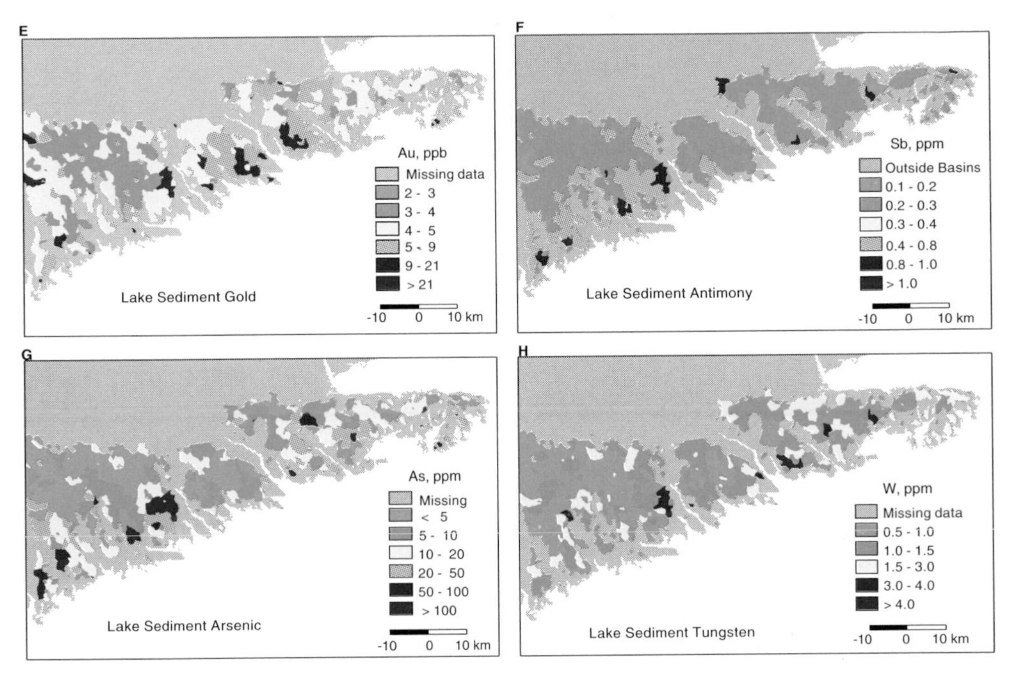

FIG. 9-3. E. Gold in lake sediments, by catchment basin. **F**. Antimony in lake sediments, by catchment basin. **G**. Arsenic in lake sediments, by catchment basin. **H**. Tungsten in lake sediments, by catchment basin.

```
: The arsenic and gold in vegetation maps are used instead of the vegetation
attribute table
        M6 = class('BIOAS') < 3
        M7 = class('BIOAU') < = 3
        M8 = class('ANTI') < 6
        M9 = class('GOLDHAL') < 4
        M10 = class('NWLINS')< =1
: Determine an overall geochemical condition by applying Boolean "OR"
        GEOCHEM = M2 OR M3 OR M4 OR M5 OR M6 OR M7
: Combine with other conditions using Boolean "AND" and generate binary map
        NEWMAP = M1 AND GEOCHEM AND M8 AND M9 AND M10
        RESULT(NEWMAP)
```

Note that for display purposes, some of the maps used in this procedure are reclassified, so there is not always agreement between the class number in the procedure and the sequence of map units shown in the legends of Figure 9-3.

The principal difference from the landfill procedure is that reference is made to an attribute table. The BASIN map is linked to the BASIN table by the basin number (class) keyfield. In

I

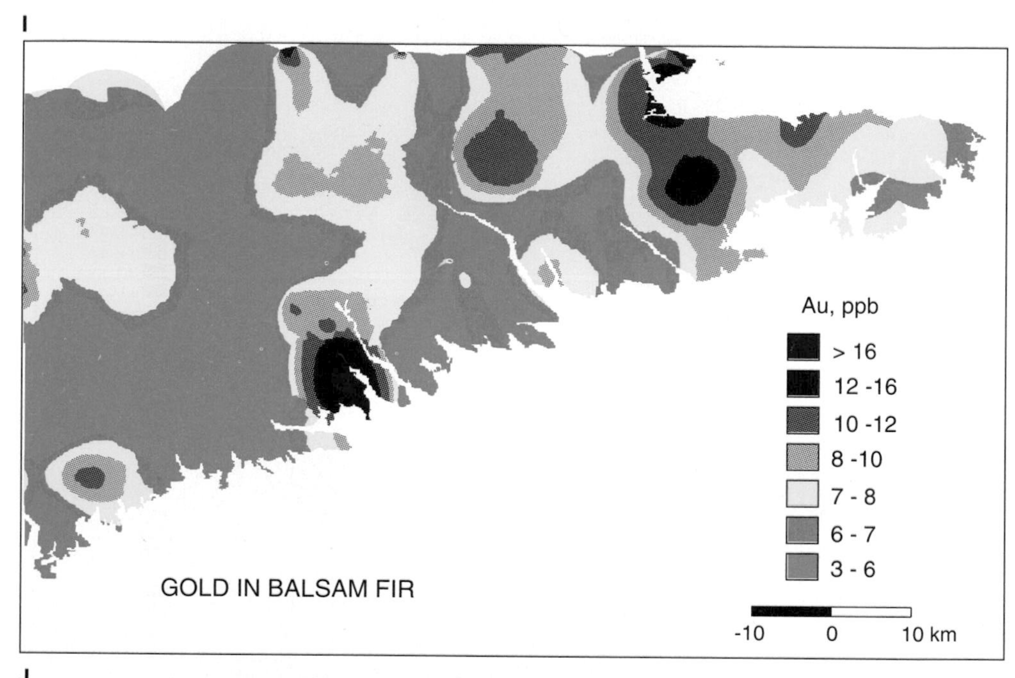

J

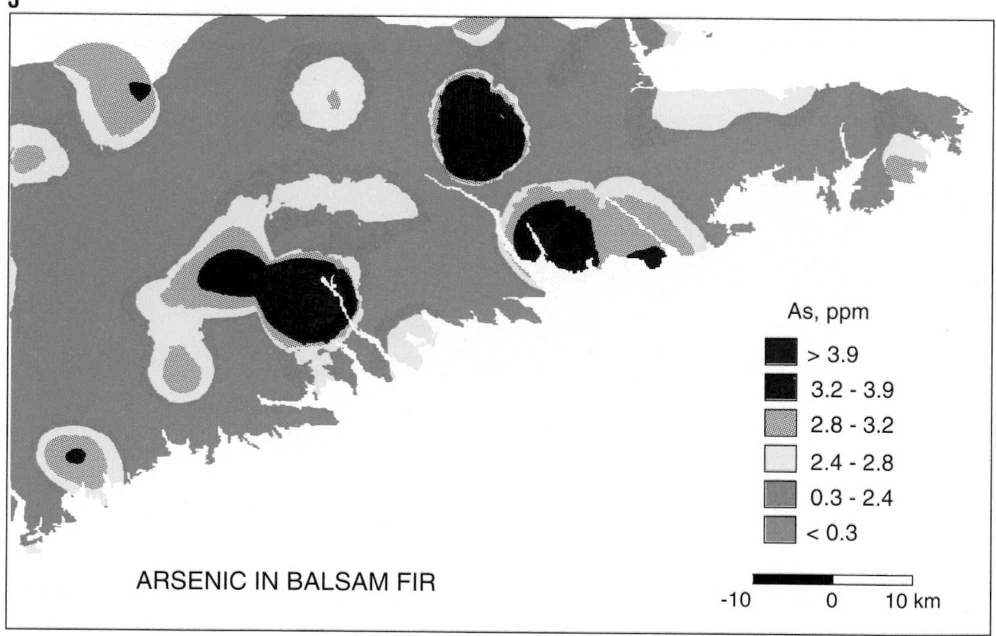

FIG. 9.3 I. Gold in twigs of Balsam Fir, interpolated from point samples by kriging. **J.** Arsenic in Balsam Fir twigs, also interpolated from point samples.

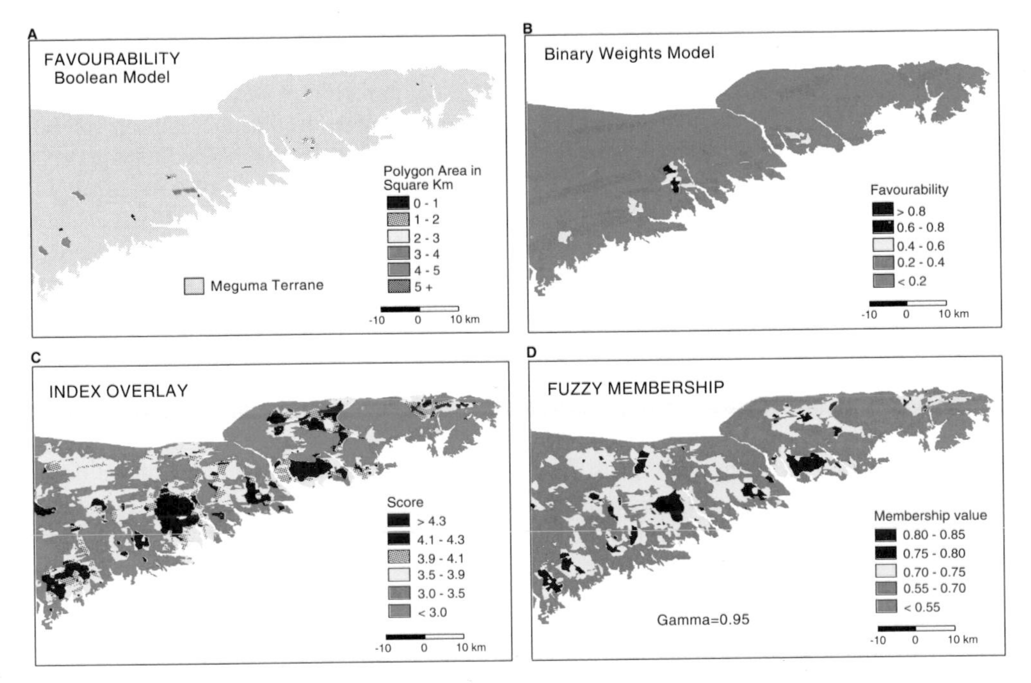

FIG. 9-4. Output maps showing gold potential calculated by various methods. **A**. Application of Boolean rules, with a binary output. The polygons that satisfy the conditions have been classified by area. **B**. Areas ranked using the weighted binary model. The red area is the Goldenville district, one of the main gold camps in the Meguma terrane. **C**. Areas ranked by an index overlay and classified. **D**. Areas ranked according to fuzzy membership values. The fuzzy set comprises those locations that satisfy the proposition "*favourable for gold exploration*", and fuzzy membership ranges from 0 to 1. Most of the main known gold occurrences are predicted by the model, and several new prospective areas are suggested.

the second line of the procedure, the class of the BASIN map is saved as a temporary variable called ROW, which saves having to repeat class('BASIN') in the next four statements. Notice that the OR operation has been employed as well as the AND operation, because there is very little spatial overlap between the various kinds of geochemical anomaly. By simply applying AND to all the geochemical anomaly zones, the output would indicate that all the geochemical criteria are never satisfied for any single location, and the output map would be zero everywhere. The OR operator produces zones where the presence of one or more of the geochemical element anomalies is present.

On the output map, see Figure 9-4A, the selected polygons have been classified according to size, although in this case the sizes of the mineral potential zones are not critical to the analysis, as for the landfill. The Boolean model can be a useful first step in defining mineral potential zones, but is not ideal because selected zones are not ranked according to the degree of favourability.

Comments on Boolean Logic Model

In effect, Boolean modelling involves the logical combination of binary maps resulting from the application of conditional operators. Each of the maps used as a condition can be thought of as a layer of evidence. The various layers of evidence are combined to support a hypothesis, or proposition. Thus the hypothesis being evaluated in the landfill siting example is "*this area is suitable for a landfill*"; in the mineral potential example, the hypothesis is "*this area is favourable for gold deposits*". Each location is tested to determine whether it belongs to the set of locations for which the criteria are satisfied. Set membership is expressed either as 1 or 0, with no possibility of "maybe". The hypothesis is evaluated repeatedly over all locations of the study region, resulting in a binary hypothesis map. In the language of sets, set membership is expressed only with binary 1 (TRUE) or 0 (FALSE).

The appeal of the Boolean approach is its simplicity. The logical combination of maps in a GIS is directly analogous to the physical "stacking" of maps on a light table, the traditional method employed by many geologists. In cases where prescriptive guidelines have been established by law or by code, Boolean combinations are a practical and easily-applied approach. In practice, however, it is usually unsuitable to give equal importance to each of the criteria being combined. Evidence needs to be weighted depending on its relative significance. For example, the presence of anomalous gold in lake sediment is a far more important indicator of a nearby gold deposit than simply being close to a lineament, but in the Boolean logic approach both kinds of evidence are treated equally.

Inferences about a hypothesis based on the application of a set of Boolean rules constitute a method of representing knowledge. Thus a set of rules for a particular type of gold deposit, based on the opinion of an exploration geologist, comprises a concise statement about the geologists knowledge of the factors important for finding a deposit. Where evidence is linked together with a set of rules, the resulting network is sometimes called a **decision tree**. Decision trees are one of the tools used to represent knowledge in an expert system. They are used in many fields, for example in the construction of keys for the optical identification of minerals, or for the identification of biological taxa from morphological and other characteristics. Decision trees not only comprise a knowledge base of rules, they also constitute a decision framework that can be represented in a modelling language, and applied to allocate objects to classes based on their attributes. Expert systems that employ deterministic decision rules use languages such as PROLOG for coding the rules that make up the decision tree.

The rules in decision trees are not always created by an expert, they can also be generated *automatically* from data. Suppose that a large dataset has been built for describing the characteristics of *Foraminifera* (a group of microfossils, widely used in stratigraphic age dating). Each record in the dataset represents a numerical description of the attributes of a foraminiferal individual. About half the records in the dataset include a positive identification of the correct species (one field contains a code indicating species), whereas the identification of the remaining half is unknown. The individuals with a known identification can be used as a **training set** to construct a decision tree, based on the attributes. At the top of the tree, all members of the training set are placed in a single group, or node. The group is partitioned into two or more subgroups, by finding the attribute that maximizes the **distinctness** of the subgroups. Each of the subgroups is, in turn, split into further sub-subgroups, using other attributes. The process is repeated until the groups become smaller than some threshold, or when the composition of a sub-group is a single species. The repeated splitting process can be represented graphically as a tree structure, or algebraically as series of decision rules. The set of automatically-generated decision rules can now be treated as a knowledge base, and can be applied to unknown individual records for the purpose of identification. Probably the best known algorithms for building decision trees are embodied in the procedures called CART (Classification and Regression Trees), see Breiman et al. (1984). CART generates binary splits at each partition. Walker and Moore (1988) have used CART procedures with a GIS for studying the factors that control the spatial distribution of kangaroo species in Australia. A number of similar procedures have been developed by others. A method that allows multi-way (as opposed to binary) splits, described by Biggs et al. (1991) was applied by Reddy and Bonham-Carter (1991) in conjunction with a GIS to analyze base metal potential in a greenstone belt.

We will now turn to the methods known as index overlays. The advantage of the index overlay procedure is that each input map to be used as evidence is assigned a different weight, depending on its significance to the hypothesis under consideration.

INDEX OVERLAY MODELS

The simplest kind of index weighting is where the input maps are binary and each map carries a single weight factor. However, where multi-class maps (either categorical or higher levels of measurement) are used, each class of every map is given a different score, allowing for a more flexible weighting system.

Binary Evidence Maps

If the evidence to be combined together is binary, each map is simply multiplied by its weight factor, summed over all the maps being combined and normalized by the sum of the weights. The result is a value ranging between 0 and 1, which can be classified into intervals appropriate for mapping. At any location, the output score, *S*, is defined as:

$$S = \frac{\sum_{i}^{n} W_i \; class(MAP_i)}{\sum_{i}^{n} W_i} \tag{9-1}$$

where W_i is the weight of the i-th map, and $class(MAP_i)$ is either 1 for presence or 0 for absence of the binary condition. The output score is between 0 (implying extremely unfavourable) to 1 (implying highly favourable). The result is to produce a map showing regions that are ranked according to the score.

Consider the map modelling procedure for the landfill case, modified for binary index weighting.

```
: Add the map weights together, for use in normalization
: Weights are not restricted to integer values, as used here
        SUMW =5+3+3+2+6+8+4+6+6+1
: At current location, set temporary variable to the map weight if condition
satisfied, else set it to 0
: The logical operator takes priority, so the weight is multiplied by either 1
(TRUE) or 0 (FALSE)
        C1 = 5 * class('OVERTHIK') > 4
        C2 = 3 * class('PERMEAB') < 2
        C3 = 3 * class('SLOPE') < 2
        C4 = 2 * class('GEOLOGY') <> 4
        C5 = 6 * class('FLOOD') <> 1
        C6 = 8 * class('ZONING') > 1
        C7 = 4 * class('SUITAB') > 1
        C8 = 6 * class('MUNIBUF') < 4
        C9 = 6 * class('ROADBUF') < 6
        C10= 1 * class{'ECOLOG') == 1
  : Calculate sum of weighted conditions and divide by normalization factor
        OUTPUT = (C1 + C2 + C3 + C4 + C5 + C6 + C7 + C8 + C9 +
        C10) / SUMW
: OUTPUT has a value in the range (0,1) and must now be divided into classes
: Apply classification table of breakpoints called 'BINWT'
        FINAL = CLASSIFY(OUTPUT, 'BINWT')
        RESULT(FINAL)
```

The weights for the 10 input maps are added together in the first statement. The sequence of weight values in the first executable statement is the same as the map order defined by the statements that follow. The Boolean procedure has been modified by multiplying each logical expression by a weight. Thus, at locations where the overburden thickness map is greater than 4, the criterion is TRUE and the weighted result is 5*1=5; where the criterion is FALSE, the result is 5*0=0. The output value is then the sum of the weighted criteria maps divided by the sum of the weights. Where all the criteria are satisfied, OUTPUT=1. Where none are satisfied, OUTPUT=0. By classifying OUTPUT into, say 10 equal classes with an interval of 0.1, a very different map for landfill site selection than the one based on Boolean operators is produced, as shown in Figure 9-2B.

Notice that the polygonal areas selected by the unweighted Boolean model (Figure 9-2A) are actually enclosed within larger zones that are ranked in the highest suitability class by the binary weighting (Figure 9-2B). Because the output map shows areas ranked according to suitability, the decision as to where a landfill should be sited can now be made with a greater degree of flexibility. Perhaps a location would be chosen although not having the very highest rank, but in consideration of other factors not easily mapped.

For the mineral potential case, index weights can also be incorporated into the modelling procedure, as follows:

```
: Calculate normalization factor
        SUMW = 3 + 4 + 5 + 3 + 2 + 4 + 5 + 4 + 2 + 1
        ROW = class('BASIN')
: For current location, determine map weights
        M1 = 3 * (class('GEOL') == 1 OR class('GEOL') == 2)
        M2 = 4 * table('BASIN', ROW, 'AS') > 30
        M3 = 5 * table('BASIN', ROW, 'SB') > 0.8
        M4 = 3 * table('BASIN', ROW, 'AU') > 9
        M5 = 2 * table('BASIN', ROW, 'W') = > 3
        M6 = 4 * class('BIOAS') < 3
        M7 = 5 * class('BIOAU') < = 3
        M8 = 4 * class('ANTI') < 6
        M9 = 2 * class('GOLDHAL') < 4
        M10 = 1 * class('NWLINS') < = 1
: Calculate normalized sum of weight factors
        NEW = (M1 + M2 + M3 + M4 + M5 + M6 + M7 + M8 + M9 +
        M10) / SUMW
: Classify and map output
        NEWMAP = CLASSIFY(NEW, 'BINWT')
        RESULT(NEWMAP)
```

The resulting map, Figure 9-4B, shows regions ranked on a scale from 0-1. The red areas for highest potential are in the neighbourhood of the Goldenville district, once one of the most important gold mining districts of Nova Scotia.

Index Overlay with Multi-Class Maps

In this case, the map classes occurring on each input map are assigned different scores, as well as the maps themselves receiving different weights as before. It is convenient to define the scores in an attribute table for each input map (some GIS provide specialized templates for inserting values in a special attribute table for all the maps being combined). The average score is then defined by

$$\bar{S} = \frac{\sum_{i}^{n} S_{ij} W_i}{\sum_{i}^{n} W_i} \tag{9-2}$$

where $\bar{S}$ is the weighted score for an area object (polygon, pixel), W_i is the weight for the i-th input map, and S_{ij} is the score for the j-th class of the i-th map, the value of j depending on the class actually occurring at the current location.

Each map must be associated with a list of scores, one per map class. Class scores can be put into an attribute table with an editor, for access by the modelling procedure. The attribute table can then be modified without changing the procedure. Attribute tables containing scores are shown in Table 9-5A to J for the landfill example. Notice that scores for some classes are set to negative 1. Areas where such a class occurs are automatically set to class 0 in the output, removing them from consideration altogether. For example, the area lying within the municipal limits are set to -1 in the "distance from municipality" map. No matter how favourable the other maps might be, this region can never be chosen as a landfill site. The modelling procedure for the landfill case follows:

```
: Calculate the sum of weights, as before
        SUMW = 5 + 3 + 3 + 2 + 6 + 8 + 4 + 6 + 6 + 1
: At the current location, determine class score from table and multiply by map weight
: The attribute tables have been assigned the same names as the corresponding maps
        C1 = 5 * table('OVERTHIK', class('OVERTHIK'), 'SCORE')
        C2 = 3 * table('PERMEAB', class('PERMEAB'), 'SCORE')
        C3 = 3 * table('SLOPE', class('SLOPE'), 'SCORE')
        C4 = 2 * table('GEOLOGY', class('GEOLOGY'), 'SCORE')
        C5 = 6 * table('FLOOD', class('FLOOD'), 'SCORE')
        C6 = 8 * table('ZONING', class('ZONING'), 'SCORE')
        C7 = 4 * table('SUITAB', class('SUITAB'), 'SCORE')
        C8 = 6 * table('MINIBUF', class('MINIBUF'), 'SCORE')
        C9 = 6 * table('ROADBUF', class('ROADBUF'), 'SCORE')
        C10= 1 * table('ECOLOG', class('ECOLOG'), 'SCORE')
: Calculate normalized score
        OUT = (C1 + C2 + C3 + C4 + C5 + C6 + C7 + C8 + C9 +
        C10) / SUMW
: Check for presence of negative scores
        NEG = MIN(C1, C2, C3, C4, C5, C6, C7, C8, C9, C10)
: Set output to zero if any score is negative
        OUT = {0 if NEG < 0, OUT}
: Use classification table called 'MULTI' to determine output map class
        NEW = CLASSIFY(OUT, 'MULTI')
        RESULT(NEW)
```

Note that the map weights are still used as before, but instead of applying conditional statements to determine a binary condition, a score is obtained from the attribute table of each map, depending on the map class present at the location under consideration. Thus in the case of the map OVERTHIK, in the second statement of the procedure, the table called 'OVERTHIK' is used, the name being the first argument following the word "table". The second argument is the row of the table, determined by the class of the OVERTHIK map for the location being processed. The third argument is the column of the table, denoted by the title of the field, in this case 'SCORE'. The result of such a table operation is to retrieve the appropriate map score for overburden thickness, depending on the map class. Suppose that the class is 6 at a particular location, the score for

Table 9-5. Attribute tables for the 10 maps used for landfill site selection. Note the score[1] and fuzzy membership fields. The rows of the tables are linked to map classes, not polygon numbers.

A. Overburden thickness (OVERTHIK)

Class	Score	Fuzzy	Legend
1	2	0.1	"1 m"
2	5	0.3	"2 m"
3	7	0.9	"3 m"
4	8	0.9	"4 m"
5	9	0.9	"5 m"
6	10	0.9	"6 m"

B. Permeability (PERMEAB)

Class	Score	Fuzzy	Legend
1	7	0.9	"low"
2	5	0.6	"med"
3	3	0.2	"high"

C. Surface slope (SLOPE)

Class	Score	Fuzzy	Legend
1	9	0.9	"low"
2	8	0.9	
3	7	0.7	
4	6	0.5	"medium"
5	5	0.3	
6	4	0.1	
7	3	0.1	
8	2	0.1	"steep"

D. Geology (GEOLOGY)

Class	Score	Fuzzy	Legend
1	6	0.8	"granite"
2	4	0.5	"sandstone"
3	6	0.9	"shale"
4	1	0.1	"limestone"
5	3	0.2	"conglomerate"

E. Zoning map(ZONING)

Class	Score	Fuzzy	Legend
0	-1	0.1	"city"
1	2	0.3	"industrial"
2	10	0.8	"agricult A"
3	5	0.7	"agricult B"

F. Distance from city limits (MUNIBUF)

Class	Score	Fuzzy	Legend
0	-1	0.0	"0 km"
1	8	0.6	"<1 km"
2	8	0.8	"<2"
3	7	0.9	"<3"
4	5	0.7	"<4"
5	3	0.5	"<5"
6	1	0.3	"<6"
7	1	0.1	"<7"
8	1	0.1	"<8"
9	1	0.1	"<9"
10	1	0.1	"<10"
11	1	0.1	">=10"

G. 100-year flood zone (FLOOD)

Class	Score	Fuzzy	Legend
1	-1	0.1	"100 yr"
2	10	0.9	">100"

H. Suitability for farming(SUITAB)

Class	Score	Fuzzy	Legend
1	2	0.1	"good"
2	4	0.4	"fair"
3	12	0.9	"poor"

I. Distance from major road (ROADBUF)

Class	Score	Fuzzy	Legend
1	8	0.6	"<1 km"
2	7	0.9	"<2"
3	6	0.8	"<3"
4	5	0.7	"<4"
5	4	0.5	"<5"
6	3	0.3	"<6"
7	2	0.1	"<7"
8	1	0.1	"<8"

J. Ecologically sensitive (ECOL)

Class	Score	Fuzzy	Legend
1	3	0.1	"sensitive"
2	8	0.9	"insensitive"

[1] Using this system of weights and scores, it is important to note that the class scores are not normalized for each map. Therefore, scores should be chosen according to a scheme that is similar for each map. For example, the scores might range from 0 (least favourable) to 10 (most favourable), with -1 meaning "impossible". The scores should not range from 0-1 for one map, and 0-100 for another map. Weights and scores are not restricted to integer values.

OVERTHIK is 10. Similar table operations are invoked for the other nine maps. Notice that the scores are multiplied in each case by the map weight. OUT is then calculated as a weighted average as before. Finally a check is made to determine whether any of the maps have a negative score, using the MIN() operator, which returns the minimum value of two or more arguments. If the minimum score is negative, then OUT is automatically set to 0. In the last step, a classification is applied to the scores before generating the output map. The assignment of scores has resulted in a very different distribution of favourability from the Boolean and binary weights models, as shown on Figure 9-2C, with a region on the NW side of town that appears to be strongly favoured, resulting from the choice of scores for the input maps. Quite different scenarios can be tested by modifying the class scores and the map weights.

In the mineral potential case, the map attribute tables with scores are shown in Table 9-6. The following modelling procedure is similar to the landfill procedure.

```
: Determine sum of weights as before
        SUMW = 3 + 4 + 5 + 3 + 2 + 4 + 5 + 4 + 2 + 1
: At current location, multiply tabled score times map weight
        M1 = 3 * table('GEOL', class('GEOL'), 'SCORE')
: Each geochemical element is now represented by a separate map
        M2 = 4 * table('LSAS', class('LSAS'), 'SCORE')
        M3 = 5 * table('LSAU', class('LSAU'), 'SCORE')
        M4 = 3 * table('LSSB', class('LSSB'), 'SCORE')
        M5 = 2 * table('LS0W', class('LS0W'), 'SCORE')
        M6 = 4 * table('BIOAS', class('BIOAS'), 'SCORE')
        M7 = 5 * table('BIOAU', class('BIOAU'), 'SCORE')
        M8 = 4 * table('ANTI', class('ANTI'), 'SCORE')
        M9 = 2 * table('GOLDHAL', class('GOLDHAL'), 'SCORE')
        M10 =1 * table('NWLINS', class('NWLINS'), 'SCORE')
: Normalize the sum of current weighted scores
        OUT = (M1 + M2 + M3 + M4 + M5 + M6 + M7 + M8 + M9 +
        M10) / SUMW
: Check for negative scores.
: The presence of a negative score automatically causes the output to be zero
        NEG = MIN(M1, M2, M3, M4, M5, M6, M7, M8, M9, M10)
: Set output to normalized sum if no negative scores present, else set to zero
        OUT = {OUT IF NEG> = 0 , 0}
: Classify and create output map
        CLASSIFY(OUT, 'MULT')
        RESULT(OUT)
```

Figure 9-4C shows mineral potential based on this index overlay scheme. The highest score is about 5, the lowest about 2. Many of the important past-producing gold deposits are indicated as having a high score, and a number of regions are picked out where no known deposits are known. This is a useful potential map, with many of the same features as the maps produced by the somewhat more sophisticated fuzzy logic and weights of evidence models, introduced next.

Comments on Index Overlays

Index overlays allow for a more flexible combination of maps than is possible with Boolean logic operations alone. The tables of scores and the map weights can be adjusted to reflect the judgment of an expert in the domain of the application under consideration. Map scores can be chosen as positive integers or real numbers, with no limit on the numerical range (except that the range should be compatible between maps), because scaling to output map classes is achieved by applying a classification table of breakpoints. The greatest disadvantage of the method probably lies in its linear additive nature. The following fuzzy logic approach is in many respects similar, but the combination rules are more flexible, and improve on the linear additive nature of the model.

FUZZY LOGIC METHOD

In classical set theory, the membership of a set is defined as true or false, 1 or 0. Membership of a **fuzzy set**, however, is expressed on a continuous scale from 1 (full membership) to 0 (full non-membership). Thus individual measurements of arsenic (As) in lake sediment might be defined according to their degree of membership in the set called "*Arsenic anomaly*". Very high values of As are definitely anomalous, with a fuzzy membership of 1; very low values at or below background have a fuzzy membership of zero; between these extremes a range of possible membership values exist. Such a membership function might be expressed analytically as

$$\mu(x) = \begin{cases} 0 & x < 50 \\ \dfrac{x - 50}{200} & 50 < x < 250 \\ 1 & x > 250 \end{cases} \qquad (9\text{-}3)$$

where x is the As concentration value in ppm and $\mu(x)$ is the fuzzy membership function. Every value of x is associated with a value of $\mu(x)$, and the ordered pairs $[x, \mu(x)]$ are known collectively as a fuzzy set. The shape of the function need not be linear, as in Equation 9-3 and shown in Figure 9-5, it can take on any analytical or arbitrary shape appropriate to the problem at hand. Fuzzy membership functions can also be expressed as lists or tables of numbers. Thus in the As case, the discrete representation of the fuzzy membership function shown in Table 9-7 (for class intervals of 50 ppm), is equivalent to the analytical expression in Equation 9-3.

Now suppose that As has been mapped, with 50 ppm class intervals, then the fuzzy membership function can be treated as an attribute table of an arsenic map, as shown in Table 9-8.

The classes of any map can be associated with fuzzy membership values in an attribute table. The level of measurement of the mapped variable can be categorical, ordinal or interval.

Table 9-6. Attribute tables for mineral potential study, showing scores for class weighting and fuzzy membership values. Only 4 out of the 10 tables used in the study are shown. Tables have been assigned the same names as their associated maps.

A. Geology (GEOL)			
Class	**Score**	**Fuzzy**	**Legend**
0	-1	0.0	'outside'
1	9	0.8	'Goldenville'
2	7	0.7	'Halifax'
3	-1	0.1	'Granite'
B. Lake Sediment Antimony (LSSB)			
0	1	0.1	'no data'
1	8	0.8	'0.9-1.3 ppm'
2	7	0.8	'0.8-0.9'
3	6	0.6	'0.6-0.8'
4	5	0.4	'0.5-0.6'
5	4	0.3	'0.4-0.5'
6	2	0.2	'0.3-0.4'
7	2	0.2	'0.2-0.3'
8	1	0.1	'<0.2'
C. Balsam Fir Gold (BIOAU)			
0	0	0.0	'no data'
1	9	0.9	'24-136 ppb'
2	8	0.8	'16-24'
3	8	0.6	'12-16'
4	7	0.4	'10-12'
5	6	0.3	'8-10'
6	5	0.2	7-8'
7	4	0.2	'6-7'
8	2	0.2	'3-6'
9	1	0.1	'<3'

D. Anticline distance (ANTI)			
Class	**Score**	**Fuzzy**	**Legend**
0	0	0.1	'> 6 km'
1	9	0.9	'<0.25'
2	9	0.9	'0.25-0.5'
3	9	0.9	'0.5-0.75'
4	9	0.9	0.75-1.0'
5	8	0.8	'1.0-1.25'
6	8	0.8	'1.25-1.5'
7	8	0.8	'1.5-1.75'
8	8	0.8	'1.75-2.0'
9	7	0.7	'2.0-2.25'
10	7	0.7	'2..25-2.5'
11	7	0.7	'2.5-2.75'
12	6	0.6	'2.75-3.0'
13	6	0.5	3.0-3.25'
14	6	0.5	'3.25-3.5'
15	5	0.5	'3.5-3.75'
16	5	0.4	'3.75-4.0'
17	4	0.4	'4.0-4.25'
18	4	0.4	'4.25-4.5'
19	3	0.3	'4.5-4.75'
20	3	0.3	'4.75-5.0'
21	2	0.3	'5.0-5.25'
22	2	0.3	'5.25-5.5'
23	1	0.2	'5.5-5.75'
24	1	0.2	'5.75-6.0'

Fuzzy membership values must lie in the range (0,1), but there are no practical constraints on the choice of fuzzy membership values. Values are simply chosen to reflect the degree of membership of a set, based on subjective judgment. Values need not increase or decrease monotonically with class number, as in the case of As above.

The presence of the various states or classes of a map might be expressed in terms of fuzzy memberships of different sets, possibly storing them as several fields in the map attribute

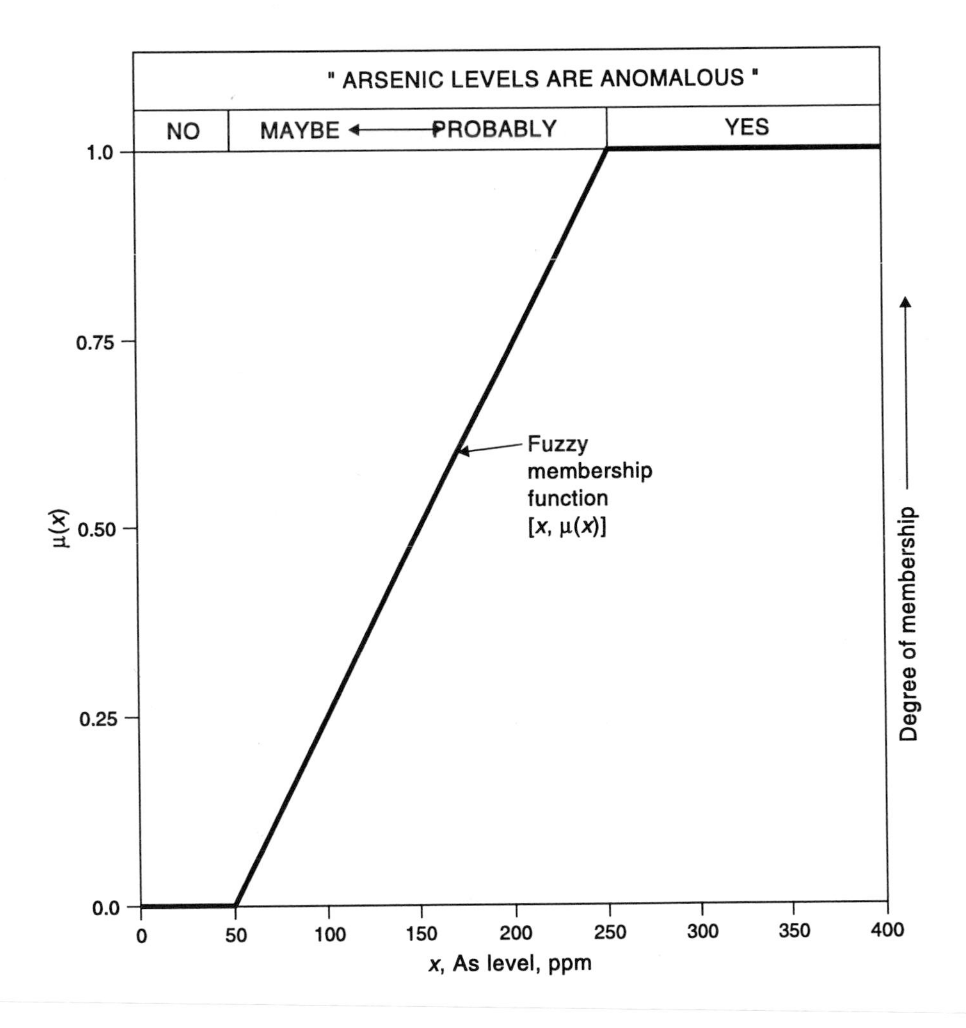

FIG. 9-5. A graph showing fuzzy membership of the set of observations for which "*arsenic levels are anomalous*". Fuzzy membership can, in some cases, be expressed as an analytical function, not necessarily linear as shown here, in other cases membership is defined more readily as a table.

table. Thus the As values on a map might be considered in terms of their fuzzy membership of a set "*favourable indicator for gold deposits*", or a second set "*suitable for drilling water wells*". The membership functions for these two sets would look very different, one reflecting the importance of As as a pathfinder element for gold deposits, the other reflecting the undesirability of drilling a water well in rocks with elevated levels of As.

Not only can a single map have more than one fuzzy membership function, but also several different maps can have membership values for the same proposition or hypothesis. Suppose that the spatial objects (polygons, pixels) on a map, are evaluated according to the proposition "*favourable location for gold exploration*": then any of the maps to be used as evidence in support

Table 9-7. Fuzzy membership function for As expressed as the ordered pairs $[x, \mu(x)]$, and organized in a table.

x	$\mu(x)$
300	1
250	1
200	0.75
150	0.5
100	0.25
50	0
0	0

Table 9-8. Attribute table for a map of As, with the fuzzy membership values shown as one field.

Map Class	Fuzzy Membership	Legend Entry
1	1.00	'> 275 ppm As'
2	1.00	'225 - 275'
3	0.75	'175 - 225'
4	0.50	'125 - 175'
5	0.25	' 75 - 125'
6	0.00	' 25 - 75'
7	0.00	'< 25'

of this proposition can be assigned fuzzy membership functions. Table 9-5 shows a series of fuzzy membership functions for the maps used to select a landfill. The membership values were chosen arbitrarily (like the index overlay scores) based on subjective judgment about the relative importance of the maps and their various states. The fuzzy membership values are in the field labelled "Fuzzy". Table 9-6 also shows fuzzy membership functions for the mineral potential maps. Note that the fuzzy memberships assigned to categorical maps (such as the geological map or the zoning map in the landfill study) do not increase or decrease monotonically with class number, but are assigned values in the range (0,1) that reflect, subjectively, the importance of individual map units. Thus limestone is assigned a value of 0.1 (highly unfavourable for a landfill), whereas a shale is assigned a very favourable value (0.9).

Note that the fuzzy membership values must reflect the relative importance of each map, as well as the relative importance of each class of a single map. The fuzzy memberships are similar to the combined effect of the class scores and the map weights of the index overlay method.

Combining Fuzzy Membership Functions

Given two or more maps with fuzzy membership functions for the same set, a variety of operators can be employed to combine the membership values together. The book by Zimmermann (1985), for example, discusses a variety of combination rules. An et al. (1991) discuss five operators that were found to be useful for combining exploration datasets, namely the fuzzy AND, fuzzy OR, fuzzy algebraic product, fuzzy algebraic sum and fuzzy gamma operator. These operators are briefly reviewed here.

Fuzzy AND

This is equivalent to a Boolean AND (logical intersection) operation on classical set values of (1,0). It is defined as

$$\mu_{combination} = MIN(\mu_A, \mu_B, \mu_C, \ldots) \quad , \tag{9-4}$$

where μ_A is the membership value for map A at a particular location, μ_B is the value for map B, and so on. Of course, the fuzzy memberships must all be with respect to the same proposition. Suppose that at some location the membership value for map A is 0.75 and for map B is 0.5, then the membership for the combination using fuzzy AND is 0.5. It can readily be seen that the effect of this rule is to make the output map be controlled by the smallest fuzzy membership value occurring at each location. Like the Boolean AND, fuzzy AND results in a conservative estimate of set membership, with a tendency to produce small values. The AND operation is appropriate where two or more pieces of evidence for a hypothesis must be present together for the hypothesis to be true.

Fuzzy OR

On the other hand, the fuzzy OR is the like the Boolean OR (logical union) in that the output membership values are controlled by the maximum values of any of the input maps, for any particular location. The fuzzy OR is defined as

$$\mu_{combination} = MAX(\mu_A, \mu_B, \mu_C, \quad ...) \tag{9-5}$$

Using this operator, the combined membership value at a location (=suitability for landfill etc) is limited only by the most suitable of the evidence maps. This is not a particularly desirable operator for the landfill case, but might in some circumstances be reasonable for mineral potential mapping, where favourable indicators of mineralization are rare and the presence of *any* positive evidence may be sufficient to suggest favourability. Note that in using either the fuzzy AND or fuzzy OR, a fuzzy membership of a single piece of evidence controls the output value. On the other hand, the following operators combine the effects of two or more pieces of evidence in a "blended" result, so that each data source has some effect on the output.

Fuzzy Algebraic Product

Here, the combined membership function is defined as

$$\mu_{combination} = \prod_{i=1}^{n} \mu_i \quad , \tag{9-6}$$

where μ_i is the fuzzy membership function for the i-th map, and i=1,2...,n maps are to be combined. The combined fuzzy membership values tend to be very small with this operator, due to the effect of multiplying several numbers less than 1. The output is always smaller than, or equal to, the smallest contributing membership value, and is therefore "decreasive". For example, the algebraic product of (0.75, 0.5) is 0.375. Nevertheless, all the contributing membership values have an effect on the result, unlike the fuzzy AND, or fuzzy OR operators.

Fuzzy Algebraic Sum

This operator is comlementary to the fuzzy algebraic product, being defined as

$$\mu_{combination} = 1 - \prod_{i=1}^{n} (1 - \mu_i) \quad . \tag{9-7}$$

The result is always larger (or equal to) the largest contributing fuzzy membership value. The effect is therefore "increasive". Two pieces of evidence that both favour a hypothesis reinforce one another and the combined evidence is more supportive than either piece of evidence taken individually. For example, the fuzzy algebraic sum of (0.75, 0.5) is 1-(1-0.75)*(1-0.5), which equals 0.875. The increasive effect of combining several favourable pieces of evidence is automatically limited by the maximum value of 1.0, which can never be exceeded. Note that whereas the fuzzy algebraic product is an algebraic product, the fuzzy algebraic sum is not an algebraic summation.

Gamma Operation

This is defined in terms of the fuzzy algebraic product and the fuzzy algebraic sum by

$$\mu_{combination} = (\text{Fuzzy algebraic sum})^{\gamma} * (\text{Fuzzy Algebraic product})^{1-\gamma}, \tag{9-8}$$

where γ is a parameter chosen in the range (0,1), Zimmermann and Zysno (1980). When γ is 1, the combination is the same as the fuzzy algebraic sum; and when γ is 0, the combination equals the fuzzy algebraic product. Judicious choice of γ produces output values that ensure a flexible compromise between the "increasive" tendencies of the fuzzy algebraic sum and the "decreasive" effects of the fuzzy algebraic product. For example, if $\gamma= 0.7$, then the combination of (0.75, 0.5) is $0.875^{0.7} * 0.375^{0.3} = 0.679$, a result that lies between 0.75 and 0.5. On the other hand, if $\gamma=0.95$, then the combination is 0.839, a mildly increasive result. If $\gamma=0.1$, then the combination is 0.408, a result that is less than the average of the 2 input function values, and is therefore decreasive. The effect of choosing different values of γ are shown in Figure 9-6. Note that although the same tendencies occur, the actual values of γ for which the combined membership function becomes increasive or decreasive vary with the input membership values. An et al. (1991) used a value of $\gamma=0.975$ to combine geophysical and geological datasets in their study of iron and base metal deposits in Manitoba, presumably because the increasive effects of large γ values best seemed to reflect the subjective decision-making of typical exploration geologists.

Returning to the internal modelling procedure, with the landfill case, the following steps can used to combine the 10 maps with the fuzzy gamma operation. Note the similarity of the modelling procedure to the index overlay, with the fuzzy membership values for each map being held in a map attribute table, similar to the class scores.

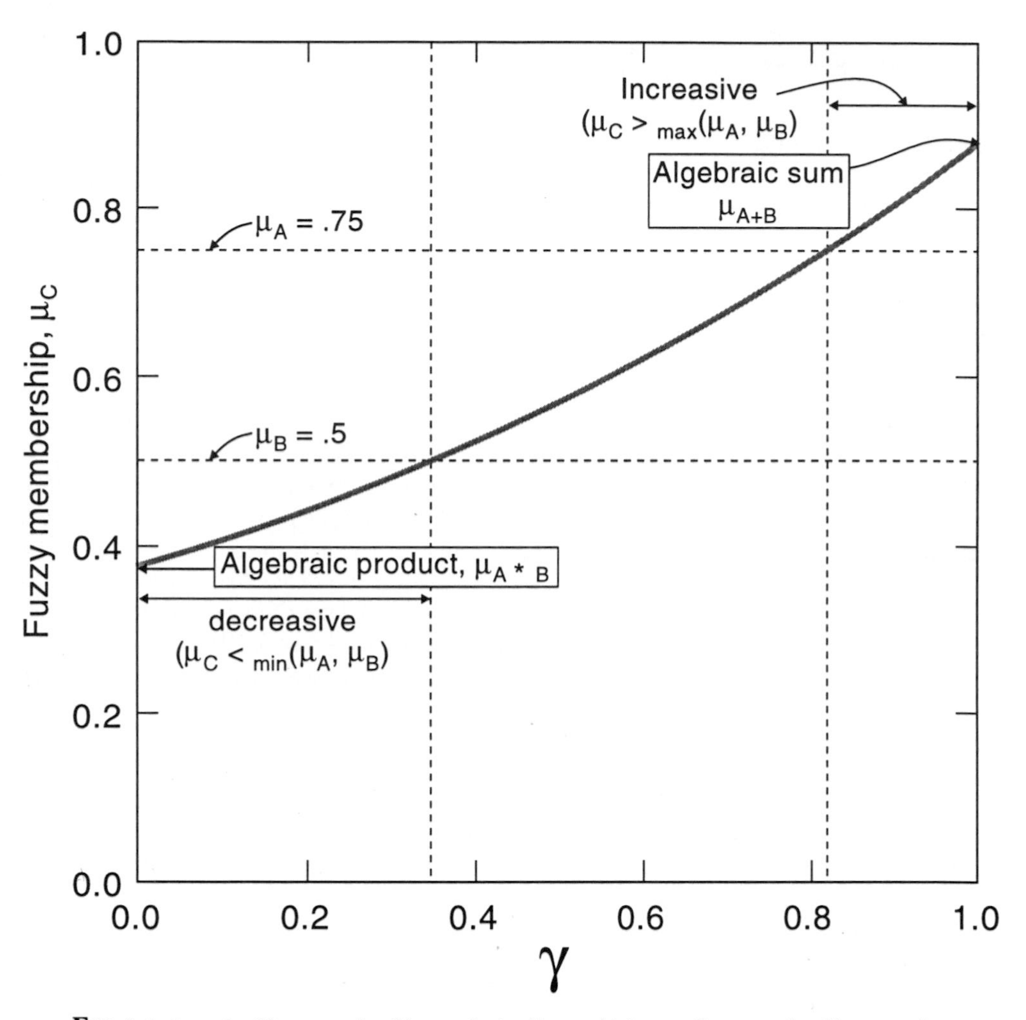

FIG. 9-6. A graph of fuzzy membership, μ_C, obtained by combining two fuzzy memberships, μ_A and μ_B, versus γ. This shows the effect of variations in γ for the case of combining two values, μ_A=0.75 and μ_B=0.5. When γ=0, the combination equals the fuzzy algebraic product; when γ=1, the combination equals the fuzzy algebraic sum. When $0.8 < \gamma < 1$, the combination is larger than the largest input membership value (in this case 0.75), and the effect is therefore "increasive". When $0 < \gamma < 0.35$, the combination is smaller than the smallest input membership value (0.5 in this case), and the effect is therefore "decreasive". When $0.35 < \gamma < 0.8$, the combination is neither increasive nor decreasive, but lies within the range of the input membership values. The limits 0.8 and 0.35 are data dependent.

```
: Set value of Gamma
        GAMMA = 0.95
: At current location, lookup fuzzy membership values for each input map
        C1 = table('OVERTHIK', class('OVERTHIK'), 'FUZZY')
        C2 = table('PERMEAB', class('PERMEAB'), 'FUZZY')
        C3 = table('SLOPE', class('SLOPE'), 'FUZZY')
        C4 = table('GEOLOGY', class('GEOLOGY'), 'FUZZY')
        C5 = table('FLOOD', class('FLOOD'), 'FUZZY')
        C6 = table('ZONING', class('ZONING'), 'FUZZY')
        C7 = table('SUITAB', class('SUITAB'), 'FUZZY')
        C8 = table('MINIBUF', class('MINIBUF'), 'FUZZY')
        C9 = table('ROADBUF', class('ROADBUF'), 'FUZZY')
        C10= table('ECOLOG', class('ECOLOG'), 'FUZZY')
: Calculate the fuzzy algebraic product and fuzzy algebraic sum
        PRODUCT = C1 * C2 * C3 * C4 * C5 * C6 * C7 * C8 * C9 * C10
        SUM  = 1 - ((1 - C1) * (1 - C2) * (1 - C3) * (1 - C4) * (1 - C5) *
        (1 - C6) * (1 - C7) * (1 - C8) * (1 - C9) * (1 - C10))
: Apply gamma operator and classify result with classification table called
'FUZTAB'
: The symbol ^ is an operator for "raised to the power of"
        OUT = (SUM ^ GAMMA) * (PRODUCT ^ (1 - GAMMA))
        MEMBER=CLASSIFY(OUT, 'FUZTAB')
        RESULT(MEMBER)
```

Notice that for each of the 10 input maps, the 'FUZZY' column is the field in the corresponding map attribute table where the fuzzy membership functions are stored, see Table 9-5. The final output makes use of the exponentiation operator ^ to raise the algebraic sum to the power GAMMA and the algebraic product to the power (1-GAMMA). The output map, after classification with a table of breakpoints called 'FUZTAB', show areas ranked according the combined fuzzy membership, see Figure 9-2D.

The procedure for the mineral potential case is similar, except for two features. First, the value of gamma is specified as keyboard input, allowing different values to be selected at run time. Second, the four lake sediment maps are combined using fuzzy OR, and the two biogeochemical maps are also combined with fuzzy OR. This means the combined effect of the lake sediment geochemical evidence will take on the maximum fuzzy membership of the four contributing maps. An anomalous value from any one of the maps is therefore sufficient to give this factor a large fuzzy score. The effect is the same for the biogeochemical combination. Finally, the gamma operator is used, as before, for the final combination step. The resulting map is shown in Figure 9-4D. Superficially it looks similar to the index overlay, but careful comparison shows some important differences.

```
: This procedure is shown graphically as an inference net in Figure 9-7
: Type input gamma value from keyboard (INPUT)
        "VALUE OF GAMMA?"
        GAMMA = INPUT
: At current location, get fuzzy membership values for each map
        M1 = table('GEOL', class('GEOL'), 'FUZZY')
        M2 = table('LSAS', class('LSAS'), 'FUZZY')
        M3 = table('LSAU', class('LSAU'), 'FUZZY')
        M4 = table('LSSB', class('LSSB'), 'FUZZY')
        M5 = table('LS0W', class('LS0W'), 'FUZZY')
        M6 = table('BIOAS', class('BIOAS'), 'FUZZY')
        M7 = table('BIOAU', class('BIOAU'), 'FUZZY')
        M8 = table('ANTI', class('ANTI'), 'FUZZY')
        M9 = table('GOLDHAL', class('GOLDHAL'), 'FUZZY')
        M10 = table('NWLINS', class('NWLINS'), 'FUZZY')
: Apply fuzzy "OR" to lake sediment maps.
: Favourable lake sed geochem is an intermediate hypothesis
        FAVLS = MAX(M2, M3, M4, M5)
: Favourable biogeochem is an intermediate hypothesis
        FAVBIO = MIN(M6, M7)
: Calculate fuzzy product, sum and gamma
: Favourable location for gold deposits is a final hypothesis
        FPROD = M1 * FAVLS * FAVBIO * M8 * M9 * M10
        FSUM = 1- ((1 -M1)*(1 - FAVLS)*(1 - FAVBIO)*(1 - M8)*(1 -
        M9)*(1 - M10))
        FAVLOC = FSUM ^ GAMMA * FPROD ^ (1 - GAMMA)
: Classify output and generate new map
        NEW = CLASSIFY(FAVLOC, 'FUZTAB')
        RESULT(NEW)
```

Comments on the Fuzzy Logic Method

In practice, it may be desirable to use a variety of different fuzzy operators in the same problem, as shown for the mineral potential example. In particular, fuzzy AND and fuzzy OR can be more appropriate than fuzzy gamma in some situations, but not in others. For example, suppose that two input maps represent evidence for a proposition that requires that the evidence occur jointly. To take a slightly contrived example, consider a map of sulphur content and a map of zinc content from lithological samples. The combination is highly suggestive evidence for the presence of zinc sulphide (sphalerite), an important mineral in many zinc deposits. Ignoring the obvious problems of concentration units and other factors for the sake of simplicity, we can deduce that because the joint presence of the two elements is needed, the importance of the evidence is limited by the lesser abundance of the two elements. In this case, fuzzy AND would be an appropriate combination operator, because at each location the combination would be controlled by the minimum of the fuzzy membership values. In other situations, fuzzy OR is more appropriate, where for example, the presence of any one of the pathfinder elements in abundance might be significant evidence for the presence of a mineral deposit, even though other pathfinder elements are not present in anomalous amounts.

Evidence maps can be combined together in a series of steps, as depicted in an inference network, Figure 9-7. Thus instead of combining all the maps in one operation, for example with the gamma operator, it may be more appropriate to link together some maps with, say the fuzzy OR to support an intermediate hypothesis, other maps with fuzzy AND to support another intermediate hypothesis, and finally to link both raw evidence and intermediate hypotheses (now in turn being used as evidence) with a fuzzy gamma operation. Many combinations are possible. The inference network becomes an important means of simulating the logical thought processes of an expert. In expert system terminology, the fuzzy

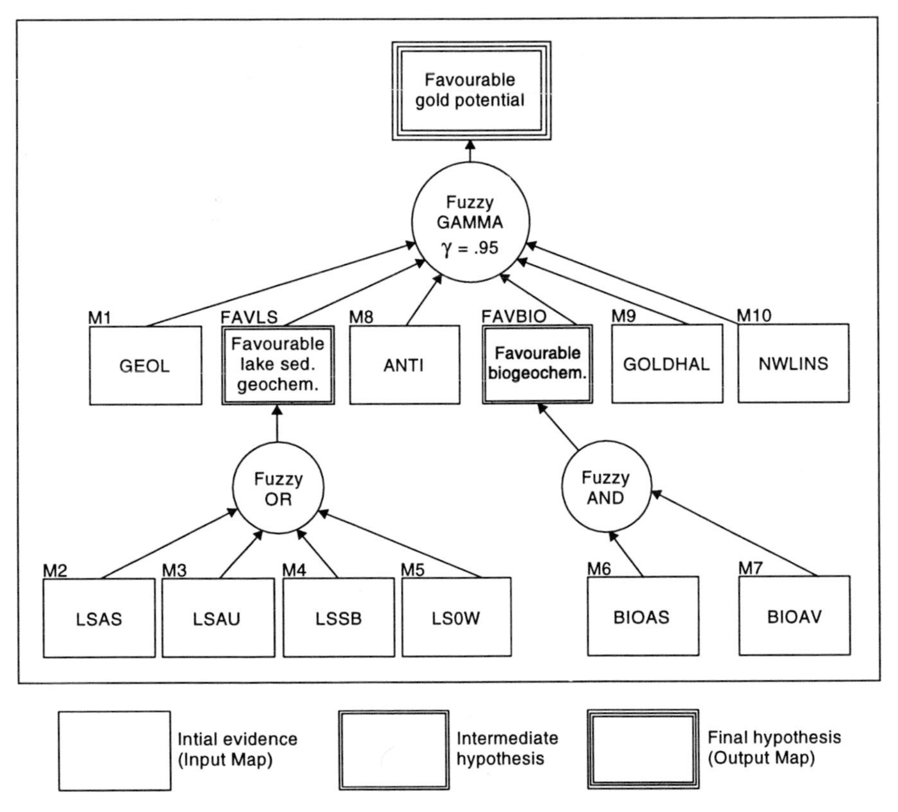

FIG. 9-7. A simple inference network for predicting gold in Meguma terrane, using fuzzy logic as the "inference engine". Each box (or "space") in the network is a map. The "knowledge base" is contained in the map attribute tables as fuzzy membership functions associated with each input map. Information is propagated through the network in the direction indicated by the arrows. The final outcome is a fuzzy membership map for the proposition "*this location is a favourable target for gold*".

membership functions are the "knowledge base" and the inference network and fuzzy combination rules are the "inference engine". Fuzzy logic is one of the tools used in expert systems where the uncertainty of evidence is important. Even quite complex inference networks can be implemented in a map modelling language. Fuzzy logic has also been applied to problems of pattern recognition in geology, see Griffiths (1987).

We now turn to methods that use probability concepts for combining maps. In general, Bayesian methods can be applied to multi-state maps, whether they are categorical or use higher levels of measurement. The weights of evidence method is a log-linear version of the general Bayesian model, normally applied where the evidence is binary. The Prospector expert system, in its original form, uses both the Bayesian model and fuzzy logic operators, with evidence that is either binary or multi-state.

BAYESIAN METHODS

The Bayesian approach to the problem of combining datasets uses a probability framework. One of the main concepts in the Bayesian approach is the idea of *prior* and *posterior* probability. To introduce the idea of *prior* and *posterior* probabilities, suppose that an individual wishes to predict the likelihood of rain for the next day. On average it rains, say, 80 days in the year. A reasonable estimate of the *prior* probability of rain the next day might therefore be the average proportion 80/365. This initial or *prior* estimate can now be modified by other sources of information. For example, one important factor that affects the likelihood of rain is the month of the year. The prior probability of rain can be modified, or updated, by multiplying by a factor that varies with the time of year. Thus

$$P\{Rain \mid Time\text{-}of\text{-}Year\} = P\{Rain\} * Time\text{-}of\text{-}Year\ Factor \quad (9\text{-}9)$$

where *P{Rain}* is the *prior* probability, and *P{Rain|Time-of-Year}* is the posterior probability, equal to the conditional probability of rain given the time of year, or the *prior* probability multiplied by a factor. The time-of-year factor varies according to the month, based on historical data. Another factor that affects the likelihood of rain is whether rain has occurred the previous day. Rain is more likely tomorrow if it rains today. These two sources of evidence for the proposition "*it will rain tomorrow*" can then be combined by the expression:

$$P\{Rain \mid Evidence\} = P\{Rain\} * Time\text{-}of\text{-}Year\ Factor * Rain\text{-}Today\text{-}Factor \quad (9\text{-}10)$$

where the Rain-Today-Factor is determined from historical data, and the combined evidence is the product of two factors. Several sources of data that provide evidence about tomorrow's weather can be combined using such a model. Some evidence will cause the *posterior* probability to increase as compared with the *prior*, i.e. the likelihood of rain is more than average. In such cases, the evidence will have a multiplying factor greater than 1. But where the multiplying factor is less than 1 (but must always be positive), the evidence causes the *posterior* probability to be less than the *prior*. The *prior* probability can be successively updated with the addition of new evidence, so that the *posterior* probability from adding one piece of evidence can be treated as the *prior* for adding a new piece of evidence.

Turning to a spatial example, consider the probability that a geochemical zinc anomaly will occur in a region, as defined by locations with values greater than 250 ppm. The *prior* probability of a zinc anomaly in the absence of any other information might be established as the proportion of samples greater than 250 ppm from extensive data records. Given information from a geological map, the *posterior* probability of a zinc anomaly will vary considerably, depending on the bedrock lithology (and other factors). Thus the geological factor might be greater than 1 for greenstone lithologies, but less than 1 for sandstone and limestone. One might collect information about the statistical relationship between lithology and zinc values to establish a geological multiplying factor that could be saved, for example, as the field of a geological attribute table, analogous to (but different from) the fuzzy membership functions discussed earlier. This geological information could be used to make a *posterior* probability map where the favourability for a zinc anomaly is greater or less than the average, depending on the lithology present. Addition of other factors, like the presence of alteration, the occurrence of particular vegetation types, the intensity of certain geophysical signatures, and others, also affect the *posterior* probability and can be combined together for a prediction based on multiple data sources.

This kind of model has been applied to the problem of combining evidence in various disciplines. In particular, it has been applied to quantitative medical diagnosis, combining evidence from clinical symptoms to predict disease, e.g. Lusted(1968), Aspinall and Hill (1983), Spiegelhalter and Knill-Jones (1984) and others. Aspinall (1992) uses a Bayesian model in an ecological GIS application. In geology, the Prospector model, originally developed in a non-spatial mode (Reboh and Reiter, 1983; McCammon, 1989), uses Bayesian "updating" to combine evidence about mineral prospects in an expert system. It was later applied in a GIS context to combine evidence in map form, and successfully predicted the extension of the Mount Tolman molybdenum deposit in Washington, Campbell et al. (1982), see also Katz (1991). Reddy et al. (1992) applied the same approach to the prediction of base-metal deposits in a greenstone belt. In the Prospector model, the importance of evidence is weighted by a mineral deposits expert. It is, therefore, a **knowledge-driven** model. The Bayesian model, in a log-linear form known as weights of evidence (Spiegelhalter, 1986), has been applied where sufficient data is available to estimate the relative importance of evidence by statistical means, and is therefore **data-driven**. Agterberg et al. (1990), Bonham-Carter et al. (1988), Watson et al. (1989) and a number of other authors have applied the method in this form for mineral potential mapping. Goodacre et al. (1993) have applied the same approach to model the spatial distribution of seismic epicentres in west Quebec.

The Bayesian approach is described here in the context of mineral potential mapping, where the goal is to predict the presence of a set of point objects. The point objects (mineral deposits, seismic epicentres, sinkholes in karst, lineament intersections, etc.) are treated as being binary, either present or absent. In fact the model requires that each point is treated as a small area object, occurring within a small unit cell. The method is not restricted to this case, however, and can be applied to the prediction of area objects, such as the presence of a rock type, or the presence of an anomaly. In the following sections, both the method used to propagate evidence in the Prospector model, and the weights of evidence model are discussed together, because they use the same principles. We begin with a general discussion of the philosophy of estimating mineral favourability.

Favourability and Conditional Probability

Consider the problem of finding a mineral deposit in a region that covers an area of 10,000 km^2. Suppose that 200 mineral deposits are known within the region, and for the purpose of the analysis, each deposit is assumed to occupy a small unit area, or cell. If the unit cell is selected as 1 km^2 for this case, then the total region, T, occupies N{T}=10,000 unit cells, where N{ } is the notation used to denote the count of unit cells. The average density of **known** deposits in the region is N{D}/N{T}, or 200/10000=0.02, where N{D} is the total number of deposits, assuming that not more than one deposit occurs per unit cell. This is the probability that a 1 km^2 cell, chosen at random (with a random number generator for example) contains a known deposit. Where no other information is available, this ratio can be used as the *prior* probability of a deposit, P{D}. Now suppose that a binary indicator map, such as an electromagnetic (EM) anomaly map, covers the same area as shown in Figure 9-8A, and that 180 out of the 200 deposits occur where the EM anomaly also occurs. Clearly the probability of finding a deposit is much greater than .02 if the EM anomaly is known to be present; conversely the probability is less than .02 if the EM anomaly is known to be absent. The favourability for finding a deposit given the presence of the evidence can be expressed by the conditional probability:

$$P\{D|B\} = \frac{P\{D \cap B\}}{P\{B\}} \tag{9-11}$$

where $P\{D|B\}$ is the conditional probability of a deposit given the presence of a binary pattern, B. But $P\{D \cap B\}$ is equal to the proportion of total area occupied by D and B together, or $P\{D|B\}=N\{D \cap B\}/N\{T\}$, and similarly $P\{B\}=N\{B\}/N\{T\}$, where P{B} and N{B} are, respectively the probability and area where pattern B is present. Substituting these expressions into Equation 9-11 satisfies

$$P\{D|B\} = \frac{N\{D \cap B\}}{N\{B\}} \quad . \tag{9-12}$$

From the Venn diagram in Figure 9-8B, we see that this conditional probability is the proportion of the binary pattern, B, occupied by deposits. In the present example, there are 180 deposits on B, and B occupies 3600 km^2, so that the conditional probability of a deposit given the presence of the EM anomaly is 180/3600=0.05, two and a half times larger than 0.02. Effectively, the exploration for new deposits of the same type now becomes more focused, because if the EM evidence is used as a critical indicator, the search area is reduced from 10,000 km^2 to 3,600 km^2.

In order to obtain an expression relating the *posterior* probability of a deposit in terms of the *prior* probability and a multiplication factor, we note that the conditional probability of being on the binary map B, given the presence of a deposit is defined as:

$$P\{B|D\} = \frac{P\{B \cap D\}}{P\{D\}} \quad , \tag{9-13}$$

which for the present case has the value 180/200=0.9. Because $P\{B \cap D\}$ is the same as $P\{D \cap B\}$, Equations 9-11 and 9-13 can be combined to solve for $P\{D|B\}$, satisfying the relationship

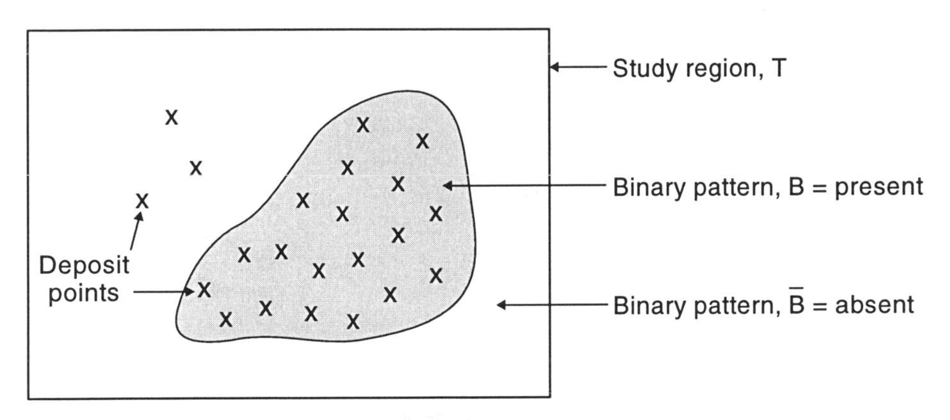

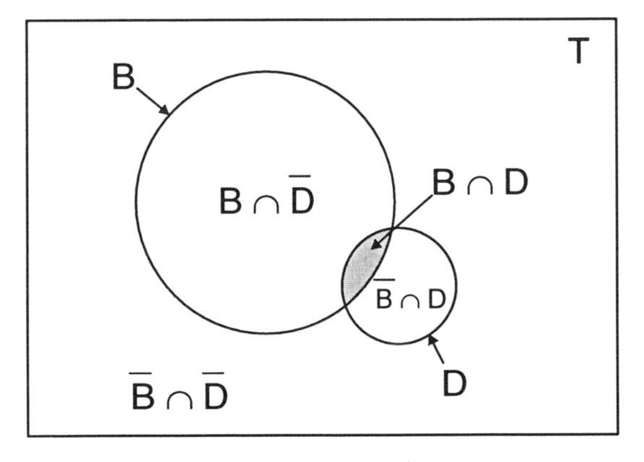

Areas in unit cells

N{T} = 10 000 (Total area)

N{B} = 3600 (Area of pattern)

N{D} = 200 (Area of deposits)

N{B ∩ D} = 180 (Area of deposits on pattern)

FIG. 9-8. Diagrams to illustrate weights of evidence calculations. **A**. Binary map, showing the locations of deposit points. **B**. A Venn diagram summarizing the spatial overlap relationships between the map pattern (area of binary anomaly) and the deposit pattern. Each deposit point is assumed to occupy a small unit area. The total study region is shown as a bounding rectangle (in reality it is usually an irregularly shaped area, bounded by geological constraints, and/or a shoreline). The areas of the circles are not drawn to scale.

$$P\{D|B\} = P\{D\} \frac{P\{B|D\}}{P\{B\}} \quad . \tag{9-14}$$

This states that the conditional (*posterior*) probability of a deposit, given the presence of the binary pattern equals the *prior* probability of the deposit P{D} multiplied by the factor P{B|D}/P{B}. The numerator of this factor is 0.9 and the denominator is 3600/10000=0.36, so the factor is 0.9/0.36=2.5. Thus given the presence an EM anomaly, the *posterior* probability of a deposit is 2.5 times greater than the *prior* probability. As Chung et al. (1992) have pointed out, the information about the numerator can often be determined from a statistical survey of deposits, to determine the proportion of them that are related to the indicator evidence, and information about the denominator comes from a knowledge of the expected occurrence of the indicator pattern in general, from regional mapping for example. We see that by applying Equation 9-14 we obtain P{D|B}=.02*2.5=.05, which is the same value determined directly by Equation 9-12, as expected.

A similar expression can be derived for the posterior probability of a deposit occurring given the *absence* of an indicator anomaly pattern. Thus

$$P\{D|\bar{B}\} = P\{D\} \frac{P\{\bar{B}|D\}}{P\{\bar{B}\}} \quad , \tag{9-15}$$

and substituting numbers for the example leads to $P\{\bar{B}\}$=(10000-3600)/10000=.64, $P\{\bar{B}|D\}$=20/200=.1, and therefore $P\{\bar{B}|D)/P\{\bar{B}\}$=0.1/0.64=0.15625. Thus the *posterior* probability of a deposit occurring per unit area at locations where the EM anomaly is absent is 0.15625 times smaller than the *prior*. In the present case, $P\{D|\bar{B}\}$=.02*0.15625=0.003125. Thus a map of the favourability for mineral exploration, based on this single source of evidence effectively reduces the area of search from 10,000 km^2 to 3,600 km^2, because the chance of finding a deposit where the EM anomaly is absent is significantly smaller (about 50 times) than where it is present.

Odds Formulation

The same model can be expressed in an ***odds*** form, as was employed in the Prospector expert system. As on the racetrack, odds are defined as a ratio of the probability that an event will occur to the probability that it will not occur. The probability of 0.5 of a horse winning a race is equivalent to odds of 0.5/(1-0.5)=1, as discussed in Chapter 8. The weights of evidence method uses the natural logarithm of odds, known as log odds or ***logits,*** but the model is otherwise very similar to the odds formulation used for Prospector. To clarify these different but related approaches, their inter-relationship is briefly outlined here.

To convert Equation 9-14 to odds, divide both sides by $P\{\bar{D}|B\}$, leading to:

$$\frac{P\{D|B\}}{P\{\bar{D}|B\}} = \frac{P\{D\} \; P\{B|D\}}{P\{\bar{D}|B\} \; P\{B\}} \quad . \tag{9-16}$$

But from the definitions of conditional probability

$$P\{\bar{D}|B\} = \frac{P\{\bar{D} \cap B\}}{P\{B\}} = \frac{P\{B|\bar{D}\}\ P\{\bar{D}\}}{P\{B\}} \quad . \qquad (9\text{-}17)$$

Substituting this expression for $P\{\bar{D}|B\}$ into the denominator of the right side of Equation 9-16, and rearranging terms yields the following:

$$\frac{P\{D|B\}}{P\{\bar{D}|B\}} = \frac{P\{D\}}{P\{\bar{D}\}} \cdot \frac{P\{B\}}{P\{B\}} \cdot \frac{P\{B|D\}}{P\{B|\bar{D}\}} \quad . \qquad (9\text{-}18)$$

The odds of a deposit $O\{D\}$ are equal to $P\{D\}/(1-P\{D\})$, or $P\{D\}/P\{\bar{D}\}$, so substituting odds into Equation 9-18 and cancelling leads to the desired expression:

$$O\{D|B\} = O\{D\} \frac{P\{B|D\}}{P\{B|\bar{D}\}} \qquad (9\text{-}19)$$

where $O\{D|B\}$ is the conditional (*posterior*) odds of D given B, $O\{D\}$ is the *prior* odds of D and $P\{B|D\}/P\{B|\bar{D}\}$ is known as the **sufficiency ratio** LS. In Prospector, the value of LS is assigned by an expert. In weights of evidence, the natural logarithm of both sides of Equation 9-19 are taken, and $\log_e$ LS is the positive weight of evidence W^+, which is calculated from the data. Then

$$logit\{D|B\} = logit\{D\} + W^+ \quad , \qquad (9\text{-}20)$$

as discussed in Chapter 8.

Similar algebraic manipulations lead to the derivation of an *odds* expression for the conditional probability of D given the absence of the binary indicator pattern, with the result being

$$O\{D|\bar{B}\} = O\{D\} \frac{P\{\bar{B}|D\}}{P\{\bar{B}|\bar{D}\}} \quad . \qquad (9\text{-}21)$$

The term $P\{\bar{B}|D\}/P\{\bar{B}|\bar{D}\}$ is called the **necessity ratio,** LN, as used in Prospector. In weights of evidence, W^- is the natural logarithm of LN, or $\log_e$ LN. Thus in *logit* form, Equation 9-21 is

$$logit\{D|\bar{B}\} = logit\{D\} + W^- \quad . \qquad (9\text{-}22)$$

LS and LN are also called **likelihood ratios**. In order to get a feel for using these equations, a sample set of calculations follows here, using the data from Figure 9-8. In practice, the more direct computing formulae for the weights, (Equations 8-20 and 8-21 in the previous chapter) can be used, but working through a numerical example may help clarify the meaning of the above equations. The following section in bold is not the map modelling language, but simply worked calculations. A short FORTRAN program is also included in the Appendix for calculating weights.

The probability of binary pattern B being present, given the presence of a deposit:

$$P\{B|D\}=P\{B \cap D\}/P\{D\} = 180/200 = 0.9$$

The probability of B being present, given the absence of a deposit:

$$P\{B|\overline{D}\}=P\{B \cap \overline{D}\}/P\{\overline{D}\} = (3600-180) / (10000-200) =3420/9800 = 0.3490$$

The probability of B being absent, given the presence of a deposit:

$$P\{\overline{B}|D\}=P\{\overline{B} \cap D\}/P(D\} = (200-180) / 200 = 20 / 200 = 0.1$$

The probability of B being absent, given the absence of a deposit:

$$P\{\overline{B}|\overline{D}\}=P\{\overline{B} \cap \overline{D}\}/P\{\overline{D}\} = (10000-3600-200+180) / (10000-200) = 6380/9800 = 0.6510$$

The sufficiency ratio for pattern B:

$$LS=P\{B|D\}/P\{B|\overline{D}\}= 0.9 / 0.3490 = 2.5789$$

The positive weight of evidence for pattern B:

$$W^{+}=\log_e LS=0.9474$$

The necessity ratio for pattern B:

$$LN=P\{\overline{B}|D\}/P(\overline{B}|\overline{D}\} = 0.1 / 0.6510 = 0.1536$$

The negative weight of evidence for pattern B:

$$W^{-}=\log_e LN=-1.8734$$

The prior probability of a deposit:

$$P\{D\}=200/10000=.02$$

The prior odds of a deposit:

$$O\{D\}=P\{D\}/(1-P\{D\})=.02/(1-.02)=.02/.98=0.020408$$

The posterior odds of a deposit given the presence of binary pattern B:

$$O\{D|B\}=O\{D\}*LS=0.05263$$

The posterior logit of a deposit, given the presence of B:

$$\text{logit}\{D|B\}=\text{logit}\{D\} + W^{+}= -2.944$$

The posterior odds of a deposit, given the absence of B:

$$O\{D|\overline{B}\}=O\{D\}*LN= 0.0031348$$

The posterior logit of a deposit, given the absence of B:

$$\text{logit}\{D|\overline{B}\}=\text{logit}\{D\} + W^{-}= -5.765$$

The posterior probability of a deposit, given the presence of B:

$$P\{D|B\}=O\{D|B\}/(1+O\{D|B\}) = 0.05263/1.005263 = 0.05$$

The posterior probability of a deposit, given the absence of B:

$$P\{D|\overline{B}\}=O\{D|\overline{B}\}/(1+O\{D|\overline{B}\}) = 0.0031348/1.0031348 = 0.003125$$

This example illustrates several important points.

1. The value of LS is greater than 1, whereas LN is in the range (0,1). This indicates that the presence of the binary pattern, B, is important positive evidence for deposits. However, if the pattern is *negatively correlated* with the deposits, LN would be greater than 1 and LS would be in the range (0,1). If the pattern is *uncorrelated* with the deposits, then LS=LN=1,

and the *posterior* probability would equal the *prior* probability, and the probability of a deposit would be unaffected by the presence or absence of the binary pattern.

2. Similarly, W^+ is positive, and W^- is negative , due to the positive correlation between the points and the binary pattern. Conversely W^+ would be negative and W^- positive for the case where fewer points occur on the pattern than would be expected due to chance. If the deposits are independent of whether the pattern is present or not, then $W^+=W^-=0$, and the *posterior* equals the *prior*, as above.

3. The values of the *posterior* probability calculated using weights or the likelihood ratios are identical to those calculated directly using the conditional probability equations.

Combining Datasets

Up to this point, it may be unclear why it is necessary to calculate weights (or likelihood ratios, or probability factors, depending on the method chosen), when the posterior probabilities can be calculated directly from the data as shown in Equation 9-11. The reason is that when the evidence from several maps is combined, the weights (or their equivalents) are calculated from each map independently, and then combined in a single equation. As will be shown, this requires an assumption of *conditional independence*, and results in predicted *posterior* probabilities that are not exactly the same as those calculated directly from the data. Direct calculation becomes unsuitable, because it yields no useful information or insight into the data.The conditional independence assumption leads to a model that, like most models, does not fit the data perfectly, but provides a simplification that, when used carefully, is useful for prediction and gives insight into the relative contributions of the separate sources of evidence.

The conditional probability of a deposit occurring, given the presence of **two** predictive map patterns, B_1 and B_2 is

$$P\{D|B_1 \cap B_2\} = \frac{P\{D \cap B_1 \cap B_2\}}{P\{B_1 \cap B_2\}} , \tag{9-23}$$

which can be written as

$$P\{D|B_1 \cap B_2\} = \frac{P\{B_1 \cap B_2|D\} \quad P\{D\}}{P\{B_1 \cap B_2\}}$$

$$= \frac{P\{B_1 \cap B_2|D\} \quad P\{D\}}{P\{B_1 \cap B_2|D\} \ P\{D\} + P\{B_1 \cap B_2|\bar{D}\} \ P\{\bar{D}\}} \tag{9-24}$$

This is Bayes' Rule. Note that there are only two mutually exclusive hypotheses, D and $\overline{D}$, with $P\{D\}+P\{\overline{D}\}=1$. This situation differs from many textbook examples of Bayes' Rule which show cases with more than two mutually exclusive hypotheses. The effects of interaction between B_1 and B_2 can be ignored by making an assumption of conditional independence. This provides a simplification, because it permits the effects of each binary map to be evaluated individually and then combined by multiplying (or adding in the log-linear case) the factors for several maps together.

The conditional independence assumption can be stated as

$$P\{B_1 \cap B_2 | D\} = P\{B_1 | D\} \; P\{B_2 | D\} \quad , \tag{9-25}$$

which allows Equation 9-24 to be simplified to the following form:

$$P\{D | B_1 \cap B_2\} = P\{D\} \; \frac{P\{B_1 | D\}}{P\{B_1\}} \; \frac{P\{B_2 | D\}}{P\{B_2\}} \quad , \tag{9-26}$$

effectively separating the multiplying factors for each map so they are independent. Note that Equation 9-26 is like Equation 9-14, except that multiplying factors for *two* maps are used to update the *prior* probability to give the *posterior* probability. Using the odds formulation, the conditional or posterior odds can be expressed for two map patterns determined from

$$O\{D | B_1 \cap B_2\} = O\{D\} * LS_1 * LS_2 \quad , \tag{9-27}$$

or with the log-linear weights of evidence from

$$logit\{D | B_1 \cap B_2\} = logit\{D\} + W^+_1 + W^+_2 \quad , \tag{9-28}$$

where the subscripts 1 and 2 refer to the likelihood ratios or weights determined independently for indicator maps 1 or 2. The weights for each of the two maps are calculated in exactly the same way as the weights for a single map, as described above. Whichever formulation of the model is used, there are now **four** different ways of combining **two** binary map patterns, the first being when both patterns are present (Equation 9-28); the other three ways are (B_1 present and B_2 absent), (B_1 absent and B_2 present), and (both B_1 and B_2 absent). In the log-linear form, they can be written as

$$logit\{D | B_1 \cap \bar{B}_2\} = logit\{D\} + W^+_1 + W^-_2 \quad , \tag{9-29}$$

$$logit\{D | \bar{B}_1 \cap B_2\} = logit\{D\} + W^-_1 + W^+_2 \quad , \tag{9-30}$$

and

$$logit\{D | \bar{B}_1 \cap \bar{B}_2\} = logit\{D\} + W^-_1 + W^-_2 \quad . \tag{9-31}$$

With 3 binary patterns as evidence, there are 2^3 or 8 possible combinations and in general with n maps there are 2^n possible different combinations. The general expression for combining i=1,2..n maps is either:

$$O\{D | B_1 \cap B_2 \cap B_3 \cap \ldots . B_n\} = O\{D\} * \prod_{i=1}^{n} LS_i \tag{9-32}$$

for the likelihood ratios or

$$logit\{D|B_1 \cap B_2 \cap B_3 \cap \ldots . B_n\} = logit\{D\} + \sum_{i=1}^{n} W^+ \quad (9\text{-}33)$$

for the weights. In these general formulas, the LS becomes LN , and W^+ becomes W^-, if the i-th map pattern is absent instead of present. Where data is missing for a particular map layer in some locations, the likelihood ratio is set to 1, or the weight is set to 0. Equations 9-32 and 9-33 are the computing formulae for combining a set of binary maps with the Bayes model.

Numerical Example

Consider the same data used for the calculations of the weights, ratios and probabilities above, where one binary pattern (now called B_1 instead of B), occupying an area 3600 km^2 out of 10000 km^2 is combined with a second pattern B_2 (lithology for example) which occupies 5000 km^2. 180 out 200 deposits occur on the first pattern, whereas 140 out of 200 occur on the second pattern. The situation is shown diagrammatically in the Venn diagram of Figure 9-9.

Using the weights of evidence formulation, and assuming a unit cell size of 1 km^2, we have already determined W^+_1=0.9474 and W^-_1=-1.8734. Similarly for the second map,

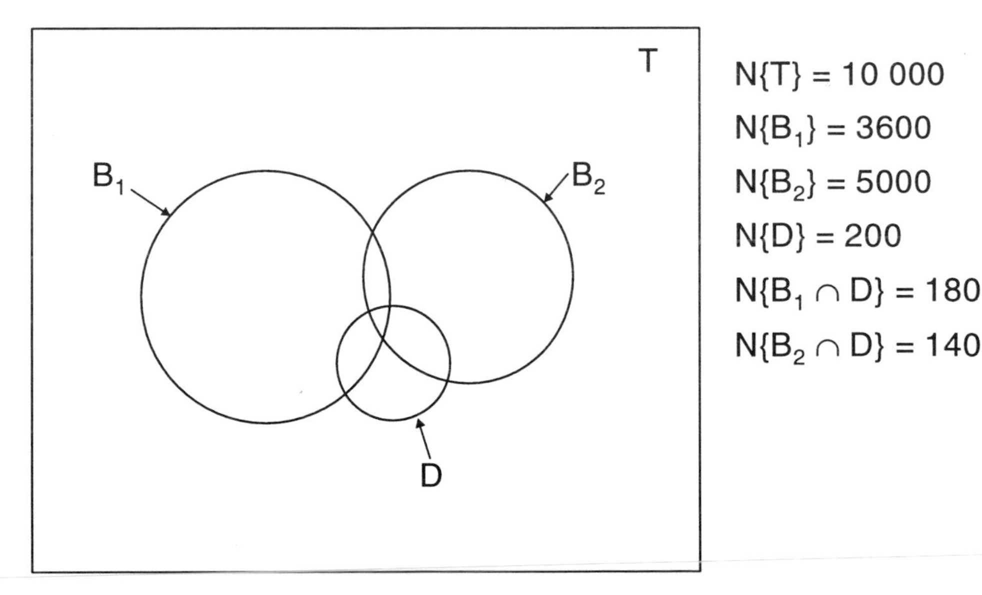

FIG. 9-9. Venn diagram to illustrate the overlap relationships of two binary map patterns (B_1 and B_2) and a set of deposit points. As in Figure 9-8, the areas of circles are not drawn to scale, and each deposit point is assumed to occupy a small unit area.

W^+_2=0.3447 and W^-_2=-0.5189 (the verification of these values is left as an exercise). Notice that the **contrast** (see Chapter 8) for the first map is 0.9474 -(-1.8734)=2.8208 and for the second map is 0.3447-(-0.5189)=0.8636, showing that the first pattern is more strongly associated with the deposit points than the second pattern. The various combinations of B_1 and B_2 are shown in Table 9-9, along with the calculated *posterior* probability, and the ratio of the *posterior* to the *prior*, to illustrate the relative changes. Notice that where one of the patterns is missing, a weight of 0 (or likelihood ratio of 1) is used, so that the probability is unaffected by missing information. This is a very useful aspect of the model in practice, because maps are seldom complete.

We now turn to the question of the meaning of the conditional independence assumption, and show how conditional independence can be tested, at least for the data-driven case.

Conditional Independence

Conditional independence (CI) is assumed to exist when combining two or more maps with the Bayesian models as described above. In practice, probably CI is always violated to some degree, and the following questions arise: How serious is the violation? Are the results distorted

Table 9-9. *Posterior* probabilities calculated from 2 binary map patterns using weights of evidence. The *prior* probability is .02. Where both binary patterns are present, the *posterior* probability is nearly 3.5 times larger than the *prior.* In case 9, both patterns are missing, so the *posterior* = the *prior.*

Case	B_1	B_2	Posterior Probability	Posterior/ Prior
1	+	+	.06915	3.458
2	+	-	.03037	1.512
3	-	+	.00441	0.220
4	-	-	.00186	0.093
5	+	?	.05000	2.5
6	?	+	.00280	1.4
7	-	?	.00031	0.156
8	?	-	.01200	0.6
9	?	?	.02000	1.0

+ = pattern present, - = pattern not present, ? = pattern missing

by the lack of CI? What can be done to mitigate the effects of violating CI? Where weights of evidence are being calculated from data, CI can be checked with statistical tests to show the magnitude of the problem, and pinpoint the maps that are causing the most difficulty. These maps can then be rejected from the analysis or modified to reduce the problem. Where likelihood ratios are being estimated subjectively by an expert, testing for CI is not possible. The modeller must simply be aware that the problem exists and interpret the results accordingly. In this section, we examine what CI means, and describe testing methods.

Pairwise test

The first method involves testing pairwise CI between all possible pairings of the binary maps being combined. Consider first the meaning of **independence**, then **conditional independence**. If two maps, B_1 and B_2, are statistically independent, it implies that

$$P\{B_1|B_2\} = P\{B_1\} \quad and \quad P\{B_2|B_1\} = P\{B_2\} \quad . \tag{9-34}$$

Stated in words, the conditional probability of B_1 being present is independent of whether B_2 is present or not, and vice-versa. Contingency tables and χ^2 can be used to characterize the degree of association between binary patterns, but cannot be used to make formal tests for statistical independence with areal data, because χ^2 varies according to the units of measurement and with the degree of spatial autocorrelation, as discussed in Chapter 8.

On the other hand, if two binary patterns are **conditionally** independent with respect to a set of deposit points, it means that the following relationship is satisfied:

$$P\{B_1 \cap B_2|D\} = P\{B_1|D\}\ P\{B_2|D\} \quad , \tag{9-35}$$

see Equation 9-25. A little algebraic manipulation shows that the following expression is equivalent to Equation 9-35:

$$N\{B_1 \cap B_2 \cap D\} = \frac{N\{B_1 \cap D\}\ N\{B_2 \cap D\}}{N\{D\}} \quad . \tag{9-36}$$

The left-hand side of Equation 9-36 is the observed number of deposits occurring in the overlap region where both B_1 and B_2 are present. The right-hand side is the predicted or expected number of deposits in this overlap zone, which should equal the number of deposits on B_1 times the number on B_2 divided by the total number of deposits, if the two patterns are conditionally independent. This is the normal (2x2) contingency table calculation for testing the independence of two factors, except that it is applied only where deposits are present. The expected frequencies are determined by multiplying the marginal frequencies together and dividing by the total. Because the deposits are considered as points, or small unit cells, the resulting values of χ^2 are unaffected by the units of area measurement, and calculated χ^2 can be compared with tabled values to test for independence. The contingency table layout is shown algebraically in Table 9-10. The four cells in the table correspond to the 4 overlap conditions between two binary patterns. The data from the numerical example above are inserted into Table 9-11.

It is clear from the example that the number of deposits occurring where both patterns are present (130) is very close to the expected number of (140*180)/200 = 126. The chi-square test is then calculated using the expression

$$\chi^2 = \sum_{i=1}^{4} \frac{(Observed_i - Expected_i)^2}{Expected_i}, \quad (9\text{-}37)$$

and then compared to the tabled value of chi-square with one degree of freedom. At a probability level of 98%, for example, there is no reason to reject the CI hypothesis unless χ^2 is greater than 5.4. Note that with small expected frequencies it is necessary to apply Yates correction (Walker and Lev, 1953). The equation including the correction is

Table 9-10. Contingency table for testing conditional independence, based on cells containing a deposit only. The four values within the table are either the expected values assuming independence, calculated from the marginal totals (Equation 9-36), or the observed values measured from maps. There is 1 degree of freedom.

	B_1 Present	B_1 Absent	Totals
B_2 Present	$N\{B_1 \cap B_2 \cap D\}$	$N\{\bar{B}_1 \cap B_2 \cap D\}$	$N\{B_2 \cap D\}$
B_2 Absent	$N\{B_1 \cap \bar{B}_2 \cap D\}$	$N\{\bar{B}_1 \cap \bar{B}_2 \cap D\}$	$N\{\bar{B}_2 \cap D\}$
	$N\{B_1 \cap D\}$	$N\{\bar{B}_1 \cap D\}$	$N\{D\}$

Table 9-11. Contingency table for the example data shown in Figure 9-9. Values in bold are the values observed on the map, those in brackets are expected under the assumption of CI. χ^2 is 4.23 (or 3.24 with Yates correction) showing that the deviation between observed and expected is small, and there is no reason to reject the hypothesis of CI at th 95% level. Tabled $\chi^2_{.95,1}$=3.8 and tabled $\chi^2_{.98,1}$=5.4, (from Walker and Lev, 1953).

	B_1 Present	B_1 Absent	Totals
B_2 Present	**130** (126)	**10** (14)	140
B_2 Absent	**50** (54)	**10** (6)	60
Totals	180	20	200

$$\chi^2 = \sum_{i=1}^{4} \frac{(|Observed_i - Expected_i| - 0.5)^2}{Expected_i}, \qquad (9\text{-}38)$$

which in this case reduces χ^2 to 3.24, making the case for rejecting the null hypothesis of conditional independence even weaker. Where the absolute value of the difference between observed and expected in any cell is less than 0.5, the correction is not made in that cell. Snedecor and Cochran (1967) state that when the expected frequencies are less than 5 in any cell, the approximation to the χ^2 distribution becomes poor. Nevertheless, this is a useful test for pairwise independence, as will be further demonstrated in the Nova Scotia application, below.

The meaning of pairwise CI can be graphically illustrated by Figure 9-10, with a Venn diagram. By focusing on the circle that represents those cells containing a deposit, we see that CI implies that the ratio of the area where both B_1 and B_2 occur to the area of B_2 equals the ratio of the area of B_1 to the area of D. Clearly there are some binary patterns that will not be independent where deposits are present. For example, geochemical elements, used as pathfinders for mineral deposits, will often be inter-correlated, even by making the CI test at the deposit sites only. In such cases, one of the correlated elements must either be discarded, or possibly combined together before the weights of evidence analysis, as discussed by Agterberg (1992). One way of doing this is to combine the dependent patterns together with Boolean operators. Thus given an anomaly map for zinc and another for lead, a combined map might be (zinc anomaly AND lead anomaly), thereby restricting the combined anomaly to those areas with *both* zinc and lead anomalies. On the other hand, a less restrictive approach would be to use (zinc anomaly OR lead anomaly), so that the combined map is anomalous where *either* of the two elements is anomalous. Another method that may be particularly suitable with geochemical data, for example, is to carry out principal components analysis on the original variables, and then to convert selected principal component maps to binary form. The principal components are statistically uncorrelated with one another, so that the likelihood of violating CI is greatly reduced. Furthermore, the technique utilizes data from many variables simultaneously, which is clearly preferable to discarding some variables. Multiple regression is another method that can be used to combine geochemical variables, prior to weights of evidence, see for example Bonham-Carter et al. (1988).

Pairwise CI tests should be run on all possible pairs of maps that are to be combined. This is only possible for Bayesian modelling in data-driven applications, where the factors for each map are being calculated from data about known deposits (or an observed response variable). Where the factors are being estimated subjectively, formal testing is not possible, and judgment must be used to evaluate likely sources of conditional dependence. Of course, conditional dependence may still be present due to three-way or multi-way interactions, and testing for these cases is also possible. However, for practical purposes, pairwise testing will reveal the most glaring problems.

Overall Test

In mineral potential mapping, where the *prior* probability is assumed to be the average known deposit point density, a simple test can be applied to determine the total number of

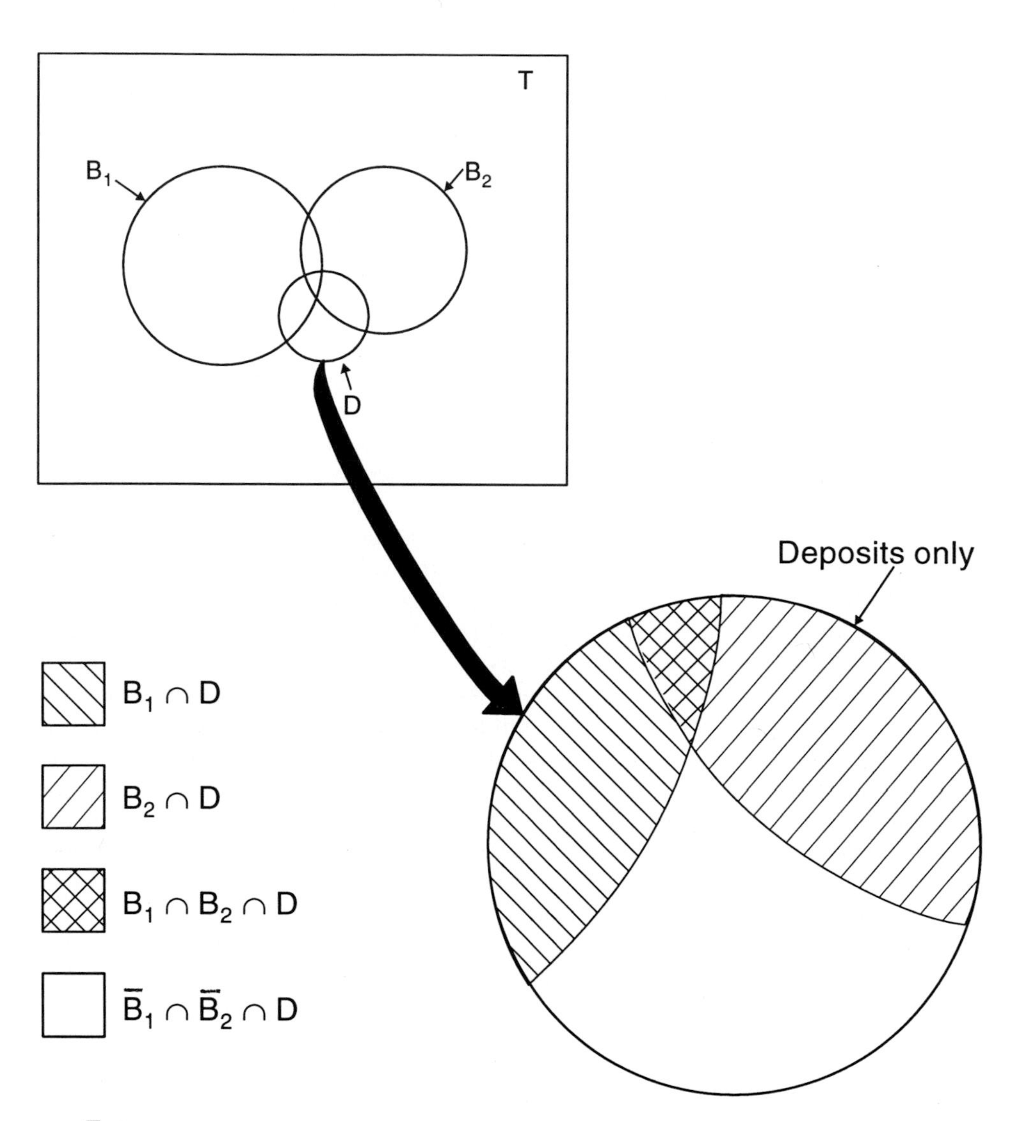

FIG. 9-10. Part of the Venn diagram in Figure 9-9 magnified to illustrate the concept of conditional independence. The enlarged circle is the set of the deposit points only. Within this circle, the patterns of B_1 and B_2 are conditionally independent with respect to the deposits if the area $(B_1 \cap D)$ multiplied by the area $(B_2 \cap D)$ equals the area $(B_1 \cap B_2 \cap D)$/area (D).

predicted deposit points. If the total predicted number of deposits is much larger than the total observed number, it suggests that CI is being violated. The predicted number $N\{D\}_{calc}$ is determined by adding together the product of the area in unit cells, $N\{A\}$, times the *posterior* probability, P, for all polygons on the map, thus

$$N\{D\}_{calc} = \sum_{k=1}^{m} P_k \; N\{A\}_k \quad , \tag{9-39}$$

where there are k=1,2...m polygons (or pixels) on the map. In practice, the predicted number is always larger than the observed number with weights of evidence. If the predicted number is more than, say, 10-15% larger than the observed number, then a serious check of the pairwise tests and remedial action is in order.

We now turn to a real application, and illustrate how the method is implemented in a GIS. We will continue to use the Nova Scotia dataset, so the problem is to predict gold deposits in the Meguma terrane, assuming that the known deposits in the region are an adequate sample of all the possible deposits, both discovered and undiscovered. If we assume that the undiscovered deposits have, on average, the same characteristics as the discovered deposits (or deposits and minor occurrences), then the values of W^+ and W^- will not be greatly affected by new discoveries. However, the prior probability is underestimated by using the average density of known deposits, and in theory can be increased to reflect the density of known plus unknown deposits. However, this will not change the ranking of regions by posterior probability, but will simply increase the posterior probability by a constant multiplication factor.

Application of Weights of Evidence to Mineral Potential Mapping

The steps in applying weights of evidence are as follows:

1. Choose a series of maps that are likely to be useful evidence for predicting mineral deposits. A particular deposit type and associated conceptual deposit model should be used to guide the selection process.

2. For each map, assumed to be multi-class (categorical or higher measurement levels), determine the optimum reclassification scheme to convert to binary form, maximizing the spatial association between the map and the deposit points. Weights of evidence calculations can be used for this optimization process, as discussed below.

3. Check for pairwise CI between the binary maps. Delete problem maps, or combine binary maps to reduce effects of CI.

4. Combine the binary maps with weights determined in step 2 with Equation 9-33. This can be carried out either with a modelling language internal to the GIS, or with an external modelling program.

5. Make new maps showing posterior probability. Optionally, calculate the effects of uncertainty in the weights, and uncertainty due to missing information, and produce an uncertainty map, as discussed below.

This process takes considerably more computation than the methods involving subjective choice of weights. However, the appeal of the method is that it produces objective estimates of weights that reflect spatial associations between map patterns and known deposit points.

Nova Scotia Maps Used for Analysis

The maps used to derive the weights are summarized in Table 9-12. They are not exactly the same ones used for the fuzzy logic model. The multi-element lake sediment geochemical data has been compressed into a single multi-element map, in order to reduce the likelihood of creating CI problems. The compression was carried out used multiple regression on the geochemical attribute table, combining the geochemical attributes with a linear additive model. The coefficients for the model were calculated by a standard multiple linear regression program, where the dependent variable is binary, being the presence or absence of known mineral deposits in the catchment basin corresponding to the geochemical sample, and the independent variables are the logarithms of the geochemical elements. The logarithms were used to stabilize the variance, and because geochemical data is often approximately lognormal. An initial stepwise regression analysis showed that the 4 strongest predictors were Au, As, Sb and W. A final map of regression scores (LKSDSIG) was created based on a combination of these four elements, eliminating possible conditional dependence that would likely be present in using the four maps independently in weights of evidence, Wright (1988), Bonham-Carter et al. (1988). In addition a map showing the proximity to Devonian granite

Table 9-12. Maps used for gold prediction, Nova Scotia study. The original multi-class maps are converted to binary form, and the cutoffs are determined by maximizing the contrast. See Table 9-4 for description of maps.

Map Name	Present	Absent	Comments
ANTICLIN	<= 1.25 km	> 1.25 km	Contrast at maximum
BIOAS	>= 5.2 ppm	< 5.2 ppm	Contrast at maximum
BIOAU	>= 8 ppb	< 8 ppb	Contrast at maximum
GEOLOGY	Goldenville and Halifax	Granite	Occurrences rare in Halifax in this study area
GOLDHAL	<= 2 km	> 2 km	Contrast at maximum
GRANCON	<= 1 km	> 1 km	Contrast at maximum
LKSDSIG	>= 0.045	< 0.045	Lake sed geochem signature regression scores
NWLINS	<= 0.5	> 0.5	Contrast at maximum

contacts (GRANCON) was constructed, because some deposit geologists have suggested that the granite might be a source of mineralizing fluids, and therefore there might be a spatial association between granite proximity and known deposits.

Binary Reclassification

The process of converting multiclass maps to a binary form can either be carried out subjectively, using geological judgment, or can be carried out statistically, so as to determine the threshold that maximizes the spatial association between the resulting binary map pattern and the point pattern. Let us illustrate this procedure with the anticline map.

The multiclass map is shown in Figure 9-11, with the mineral occurrence points superimposed. Even without statistical confirmation, it is clear that the points tend to lie close to the axes of anticlinal folds, a fact widely recognised by geologists who have studied this region. The question is, in determining how to make a binary anomaly map for this evidence, what is the best distance to choose as a cutoff? By making the distance too short, the effect is to reduce the binary pattern to a very small area, and many of the occurrences are missed. By making the distance too long, the binary pattern becomes too large, and effectiveness of this pattern for narrowing the search area is reduced. This map has 24 distance buffers spaced at 250 m intervals, so by calculating the weights (and contrast) for each distance corridor

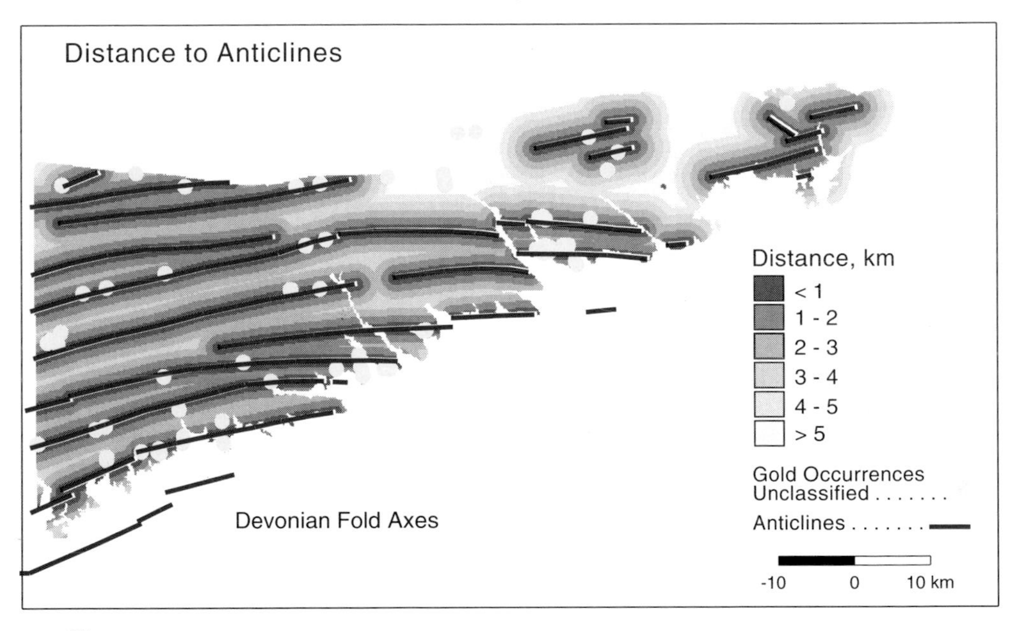

FIG. 9-11. Proximity to anticlines map, with the fold axes and gold occurrence points superimposed.

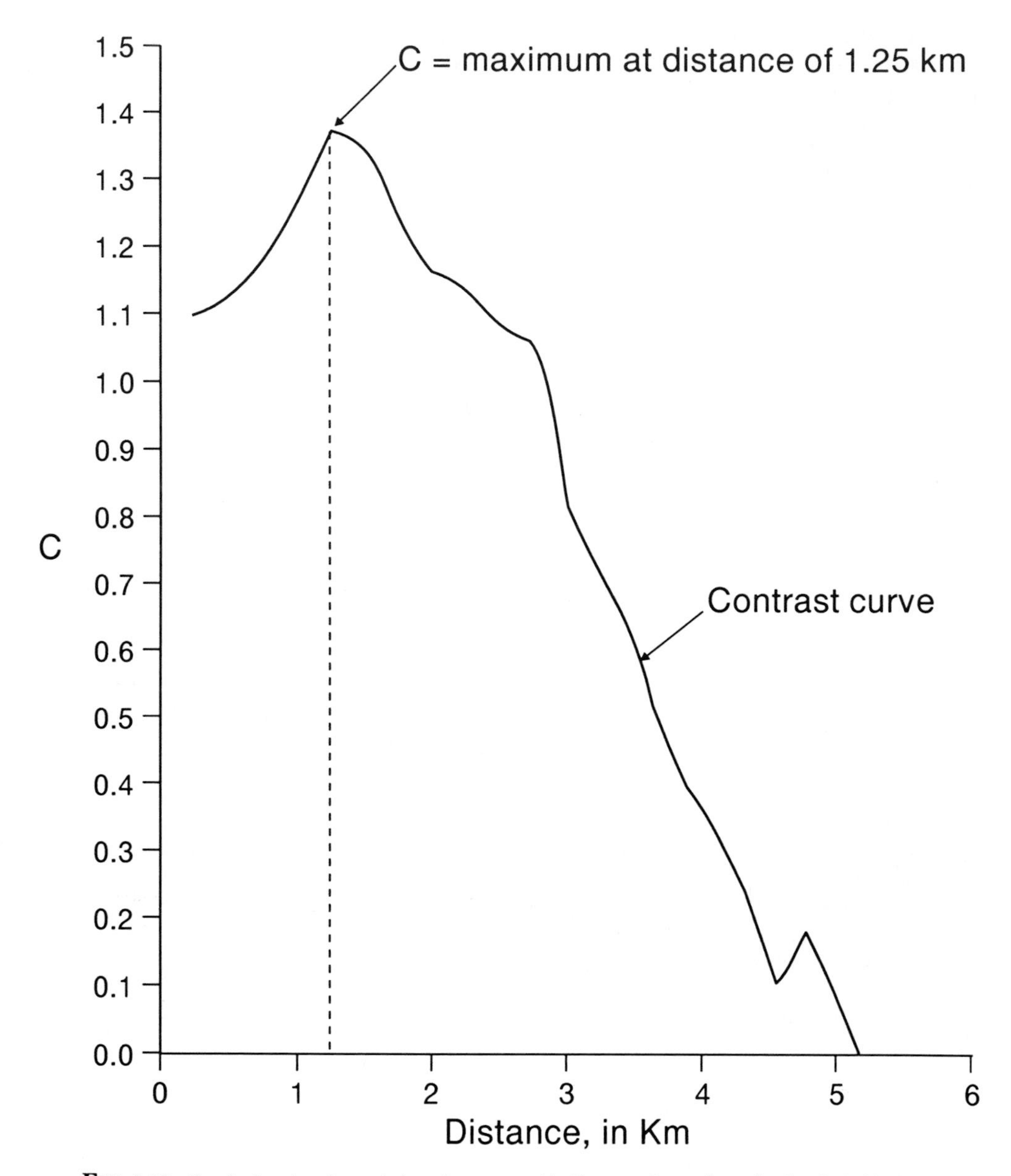

FIG. 9-12. Graph showing the variation of contrast with distance, drawn from the data in Table 9-13. The contrast is a measure of the spatial association between the gold occurrence points and the anticlinal fold axes (Figure 9-11). Notice that the contrast reaches a maximum at a distance of 1.25 km, then decays at greater distances.

successively, the variation in contrast with distance can be determined. The problem with this approach is that many of the classes have either no points occurring in them, or a very small number. This gives rise to a very noisy and unreliable curve, with poor estimates. The alternative is to calculate weights for *cumulative* distances, and to examine the variation of the weights and contrast at successive cumulative distance intervals, as shown in Figure 9-12, and Table 9-13. In this example, there is a particularly clear relationship between the anticline map and the gold occurrences. At a distance of 1.25 km, 51 out of the 68 occurrences are present, yet the area of the region lying within 1.25 km of the nearest anticline is 1280 km^2 (or 1280 unit cells, assuming that the unit cell is 1 km^2) out of a total study area of 2942 unit cells. The contrast is at a maximum at this distance. Note, however, that W^+ is at a maximum at 0.25 km, the closest buffer zone. At this distance, 15 of the 68 deposits occur within a region of only 260 unit cells, indicating a very strong affinity of the occurrences with the region immediately at the axis. However, because 53 of the occurrences lie beyond 0.25 km from the fold axes, the W^- value is not strongly negative, and the downweighting effect of **not** being within 0.25 km is therefore not great, and the combined spatial association arises from the total effect of the upweighting and downweighting, where the contrast is at a maximum.

The anticline map "thresholded" at 1.25 km, therefore, is the "best" binary predictor of the known gold occurrence points. It should be pointed out, however, that making this decision is often not so straightforward, because there is not always a clear maximum on the contrast curve. In such cases, subjective judgement, supported by geological reasoning, comes into play. Further, the actual values of the weights and contrast, although not very sensitive to the choice of the unit cell size, can be very sensitive to the choice of the study region. By changing the boundaries of the region, the values of the weights can be drastically changed, as indicated in Chapter 8. This does not negate the usefulness of the approach, but does indicate that the results are region-dependent and this should be considered in the interpretation.

Another guide to the interpretation of the contrast curve comes from estimates of the uncertainty of the weights and contrast. If the total number of occurrences is large, the variances of the weights are approximately given by expressions from Bishop et al. (1975):

$$s^2(W^+) = \frac{1}{N\{B \cap D\}} + \frac{1}{N\{B \cap \bar{D}\}} \quad \text{and} \tag{9-40}$$

$$s^2(W^-) = \frac{1}{N\{\bar{B} \cap D\}} + \frac{1}{N\{\bar{B} \cap \bar{D}\}} \quad . \tag{9-41}$$

The variance of the contrast is the sum of the variances of the weights. The problem with these expressions in a spatial context is that they vary with the units of areal measurement, so they are only useful in a relative sense. Fortunately this problem is minor with deposit data because $1/N\{B \cap \bar{D}\}$ and $1/N\{\bar{B} \cap \bar{D}\}$ are nearly zero with a small unit cell; and $1/N\{B \cap D\}$ and $1/N\{\bar{B} \cap D\}$ are unaffected by the size of the unit cell. From Table 9-13 we notice that the standard deviation (square root of variance) of W^+ is relatively large at short distances, and decreases with

Table 9-13. Variation of weights of evidence with cumulative distance from anticline axes. The contrast, C, reaches a maximum value at class 5 (1.25 km). The area of the unit cell is 1 km^2, and the area units are in terms of unit cells. The point column is the number of cells containing a point. $s(W^+)$ is the standard deviation of W^+, and $s(W^-)$ is the standard deviation of W^- (see Equations 9-40 and 9-41). Addition of an extra column (not shown here) containing the Studentized contrast (C/s(C)) is also helpful for choosing the cutoff distance, because it shows the contrast relative to the uncertainty due to the weights.

Class	Distance	Area	Points	W^+	$s(W^+)$	W^-	$s(W^-)$	C
1	<0.25	260	15	.949	.266	-.160	.139	1.109
2	0.50	617	30	.770	.187	-.353	.164	1.123
3	0.75	813	37	.700	.168	-.471	.181	1.171
4	1.00	998	43	.643	.156	-.596	.201	1.237
5	*1.25*	*1280*	*51*	*.561*	*.143*	*-.820*	*.244*	*1.381*
6	1.50	1497	54	.458	.139	-.883	.269	1.341
7	1.75	1643	55	.381	.137	-.850	.279	1.231
8	2.00	1848	57	.296	.135	-.845	.303	1.142
9	2.25	2009	59	.246	.132	-.887	.335	1.133
10	2.50	2133	60	.202	.131	-.862	.355	1.064
11	2.75	2229	61	.173	.130	-.869	.380	1.042
12	3.00	2343	61	.122	.130	-.693	.380	0.815
13	3.25	2403	61	.096	.130	-.586	.380	0.682
14	3.50	2444	61	.079	.130	-.505	.381	0.584
15	3.75	2495	61	.057	.130	-.396	.381	0.453
16	4.00	2531	61	.043	.130	-.310	.381	0.352
17	4.25	2560	61	.031	.130	-.235	.382	0.265
18	4.50	2607	61	.013	.130	-.103	.382	0.116
19	4.75	2633	62	.019	.129	-.176	.412	0.194
20	5.00	2654	62	.011	.129	-.103	.413	0.114
21	5.25	2695	62	-.005	.129	.054	.413	-0.060
22	5.50	2715	62	-.013	.129	.140	.414	-0.153
23	5.75	2728	62	-.017	.129	.202	.414	-0.219
24	6.00	2757	63	-.012	.127	.166	.453	-0.178
25	>6.00	2942	68					

distance, whereas the standard deviation of W^- behaves in the opposite fashion, increasing with distance.

The variances of the weights and contrast are useful for 1) for helping to determine the cutoff level at which to convert multiclass maps to binary form, and 2) for mapping the uncertainty of the posterior probability due to uncertainty in the weights.

In the first case, a useful measure is to calculate the Studentized value of the contrast, a measure of the certainty with which the contrast is known. The Studentized value is calculated as the ratio of C to its standard deviation, $C/s(C)$. This ratio serves as an informal test of the hypothesis that C=0, and as long as the ratio is relatively large, implying that the contrast is large compared with the standard deviation, then the contrast is more likely to be "real". Ideally it is nice to see a Studentized value larger than 1.5, or even 2. Because of the assumptions required for a formal statistical test, it is best to use this ratio in a relative, rather than an absolute, sense. Thus in examining the variation of C with distance in Table 9-13, the variations in $s(C)$ indicate the errors associated with particular C values due to the small number of points, or due to small areas.

In the second case, the variances of the weights can be used to calculate the variance of the posterior probability at each location, and to generate an uncertainty surface that accompanies the posterior probability surface. Again, the Studentized posterior probability acts as a measure of the relative certainty of the posterior probability. Regions where the Studentized value falls below some threshold can be masked out, due to lack of confidence in the results. Bonham-Carter et al. (1989) discuss these calculations in some detail, and illustrate them with uncertainty maps for the Nova Scotia Meguma gold data.

Returning to the procedure for picking optimal cutoffs for converting multistate to binary maps, it should be noted that actual real world results often call for the use of subjective judgment; therefore it is unwise to apply a completely automated "black box" to the problem. For the Nova Scotia data, the weights are calculated for each map, using a cumulative form for the maps with ordinal (and higher) measurement levels (as in Table 9-13), and a noncumulative form for the categorical maps. The proximity to the Goldenville-Halifax contact, the proximity to granite contact, and the proximity to NW lineaments are all evaluated by measuring the contrast as a function of cumulative distance, and finding the distance at which C is a maximum, tempered by inspection of the Studentized C values, and by the geological interpretability of the result. A similar procedure is followed for the geochemical maps, in order to find the element concentration level for each map at which C is maximized. In this way the cutoff between background and anomaly is determined that maximizes the spatial association between the known mineral occurrences and the geochemical anomaly maps. Note that this is not the common procedure for determining anomaly levels in exploration geochemistry, and a comparison of conventional anomaly finding methods and this method can be instructive. For categorical maps, it is sometimes desirable to use a separate weight for each class (this is actually the W^+ value calculated for each class versus all other classes), as described by Agterberg et al. (1990). Alternatively, if some classes have only small areas, or contain only a small number (or no) points, then it is prudent is combine classes. This can be done on geological grounds, lumping map units that can be treated together from the point of view of the deposit model. In the present case, the geology was reduced to two classes, one comprising the combined Halifax and Goldenville Fms, the other being the Devonian granite.

These procedures for determining the optimal weights were carried out on the Nova Scotia data, and the weights are summarized in Table 9-14.

Table 9-14. Weights for Nova Scotia maps. The prior probability is 0.02309. The values for the geological map are based on combining the Halifax and Goldenville Fms.

Map	W^+	$s(W^+)$	W^-	$s(W^-)$	C
ANTICLIN	0.561	0.143	-0.820	0.244	1.381
BIOAS	0.952	0.243	-0.200	0.143	1.152
BIOAU	0.280	0.233	-0.090	0.144	0.370
GEOLOGY	1.744	0.709	-0.154	0.125	1.899
GOLDHAL	0.367	0.174	-0.268	0.173	0.635
GRANCON	0.223	0.306	-0.038	0.134	0.261
LKSDSIG	1.423	0.343	-0.375	0.259	1.798
NWLINS	0.041	0.271	-0.010	0.138	0.051

Table 9-15. Table of χ^2 values for testing conditional independence between all pairs of 8 binary maps. With 1 degree of freedom and a probability level of 98%, tabled χ^2 = 5.4, larger than any of the observed values, so the null hypothesis of CI is not rejected at this level.

MAP	BIOAS	BIOAU	GEOLOGY	GOLDHAL	GRANCON	LKSDSIG	NWLINS
ANTICLIN	0.40	0.02	2.75	5.02*	0.33	1.09	0.48
BIOAS		2.29	0.00	3.70	0.00	5.00*	0.04
BIOAU			0.01	0.07	3.57	0.66	0.08
GEOLOGY				0.52	0.40	0.63	3.73
GOLDHAL					0.43	1.14	2.25
GRANCON						0.42	0.39
LKSDSIG							0.03

* Significant at 95% level of probability, for which tabled χ^2= 3.8.

Conditional Independence

The results of carrying out pairwise CI tests are summarized in Table 9-15, and show that there is little evidence to suggest that any of the maps show a serious conditional dependence problem on a pairwise basis. The cases with the largest χ^2 values are the (anticlines- Goldenville/Halifax contact) pair, and the (Balsam Fir As-Lake sediment geochemical signature) pair. In the former case, the fold axes and contact are indeed often spatially related, but the value of 5.02 is not large enough to warrant rejecting the null hypothesis at the 98% probability level. In the latter case, it is again not surprising that two kinds of geochemical indicators are weakly dependent, but not sufficiently so to reject the null hypothesis. The actual contingency tables for two cases are shown in Table 9-16. Notice that in the anticline-G-H contact case, the number of deposits occurring where both patterns are present (21) is actually smaller than the number expected under CI (25.5). On the other hand, in the test on As and Au in Balsam Fir, the observed number of deposits where both patterns are present is 8, whereas only 5.0 are predicted under CI.

Table 9-16. Tests of conditional independence, using (2x2) contingency tables. Values in round brackets are the expected number of deposits, and are preceded by the observed number.
A. The two maps are ANTICLIN (B_1) with GOLDHAL (B_2). Calculated χ^2=5.02. The tabled value of chi-squared with 1 df and probability level of 0.98 is 5.4. Therefore the null hypothesis of CI is not rejected at this level, but is rejected at the 95% level because tabled $\chi^2_{.95,1}$=3.8.

	$N\{B_1\}$	$N\{\bar{B}_1\}$	TOTAL
$N\{B_2\}$	21 (25.5)	30 (25.5)	51
$N\{\bar{B}_2\}$	13 (8.5)	4 (8.5)	17
TOTAL	34	34	68

B. The two maps are BIOAS (B_1) with BIOAU (B_2). Calculated χ^2=2.29. The null hypothesis of CI is not rejected at the 95% level of confidence.

	$N\{B_1\}$	$N\{\bar{B}_1\}$	TOTAL
$N\{B_2\}$	8 (5.0)	10 (13.0)	18
$N\{\bar{B}_2\}$	11 (14.0)	39 (36.0)	50
TOTAL	19	49	68

Generating Posterior Probability Maps

Creating a map of posterior probability now becomes a matter of applying Equation 9-33, and this may be done readily with a GIS modelling language. Using the actual map names and weights for the 8 Nova Scotia maps, the following map modelling statements define the steps needed to generate a posterior probability map:

```
: Calculate prior logit of a deposit given prior probability of .023
        prilogit = log(0.023/(1-0.023))
: At current location, choose positive weight if pattern is present, negative
weight if absent
        wt1 = {0.561 if (class('ANTICLIN') < 6 AND class ('ANTICLIN')
        > 0) , -0.820}
        wt2 = {0.952 if class('BIOAS') < 5 , -0.200 }
        wt3 = {0.280 if class('BIOAU') < 6 , -0.090 }
        wt4 = {0.154 if class('GEOLOGY') == 1 OR class('GEOLOGY')
        == 2, -1.744}
        wt5 = {0.367 if (class('GOLDHAL') < 5 AND class('GOLDHAL')
        > 0) , -0.268}
        wt6 = {0.223 if (class('GRANCON') < 3 AND class('GRANCON')
        > 0), -0.038}
        wt7 = {1.423 if (class('LKSDSIG') > 3 , -0.375 if (class('LKSDSIG')
        <= 3 AND class('LKSDSIG')0), > 0}
        wt8 = {0.041 if (class('NWLINS') < 2 AND class('NWLINS'
        > 0), -0.010}
: Posterior logit equals prior logit plus sum of weights
        pstlogit = prilogit + wt1 + wt2 + wt3 + wt4 + wt5 + wt6 + wt7 + wt8
: Convert to posterior odds
        pstodds = exp(pstlogit)
: Convert to posterior probability
        pstprob = pstodds / (1 + pstodds)
: Generate output class after applying classification table called 'PROB'
        OUT = CLASSIFY(pstprob, 'PROB')
        RESULT(OUT)
```

In the first line, the prior logit is calculated from the prior probability, inserted as 0.023 (=68/2942). In the second line, the value of **wt1** equals 0.561 (the value of W^+ for the anticline map) if the anticline map class is less than 6 and greater than 0, else it equals -0.820 (the value of W^-). Similar operations are carried out for the remaining 7 maps. Notice that **wt7** for the lake sediment signature map is either set to W^+, or to W^-, or if the class value is zero (meaning no data, because the location is outside the sampled basin) then it is set to 0. Thus, for missing data, no weight is contributed from that input map and the posterior probability is neither upweighted nor downweighted. However, class "zero" for the buffer maps does not mean "missing data", but simply that the location is beyond the outermost buffer zone. The final three statements determine the posterior logit, convert it to posterior odds, and finally to posterior probability. The posterior probability values are then classified and a map produced, as shown in Figure 9-13A.

The check of the total number of predicted deposits in this case is 76.7 (calculated using Equation 9-39), as compared with the actual number of 68, and is therefore about 12% greater than it should be, indicating the presence of some conditional dependence, despite the apparent lack of serious CI problems in the pairwise tests. Therefore, to check the possible effects of this dependence, one of the maps that was suspect in the CI tests, namely the Goldenville-Halifax contact map, was removed from the analysis and a new posterior probability map generated. The result is shown in Figure 9-13B, a map that shows almost no difference from A, and the total number of predicted deposits is reduced to 72.4, indicating that conditional dependence can now be considered at an acceptably low level. There are minor differences in the 7- and 8-map cases, of course, as shown by the difference map (calculated before classification), Figure 9-13C, but at the post-classification stage, the

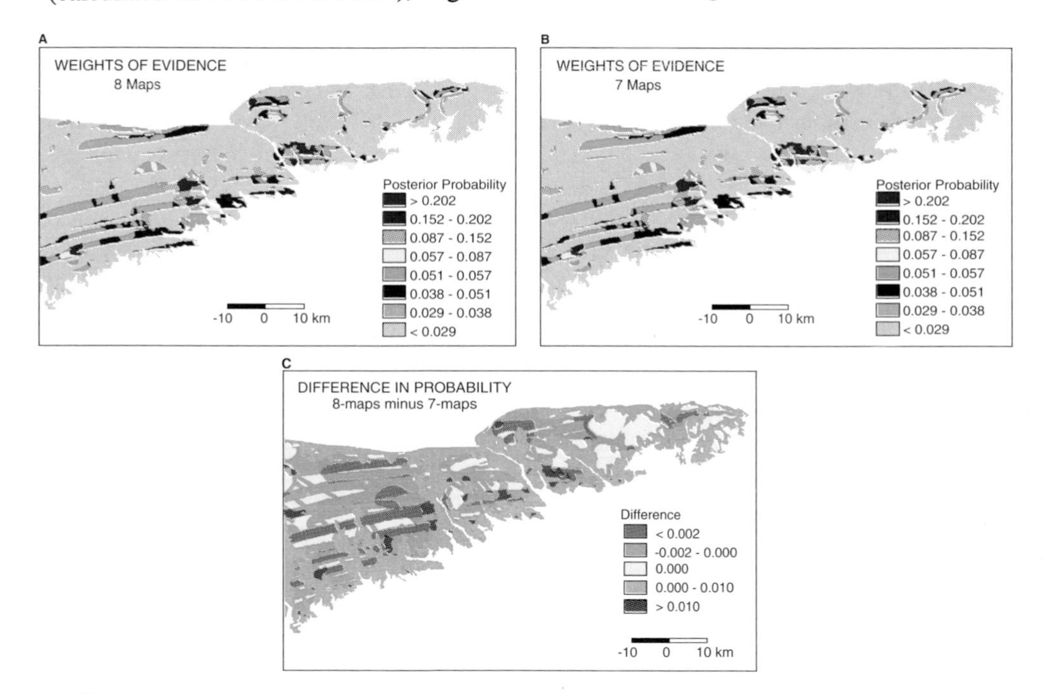

FIG. 9-13. Maps of gold potential, estimated as posterior probabilities by the weights of evidence method. **A**. Map showing the probability of finding a gold occurrence per km^2 (unit cell=1 km^2), using 8 input maps. **B**. Same as A, using 7 maps (Goldenville-Halifax contact map removed), in order to check the effects of conditional dependence introduced by the Goldenville-Halifax map). **C**. Map showing differences in posterior probability between A and B. The differences are small, and have been accentuated on this map by the classification scheme.

differences between the maps are negligible. This does not mean that conditional dependence can be ignored in all cases, simply that for this dataset the problem is judged as not serious.

Discussion of weights of evidence

Let us summarize some of the good and bad points about using weights of evidence as a method of combining maps. The principal advantages of the method are:

1. The method is objective, and avoids the subjective choice of weighting factors, as in the index overlay method, for example.

2. Multiple map patterns can be combined with a model that is straightforward to program with a modelling language.

3. Conversion of multistate maps to binary map patterns so as to optimize the contrast is a process that gives insight into spatial data relationships. This is an important *inductive* process.

4. Input maps with missing data (incomplete coverage) can be accommodated in the model.

5. Uncertainty a) due to variances of weights, and b) due to missing data can be modelled to show the effect on the *posterior* probability. The result can be mapped separately, or combined with the *posterior* probability map in various ways. For example, Bonham-Carter et al. (1989) combined the *posterior* probability map for gold in the Nova Scotia study with a binary mask produced from the uncertainty map, so that areas that were relatively uncertain were masked out.

6. Not discussed here, the Bayesian CI model can be applied to multiclass input maps, where each map class is associated with a weight (or likelihood ratio). For example, Agterberg and Bonham-Carter(1990) discuss the "grey-scale" weights of evidence model, as applied to the gold occurrences versus anticline data, showing how individual weights can be derived from the weights based on cumulative distance classes. This same approach is used in a study of seismic epicentres in western Quebec, showing how they are related to drainage patterns and geophysical maps, Goodacre et al. (1993). In this paper the binary model is compared with the "grey-scale" model, and shows that the predictive map of epicentre density is very similar to the map generated using the "grey-scale" model. An (1992) has successfully applied the grey-scale Bayesian model to predict iron and base-metal deposits in part of Manitoba, but with weight functions that were estimated subjectively.

Some of the disadvantages of weights of evidence modelling are:

1. The combination of input maps assumes that the maps are conditionally independent of one another with respect to the response variable. The testing for conditional independence is only possible where the method is applied in a data driven mode. Tests for CI require overlay data between pairs of binary maps (and the overlap of deposit points), and involve contingency table calculations, generally carried out outside the GIS environment, using data files generated by the GIS.

2. Weights of evidence, in common with other data-driven methods, is only applicable in regions where the response variable (e.g. distribution of known mineral occurrences in this case) is fairly well known. In the mineral deposits case, unit areas containing a deposit are known (almost) without error, but unit areas with no known deposit may or may not contain one or more

undiscovered deposits, i.e. the true mineral deposit population is under-represented. In frontier regions, where only a few occurrences are known, then the estimates of weights will be in error, and the weights will have large variances. Therefore the weights of evidence model is not always applicable in poorly explored regions, and if it is used with only a small number of known mineral occurrences, the results must be interpreted with caution. However, in partly explored regions, where a reasonable sample of deposits and showings are known, the estimates of weights may be sufficiently stable that adding new as yet undiscovered deposits (with essentially the same characteristics, on average, as the known deposits), will not greatly alter the weights. Under these circumstances, the addition of undiscovered deposits can be modelled by increasing the *prior* probability, which does not change the ranking of areas by *posterior* probability. Weights of evidence should not be used to estimate the *number* of undiscovered deposits, or the total *mineral endowment* of a region. A different approach, such as the "one-level" prediction method by McCammon and Kork (1992) should be applied for estimating total endowment. The resulting estimate of the total number of undiscovered deposits might then be chosen as the *prior* probability for weights of evidence.

Prospector model

The name Prospector was given to an expert system, developed at the Stanford Research Institute in the 1970s, for evaluating mineral prospects (Duda et al., 1978). The goal of the system was to provide an exploration geologist with a method to determine the deposit type or deposit model appropriate for an unknown mineral prospect, a task normally carried out by an exploration expert. Prospector thus attempts to emulate the performance of a deposit model expert. The interaction with the system consists of a dialogue or consultation, during which the computer asks questions about the characteristics of the prospect - the mineralogy, host rock geology, structural setting, alteration, and so on. Within the system, the data are used to draw inferences about hypotheses, using a series of inference networks, built into the program by the designers of the system working with deposit geologists. The consultation ultimately generates an estimate of the most likely deposit type, and an explanation of how the result was obtained.

Although the original Prospector did not deal with regional spatial data, like geophysical surveys or geological maps, later versions of the program were modified to make regional assessments of mineral favourability, Duda et al. (1978). For example, Campbell et al. (1982) published the results of applying Prospector to map the potential for molybdenum deposits in the Mt. Tolman area of Washington State. In that study, digitized maps of geology, structure, geochemistry and alteration were combined, using the inference network applicable to the porphyry molybdenum model. The output consisted of a map showing favourability for molybdenum and outlined not only the area of known mineralization, but also an extension to the known deposit. Duda et al. (1978) published a study of the Island Copper deposit in British Columbia, that was also modelled by combining map patterns, in this case using a porphyry copper model. Some illustrations based on that report are shown here to show the general principles of inference networks and how they can be applied for regional mineral potential mapping. Katz (1991) emulated the Island Copper study with a GIS called MAP.

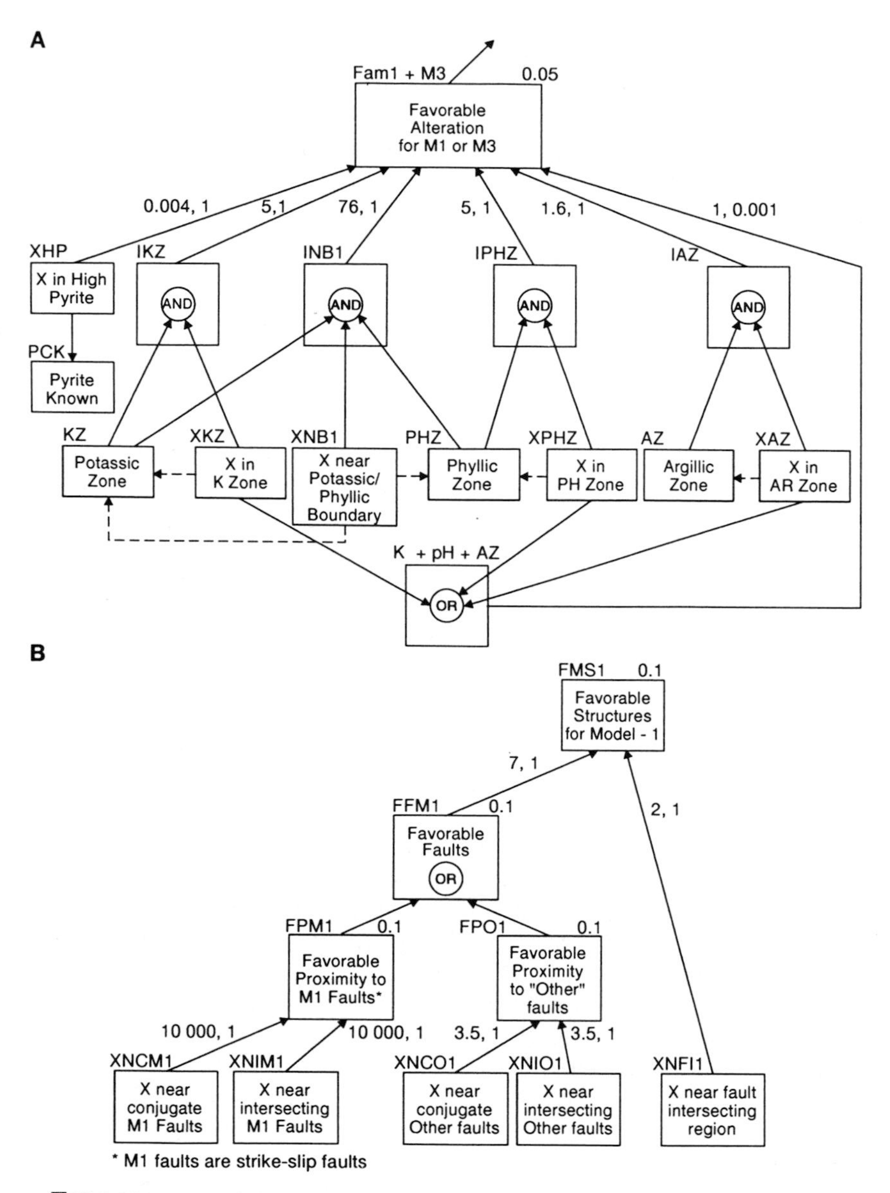

FIG. 9-14. Prospector inference networks for part of the Island Copper study, British Columbia, modified from Duda et al. (1978). **A.** The network for alteration. The pairs of numbers alongside each arrow are the likelihood ratios, LS and LN, estimated by the deposit model expert. The numbers above the boxes are prior probabilities. The three kinds of combination operations are 1) Bayesian "updating", using LS and LN values, 2) fuzzy AND operations, and 3) fuzzy OR operations. **B.** The inference network for structures. The full network is not shown.

Reddy et al. (1992) applied similar principles to model base-metal volcanogenic massive sulphides in Manitoba (same data as Chapter 1, but different inferencing mechanism). Yatabe and Fabbri (1988) provide a review of artificial intelligence applications in earth science, including a discussion of the Prospector expert system.

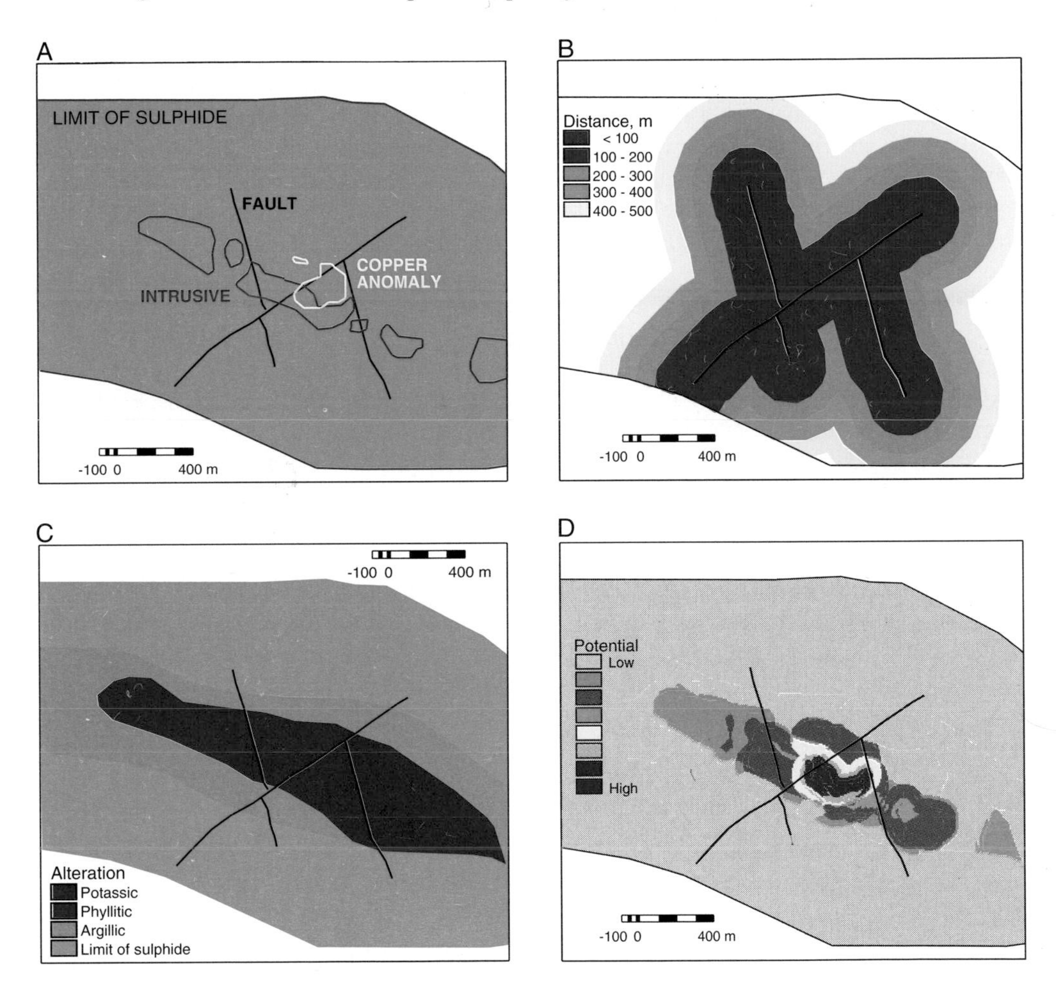

FIG. 9-15. Maps from Prospector study of Island Copper deposit, based on data from Duda et al. (1978). The calculations have been repeated, and new maps drawn, so there are minor differences from the original. **A**. Basic input map, showing alteration types, structures, intrusive contacts, and a line showing the limit of the sulphide zone, which is taken as the boundary of the study region. **B**. A buffer map showing the proximity to conjugate faults, used in the evaluation of structural evidence. **C**. Map showing types of alteration, used in the evaluation of the hypothesis "favourable alteration". **D**. Ouput map showing copper potential ranked on the basis of favourability of the hypothesis "favourable location for copper deposits". The most favourable zone is the outcome of the overlap of structural, alteration, geochemical and geophysical evidence.

Island Copper study

The principal types of evidence used in the copper porphyry model are derived from maps showing alteration and proximity to structures. Part of the inference net dealing with alteration is shown in Figure 9-14A. The map showing some of the evidence as vectors is in Figure 9-15A. The boxes (also called "spaces") in the inference network contain evidence maps and hypothesis maps of various types; each box can be thought of as either an input map, or a derived map forming part of the sequence of operations involved in combining and propagating spatial information through the inference network. Arrows represent rules, and broken arrows are context-constraint rules (one hypothesis must be evaluated before another one can be assessed). The numbers beside the arrows are the likelihood ratios, LS and LN. The numbers at the top of boxes are *prior* probabilities. For example the *prior* probability assigned to the "favourable alteration" hypothesis in the upper space is 0.05, implying that in the absence of other information, the probability of favourable alteration being present is 5%. The *prior* probabilities and likelihood ratios are assigned by the deposit model expert, and are part of the system.The "AND" and "OR" symbolize fuzzy logic operations. Note that the various kinds of alteration on the map ("potassic", "phyllitic", and "argillic"), are either present or absent, but one space contains the evidence "near the potassic-phyllic boundary". This evidence is expressed on a scale between 0 and 1, 1 for distances $< a$, 0 for distances $> b$, with values varying linearly between 0 and 1 for distances between a and b. The parameters a and b are specified in metres by the deposit expert. The sequence with which these various types of mapped evidence are combined is determined by the inference network.

In applying the LS and LN values, the same logic discussed for Bayesian weights of evidence are used. Thus note that the evidence space called INB1 (obtained by combining 3 binary alteration maps with an AND) is propagated to the "favourable alteration" space as follows. When INB1 is present, the *posterior* odds of the hypothesis are equal to the *prior* odds (=0.05/(1-.05)) multiplied by LS (=76); but when INB1 is absent, the *prior* odds are multiplied by 1. In other words, the *posterior* odds are increased by a large factor when all 3 types of alteration are present together, but when the evidence is absent, the *posterior* odds are unchanged and remain the same as the *prior*. Conversely, the effect of the presence of pyrite has a depressing effect on the *posterior* odds, (LS<1), and the absence of pyrite has no effect, (LN=1).

The network dealing with proximity to structures of various kinds is shown in Figure 9-14B. Note that the relative weighting of evidence is controlled by the choice of LS and LN values, as well as the *prior* probability assignments.

Prospector deals with uncertain evidence (i.e. where the evidence is between 0 and 1), by using a piecewise interpolation method to calculate an "effective" likelihood ratio, LE, that lies between the values of LS and LN. The actual value of LE is determined by 4 parameters: LS, LN, P{H} and P{E}, where the last two are the prior probabilities for the hypothesis and evidence, respectively. The details of this calculation are given in various Prospector reports, now often difficult to obtain. More accessible publications that discuss Prospector calculations are Harris (1984), or Reddy et al. (1992), see also Agterberg (1989a).

In Figure 9-15B, C and D, the results of modelling are shown. The "*favourable alteration*" and "*favourable structure*" hypotheses are produced as intermediate maps that themselves become evidence for the final combination to evaluate the "*favourability for copper deposits*" map.

In summary, we see that Prospector uses two methods for propagating information through the inference network. One is the Bayesian "updating" of *prior* probabilities to *posterior* probabilities. The other is by applying fuzzy logic AND and OR operators. In a GIS environment, small inference networks of this type can be implemented directly in the modelling language. However, very large inference networks are better handled outside the GIS (possibly in an expert system shell), in an external program linked to the GIS by shared data files. Weights of evidence is the same as the odds formulation for the Bayesian model used in Prospector (except it is in a log-linear form), where the evidence is binary. The major difference is that in Prospector, the likelihood ratios and prior probabilities are estimated by a deposit model expert, whereas in weights of evidence they are normally estimated statistically from data.

Dempster-Shafer belief theory is an alternative method for knowledge representation and map combination that has been applied by An (1992) and Moon (1990) for modelling base metal deposits in Manitoba, and by other authors.

SUMMARY

The methods of combining multiple maps in a GIS encompass a wide variety of models. This chapter has focused on empirical models in which map weighting is controlled either subjectively (knowledge-driven) or is determined from measured associations (data-driven). In some geological fields, models based on theoretical or semi-empirical principles (such as modelling heat flow or sediment transport) provide the rules for map combination. Such models are not discussed in this book, mainly because they often require special purpose software outside the normal GIS environment.

Boolean rules, index weighting and fuzzy logic are three basic and important approaches for GIS map combination. They are suitable for site-selection problems, and for the closely-related procedure of ranking areas according to potential. Potential mapping not only applies to mineral potential, the main application described here, but to other kinds of potential, such as earthquake damage potential, landslide potential, damage pollution due to pollution, and to many other geological problems.

Weights of evidence is one type of statistical model that has been shown to be suitable for combining maps to evaluate mineral potential. It has also bee used for slope stability mapping, van Westen (1993), and prediction of earthquake epicentre densities, Goodacre et al. (1993). Many other statistical models are applicable for combining maps in a GIS, but are beyond the scope of this book. Logistic regression is a particularly useful multivariate statistical model for map problems, because it is able to handle categorical map variables, see for example Bonham-Carter and Chung (1983), Chung and Agterberg (1980), Agterberg(1989b), Reddy et al. (1991).

Expert systems, using various types of knowledge representation (such as fuzzy memberships, probability distributions, Dempster-Shafer belief functions) and various methods for propagating evidence through an inference network (such as fuzzy logic operators, Bayesian updating of *prior* probabilities, Dempster's rule for combining belief functions), are likely to be used increasingly with GIS datasets. These methods bring the knowledge of a domain expert to bear on the problem of how best to combine spatial datasets.

REFERENCES

Agterberg, F.P., 1989a, Computer programs for mineral exploration: *Science*, v.245, p. 76-81.

Agterberg, F.P., 1989b, LOGDIA - FORTRAN 77 program for logistic regression with diagnostics: *Computers & Geosciences*, v. 15(4), p. 599-614.

Agterberg, F.P., 1992, Estimating the probability of occurrence of mineral deposits from multiple map patterns: *In The Use of Microcomputers in Geology*, Editors: Merriam, D.F. and Kurzl, H., Plenum Press, New York, p. 73-92.

Agterberg, F.P. and Bonham-Carter, G.F., 1990, Deriving weights of evidence from geoscience contour maps for the prediction of discrete events: *Proceedings 22nd APCOM Symposium*, Berlin, Germany, Sept. 17-21, 1990, Technical University of Berlin, v. 2, p. 381-395.

Agterberg, F.P., Bonham-Carter, G.F. and Wright, D.F., 1990, Statistical pattern integration for mineral exploration: In *Computer Applications in Resource Estimation Prediction and Assessment for Metals and Petroleum*, Editors: Gaál, G. and Merriam, D.F., Pergamon Press, Oxford-New York, p. 1-21.

An, P., 1992, *Spatial Reasoning and Integration Techniques for Geophysical and Geological Exploration Data*, Unpublished Ph.D. Thesis, University of Manitoba, 280 p.

An, P., Moon, W.M., and Rencz, A., 1991, Application of fuzzy set theory for integration of geological, geophysical and remote sensing data; *Canadian Journal of Exploration Geophysics*, v. 27, p. 1-11.

An, P., Moon, W.M. and Bonham-Carter, G.F., 1992, On a knowledge-based approach of integrating remote-sensing, geophysical and geological information: *Proceedings IGARSS'92*, p. 34-38.

Aspinall, R.J., 1992, An inductive modelling procedure based on Bayes' theorem for analysis of pattern in spatial data: *International Journal of Geographical Information Systems*, v. 6(2), p. 105-121.

Aspinall, P.J. and Hill, A.R., 1983, Clinical inferences and decisions-I. Diagnosis and Bayes' theorem: *Opthalmic and Physiological Optics*, v. 3, p. 295-304.

Biggs, D., de Ville, B. and Suen, E., 1991, A method of choosing multi-way partitions for classification and decision trees: *J. Applied Statistics*, v. 18(1), p. 49-62.

Bishop, M.M., Fienberg, S.E. and Holland, P.W., 1975, *Discrete Multivariate Analysis: Theory and Practice*: MIT Press, Cambridge Massachusetts, 587 p.

Bonham-Carter, G.F. and Chung, C.F., 1983, Integration of mineral resource data for Kasmere Lake area, Northwest Manitoba, with emphasis on uranium: *Computers & Geosciences*, v. 15, p. 25-45.

Bonham-Carter, G.F., and Thomas, J.H., 1973, Numerical calculation of steady wind-driven currents in Lake Ontario and the Rochester embayment: *Proceedings 16th Conference on Great Lakes Research*, p. 640-662.

Bonham-Carter, G.F., Agterberg, F.P. and Wright, D.F., 1988, Integration of geological datasets for gold exploration in Nova Scotia: *Photogrammetric Engineering and Remote Sensing*, v. 54 (11), p. 1585-1592.

Bonham-Carter, G.F., Agterberg, F.P. and Wright, D.F., 1989, Weights of evidence modelling: a new approach to mapping mineral potential: In *Statistical Applications in the Earth Sciences*, Editors: Agterberg, F.P. and Bonham-Carter, G.F., Geological Survey of Canada Paper 89-9, p. 171-183.

Breiman, L., Friedman, J.H., Olshen, R.A. and Stone, C.J., 1984, *Classification and Regression Trees*: Wardsworth and Brooks, Cole Advanced Books and Software, Monterey, California, 358 p.

Campbell, A.N., Hollister, V.F. and Duda, R.O., 1982, Recognition of a hidden mineral deposit by an artificial intelligence program: *Science*, v. 217 (3), p. 927-929.

Chung, C.F. and Agterberg, F.P., 1980, Regression models for estimating mineral resources from geological map data: *Mathematical Geology*, v. 12 (5), p. 473-488.

Chung, C.F. and Fabbri, A.G., 1993, The representation of geoscience information for data integration: *Nonrenewable Resources*, v. 2(2), p. 122-139.

Chung, C.F. and Moon, W.M., 1991, Combination rules of spatial geoscience data for mineral exploration: *Geoinformatics*, v.2 (2), p. 159-169.

Chung, C.F., Jefferson, C.W. and Singer, D.A., 1992, A quantitative link among mineral deposit modelling, geoscience mapping, and exploration-resource assessment: *Economic Geology*, v. 87, p. 194-197.

Duda, R.O., Hart, P.E., Nilsson, N.J. and Sutherland, G.L., 1978, Semantic network representations in rule-based interference systems: In *Pattern-Directed Inference Systems*, Editors: Waterman, D.A. and Hayes-Roth, F., Academic Press, p. 203-221.

Goodacre, A., Bonham-Carter, G.F., Agterberg, F.P. and Wright, D.F., 1993, A statistical analysis of the spatial association of seismicity with drainage patterns and magnetic anomalies in western Quebec; *Tectonophysics*, v. 217, p. 205-305.

Griffiths, C.M., 1987, An example of the use of fuzzy-set based pattern recognition approach to the problem of strata recognition from drilling response: In *Handbook of Geophysical Exploration, Section I: Seismic Prospecting, Volume 20: Pattern Recognition and Image Processing*, Editor, Aminzadeh, F., Geophysical Press, London-Amsterdam, p. 504-538.

Harris, D.P., 1984, *Mineral Resources Appraisal*: Clarendon Press, Oxford, 445 p.

Katz, S.S., 1991, Emulating the Prospector expert system with a raster GIS: *Computers & Geosciences*, v. 17, p. 1033-1050.

Lusted, L.B., 1968, *Introduction to Medical Decision Making*: Charles Thomas, Springfield, 271 p.

McCammon, R.B., 1989, Prospector II - The redesign of Prospector: *AI Systems in Government*, March 27-31, 1989, Washington, D.C., p. 88-92.

McCammon, R.B. and Kork, J.O., 1992, One-level prediction - A numerical method for estimating undiscovered metal endowment: *Nonrenewable Resources*, v. 1(2), p. 139-147.

Moon, W.M., 1990, Integration of geophysical and geological data using evidential belief function: *IEEE Transactions, Geoscience and Remote Sensing*, v. 28, p. 711-720.

Reboh, R. and Reiter, J., 1983, *A knowledge-based system for regional mineral resource assessment*: Final Report, Contract No. 14-18-0001-20717, SRI Project 4119, 267 p.

Reddy, R.K.T. and Bonham-Carter,G.F., 1991, A decision-tree approach to mineral potential mapping in Snow Lake area, Manitoba: *Canadian Journal of Remote Sensing*, v. 17, p.191-200.

Reddy, R.K., Agterberg, F.P. and Bonham-Carter, G.F., 1991, Application of GIS-based logistic models to base-metal potential mapping in Snow Lake area, Manitoba: *Proceedings of Canadian Conference on GIS*, Ottawa, p. 607-618.

Reddy, R.K., Bonham-Carter, G.F. and Galley, A.G., 1992, Developing a geographic expert system for regional mapping of Volcanogenic Massive Sulphide (VMS) deposit potential: *Nonrenewable Resources*, v. 1(2), p. 112-124.

Robinove, C.J., 1989, Principles of logic and the use of digital geographic information systems: In *Fundamentals of Geographic Information Systems: a Compendium*, Editor: Ripple, W.J., American Society for Photogrammetry and Remote Sensing, p. 61-80.

Simons, T.J., 1971, Development of numerical models of Lake Ontario, *Proceedings 14th Conference on Great Lakes Research*, p. 654-669.

Snedecor, G.W. and Cochran, W.G., 1967, *Statistical Methods*: 6th Edition, Iowa State University Press, Ames, Iowa, 593 p.

Spiegelhalter, D.J., 1986, Uncertainty in expert systems: In *Artificial Intelligence and Statistics*, Editor: Gale, W.A., Addison-Wesley, Reading, Massachusetts, p. 17-55.

Spiegelhalter, D.J. and Knill-Jones, R.P., 1984, Statistical and knowledge-based approaches to clinical decision-support systems, with an application in gastroenterology: *Journal of the Royal Statistical Society*, A, Part 1, p. 35-77.

Tetzlaf, D.M. and Harbaugh, J.W., 1989, *Simulating Clastic Sedimentation*: Van Nostrand Reinhold, New York, 202 p.

van Westen, C.J., 1993, *GISSIZ: Training package for geographic information systems in slope instability zonation. Volume 1 - Theory*: International Institute for Aerospace Survey and Earth Sciences (ITC) Publication no. 15, 245 p.

Varnes, D.J., 1974, The logic of maps with reference to their interpretation and use for engineering purposes: *United States Geological Survey Professional Paper 837*, 48 p.

Wadge, G., 1988, The potential of GIS modelling of gravity flows and slope instabilities: *International Journal of Geographical Information Systems*, v. 2 (2), p. 143-152.

Walker, H.M. and Lev, J., 1953, *Statistical Inference*: Holt, Rinehart and Winston, New York, 510 p.

Walker, P.A. and Moore, D.M., 1988, SIMPLE-An inductive modelling and mapping tool for spatially-oriented data: *International Journal of Geographical Information Systems*, v.2, p.347-363.

Watson, G.P., Rencz, A.N. and Bonham-Carter, G.F., 1989, Computers assist prospecting, *Geos*, v. 18 (1), p. 8-15.

Wright, D.F., 1988, *Data Integration and Geochemical Evaluation of Meguma Terrane, Nova Scotia, for Gold Mineralization*, Unpublished M.Sc. Thesis, University of Ottawa., 82 p.

Yatabe, S.M. and Fabbri, A.G., 1988, Artificial intelligence in the geosciences: a review: *Sciences de la Terre, Ser. Inf., Nancy*, v. 27, p. 37-67.

Zimmermann, H.J., 1985, *Fuzzy Set Theory - and Its Applications*, Kluwer-Nijhoff Publishing, Boston-Dordrecht-Lancaster, 363 p.

Zimmermann, H.J. and Zysno, P., 1980, Latent connectives in human decision making: *Fuzzy Sets and Systems*, v. 4, p. 37-51.

APPENDIX I

Syntax for Modelling Pseudocode

DATA TYPES

Constants may be numeric (either real or integer), character strings (enclosed in single quotes), or text strings (enclosed in double quotes). *Character strings* are used for names of maps, fields in tables, and names of tables. *Text strings* are for messages to the keyboard. ***Variables*** are either numeric (integer or real), or character strings. Names of variables must start with a letter and be < 25 characters long, case sensitive. Character variables end with a $.

OPERATORS

Arithmetic operators are +, -, *, /, and ^ for add, subtract, multiply and exponentiate, respectively. The ***assignment*** operator is =. The ***logical*** operators are AND, OR, NOT, and XOR. The ***relational*** operators are >, >=, <, <=, ==, and <> for greater than, greater than or equal to, less than, less than or equal to, equal to, and not equal to, respectively. The result of a logical operation is either 1 (TRUE) or 0 (FALSE).

EXPRESSIONS

A *numeric* expression is a combination of constants, variables and operators, leading to a single numeric value, such as A >= 51.7. ***Conditional*** expressions are enclosed in curly brackets and take the form $\{e_1 \text{ if } l_1, e_2 \text{ if } l_2, e_3\}$. The result of this expression is e_1 if l_1 is true, else it is e_2 if l_2 is true, else e_3 where e_1, e_2, and e_3 are numeric expressions, and l_1, and l_2 are Boolean expressions with a value either 1 or 0. The evaluation of the expression stops at the first true phrase.

STATEMENTS

Assignment statements take the form <numeric variable> = <expression>, such as

XXa = p * 3 + {4.1 if alpha <> beta, 0}

Precedence of operations is similar to FORTRAN, with round brackets to force precedence. The **classify** statement results in a table lookup operation with a classification table of breakpoints. For example

out = classify(total, 'breakpnt')

results in applying the classification table called **'breakpnt'** to be used with a numeric value called **total** to produce a numeric variable called **out**. The breakpoint table is a one-dimensional list of breakpoints between classes. The value of **out** is a non-negative integer, suitable to be used as a class value of a map. The ***result*** statement is the last statement in a procedure, and in map modelling is the class value of a new map. Thus

result **(x + y)**

causes the numeric expression to be truncated to an integer, and assigned to the current location of the new map.

Comment statements begin with a colon.

The expression **class('GEOLOGY')** fetches the class value of the map called GEOLOGY for the current location. Map names are character strings.

The expression **table('name', row,'column')** is an expression that takes the value located in a 2-dimensional table (like a FORTRAN array) called **'name'**, using the row number governed by the numeric expression **row** and column whose name is the character string **'column'**. For example the statement

NEW = table('GEOCHEM', class('BASIN'), 'ZINC')

causes the class value of the map called 'BASIN' at the current location to be used as the row number of the table called 'GEOCHEM'. The result is to fetch from that row of the table the numeric value in the column (field) called 'ZINC' and assign it to the variable called NEW. The row of the table is normally the same as the record number, numbered sequentially, but can also be determined by setting a keyfield, as in a relational database system.

FUNCTIONS

Functions are similar to the built-in library functions of FORTRAN, such as **max, min, exp, log, cos, tan,** etc.

APPENDIX II

Fortran Program for Calculating Weights of Evidence

```
      program contrast
      real ls,ln,or
      character*1 q
1     print *, ' area of study region ?'
      read *, s
      print *, ' area of binary map pattern ?'
      read *, b
      print *, ' area of unit cell?'
      read *, unit
      print *, ' no of deposits on pattern?'
      read *, db
      print *, ' total no of deposits ?'
      read *, ds
      if (db.le.ds) go to 6
      print *, 'error '
      stop
6     s=s/unit
      b=b/unit
      pbd=db/ds
      pbdb=(b-db)/(s-ds)
      or=db*(s+db-ds-b)/(b-db)/(ds-db)
      ls=pbd/pbdb
      wp=alog(ls)
      vp=1./db+1./(b-db)
      sp=sqrt(vp)
      pbbd=(ds-db)/ds
      pbbdb=(s-b-ds+db)/(s-ds)
      ln=pbbd/pbbdb
      wm=alog(ln)
      vm=1./(ds-db)+1./(s-b-ds+db)
```

```
      sm=sqrt(vm)
      c=wp-wm
      sc=sqrt(vp+vm)
      priorp=ds/s
      vprip=priorp/s
      sprip=sqrt(vprip)
      sprilo=sprip/priorp
      prioro=priorp/(1-priorp)
      prilo=alog(prioro)
      cpp=exp(prilo+wp)
      cpp=cpp/(1.+cpp)
      cpm=exp(prilo+wm)
      cpm=cpm/(1.+cpm)
      print 2, db,ds,b,s,ls,wp,sp,ln,wm,sm,c,or,sc,c/sc,priorp,
     + sprip,prilo,sprilo,cpp,cpm
2     format(//////' Number of deposits on pattern          ',f10.2/
     +  ' Total number of deposits              ',f10.2/
     +  ' Area of binary pattern (unit cells)  ',f10.2/
     +  ' Total area (unit cells)               ',f10.2/
     +  ' LS                                   ',f15.4/
     +  ' W+                                   ',f15.4/
     +  ' Standard deviation of W+              ',f10.4/
     +  ' LN                                   ',f15.4/
     +  ' W-                                   ',f15.4/
     +  ' Standard deviation of W-              ',f10.4/
     +  ' Contrast                             ',f15.4/
     +  ' Odds ratio                           ',f15.4/
     +  ' Standard deviation of Contrast        ',f10.4/
     +  ' Contrast/Standard deviation           ',f10.4/
     +  ' Prior probability                    ',f15.4/
     +  ' Standard deviation of prior prob      ',f10.4/
     +  ' Prior log odds                        ',f10.4/
     +  ' Standard deviation of prior log odds ',f10.4/
     +  ' Cond prob of deposit given pattern   ',f15.4/
     +  ' Cond prob of deposit given no pattern',f15.4///)
      print *, 'another case ?'
      read(*,'(a)') q
      if (q.eq.'Y'.or.q.eq.'y') go to 1
      stop
      end
```

EXAMPLE OF OUTPUT (input values are underlined)

Number of deposits on pattern	80.00
Total number of deposits	100.00
Area of binary pattern (unit cells)	500.00
Total area (unit cells)	1000.00
LS	1.7143
W+	.5390
Standard deviation of W+	.1220
LN	.3750
W-	-.9808
Standard deviation of W-	.2282
Contrast	1.5198
Odds ratio	4.5714
Standard deviation of Contrast	.2588
Contrast/Standard deviation	5.8732
Prior probability	.1000
Standard deviation of prior prob	.0100
Prior log odds	-2.1972
Standard deviation of prior log odds	.1000
Cond prob of deposit given pattern	.1600
Cond prob of deposit given no pattern	.0400

APPENDIX III

Glossary of GIS Terms

This glossary[1] is reproduced with minor modifications from the Lexicon of Terms for Users of Geographic Information Systems, published in the 1994 International GIS Sourcebook. Citations to references are indexed by number, for brevity. The American spelling (as opposed to Canadian spelling, used in the rest of the book) is retained from the original.

ORGANIZATION OF LEXICON ENTRIES

An *ENTRY* in the lexicon consists of a *term* or an abbreviation or acronym for a term, followed by a *commentary* or explanation. When there are several possible meanings, each constitutes a separate item in the commentary.

A ***LEXICON TERM*** is a single word (a singular form of a noun or an infinitive of a verb) or a group of words. When the term is an acronym, its full explanation (in parenthesis) follows immediately.

A ***COMMENTARY*** consists of one or more *items*. Words in a commentary that are terms in the lexicon are in capital letters. When commentary provides the items qualified as "commonly" are always given as the last item.

Each ***ITEM*** may include an *ITEM NUMBER*, *USAGE LABEL*, *DESCRIPTIVE PHRASE*, *ANNOTATION*, *REFERENCE SOURCE*, and *ADDITIONAL INFORMATION POINTERS.*

- ***ITEM NUMBERS*** start with 1 and continue sequentially but if there is only one item, the number one is omitted.
- ***USAGE LABEL*** indicates the area or a manner of usage of the term; "in remote sensing" would indicate the area of usage. "Commonly" indicates the manner of usage. Usage label is followed by comma.
- ***DESCRIPTIVE PHRASE*** provides the definition of the term.
- ***ANNOTATION*** is an additional explanation that may be deemed necessary for the full understanding of term.
- ***REFERENCE SOURCE*** points to the reference material (listed) where the term is used, defined, or explained in detail.

---

[1] Copyright © 1993 Krzanowski, Palylyk and Crown and GIS World, Inc. Reprinted by permission. *1994 International GIS Sourcebook*, Ft. Collins, CO: GIS World, 1993. All rights reserved. The lexicon cannot be reproduced or reprinted without permission.

ADDITIONAL INFORMATION POINTERS direct the reader to other terms in the lexicon and include:

- ***SEE***, referring to another entry in the lexicon where the given term is defined.
- ***SYNONYM FOR***, referring to a word or term with the same meaning.
- ***SEE ALSO***, referring to a related term or terms.

If additional information pointers occur, they do so in the above order.

LIST OF ENTRIES

ACCEPTANCE TEST - A test performed on software or a hardware system to evaluate its functional and performance capabilities with reference to predefined requirements (7). See also BENCHMARK.

ACCESS TIME - 1. The time required to access a record on a disk. Access time is the sum of the time required to position a disk head to the designated cylinder plus the time required to rotate a particular track to the desired position. The following types of access time have been defined:

- maximum access time - the longest time of access of the record on the disk,
- minimum access time - the shortest possible time required to access a record,
- average access time - the expected time to access a record,
- track-to-track access time - time required to move a disk head to the adjacent cylinder (2).

See also HARD DISK. 2. Commonly, the time required to access a file by the application program.

ACCURACY - 1. The degree to which a measurement approximates a given value (42). Synonym for CORRECTNESS. 2. The degree of consistent conformity to a standard (7)(42).

ADDITIVE COLOR PROCESS - 1. In cartography, the mixing of three additive primary colors (red, green, blue) in various proportions to create an arbitrary color (9).

ADDITIVE PRIMARY COLORS - The colors red, green, and blue that can be mixed to produce other colors using light sources.

ADDRESS RANGE - In street network data bases, the lowest and highest address numbers along the street, usually within a given area (29). See also TIGER, AMF, DIME.

ADDRESS CODING - An operation of relating street address to census blocks, census tracks, or other administrative units (29)(31). See also CENSUS BLOCK, CENSUS TRACK.

ADJACENCY ANALYSIS - 1. Grouping of continuous geographic areas by a selected common attribute (28). 2. Analysis performed to determine if a set of areas has a common boundary (16). Synonym for CONTIGUITY ANALYSIS.

AERIAL PHOTOGRAPH - A photograph taken from the air (20). See also VERTICAL AERIAL PHOTOGRAPH.

AFFINE TRANSFORMATION - A combination of LINEAR TRANSFORMATIONS followed by translation. The basic invariant of an affine transformation is the coincidence of the point on line (44).

AGGREGATION - A process of grouping distinct data (attributes and/or spatial data). The aggregated data set has a smaller number of data elements than the input data set (16). See also DISSOLVE.

AI - See ARTIFICIAL INTELLIGENCE.

ALIASING - The "staircase" appearance of the edges or lines on a raster display (42).

ALGORITHM - 1. A computer method or procedure devised to solve a certain problem. The algorithm possess the following characteristics:

- its application to a particular problem results in a finite sequence of steps;
- there is a unique initial step;
- each step has a unique successor;
- an algorithm terminates with the solution or with an error message stating that the problem has no solution (2).

2. Commonly, a set of steps designed to solve a problem.

ALPHANUMERIC SYMBOL - A STRING consisting of printable symbols (letters, numbers, punctuation symbols) (31).

ALTITUDE MATRIX - A rectangular grid containing elevation data. An altitude matrix can be obtained from the stereoscopic study of overlapping aerial photographs on analytical stereoplotters, or from interpolation of irregularly distributed elevation data points (7). See also DEM, DTM.

AM/FM - (Automated Mapping/Facilities Management). A Geographic information system designed for the optimal processing of information about utilities and infrastructures such as power lines, and water and telephone networks.

AMF - See AREA MASTER FILE.

ANAGLYPH - A stereoscopic image with the right component portrayed in a red color superimposed on the left component portrayed in another color (usually bright green). When viewed through the correspondingly colored filters, a 3-dimensional image is rendered (5).

ANALOG - The representation of a numerical quantity by a physical quantity (electrical voltage, light intensity) (31)(42). See also ANALOG MAP.

ANALOG MAP - Commonly, a map plotted on a permanent media such as paper or MYLAR. See also HARD COPY.

ANCILLARY DATA - In remote sensing, the extra data for an area of interest used to enhance the analysis of the primary remote sensing data (spectral data). Ancillary data frequently include topography, land cover, or climatic data (5).

ANNOTATION - Textual information used to describe an object on a map (or image) or to provide additional information related to the map (or image) content. See also LABEL.

ARC - 1. A curve defined by an analytical equation. A coding method used in graphic exchange formats is to represent a curve by its analytical description instead of by a list of its vertices. 2. Used as a synonym for CHAIN.

ARCHIVAL STORAGE - The storage of information for archiving purposes. The stored information is usually not accessible ON-LINE. Archiving media include MAGNETIC TAPES, magnetic cartridges, DISKETTES, and OPTICAL DISKS (7). See also ARCHIVING, ON-LINE.

ARCHIVING - The process of storing information for security or safety on permanent storage media, usually OFF-LINE (28). See also ARCHIVAL STORAGE.
AREA - An extent of surface or space. The size or extent of an area is expressed in length units squared.
AREA MASTER FILE - (AMF). A digital data file which contains a complete street network and physical and cultural features for a specific geographical area. Area Master Files contain information on every street address range, block-face, and centroid number (53).
ARRAY - A data structure in which each element has ben assigned a unique index. An array with a one dimensional index is called a VECTOR. An array with a two dimensional index is called a table. A mathematical representation of an array is a MATRIX.
ARRAY PROCESSOR - A processor designed to speed up specific arithmetical operations (2). Synonym for PIPELINED ARITHMETIC PROCESSOR.
ARTIFICIAL INTELLIGENCE - (AI). A branch of computer science dealing with the formalization of cognitive processes.
ARTIFICIAL NEURAL NETWORKS - Hardware or software systems that simulate certain functions and topology of a central nervous system. Artificial neural networks, because of their capability to generalize and make abstractions, are gaining acceptance as pattern recognition tools in REMOTE SENSING and SPATIAL INFORMATION SYSTEMS (49).
ASCII - (American Standard Code for Information Interchange). A seven bit code representing a character set for modern written English (2).
ASPECT - The AZIMUTH of maximum slope (GRADIENT) (7). Synonym for EXPOSURE, DIRECTION OF STEEPEST SLOPE. See also SLOPE, GRADIENT.
ATEMPORAL DATA BASE - A data base storing information not indexed by time. See also TEMPORAL DATA BASE.
ATTRIBUTE - 1. Descriptive information characterizing a geographical feature (point, line, area) (2). 2. Commonly, a fact describing an ENTITY in a relational data model, equivalent to the column in a relational table (1).
ATTRIBUTE ACCURACY - The degree of ACCURACY of ATTRIBUTEs in spatial data bases. The concept of attribute accuracy encompasses accuracy of measurements and accuracy of matching attributes to specific spatial objects (34).
ATTRIBUTE MATCHING - Synonym for ATTRIBUTE TAGGING.
ATTRIBUTE TAGGING - An assignment of attributes to objects (points, lines, polygons) in SPATIAL INFORMATION SYSTEMS (16). Synonym for ATTRIBUTE MATCHING.
AUTOMATIC NAME PLACEMENT - The automated LETTERING of objects on a map. Simple automatic name placement systems locate the text in certain predefined positions, relative to graphic objects. Complex automatic name placement systems resolve LABEL CONFLICTS and overlaps to improve the readability of textual information and the aesthetic quality of maps (29). See LABELLING.
AUXILIARY MEMORY - Memory from which instructions are not taken for execution, in contrast to main memory. Auxiliary Memory usually uses magnetic or optical digital storage technology (2).

AVERAGE ACCESS TIME - See ACCESS TIME.
AVHRR - Advanced Very High Resolution Radiometer, a multispectral imaging system on the TIROS-NOAA meteorological satellites.
AZIMUTH - A direction measured in degrees of angle east or clockwise from North. For example, South has an Azimuth of 180 degrees (31). Synonym for COMPASS DIRECTION.
AZRAN - (Azimuth and Range). Abbreviation for the components of a polar coordinate system (21).

BACK-UP - 1. The archived copy of a file. 2. A process of archiving a file. See also ARCHIVING.
BACKUS-NAUR FORM - A system to describe the SYNTAX of a language and used to represent data formats and computer languages (32).
BAND - 1. A range of wavelengths or frequencies (spectral band), often used to characterize (a) the spectral sensitivity or character of a remote sensing device (radar, thermal scanner), or (b) the special phenomena from the process of either emitting, absorbing or receiving waves (27). 2. An image file containing the reflectance or emittance values for a particular wavelength. 3. A synonym for a layer in a raster data base (32).
BAND SEQUENTIAL FORMAT - A method of organizing a raster data base such that each complete LAYER (often all the data for one spectral band) is stored in contiguous sequence (32).
BAND INTERLEAVING FORMAT - A method of organizing a raster data base such that the data for different LAYERS are stored together in specific patterns such as by row or by pixel (32).
BAR SCALE - 1. A graduated line on a MAP, plan, photograph, or MOSAIC used to relate distances on the former to actual ground distances. Synonym for GRAPHIC SCALE.
BARRIER - A feature impeding the flow of entities through a NETWORK (14).
BASE MAP - A map portraying background reference information onto which other information is placed. Base maps usually show the location and extent of natural Earth surface features and permanent man made objects.
BATCH PROCESSING - Non-interactive, automatic execution of a COMMAND PROCEDURE. See also SCRIPT FILE.
BATHYMETRIC MAP - A MAP portraying the shape of a water body or reservoir using ISOBATHS (27).
BAUD - 1. A unit of signaling speed based on the number of times that the state of a transmission line changes per second. 2. Commonly, baud is used to denote the number of transmitted bits per second, which is correct only for the specific case of two-state signaling (41).
BAUD RATE - See BAUD
BCD - See BINARY-CODED DECIMAL NOTATION.
BEARING - Horizontal angle measured East from a reference meridian (14).
BELIEF - In AI, an hypothesis about a process or situation.

BENCHMARK - 1. A controlled test of capabilities of a software or hardware system. 2. A permanent natural or artificial object of known elevation or other characteristic taken as a standard (27).
BINARY ARITHMETIC OPERATION - An arithmetic operation using PURE BINARY NUMERICAL SYSTEM (2).
BINARY-CODED DECIMAL NOTATION - A code for decimal numbers in which each of the decimal digits (1,2,...,0) is represented by a binary numeral equivalent to four-bits. The four bits in each BCD have weights of 8,4,2,1, such that in BCD the number "fifty two" is represented as "01010010". In a pure binary numeration system it is represented as "110100" (2)(41). Synonym for BINARY-CODED DECIMAL CODE, BINARY-CODED DECIMAL REPRESENTATION. See also BINARY NOTATION, DECIMAL NOTATION, BINARY NUMERATION SYSTEM, EBCDIC, ASCII.
BINARY-CODED DECIMAL REPRESENTATION - Synonym for BINARY-CODED DECIMAL NOTATION.
BINARY DIGIT - See BIT.
BINARY ELEMENT - An element of data that can take two values or two states. The term BIT should not be used as synonym for BINARY ELEMENT (2).
BINARY NOTATION - A notation that uses two different characters. Most frequently, the term binary notation refers to the PURE BINARY NUMERATION SYSTEM (2).
BINARY NUMBER - See BINARY NUMERAL (2).
BINARY NUMERAL - 1. A numeral in the PURE BINARY NUMERATION SYSTEM (ie. the binary numeral equivalent to the number "5" is "101") (2).
BINARY NUMERATION SYSTEM - See PURE BINARY NUMERATION SYSTEM.
BINARY OPERATION - Synonym for BINARY ARITHMETIC OPERATIONS (2).
BIT - 1. In a PURE BINARY NUMERATION SYSTEM, either of the digits "0" or "1" (2). Synonym for BINARY ELEMENT. 2. Synonym for BINARY DIGIT (2).
BITFIELD - A group of bits treated as a unit.
BITMAP - 1. In computer graphics, a PIXMAP having a depth of 1 (48). Synonym for BIT PLANE. 2. Bit representation in computer memory of a data file or a hardware component (4).
BITMASK - An ordered list indicating the position of active or inactive bits in a BITFIELD (32).
BIT PLANE - Synonym for BITMAP.
BLIND-DIGITIZING - DIGITIZING without immediate graphic feedback on the results of the digitizing process (34).
BLOCK - In computer systems, a group of records treated as a unit.
BLOCK-FACE - One side of a city street between two distinct features such as intersections, end or beginning points, etc. This term used in relation to AREA MASTER FILES (AMF) (53).
BOOLEAN ALGEBRA - Finite or infinite set of elements with the three defined operations of negation, addition, and multiplication. These operations correspond to the set operations of complementation, union, and intersection (41). See also BOOLEAN OPERATIONS.

BOOLEAN OPERATIONS - Operations based on BOOLEAN ALGEBRA.
BOOT-UP - The initialization of a computer system. Cold boot-up (cold start) implies a loss of any data occurring in the system. Warm boot-up (warm start) implies that data occurring in the system are not lost.
BOUNDARY - A line or set of lines defining the extent of an area having specific characteristics. A "logical" boundary is defined by human interpretation of geographical features (the boundary of ecosystems), while a "physical" boundary is defined by physical objects such as rivers, shorelines, etc.
BPI - (Bytes per inch). The unit used to measure or describe the density of information storage on gnetic media.
BREAK - 1. See SURFACE SPECIFIC POINTS. 2. The interruption of the execution of a computer process by intention of the operator, error condition, or a "break" command in the program. The break-ending to a procedure implies that there will be no possibility of returning to the process at the point of interruption. See also PAUSE.
BREAK LINES - See SURFACE SPECIFIC POINTS.
BROWSING - A function of information systems allowing for a general search through a DATA BASE without an operator having to specify a particular search criteria.
BUFFER - 1. A storage area which temporarily holds data transferred from one device to another (41). See also CACHE. 2. In spatial information systems, a polygon enclosing an area within specified distance from a point, line, or polygon. Accordingly, there are point buffers, line buffers, and polygon buffers (7). See also CORRIDOR.
BUFFER ZONE - An area enclosed by BUFFER.
BUG - An error in a system or in the logic of a computer program (41). Synonym for malfunction or mistake.
BUS - 1. A hardware component for transmitting signals. 2. Synonym for DATA BUS (2).
BYTE - A group of binary digits treated as a unit. In current computer implementations, the most common are 8-bit bytes (2) (41). See also WORD, LONGWORD, NIBBLE.

CACHE - The mechanism improving transfer rates between the processor (CPU) and storage devices (memory, disk). CACHE can be implemented as a high speed semiconductor device with a speed similar to that of the CPU or as an organization of a memory (41).
CADASTRE - A public register or survey that defines or re-establish boundaries of public and/or private land for purposes of ownership and/or taxation (14). See also MULTIPURPOSE CADASTRE.
CADASTRAL MAP - A MAP showing the boundaries of the subdivisions of land for purposes of describing and recording ownership or taxation (27).
CAD/CAM - (Computer Aided Design and Computer Aided Manufacturing). A complete system for the manipulation of graphic information related to industrial design and manufacturing (2).
CALIBRATION - The process of comparing certain specific measurements from an instrument to those from a standard instrument (9).

CALIBRATION DATA - Data from a standard instrument or device. In remote sensing, these are data pertaining to spectral and/or geometric characteristics of a sensor or radiation source (21). See also TRAINING DATA.
CAMERA - A lightproof chamber in which an image of an external object is focussed upon a photo-sensitive plate or film through an opening in the chamber that is equipped with a lens, shutter and aperture (21).
CARDINAL POINTS - The main, chief, or principal directions (north, east, south, and west) (21).
CARTESIAN COORDINATE SYSTEM - The COORDINATE SYSTEM in which the location of a point in n-dimensional space is defined by distances from the point to the reference plane. Distances are measured parallel to the planes intersecting a given reference plane. If reference planes are mutually orthogonal, the coordinate system is called a rectangular Cartesian coordinate system. See also COORDINATES, COORDINATE SYSTEM (21).
CARTOGRAPHIC ENHANCEMENT - The addition of information to a map or photo-image thus increasing its readability, aesthetics, or information content (21).
CARTOGRAPHIC MODELING - The process by which CARTOGRAPHIC MODELS are generated (7).
CARTOGRAPHIC MODELS - A sequence of primitive spatial operations resulting in complex SPATIAL MODELS (7). See also MAP ALGEBRA, SPATIAL MODELS.
CARTOGRAPHY - The art and science of making MAPS and CHARTS (42)(7).
CATHODE RAY TUBE - See CRT.
CCT - (Computer Compatible Tape). A magnetic tape holding data stored in a computer readable format (5).
CD-ROM - (Compact Disk - Read Only Memory). Read-only, optical, data storage media similar to a commercial audio compact disk. The storage capacity of some CD-ROMs is in excess of 650 Mb.
CELL - Synonym for RASTER.
CELL CODE - The value of an ATTRIBUTE assigned to a cell.
CENSUS BLOCK - The smallest geographical area, bounded by visible boundaries,.for which census data are collected (29).
CENSUS TRACT - A small, permanent, subdivision of a county with homogeneous population characteristics, status, and living conditions (29).
CENTERLINE - A line drawn from the center point of a vertical aerial photograph through the transposed center point of an overlapping aerial photograph (21).
CENTERPOINT - The point at the exact center of a photograph that corresponds in position to the optical axis of the camera.(21).
CENTROID - 1. The geometric center of a POLYGON. The location of the centroid can be calculated as the average location of vertices defining polygon boundaries (38). 2. In spatial information systems, the centroid is a point in a POLYGON to which attribute information about that specific area is linked. 3. Mid-point between the beginning and the end of a valid ADDRESS RANGE (53).

CHAIN CODES - A method of coordinate compaction. Chain codes are used to represent lines on a regular grid. Each line segment in a chain code is represented by its length (in grid units) and a unique directional code (37).
CHAIN - In spatial information systems, a chain is an ordered set of points having a beginning and an end. A chain represents the location of ENTITIES such as linear features like roads or streets, or the border or boundary of an area (polygon). The beginning and the end of chains are called NODES; while intermediate points are called VERTICES (16)(31)(37). See also ARC.
CHANNEL - Synonym for BAND.
CHARACTER SET - A set of letters, numerals, punctuation marks, mathematical, and/or other symbols (34).
CHARACTER STRING - See STRING.
CHART - A MAP used for navigation (nautical or aeronautical) (27).
CHORD - Synonym for SEGMENT.
CHOROPLETH MAP - A map portraying properties of a surface using area symbols. Area symbols on a CHOROPLETH MAP usually represent categorized classes of the mapped phenomenon (42)(14).
CHROMA - Color dimension that refers to the strength or purity of the dominant color or, the degree of SATURATION of the dominant color (21).
CLASSIFICATION - A method of GENERALIZATION. In the process of classification, an attempt is made to group data into classes according to some common characteristics thereby reducing the number of data elements or their variety. Classification tends to be based on the attributes or characteristics of data rather than their geometry (42)(53).
CLEAN DATA - 1. Data that are devoid of errors. 2. A COVERAGE that is without topological errors.
CLIPPING - The process of extracting a subset of spatial data from a larger data set by selecting only those data located inside(or outside) a selected boundary. The area enclosed by this boundary may be referred to as a CLIPPING WINDOW.
CLIPPING WINDOW - A POLYGON used to perform CLIPPING.
CLOBBERING - Accidental overwriting of a file.
COGO - (Coordinate Geometry). A set of procedures for encoding and manipulating bearings, distances and angles of survey data into a graphic representation. COGO is frequently a subsystem of GIS (5).
COLOR - Property of an object that depends upon the wavelength of light that it reflects or emits (21).
COLOR COMPOSITE - A color image produced by assigning a different primary color to each of a number of registered, black and white, positive images of a SCENE and displaying the superimposed result (20).
COLOR TABLE - A LOOK-UP TABLE relating a color code to its computer representation scheme such that for any particular color, the intensities for the mix of red, green, and blue are given.
COMMAND PROCEDURE - Synonym for SCRIPT FILE.

COMPACTION - See COMPRESSION.
COMPASS DIRECTION - Synonym for AZIMUTH.
COMPILATION - 1. The translation of a computer program, written in a problem-oriented language, into a computer-oriented language (2). 2. Commonly, the translation of a computer program from a high-level language into machine readable instructions (14).
COMPILER - A computer program performing COMPILATION (2). Synonym for COMPILING PROGRAM.
COMPILING PROGRAM - Synonym for COMPILER.
COMPLEX POLYGON - A POLYGON with one or more ISLANDS (32).
COMPLEX OBJECT - An OBJECT that can be decomposed into other OBJECTS (32).
COMPLEX SURFACE - A SURFACE that cannot be described by a deterministic function.
COMPOSITE MAP - A map on which the combined information from different thematic maps is presented. A composite map can be create in the process of geographical analysis.
COMPRESSION - The operation of reducing the size of a FILE. Compression techniques eliminate repetitious groups of BYTES and/or BITS, or introduce a different coding schemes. Compression may be reversible or non-reversible (17)(41). See also DECOMPRESSION, RUN-LENGTH CODING.
CONCAVITY - Negative CONVEXITY (7).
CONFLATION - A process by which digital map files may be matched and merged into one by selecting the best features from each individual MAP to be added to the composite new MAP (26).
CONJOINT BOUNDARY - A boundary shared by two adjacent geographical areas (polygons, map sheets).
CONNECTIVITY ANALYSIS - An analysis which is performed in a SPATIAL INFORMATION SYSTEM to determine which features in a set are connected to each other (7).
CONTACT PRINT - A photographic print made with a negative or a diapositive in direct contact with the photosensitive material (21).
CONTIGUITY ANALYSIS - Synonym for ADJACENCY ANALYSIS.
CONTINUOUS DATA BASE - See SEAMLESS DATA BASE.
CONTOURING - The process of generating CONTOUR LINES (7).
CONTOUR LINE - A logical (imaginary) line connecting points with the same value for a selected characteristic of a surface. The term "contour line" is most commonly used for lines connecting points on the ground having the same elevation (16). See ISARITHM.
CONTOUR TAGGING - An automated process of assigning values to scanned CONTOUR LINES.
CONTRAST STRETCHING - An IMAGE ENHANCEMENT procedure in which the original range of grey scale presented on an image is expanded to the full range of the display device or recording media (20).

CONTROL CHARACTER - A character that is used to cause a change in computer operation such as line feed, escape, or carriage control. CONTROL CHARACTERs are not graphic characters although they may have a graphic representation (2).
CONTROL POINT - A point or location with a known position and/or magnitude (2). See also TIC, BENCHMARK.
CONVERSATIONAL MODE - Synonym for INTERACTIVE MODE.
CONVEXITY - 1. The outward (as on the outside of a sphere or circle) rate of change of SLOPE. 2. The second derivative of a HYPSOMETRIC CURVE. CONVEXITY may be expressed as vertical convexity or horizontal convexity (7). See also CONCAVITY.
COORDINATE SYSTEM - A reference system for the unique definition of a location of a point in n-dimensional space.
COORDINATES - An n-tuple of values uniquely defining a point within an arbitrary n-dimensional reference (coordinate) system.
COORDINATES REPRESENTATION - Format and type of coordinates used to define the location of a cartographic object (32).
CORRECTNESS - Synonym for ACCURACY.
CORRIDOR - A BUFFER along a line feature (5). See also PROXIMITY ANALYSIS.
COVERAGE - 1.An OBJECT in a SPATIAL DATA BASE. The representation of a map composed of graphic and attribute files in a digital mapping system. 2. In remote sensing, often used to describe the extent of the Earth's surface represented on an image or a set of images, as in areal coverage. See also SCENE.
CPU - (Central Processing Unit). The central hardware component of a computer with which essential arithmetic and logical operations are performed (41).
CPU TIME - Time required for the execution of the instructions in the CPU (11). See also RESPONSE TIME.
CRITICAL POINTS - 1. Points on the LINE that remain stable under orthographic projection. CRITICAL POINTS include curvature minima and maxima, end points, and points of intersections. CRITICAL POINTS are the bases for efficient line generalization algorithms (24). 2. Commonly, used to refer to those points on a LINE that capture its basic character.
CROSS-HATCHING - The filling of POLYGONS with line symbols.
CROSS-SECTION - 1. A graph portraying elevations along a specific direction. Synonym for PROFILE. 2. A section showing variations in an arbitrary or selected parameter in a direction normal to surface.(42).
CRT - (Cathode Ray Tube display). A rarely used abbreviation for a device for the visual displaying of data (2). See also MONITOR.
CULTURAL FEATURES - Items in a SPATIAL DATA BASE representing man-made objects.
CURRENCY - The temporal quality of spatial information reflecting the agreement between information stored in spatial data base and the actual phenomena this information describes, with respect to changes over time (34).

CURSOR - 1. A visible mark indicating a location of a next display operation on a CRT.(2). 2. A transducer used to digitize and control a digitizing process. Synonym for PUCK.
CURVE FITTING - The generation of a smooth curve passing through, or close to, a number of existing points (34). See also SURFACE FITTING.

DANGLE - A TOPOLOGICAL ERROR that occurs when an ARC has a NODE with a VALENCY of 1. In contrast with UNDERSHOOT, DANGLE is not related to a LEAKING POLYGON (7).
DANGLING NODE - A node of VALENCY 1. A DANGLING NODE is considered to be a TOPOLOGICAL ERROR in graphic files containing area features (7).
DATA BANK - Synonym for DATA BASE.
DATA BASE - An organized collection of information (41)(2).
DATA BUS - A BUS for the transfer (internal or external) of data between processing unit(s), storage devices, and/or peripheral devices (2).
DATA CAPTURE - The encoding of data into a computer compatible format (34). See also SCANNING, DIGITIZING, REMOTE SENSING.
DATA COVERAGE - The EXTENT of available data with respect to the selected theme or target area.
DATA DICTIONARY - A repository of information about data in which is stored information on all of the objects in a DATA BASE and their relationships (43).
DATA FORMAT - See FORMAT.
DATA LAYER - See LAYER.
DATA MODEL - Collection of concepts allowing for the representation of an environment according to arbitrary requirements (1).
DATA PROCESSING - Operations performed on data using a computer system (21).
DATA REDUCTION - See COMPRESSION.
DATA REPRESENTATION - Methods of representing spatial objects in a SPATIAL INFORMATION SYSTEM, the most common of which being VECTORS and TESSALATIONS.
DATA SET - Ensemble of data with common characteristics.
DATA STRUCTURE - The organization of simple DATA TYPES directly represented by computer as in TREE or QUEUE (38).
DATA TYPE - A representation of data by a computer including INTEGER, FLOAT, BOOLEAN.
DATUM - 1. Any point, line, or surface used as a reference for a measurement of another quantity. 2. A model of the earth used for geodetic calculations (6). See also GEODETIC DATUM, SPHEROID, ELLIPSOID.
DECOMPRESSION - The operation of expanding a compressed file into its original form. See also COMPRESSION.
DECIMAL NOTATION - A numeration system having radix of 10. Decimal notation uses the numbers 0,1,..,9.

DEBUGGING - The process of correcting errors (BUGS) in computer software or hardware. See also DEBUGGER.
DEBUGGER - A software system used to perform DEBUGGING.
DEDICATED SYSTEM - A computer system allocated solely to a single pre-selected task.
DEFAULT - A value assigned to a parameter of a variable by the system.
DEM - (Digital Elevation Model). A model of terrain relief in the form of a MATRIX. Each element of the DEM is regarded as a node of an imaginary grid. The grid is defined by identifying one of its corners (lower left usually), the distance between nodes in both the X and Y directions, the number of nodes in both the X and Y directions, and the grid orientation (7). See also ALTITUDE MATRIX.
DERIVATIVE MAP - A map created by altering, combining, or through the analysis of another maps(s). See also CARTOGRAPHIC MODEL.
DIAPOSITIVE - A positive photographic image on a transparent medium that is used in stereoplotting instruments (21).
DIGITAL DATA - Data represented in a computer compatible format.
DIGITAL IMAGE - A digitally encoded record of spectral reflectance or emittance intensity for a selected object or area. Each element of a digital image (referred to as a picture element or PIXEL) has a unique value for spectral intensity (21).
DIGITAL NUMBER - A positive integer representing the relative reflectance or emittance of an object in a DIGITAL IMAGE.
DIGITIZER TABLET - Synonym for DIGITIZER.
DIGITIZER - A device for a manual DIGITIZING. It consists of a flat surface (TABLE) and a CURSOR. See also SCANNER.
DIGITIZING - 1. An automated process of converting information in analog form (photograph, map, graph) into a digital representation, directly readable by a computer system using scanners (7). See also SCANNING. 2. Commonly, a process of manual encoding analog maps into computer readable format using a DIGITIZER.
DIGITIZING THRESHOLD - FUZZY DISTANCE used in DIGITIZING (45). See also WEEDING DISTANCE.
DIME - (Dual Independent Map Encoding). A data format used by the U.S. Census Bureau to code street network and related data. See also TIGER.
DIRECT ACCESS - An organization of data in a file or on storage media such that each element has a unique address and may be independently retrieved. Synonym for RANDOM ACCESS.
DIRECTORY - 1. A table of pointers to the corresponding data items (2). 2. Commonly, a look-up table with the addresses of data files on disk.
DIRICHLET TESSALATION - See THIESSEN POLYGONS.
DISK - See MAGNETIC DISK, OPTICAL DISK.
DISSOLVE - The AGGREGATION of polygons on the basis of selected common attributes, resulting in the generation of a simplified polygon COVERAGE.(7)(28). See also DROPLINE.

DISTRIBUTED DATA BASE - A data base for which different parts are located on different nodes in a computer network.
DN - See DIGITAL NUMBER.
DOUBLE PRECISION - The use of two computer WORDS to represent a number (2). See also SINGLE PRECISION.
DRAPING - The display of selected two-dimensional data on a perspective view of terrain relief or any other spatially distributed variable. For example, a map of a road network may be draped over a perspective view of surface elevations (16).
DROPLINE - A process of graphic elimination of boundaries between polygons having common selected attributes. This results in the generation of simplified polygon maps (7)(28). See also DISSOLVE.
DRUM PLOTTER - A PLOTTER that draws graphics on a display surface that is mounted on a rotating drum (2)(14).
DRUM SCANNER - A SCANNER in which the material to be scanned is mounted on a rotating drum.
DTM - (Digital Terrain Model). A representation of terrain relief in a computer readable format.

EBCDIC - (Extended Binary Coded Decimal Interchange Code). A code developed by IBM to code text with special and control characters using 8 bits (41).
EDITOR - Program allowing the user to interchange or enter data in a permanent computer file.
EDGE ENHANCEMENT - Analytical image processing techniques which emphasize tonal transition between features in an image (9).
EDGEJOIN - Synonym for EDGEMATCHING.
EDGEMATCH - Synonym for EDGEMATCHING.
EDGEMATCHING - A process of joining line features located on two separate but adjacent map sheets, across the map edge. Not synonymous with MAPJOIN. Synonymous with EDGEMATCH, EDGEJOIN. See also MOSAICKING, MAPJOIN.
ELAPSED TIME - The actual time that passes between the beginning and the end of a computer process. See also CPU TIME.
ELECTROSTATIC PRINTER - A device that prints by placing small electrostatic charges on the paper to which dark toner is attracted and adheres.
ELEVATION - A vertical distance below or above a reference surface. Terrain elevation is expressed with reference to mean sea level (msl).
ELLIPSOID - A solid, often used to represent Earth, having all its planer sections as ellipses or circles. See also REFERENCE ELLIPSOID, NAD.
ELLIPSOID OF REVOLUTION - An ELLIPSOID generated by the rotation of an ellipse around one of its axes.
EMULATION - The use of an EMULATOR to perform selected tasks.
EMULATOR - A computer program which allows an operator to perform certain operations on one kind of computer or computer component when those operations were originally

designed for a different kind of computer or computer component. For example, a terminal emulator allows an operator to execute programs designed for one type of a terminal on another type of terminal (11).

ENTITY - 1. In cartography, a real world phenomenon that is not further divisible into phenomena of the same kind. For example, a road (32). 2. In relational data bases, an object and its attributes (1).

ENTITY CLASS - 1. In cartography, a group of ENTITIES with a common feature. For example, road network (32). 2. In relational data bases, a set of object classes (1).

ENUMERATION AREA - An area canvassed by a census officer. Term related to AMF (52).

ENUMERATION DISTRICT - A geographic area regarded as a basic unit for data collection and tabulation in a census (29)(52).

ERROR BAND - An area around a point or a line having a width (or radius) equalling twice the EPSILON ERROR (50). The locus of points around the line which should contain true line

EPSILON ERROR - The tolerance of a measurement of a cartographic line or point (50).

ETHERNET - A type of computer network established using coaxial cable (thin or thick), twisted-pair cable, or fiber-optic cable (35).

EXCLUSION AREAS - Areas within a DEM model excluded from interpolation or contouring (55).

EXPERT SYSTEM - A computer system that performs tasks, that if performed by human beings, would require high levels of knowledge and several years of specialized training.

EXPOSURE - In surface modeling, a synonym for ASPECT.

EXPORT - The process of transferring data from a computer system to another system or storage media. See also IMPORT.

EXTENT - A rectangle bounding a map, the size of which being determined by the minimum and maximum map coordinates. Synonym for MAP EXTENT.

EXTRAPOLATION - The estimation of the value of a function outside its sampling domain. See also INTERPOLATION.

FACT - In AI, an accepted datum stored in a KNOWLEDGE BASE.

FEATURE - An object in a SPATIAL DATA BASE with a distinct set of characteristics.

FEATURE CODE - A unique alphanumeric string representing a category of an OBJECT in a SPATIAL DATA BASE (31).

FIDUCIAL MARKS - Index marks which are formed on the negatives of aerial photographs within the aerial camera at the time of exposure . The intersection of lines drawn between opposite fiducial marks defines the principal point or centerpoint of the photograph.

FILE - A collection of data stored on a permanent medium such as disk, MAGNETIC TAPE or OPTICAL DISK (13). See also FILE STRUCTURE.

FILE STRUCTURE - The organization imposed on the FILE to facilitate file processing (13).

FILTERING - As used in image analysis, the selective process of removing certain spectral or spatial frequencies to highlight or enhance features in an image (9).
FISHNET - A method of perspective representation of surface by lines drawn in X and Y directions (42).
FLAT - See SURFACE SPECIFIC POINTS.
FLAT FILE - A two-dimensional array of data.
FLATBED PLOTTER - A PLOTTER that draws graphics on a flat display surface (2).
FLOATING POINT NUMBER - Synonym for FLOATING POINT REPRESENTATION SYSTEM.
FLOATING POINT REPRESENTATION SYSTEM - A numeration system in which a real number is represented as the product of two numerals, a mantissa and an implicit base raised to integer power (30)(39). Synonym for FLOATING POINT NUMBER, REAL NUMBER.
FLOPPY - Synonym for FLOPPY DISK.
FLOPPY DISK - A flexible MAGNETIC DISK. Synonym for FLOPPY, DISKETTE (2).
FLOWCHART - A graphical representation (a model) of a total process in which symbols represent data flow, operations, or equipment used (2).
FORMAT CONVERSION - The conversion of data from one format to another.
FRACTAL - A shape made up of parts, each one being similar to the whole in some way. Fractals are used in the description of complex geometric objects in digital cartography (12).
FONT - One of the attributes of a text STRING. Font designates a set of characters in a particular style (42).
FORMAT - The arrangement of data in a FILE.
FREQUENCY DIAGRAM - Synonym for HISTOGRAM.
FUZZY DISTANCE - The distance within which two points are treated as one by a digital graphic/mapping system (a concept of spatial resolution). Synonym for FUZZY TOLERANCE.
FUZZY TOLERANCE - Synonym for FUZZY DISTANCE.

GENERALIZATION - A process of simplifying the thematic or geometric content of a MAP (42). See also SIMPLIFICATION, CLASSIFICATION, SYMBOLIZATION, INDUCTION.
GEOCODING - 1. Synonym for ADDRESS CODING. 2. Process of assigning geographic locations to OBJECTS.
GEODETIC DATUM - A model of the Earth's shape (6). See also NAD.
GEOID - A surface around the Earth such that each point on the surface is normal to the direction of gravity, coincides with mean sea level of the oceans (6). See also REFERENCE ELLIPSOID.
GEOMETRIC CORRECTION - The adjustment of the geometry of a DIGITAL IMAGE for scaling, skewing, and other spatial distortions (9).
GEOMETRIC TRANSFORMATION - 1. Adjustments made in IMAGE data to improve geometrical consistency for cartographic purposes. 2. Transformation of IMAGE data from

its original sensor coordinate system to another selected coordinate system (ie. UTM) (9). See also GEOMETRIC CORRECTION.

GEOPROCESSING - The manipulation and analysis of geographically referenced data.

GEOREFERENCE SYSTEM - A coordinate system with which the location of a point on the Earth's surface may be identified.

GIS - (Geographic Information System). A computer software system (often including hardware) with which spatial information may be captured, stored, analyzed, displayed, and retrieved (7).

GPS - (Global Positioning System). A satellite-based navigational system permitting the determination of the position of any point on the Earth with high accuracy (19).

GRADIENT - 1. One of the morphological parameters used in spatial analysis systems and refers to the degree or percent of slope (7). 2. Maximum slope change. See also ASPECT, DEM, ORTHOCONTOURS, SLOPE.

GRAPH - A set of junction NODES with certain nodes joined by branches (18). See also NETWORK, NETWORK ANALYSIS.

GRAPHIC SCALE - Synonym for BAR SCALE.

GRATICULE - Network of parallels and meridians plotted on the map in map projection (31).

GRAY SCALE - Monochrome ordering of shades of gray from white to black with intermediate shades of gray (20)(31).

GRID - 1. A DATA STRUCTURE composed of points located at the nodes of an imaginary grid, with the spacing between nodes being constant in both the horizontal and vertical directions. 2. Commonly, RASTER DATA STRUCTURE.

GRID CELL - An element of a RASTER DATA STRUCTURE (32).

GRIDDING - 1. Process of converting spatial information into a GRID data structure. 2. Interpolation of a spatial function, sampled in irregular locations, to regularly distributed points. 3. Commonly RASTERIZATION.

GRID RESOLUTION - Distance between neighboring grid nodes.

GROUND CONTROL POINT - (GCP). 1. A point of known location that can be recognized on an IMAGE or a MAP and that can be used to calculate the transformation needed for the REGISTRATION of images or maps. Ground Control Points are related to a known projection for use in geometric transformation (9). See also GEOMETRIC CORRECTION, GEOMETRIC TRANSFORMATION.

HALFTONE - A technique to represent a continuous tone image by varying the density of discrete lines or points (42).

HARD COPY - A copy of a digital file (map, table) on a permanent media (paper, transparency) (31).

HARD DISK - See MAGNETIC DISK.

HARDWARE - Physical components or peripherals of a computer system.

HARDWARE TRANSPARENCY - A property of a software system allowing it to function on different hardware platforms.

HEADER RECORD - The record at the start of the physical unit of storage that contains identifying information about a file or data set (34).
HEADER FILE - The file that contains identifying information about a following data set or group of files.
HEURISTIC RULE - In AI, a non-algorithmic method for performing a task.
HEXADECIMAL NOTATION - A numeration system having a radix of 16. Hexadecimal notation uses numbers from 1 to 0 and letters from A to H (2).
HIDDEN LINE REMOVAL - The process of eliminating from a graphic display, line segments that are obscured from a view in a two-dimensional projection of a three-dimensional object (2).
HIGH-LEVEL LANGUAGE - A programming language that does not reflect, nor is tied to, the unique structure of any particular computer system (2).
HILL SHADING - The shading (varying pattern of light and dark) applied to a graphical representation of a terrain unit or landscape according to the relative amounts of light reflected from different terrain locations for a particular position (azimuth and elevation) of the sun (31).
HISTOGRAM - 1. A function relating values of data to their frequency of occurrence.(42). 2. A graphic display of HISTOGRAM. Synonym for FREQUENCY DIAGRAM.
HOLE - Synonym for ISLAND.
HORIZONTAL INTEGRATION - The combination of two or more adjacent computer map files so the resulting map file is topologically consistent (29). See also EDGEMATCHING.
HUE - The attribute of color associated with wavelength or dominant spectral color (31).
HYBRID DATA STRUCTURE - Data structure integrating elements of TESSELLATION and VECTOR SPATIAL DATA STRUCTURES (18)(59)(61).
HYPSOMETRIC CURVE - 1. An elevation surface (7). 2. A curve relating a mean land height to an area (42).
HYPSOMETRY - Vertical control in an elevation model with reference to an established datum (usually mean sea level).

IMAGE - The representation of an object that results from its reflection or emission of energy being recorded by chemical, mechanical, optical, or electronic means (9). A photograph is one kind of image but not all images are photographic. See DIGITAL IMAGE.
IMAGE PROCESSING SYSTEM - The system of software and hardware components designed for the processing of DIGITAL IMAGES or other compatible data.
IMAGE ENHANCEMENT - The application of any of a group of operations that alter or exaggerate the tonal differences in an image. These operations may include CONTRAST STRETCHING, EDGE ENHANCEMENT, FILTERING, and/or smoothing to improve the detection or identification of features of interest(9).
IMAGERY - The products of image forming instruments.
IMPORT - A process of loading data to a computer system from another system or storage media. See also EXPORT.

INDEFINITE CONTOURS (AREAS OF) - CONTOURS compiled from low accuracy elevation data and explicitly distinguished from other contours (55).
INDEX - Identifier, that is not a part of a data base, used to access stored information in contrast with KEY (46).
INDEX CONTOUR - A CONTOUR LINE labelled with its value.
INDEX MAP - A reference map that outlines the mapped area, identifies all the component maps for the area if several map sheets are required, and identifies all adjacent map sheets (29).
INDUCTION - One of the methods used for map GENERALIZATION involving the extension of data based on the association between data elements or data characteristics (interpolation) (42).
INK JET PRINTER - A printer in which characters are formed by projecting a jet of ink onto paper (2).
INPUT - 1. The physical process of entering data into a computer. 2. The data to be entered into a computer.
I/O - (Input and output). A term referring to the operations of input and output (2).
INSET MAP - A map that is an enlargement of some congested area of a smaller scale map, and that is usually placed on the same sheet with the smaller scale main map (29).
INTERACTIVE TOPOLOGY - A function of a GIS enabling real-time update of the TOPOLOGY of a SPATIAL DATA BASE.
INTERMEDIATE CONTOUR LINE - An unlabelled contour line.
INTERPOLATION - The estimation of the value of the function inside the sampling domain. See also EXTRAPOLATION.
INTERSECTION - 1. A common point between two or more lines or line segments. 2. A set operation of multiplication. The outcome of set intersection is a set containing common elements to all the input sets.
INTERACTIVE MODE - A mode of operation of a computer system in which the dialog takes place between the user and the system in a sequence of request-response (2). See also BATCH PROCESSING. Synonym for CONVERSATIONAL MODE, INTERACTIVE PROCESSING.
INTERACTIVE PROCESSING - Synonym for INTERACTIVE MODE.
INTERNATIONAL ELLIPSOID - The official REFERENCE ELLIPSOID used for the primary geodetic network in Europe (6). See also NAD.
ISARITHM - A line connecting points of the same value for any of the characteristics used in the representation of surfaces. An isarithm is created at the intersection of the mapped surface with a plane parallel to the datum (42). Synonym for ISOLINE. See also PROFILE, ISOLINE, ISARITHMIC LINE, CONTOUR.
ISARITHMIC LINE - Synonym for ISARITHM.
ISARITHMIC MAP - A map that uses isarithms to represent a STATISTICAL SURFACE (42).
ISLAND - A polygon within another, not sharing a common boundary. Islands require special handling in polygon oriented spatial data bases. Synonym for HOLE.

ISOBATHS - Contours of depth (of the water) describing the shape of the bottom of a water reservoir (27). See also BATHYMETRIC MAP.
ISOLATED CONTOUR LINE - A contour line represented by special symbols to designate the value of the enclosed area, often depressions or peaks. See SURFACE SPECIFIC POINTS.
ISOLINE - Synonym for ISARITHM.
ISOMETRIC MAP - A map on which isolines are used to represent the surface (42). See also ISOLINE.
ISOPLETH - A line representing quantities that cannot exist at a point, such as population density (42).
ISOPLETH MAP - A map on which isopleths are used to represent some selected quantity (42).
ITEM - Element of a record in a relational data base.

JOIN - The fusion or union of two or more tables in a relational data base on the basis of a common item. See also RELATIONAL DATA BASE.
JOURNAL FILE - A file containing a record of an interactive session for future reference. Often used to allow an operator to recover from a system failure.
JOYSTICK - A location device in form of a level with at least two degrees of freedom (2).
JUSTIFICATION - The adjusting of the position of characters or numbers to meet the requirements of a particular format. Left-justification denotes a shift to and alignment at the left, right-justification denotes a shift to and alignment at the right, and center-justification denotes alignment with the center of a particular format.

KEY - A field in data base used to obtain access to stored information.(46). See also INDEX.
KEYWORD - Synonym for RESERVED WORD.
KILOBYTE - 1000 BYTES.
KNOWLEDGE - In AI, a term denoting FACTS, BELIEFS, and HEURISTIC RULES.
KNOWLEDGE BASE - The computer data base storing KNOWLEDGE.
KRIGING - An interpolation method based on a generalized least-squares algorithm, using VARIOGRAMS as weighting functions (23).

LABEL - A textual description of the geographic object placed on a map. See also ANNOTATION.
LABEL CONFLICT - Coincidence of two or more labels on a map. See also AUTOMATIC LABEL PLACEMENT.
LABELLING - The placement of descriptive text on a map, specifically the placement of text describing particular feature characteristics. See also LETTERING.
LAND INFORMATION SYSTEM - A spatial information system containing data on land and land use (34).

LANDSAT - An Earth orbiting satellite, operated by the USA, that collects, records, and transmits the intensity of reflected and emitted energy from the Earth's surface through the use of non-photographic, multi-spectral imaging systems. LANDSAT digital data, often in the form of digital images, are useful for the extraction of temporal information on land cover for input into a GIS.

LANDSAT MSS - A multispectral scanner system that collects electromagnetic radiation in four bands (0.5 - 0.6 micrometers, 0.6 - 0.7 micrometers, 0.7 - 0.8 micrometers, and 0.8 - 1.1 micrometers). The nominal ground resolution cell size for MSS data is 80 meters.

LANDSAT TM - Thematic mapper system that collects electromagnetic radiation from a ground resolution cell of 30 meters in the following seven channels:

- Channel 1 (0.45 - 0.52 micrometers)
- Channel 2 (0.52 - 0.60 ")
- Channel 3 (0.63 - 0.69 ")
- Channel 4 (0.76 - 0.90 ")
- Channel 5 (1.55 - 1.75 ")
- Channel 6 (10.4 - 12.5 ")
- Channel 7 (2.08 - 2.46 ")

LATITUDE - North/South position of a point on the Earth defined as an angle between the normal to the Earth's surface at that point and the plane of the equator (42).

LAYER - A subset of a spatial data set dealing with one thematic topic.

LEAKING POLYGON - A TOPOLOGICAL ERROR denoting a POLYGON with a boundary that is not closed. A leaking polygon is usual associated with undershoot in which case it is called a proper leaking polygon, otherwise is called a specific leaking polygon (7).

LEAST-SQUARES ADJUSTMENT - A method for transformation of coordinates using a LEAST-SQUARES ALGORITHM.

LEAST-SQUARES ALGORITHM - A method of approximation based on the minimization of the squared distance between two sets of variables.

LEGEND - An explanation of the symbols, codes, and other information appearing on a map. A legend is usually constructed as a table or text (27).

LETTERING - The process of a designing and placing TEXT on a map (42).

LEVEL - See LAYER.

LEVEL SLICING - The mapping of a continuous valued function into a discrete classes.

LINE - A geometric object represented by a series of points. A line is a computer analog of such geographic ENTITIES as roads, rivers, and shorelines.

LINE COVERAGE - A COVERAGE containing information and related attributes for geographic ENTITIES represented by lines.

LINE SIMPLIFICATION - See WEEDING.

LINEAGE - Information about the source of data, its origin, accuracy, and scale (15).

LINE-OF-SIGHT MAP - Synonym for VIEWSHED MAP.

LINE PATTERN - A graphic design representing linear features.

LINE PRINTER - A computer-driven device that produces a line-by-line printed record of the data it receives (21).

LINE SEGMENT - See SEGMENT.
LINE WEIGHT - The width of the line pattern used for the graphic representation of line features.
LINEAR TRANSFORMATION - A transformation which maps a linear combination of vectors into the same combination of transformed vectors. Linear transformations are shearing, scaling, rotation, reflection, translation, perspective, and overall scaling (22).
LINK - The directed SEGMENT.
LOCATIONAL ACCURACY - The accuracy with which spatial objects are positioned in SPATIAL DATA BASES (34).
LONGITUDE - East/West position of a point on the Earth, defined as the angle between the plane of a reference meridian and the plane of a meridian passing through an arbitrary point (42).
LONGWORD - A group of WORDS treated as a unit.
LOOK-UP TABLE - A TABLE containing a KEY and/or additional information about the KEY.
LOXODROME - A Line of constant compass direction on Earth. A loxodrome intersects all meridians at equal oblique angles and appears as a straight line on a Mercator projection (51). Synonym for RHUMB LINE.

MACRO - An instruction in a computer language that can be replaced by a set of instructions in the same language (2).
MAGNETIC DISK - A flat circular plate with a magnetizable surface on which data can be stored by magnetic recording (2). Synonym for DISK, HARD DISK.
MAGNETIC TAPE - A tape with magnetizable surface on which data can be stored and retrieved by magnetic recording (2).
MAINFRAME - 1. A unit consisting of one or more processors and their internal storage (2). 2. Commonly, a large central computer in network or terminal system.
MAP - 1. A graphic representation of spatial relationships and spatial forms (31). 2. A graphic representation on a plane of selected features of a part of or the whole Earth's (or any other celestial body) surface (27).
MAP ALGEBRA - A set of operations defined on spatial data sets for the analysis and synthesis of spatial information (7)(57).
MAP EXTENT - Synonym for EXTENT.
MAP GENERALIZATION - Synonym for GENERALIZATION.
MAPJOIN - An automated process for joining separate but adjacent map sheets into one. The result is a topologically consistent and physically continuous map.
MAP LEGEND - Synonym for LEGEND.
MAP LIBRARY - A data base designed for the storage and manipulation of COVERAGES. See also TILE.
MAPPING UNIT - A defined geographical area feature to which attribute information is related. For example, a parcel of land, a census block, a county, or a country.

MAP POSITIONAL FILE - A file containing positionally correct and complete digital map data (55). Synonym for positional file.
MAP PROJECTION - A representation on a plane of the surface of a round body (51).
MAP REPRESENTATION FILE - A file containing cartographically correct and complete digital map data that have been enhanced to improve the map readability or aesthetic quality. Map representation files are generated from MAP POSITIONAL FILES (55).
MAP UNITS - Units for which the locations of features on a map are defined. For example, the map units for maps compiled using the UTM projection are "meters".
MARKER - A graphic symbol used to represent a point feature. Synonym for point symbol.
MASKING - A process of excluding of a part of an image or map from further processing.
MASS STORAGE - 1. A storage media with very large storage capacity (2). 2. Commonly, MAGNETIC TAPES, CARTRIDGES, CD-ROMS.
MATRIX - A rectangular array of elements arranged in rows and columns that can be manipulated according to rules of matrix algebra (2).
MAXIMUM ACCESS TIME - See ACCESS TIME.
MEGABYTE - 1,000,000 bytes. See also BYTE, KILOBYTE.
MERIDIAN - A line of a great circle perpendicular to the plane of the equator.
MENU - A type of computer interface in which a list of options is displayed allowing the operator to select the next operation, or the parameters of the operation, by pointing to the option on the list with a pointing device.
MINIMUM ACCESS TIME - See ACCESS TIME.
MIPS - (Million Instructions Per Second). The number of instructions that a computer can perform in one second on a given task. MIPS are used as a measure of computer system power (11).
MODEL - A representation of a set of objects and their relationships.
MODEM - 1. A device that allows for the transmission of digital signals over analog transmission lines (2). 2. Commonly, a device allowing for communication between computers over telephone lines.
MONITOR - The screen of a CRT, especially with a microcomputer (11).
MONOCHROMATIC IMAGE - An image pertaining to a single wavelength or to a narrow range of wavelengths (21).
MONUMENT - A surveyed marker with known, precise location (horizontal control monument) and elevation (vertical control monument).(55).
MORPHOMETRIC MAP - A map representing morphological features of the Earth's surface (42).
MORPHOMETRY - The science of the quantitative characterization of terrain forms.
MORTON NUMBER - A numbering scheme to index tiles of tessellated space (37). Synonym for Morton key or Morton index.
MOS - Marine Observation Satellites launched by the Space Development Agency of Japan (MOS-1 on February 19, 1987 and MOS-1b on February 7, 1990).

MOSAICKING - 1. In spatial information systems, a synonym for MAPJOIN. 2. The process of assembling several individual aerial photographs or other kinds of images into a composite view of an area (5).
MOUSE - A hand-held location device operated by moving it on a flat surface (2).
MTBF - (Mean Time Between Failures). One of the parameters of computer system components. MTBF is a mean length of time between consecutive failures of system components (2).
MULTIPLE CENTROIDS - Error in a POLYGON COVERAGE occurring when one or more polygons have more than one CENTROID.
MULTIPURPOSE CADASTRE - A framework supporting comprehensive, land-related information at parcel level composed of:
- a reference frame consisting of the geodetic framework,
- a series of current large-scale maps,
- a cadastral overlay delineating all cadastral parcels,
- a unique index number assigned to each parcel;
- a series of land related data files, each including a parcel identification index (33).

See also CADASTRE.
MULTISPECTRAL DATA SET - A data set containing a record of more than one spectral band.
MULTITEMPORAL DATA SET - A data set containing information pertaining to more than one time or period.
MULTISPECTRAL SCANNER - A line-scanning sensor (optical-mechanical scanner) that uses an oscillating or rotating mirror, a wavelength-selective dispersive mechanism, and an array of detectors to simultaneously measure the energy available in several wavelengths.(21).
MYLAR - A thin translucent plastic drafting material, used as a stable base in the production of analog maps. See also HARD COPY, ANALOG MAP.

NAD - (North American Datum). The official REFERENCE ELLIPSOID used for primary geodetic network in North America (6). See also INTERNATIONAL ELLIPSOID.
NAD27 - (North American Datum 1927). A datum with its origin at Meales Ranch, Kansas (51).
NAD83 - (North American Datum 1983). An Earth-centered datum defined on satellite and terrestrial data (51).
NADIR - A point on the ground vertically below the observer, camera lenses, or sensor system (27).
NEAT LINE - A line that separates the map from the margin (27).
NEGATIVE PHOTOGRAPH - 1. A photographic image on which tones are reversed. 2. A film, plate, or paper containing such a reversed image (21).
NEIGHBORHOOD ANALYSIS - The analysis of the relationships between an object and similar surrounding objects, used in image processing, etc.

NETWORK - 1. A GRAPH with a flow of some type in its branches.(18). See also NETWORK ANALYSIS. 2. In SPATIAL INFORMATION SYSTEMS, a network is a data structure composed of nodes, arcs, and information on the relationships between them (topology). 3. A data structure to hold information about geographic objects represented by lines and allowing for NETWORK ANALYSIS.
NETWORK ANALYSIS - A set of techniques for the analysis of systems represented by networks. These techniques include maximum flow analysis, travelling salesman problem, allocation of resources, and optimum path.
NEURAL NETWORKS - Synonym for ARTIFICIAL NEURAL NETWORKS.
NIBBLE - A group of bits, smaller that byte, that can be regarded as a unit (a 4 bit nibble).
NODE - 1. A beginning or ending point of an ARC. 2. An element of a TREE (56). 3. An element of a GRID.
NORMAL FORM - A set of guidelines for the design of records in relational data base theory. Normal forms, from one to five, are designed to prevent update anomalies, data redundancies, and data inconsistencies (23).

OBEL - Synonym for VOXEL.
OBJECT - 1. In object oriented programming, primitive elements combining properties of procedures and data (53). 2. In relational data base, a phenomenon characterized by a set of attributes (1). 3. In cartography, a digital representation of an ENTITY (32).
OBJECT CLASS - Synonym for OBJECT TYPE.
OBJECT CODE - Output of a compiler which is in the form of an executable machine code (2).
OBJECT TYPE - In relational data bases, a set of OBJECTS characterized by a set of attributes common to them all (1). Synonym for OBJECT CLASS.
OCTAN - A search region in interpolation. An OCTAN is a 45 degree segment of a circle centered on an interpolated point. See also QUADRANT.
OCTAL NOTATION - A fixed radix numbering system with a radix of eight. Octal notation uses numbers from 0 to 7 (2).
OCTREE - An extension of a quadtree structure into three dimensions (37).
OFF-LINE - The state of a computer system component when it is not under the control of the computer (2).
ON-LINE - The state of a computer system component when it is under the control of the computer. See also OFF-LINE (2).
OOP - Object Oriented Programming (53).
OOPL - Object Oriented Programming Language. A programming language that provides a high-level structure at the level of objects, classes, and class hierarchies (62).
OOPS - (Object Oriented Programming Systems). Commonly, a minor error in the execution of an interactive computer task.
OPERATING SYSTEM - Software that controls the operation of the computer system and the execution of the programs. Popular operating systems include DOS, OS/2, UNIX, VMS, ULTRIX (2).

OPTICAL DISK - See CD-ROM.
ORTHOCONTOURS - ISOLINES of the GRADIENT of a SPATIAL FUNCTION. Orthocontours show a rate and direction of change in a mapped phenomena (10). See also GRADIENT.
ORTHODROME - A line of a great circle. The shortest distance between two points on Earth. An orthodrome only appears as a straight line in Gnomonic projection (51). See also LOXODROME.
ORTHOPHOTOGRAPH - A photographic copy of a perspective photograph with distortions due to tilt and relief removed (5).
ORTHOPHOTOQUAD - A standard map sheet overprinted with an orthophotograph(5).
OUTLINE MAP - A map showing the limits of a specific set of mapping entities such as counties, NTS quads, etc. Outline maps usually contain a very small number of details over the desired boundaries with their descriptive codes (29).
OVERLAP - That part of a map sheet that is duplicated on the adjacent map sheet (29).
OVERLAY - 1. The process of superimposing or registering two or more maps such that the resulting maps contain the information from both maps for selected data items. Overlay is usually performed using choropleth maps (3). 2. A subset of a spatial data set dealing with one theme. See also LAYER.
OVERSHOOT - A TOPOLOGICAL ERROR denoting a DANGLE.

PAGE - 1. A BLOCK of a computer memory of fixed length and unique address, that is accessible as a unit (2). 2. Extent of the surface on which a graphic file may be displayed or plotted.
PAGE UNITS - Units that define the position and size of graphic elements on a graphic PAGE. See also MAP UNITS.
PAGING - 1. The transfer of PAGES between storage systems (2). 2. An organization of computer storage into fixed length BLOCKS and PAGES (2).
PAN - The capability of a system to display different parts of an image without changing the scale of display (roam). See also ZOOM.
PARALLEL - The line of intersection between a plane parallel to the plane of the equator and the Earth's surface.
PARTITION - The fragmentation of a data set into manageable subsets (34).
PASS - See SURFACE SPECIFIC POINTS.
PATCH - Information related to a small area (usually a subset of a large SPATIAL DATA BASE of greater extent) added to a data base to upgrade or densify its data content (16).
PATTERN - 1. The graphic design used to infill areas on a graphic display or represent lines or points. See also SHADING or CROSS-HATCHING. 2. An object with defined characteristic and assigned class.
PATTERN RECOGNITION - The process of classifying objects into discrete classes on the basis of comparisons with members of other classes, or on predefined values for characteristics (21).

PAUSE - Temporary interruption in the execution of a process, with the possibility of the process being continued afterwards. See also BREAK.
PEAKS - See SURFACE SPECIFIC POINTS.
PEANO CURVE - Family of curves that transform n-dimensional space into line (37).
PEN PLOTTER - Plotters which use pens as drawing elements.
PEN - The drawing element of a plotter similar in construction to ballpoint pens or ink pens.
PHOTOGRAMMETRY - The art and science of obtaining reliable measurements through use of photographs (21).
PHOTOGRAPH - A image formed by the action of light on a photosensitive medium such as film (21).
PHOTOGRAPH, LARGE-SCALE - Aerial photograph with a representative fraction of 1:500 to 1:10 000 (21).
PHOTOGRAPH, MEDIUM-SCALE - Aerial photograph with a representative fraction of 1:12 000 to 1:30 000 (21).
PHOTOGRAPH, SMALL-SCALE - Aerial photograph with a representative fraction smaller than 1:40 000 (21).
PICTURE ELEMENT - See PIXEL.
PIPELINE ARITHMETIC PROCESSOR - Synonym for ARRAY PROCESSOR.
PIT - See SURFACE SPECIFIC POINTS.
PIXEL - Smallest, nondivisible element of a display surface, or an image, that can be independently accessed (addressed) (2). Acronym for PICTURE ELEMENT.
PIXMAP - A three-dimensional array of bits represented as a stack of "N", two-dimensional, one bit arrays, where "N" is the depth of each array element expressed in the number of bits. Akin to a stack of "N" bitmaps (48). See also BITMAP.
PLANIMETRIC MAP - A map showing only the horizontal position of geographic objects without topographic features (21).
PLOTTER - An output device that generates HARD COPIES of computer graphic files (2). See also PEN PLOTTER, DRUM PLOTTER, FLATBED PLOTTER.
POINT - An object represented by an n-tuple of coordinates, with zero length and zero area. In cartography ,points or spot symbols (depending on scale) represent such geographic objects as benchmarks, wells, sampling sites, etc. (52).
POINT COLLISION - Points or lines located closer to each other than the FUZZY TOLERANCE for a given process (LINE SIMPLIFICATION, digitizing of lines) (60).
POINT COVERAGE - A COVERAGE containing information on geographic ENTITIES represented by POINTS and related attributes. See also LINE COVERAGE, POLYGON COVERAGE.
POINT MODE - A mode of digitizing in which each digitized point must be explicitly entered by the operator. See also STREAM MODE.
POINT SYMBOL - Synonym for MARKER.
POLAR COORDINATE SYSTEM - A coordinate system in which the position of a point is defined by its distance and angle from a reference point and line.

POLYGON - A class of spatial objects having nonzero area and perimeter, and representing a closed boundary region of uniform characteristics (47).
POLYGON COVERAGE - A COVERAGE containing information on geographic ENTITIES represented by POLYGONS and related attributes.
POLYGON RETRIEVAL - A function of SPATIAL INFORMATION SYSTEMS by which polygons falling within a query window or meeting a specific set of search conditions are selected (28).
POLYGONIZATION - A process of generating a POLYGON COVERAGE from digitized line data. The process of polygonization may include:

- calculation of line intersections;
- elimination of under- and overshoots;
- calculation of centroids;
- chaining of arcs;
- delineation of islands (7).

POSITIONAL ACCURACY - Reliability of the locating of cartographic features relative to their true position. See also LOCATIONAL ACCURACY.
PRIMARY COLORS - Colors from which other colors can be produced.
PRECISION - 1. A measure of the ability to distinguish between nearly equal values (2). 2. A degree of discrimination to which a quantity is stated (two digit number distinguishes 100 quantities) (2).
PROFILE - Synonym for CROSS-SECTION.
PROJECTION - See MAP PROJECTION.
PROJECTION CHANGE - The transfer of digital map data from one MAP PROJECTION to another (16).
PROOF PLOT - A test printout of a digitized map used to check its registration to a source map.
PROXIMAL ANALYSIS - The generation of VORONOI POLYGONS for a particular set of spatial features (5).
PROXIMITY - A measure of inter-object distance (16).
PROXIMITY POLYGON - See VORONOI POLYGON.
PRUNING - The deletion of extraneous points in stream digitizing.(45). See also WEEDING.
PSEUDOSCOPIC VIEW - A reversal of the normal stereoscopic effect in photographs, causing valleys to appear as ridges and ridges as valleys (21).
PUCK - Synonym for CURSOR (42).
PURE BINARY NUMERATION SYSTEM - A numeration system that uses a radix of 2 and binary digits.such that the numeral "5" is represented as 101 (2). Synonym for BINARY NOTATION, BINARY NUMERATION SYSTEM. See also OCTAL NOTATION, HEXADECIMAL NOTATION, DECIMAL NOTATION.

QUADTREE - (Q-TREE). 1. Hierarchical data structures based on the principle of recursive decomposition of space into square tiles, resulting in a balanced tree structure of degree 4 (37)(47). See also SPATIAL DATA STRUCTURES.

QUADRANT - A search region in interpolation. A quadrant is a 90 degree segment of a circle centered on an interpolated point. See also OCTAN.

QUERY - 1. A set of conditions or questions that form the basis for the retrieval of information from a data base. 2. The retrieval of information based on such conditions or in response to such questions. See also BROWSING.

QUERY WINDOW - Area within which the retrieval of spatial information and related attributes is to be performed. A query window may have a regular or an irregular shape (28).

QUEUE - 1. A list of variables whose content may be changed by adding or removing elements at either end (2). 2. Commonly, a list of jobs to be executed, in order, by a computer system.

RADAR - (Radio Detection and Ranging). 1. A system, method or technique, including equipment components, to detect, locate and/or track objects, to measure altitudes, and/or to acquire image data using microwave energy. Electromagnetic radiation, in the spectral region from 1mm to 0.8m in wavelength, is beamed out to illuminate target areas and the intensity and time delay of the returning energy are measured (9).

RADIX - Base used for exponentiation in numeration systems. For example, a decimal numeration system has radix of 10 (2).

RANDOM ACCESS - Synonym for DIRECT ACCESS.

RASTER - 1. An element of a space that has been sub-divided into regular tiles by TESSELLATION (47). 2. Commonly, a data set, as for an IMAGE or DEM, composed of rasters. 3. Often used as a synonym for GRID.

RASTER DATA BASE - Data base storing spatial information in raster format.

RASTER DATA STRUCTURE - A class of SPATIAL DATA STRUCTURES in which the basic units of spatial information are represented as RASTERS. Raster Data Structures do not provide specific provisions to represent points, lines and polygons (37). See also TESSELLATION DATA STRUCTURES.

RASTERIZATION - 1. A process of coding a spatial information into a RASTER DATA STRUCTURE. 2. Commonly, the conversion of data from a VECTOR DATA STRUCTURE to a RASTER DATA STRUCTURE.

RAVINE - See SURFACE SPECIFIC POINTS.

REAL NUMBER - See FLOATING POINT REPRESENTATION SYSTEM.

RECORD - 1. A set of data treated as a unit (2). 2. An element of a table in relational data bases. Synonym for TUPLE. 3. An organizational unit of a file.

RECTIFICATION - A set of techniques for the elimination of errors in data. Rectification may be used to correct aerial photographs, remotely sensed data or analog maps. See also RUBBER-SHEETING, REGISTRATION.

RECURSION - A process, function, or routine that can be execute continuously until a specified condition is met (2).

REFERENCE ELLIPSOID - The mathematical model of the Earth used in geodetic computations (6). Synonym for SPHEROID, ELLIPSOID, ELLIPSOID OF REVOLUTION. See also NAD.
REFLECTANCE - The ratio of a radiant energy reflected by an object to the radiant energy received.
REFORMAT - See FORMAT CONVERSION.
REFRESH - A utility allowing for the redrawing of the content of a graphic screen from screen memory.
REGISTRATION - The process of geometrically aligning two or more cartographic data sets or digital images, in vertical juxtaposition, while maintaining true geographic referencing.
REGION - A continuous area with some degree of uniformity in selected characteristics (47). See also POLYGON.
RELATION - Used as a synonym for TABLE in relational data models (1).
RELATIONAL ALGEBRA - A collection of operations on RELATIONS including Cartesian product, union, intersection, difference, projection, restriction, join, and division (1).
RELATIONAL DATA BASE - 1. A set of RELATIONS whose structure is specified by the relational scheme (1). 2. Commonly, a data base where information is arranged into tables and the dependencies between information is mapped by dependencies between two or more TABLES.
RELIABILITY DIAGRAM - An inset map providing an areal index of the expected accuracy of sources used to compile a map.
RELIEF MAP - Any map that appears to be or is in 3-dimensions (27).
REMOTE SENSING - 1. Acquisition of information about the properties of an object or phenomenon using a recording device that is not in physical contact with the object of the study. 2. Commonly, information gathered using airborne or satellite platforms.
RESAMPLING - 1. In digital image processing, the interpolation of PIXELS on a source digital image to new locations of transformed PIXELS, usually coinciding with a georegistered grid (5). 2. In modeling of spatial functions, generation of a new data set by subsetting the input data set or by creating a new set through INDUCTION.
RESERVED WORD - Part of a computer language definition set that cannot be used as a user-defined variable or constant name. Synonym for KEY WORD.
RESOLUTION - A minimum distance between two objects that can be distinguished by a sensor (9). Most often a synonym for SPATIAL RESOLUTION but also applies to spectral and temporal aspects of remote sensing imaging systems.
RESPONSE TIME - The elapsed time between the end of a request on a computer system and the beginning of the response. (2).
RGB MONITOR - (Red-Green-Blue Monitor). A color computer MONITOR that uses the three primary colors in various mixtures to generate all colors
RIDGE - See SURFACE SPACING POINTS.
RIDGE-LINE - See SURFACE SPECIFIC POINTS.
RING - A closed sequence of nonintersecting CHAINS, STRINGS, or ARCS (32).

RUBBER-SHEETING - 1. A map transformation bringing a map into coincidence or near-coincidence with another reference map. 2. A deformation algorithm forcing a registration of control points, distributed over a map, with their counterparts on the stable, better quality, map (63). See also RECTIFICATION.
RHUMB LINE - See LOXODROME.
RUN-LENGTH CODING - A method of compressing data based on the reduction in the number of any type of repeating sequences of characters. Used frequently to compress image data (17).

SATURATION - The degree of intensity difference between a color and an achromatic light-source color of the same brightness (21).
SCALE - The ratio of the distance on a map, photograph, or image to the corresponding distance on the ground, all expressed in the same units (42)(5).
SCAN LINE - 1. The segment of an image produced as a result of a single sweep of a light source across a recording medium. 2. The electronic signal of a single scan (21). 3. A list of PIXELS forming a horizontal row of an image with a constant Y-coordinate, and ordered by increasing X coordinate (48).
SCANNER - A device that produces a digital image from analog data input (21).
SCANNING - 1. An automated process of converting analog information into digital format. 2. The conversion of analog maps into digitally readable format using SCANNER (42).
SCENE - The total image area presented on the output from a data gathering system.
SCRATCH FILE - A temporary file created during some computer process to store intermediate results, and which is deleted after the process has been completed. Synonymous with TEMPORARY FILE. See also SCRATCH SPACE.
SCRATCH SPACE - An amount of storage space allocated to scratch files. Synonym for TEMPORARY SPACE.
SCRIPT FILE - A file with a series of commands to be executed automatically. Synonym for COMMAND PROCEDURE. See also MACRO.
SEAMLESS DATA BASE - A digital data base storing, as one continuous data structure, spatial information spanning two or more disjointed map sheets (8). Synonym for CONTINUOUS DATA BASE.
SEED - See CENTROID (34).
SEGMENT - An element of line defined by two end points. Synonym for LINE SEGMENT, CHORD.
SEMI-VARIOGRAM - A measure of the continuity of spatial phenomena expressed as an average squared difference between measured quantities at different locations. Used in KRIGING (22).
SERVER - A station on a computer network providing service such as making a file or printer available (35).
SET FUNCTION - An operation that acts on an ensemble of values (sets) instead of on a single value. Operations on spatial data bases are frequently represented by set operations such as UNION, INTERSECTION, and SUBTRACTION.

SHADING - 1. A type of graphic pattern used to represent regions, composed of lines, characters,or symbols. 2. A process of filling an area with a shading pattern. See also CROSS-HATCHING.
SIMPLE OBJECT - An OBJECT that cannot be subdivided into other objects. See also COMPLEX OBJECT.
SIMPLIFICATION - One of the elements of cartographic generalization involving the determination of important characteristics of data, elimination of unimportant details, and the retention of important characteristics (42). See also GENERALIZATION. Synonym for TYPIFICATION.
SIMULATION - The modeling of the dynamic behavior of a system by moving it from state to state in accordance with a defined set of rules (40).
SINGLE-PRECISION - Use of one computer word to represent a number. See also DOUBLE PRECISION.
SKEW - 1. Distortion from a true or symmetrical form (21). 2. In satellite imagery, distortion of the IMAGE is due to the movement of the Earth below the satellite, and movement of the satellite along its path.
SLIVER POLYGONS - Polygons created in an OVERLAY process of two vector graphic files when boundaries are not in perfect registration. Such polygons are usually regarded as artifacts of the OVERLAY process and not legitimate data. The generation of SLIVER POLYGONS is often attributed to errors in the polygon boundaries of the original files. Synonym for SLIVERS, SPURIOUS POLYGONS.
SLIVERS - Synonym for SLIVER POLYGONS.
SLOPE - 1. A morphometric parameter expressing the ratio of the change of the terrain elevation over the horizontal distance and expressed as a ratio, percentage, or tangent of the hypsometric curve. 2. Commonly, the rate of change of quantity. See also GRADIENT.
SMOOTHING - The reduction of the local variability of data and, when applied to a spatially distributed variable, results in a reduction of local variance. Smoothing, applied to a LINE, results in a reduction in the sharpness of angles between line SEGMENTS. See also FILTERING, EDGEMATCHING, WEEDING, THINNING.
SNAPPING - The automatic intersecting of disjoint lines. See also SNAPPING DISTANCE.
SNAPPING DISTANCE - The distance over which lines are intersected in SNAPPING (45).
SNAPSHOT - A query in SPATIO-TEMPORAL DATA BASE characterizing data at a given time (25). Synonym for TIME SLICE.
SOFTWARE - Programs, procedures, and rules for the execution of specific tasks on a computer system (2).
SORT - The operation of arranging of a set of items according to a KEY that determines the sequence (precedence) of items.
SPAGHETTI - 1. Data model for the storage of spatial information in VECTOR DATA STRUCTURES. In spaghetti models, points, lines and polygons are stored as simple lists of coordinates (37). 2. The approach of digitizing lines in an arbitrary sequence.

SPATIAL DATA BASE - A DATA BASE containing information indexed by location.
SPATIAL DATA STRUCTURES - A class of data structures designed to store spatial information and facilitate its manipulation (47).
SPATIAL DATA MODEL - A model of how objects are located in a spatial context.
SPATIAL FUNCTION - A function defined over space.
SPATIAL INFORMATION SYSTEM - An information system with the capability to manage spatially referenced information. See also GIS.
SPATIAL QUERY - A query that includes criteria for which selected features must meet location conditions.
SPATIO-TEMPORAL DATA - Data defined with reference to both space and time (25).
SPATIO-TEMPORAL DATA BASES - Data bases designed for the storage and management of SPATIO-TEMPORAL DATA (25).
SPATIO-TEMPORAL QUERIES - Queries performed on spatio-temporal data where one of the selection conditions is time.(25).
SPECTRAL SIGNATURE - A record of the spectral distribution and intensity of energy reflected or emitted from an object, or class of objects, by which they may be identified
SPHEROID - A solid that resembles a sphere in geometry. One of the terms used to describe the shape of the Earth. See also GEOID, ELLIPSOID.
SPIKE - Erroneous elevation data in a DEM, characterized by a value several magnitudes greater than surrounding data values.
SPLINE - Piecewise polynomial of k-degree with continuity of derivatives of order k-1 at the common joints between segments. Splines are used to calculate a line through a series of points (22).
SPOT - Satellite Probatoire pour l'Observation de la Terre, a French, Earth-orbitting, remote sensing satellite.
SPURIOUS POLYGONS - Synonym for SLIVER POLYGONS.
SQL - (Standard Query Language). Non-procedural query language used to access relational data bases, having an English-like syntax allowing for data insert, update, query and protection (1)(58).
STATISTICAL SURFACE - A surface described by some probability law.
STEREOMETER - A device for measuring parallax difference (21)
STEREOSCOPIC PLOTTER - A device for making topographic maps using stereo pairs of aerial photographs(5).
STREAM MODE - The process of manual DIGITIZING in which points are digitized automatically at preset time or distance intervals (42). See also POINT MODE. Synonym for LINE MODE.
STRING - 1. The sequence of SEGMENTS without specified end points and direction (32). 2. The set of characters treated as a unit for the purpose of display or analysis.
SUBTRACTION - A set operation. The outcome of the subtraction of two sets is a set with elements belonging to one input set and not existing in a second input set.
SUBTRACTIVE PRIMARY COLORS - The complimentary colors (yellow, magenta, and cyan) used to produce other colors in the printing process.

SURFACE FITTING - The generation of a surface passing through, or close to, a number of existing points. See also LINE FITTING.
SURFACE SPECIFIC POINTS AND LINES - Points and lines that describe surface specific local features including:

- peaks (local maxima);
- pits (local minima);
- passes (maxima in one direction minima in others);
- ridges or ridge-lines (maxima connecting peaks with passes);
- ravines or ravine-lines (minima connecting pits with passes);
- breaks or break-lines (topographical features causing abrupt change in contour direction and/or density, there are sharp break-lines and soft break-lines (55))
- flats.(areas of near zero slope) (36).

SYMBOL - A graphic representation of a geographic entity. There are three types of symbols; points, lines, and areas (42)(29).
SYMBOLIZATION - An element of cartographic generalization. A process of assigning symbols to map elements to convey their characteristics or hierarchy, and help to enhance the information contained in a map (42).
SYNTAX - The rules governing the structure of a language (2). See also BACKUS-NAUR FORM.

TABLE - 1. An object in a relational data base made up of records.(1). Synonym for RELATION. See also RECORD. 2. Part of a digitizer.
TAG - The identifier for or pointer to a cartographic object (32).
TAPE DECK - Synonym for TAPE DRIVE.
TAPE DRIVE - A system unit for controlling of operations on magnetic tapes. Tape drives allow the user to read and to write to tape (2). Synonym for TAPE DECK, TAPE UNIT.
TAPE UNIT - Synonym for TAPE DRIVE.
TCP/IP - Transmission Control Protocol/ Internet Protocol (35).
TEMPORAL DATA BASE - A DATA BASE containing information indexed by time. See also SPATIAL DATA BASE, ATEMPORAL DATA BASE.
TEMPORAL GRID - A data structure (similar to GRID) providing a framework for the storage of temporal data. The addresses of CELLS in a temporal grid are defined using the time domain. GRID cells store data characterizing an arbitrary phenomena at different times (25).
TEMPORARY FILE - Synonym for SCRATCH FILE.
TERMINAL - Input - output unit with which a user communicates with a data processing system (2).
TELNET - A TCP/IP network VIRTUAL TERMINAL interface. (35).
TESSELLATION - Division of space into regular or irregular polygons (2-D) or polyhedra (3-D). See also TESSELLATION DATA STRUCTURE.
TESSELLATION DATA STRUCTURE - A SPATIAL DATA STRUCTURE in which space is represented through TESSELLATION (37).

TEXT - Information, composed of printable characters, that are placed on a map. See also LABEL, ANNOTATION.
TEXT ATTRIBUTES - A set of parameters defining the characteristics of the text displayed on the map. Text attributes include color, style, font, size, location, and angle.
TEXT ENVELOPE - A rectangle bounding a text string. The height of the text envelope is equal to the difference between the maximum and minimum values of the Y-coordinates of the text rectangle. The width of the Text Envelope is equal to the difference between the maximum and minimum values of the X -coordinates of text rectangle (29).
TEXT POSITION - A variable describing how a text string is to be positioned. For example, lower left text position means that the origin of the text string is located in the lower left corner of the text rectangle (42).
TEXT RECTANGLE - A rectangle bounding a text string. The height of the text rectangle is equal to the height of the text string. The length of the text rectangle is equal to the sum of the lengths of all the characters and intercharacter spaces in the text string (29).
TEXT STRING ATTRIBUTES - See TEXT ATTRIBUTES.
TEXT STYLE - The graphic design for characters. Some of the more common styles are: Roman, Italic, and Gothic (42). See also FONT.
THEMATIC MAP - A map illustrating the class characteristics of a particular spatial variable (29). See also MAP.
THEMATIC MAPPER - A multispectral imaging system on some of the Landsat satellites. See LANDSAT TM.
THINNING - The process of reducing the number of grid coordinates defining a line, most often used during the conversion of lines from grid to vector format. A line is "thinned" if each grid cell defining a line is adjoined to only two other cells that are defining the same feature (31).
THIESSEN POLYGONS - Polygons constructed by bisecting, at right angles, the sides of each of the triangles of a TIN. Thiessen Polygons are logical duals of TIN mesh (37). Synonym for DIRICHLET TESSELLATION, VORONOI DIAGRAMS, VORONOI POLYGONS.
TICS - Points on a map with known locations used to define the transformation of map coordinates from one system to another. TICS are used in the process of transforming map data from digitized coordinates to the desired projection system. Often used as a synonym for CONTROL POINTS. See also DIGITIZING.
TIGER - (Topologically Integrated Geographic Encoding and Referencing System). An automated cartographic and geographic support system compiled by the Geographic Division of the US Census Bureau (39).
TILE - 1. Element of a TESSELLATION DATA BASE. 2. A unit of organization in a MAP LIBRARY.
TIN - (Triangulated Irregular Network). A SPATIAL DATA STRUCTURE generated by the TESSELLATION of space into irregular, exclusive triangles (37). See also VORONOI DIAGRAMS, THIESSEN POLYGONS, TRIANGULATION.
TONE - A distinguishable shade of gray on an IMAGE between black and white (9).

TOPOGRAPHIC MAP - Map depicting terrain relief (27).
TOPOLOGICAL ERROR - Term used in spatial information systems to denote inconsistencies in the geometry of a map file as defined by the requirements of a particular software system. The following topological errors are commonly distinguished: DANGLE, OVERSHOOT, UNDERSHOOT, LEAKING POLYGON, MULTIPLE CENTROID (42)(7).
TOPOLOGICAL VERIFICATION - The verification of topological relations between data elements (16).
TOPOLOGY - 1. Relationship between geometric objects remaining constant through any one-to-one continuous transformation. 2. Relationship in space between OBJECTS in a spatial data base. 3. Explicit coding of spatial relationships between map elements (7). See also TOPOLOGICAL ERROR.
TRACK-TO-TRACK ACCESS TIME - See ACCESS TIME.
TRAINING AREA - A portion of a digital image for which ground conditions or characteristics are known and which therefore provides the basis for defining the characteristics of a known class. The data taken from the TRAINING AREA are used as the training sample.
TRAINING DATA - Data used to calibrate a data acquisition device or data classification procedure.
TRAINING SAMPLE - See TRAINING AREA
TRANSACTION - The addition, retrieval or deletion of an element of a data base.
TRANSACTION CONTROL - The management of TRANSACTIONS to support transaction tracking, transaction recording, and to avoid transaction collision.
TRANSACTIONAL DATA BASE - A DATA BASE management system that supports TRANSACTION CONTROL.
TRANSFER FORMAT - A FORMAT in which data are transferred between systems (34). See also EXPORT, INPORT.
TRANSFER MEDIA - A storage media used for the transfer of data.
TRANSFORMATION - Changing the coordinate system in which data are defined, as from digitized coordinates to map coordinates.
TRANSECT - Synonym for PROFILE.
TREE - 1. A form of a directed graph having a distinctive vertex (root vertex) without predecessor, and with every other vertex having a unique predecessor (56). 2. Commonly, a kind of data structure organized as a TREE, used to speed up data access.
TRIANGULATION - A process of subdividing a 2-D space into bounding regions that are triangles (38). See also TIN.
TRIPLET - A datum with two location coordinates (X,Y) and an elevation (Z).
TUPLE - See RECORD.

UNCONTROLLED MOSAIC - A photomosaic made without correction for distortions of any type (21).

UNDERSHOOT - A kind of topological error in which an ARC with a NODE of VALENCY one causes an unclosed or LEAKING POLYGON (7).
UNION - A set operation. The outcome of the union of sets is a set with all the elements of all input sets.
USER INTERFACE - A method by which the user controls operation of a computer system. Typical user interfaces include MENU command language and programmable interface (16).
USER REQUIREMENT ANALYSIS - A study of the needs of a user of a system conducted prior to the system design (16).
UTM - (Universal Transverse Mercator). A grid system based on the Transverse Mercator projection. The UTM grid extends North-South from 84°N to 80°S latitude and, starting at the 180° meridian, is divided Eastwards into 60, 6° zones with Zone 1 beginning at 180° longitude. The UTM grid is used for topographic maps and geo-referencing of satellite images (42).

VALENCY OF NODE - The number of arcs beginning or ending at a NODE. A node with a valency of one is referred to as a "dangling node".
VALENCY TABLE - A list of all NODES for a given COVERAGE together with their respective VALENCY. See also TOPOLOGY, ARC,and/or VALENCY OF NODE.
VARIOGRAM - See SEMI-VARIOGRAM.
VECTOR - 1. A quantity which has magnitude and direction. 2. Commonly, the notation used to represent spatial information.
VECTOR DATA STRUCTURES - A term describing a class of spatial data structures in which spatial information is represented as VECTORS. In vector data structures, the basic units of spatial information are points (coded as vector) and lines (coded as groups of points) organized into CHAINS, ARCS, or POLYGONS.
VECTORIZATION - 1. The conversion of SPATIAL DATA from any other data structure to VECTOR DATA STRUCTURE. 2. Commonly, the conversion of the RASTER DATA STRUCTURE of some data set to a VECTOR DATA STRUCTURE (7).
VERTICAL AERIAL PHOTOGRAPH - An aerial photograph taken with the optical axis of the camera perpendicular to the Earth's datum plane.
VERTICAL INTEGRATION - The integration of several overlapping map files into one map containing the combined information of the input map layers in the correct format for a given system (29). See also OVERLAY.
VERTICES - Intermediate points forming a LINE. See also CHAIN, NODE.
VIEWPOINT - 1. Synonym for VISTA POINT. 2. A position of an imaginative viewer from which a 2-D perspective view of a 3-D object may be drawn.
VIEWPORT - A predefined portion of a display area (2).
VIEWSHED MAP - 1. A map outlining those areas visible from a particular vista point (or set of vista points) (7). Synonym for VISIBILITY MAP, LINE-OF-SIGHT MAP. 2. A map outlining areas obscured from a particular vista point for a particular communication device and medium. For example, microwave visibility, radar visibility.
VISIBILITY MAP - See VIEWSHED MAP.

VISIBLE WAVELENGTHS - The electromagnetic radiation range to which the human eye is sensitive, approximately 0.4 to 0.7 micrometers (9).
VIRUS - A computer program that disrupts the normal operation of the computer system.
VIRTUAL TERMINAL - See VIRTUAL MACHINE.
VIRTUAL MACHINE - A hardware or software that designs how the elements of a computer system appear to behave to the user.
VISTA POINT - A point from which a visibility map is compiled. See also VIEWSHED MAP, VISIBILITY MAP.
VORONOI DIAGRAMS - Synonym for THIESSEN POLYGONS.
VOXEL - A unit cube forming a 3-D object. The smallest addressable element of a 3-D array (47). Synonym for OBEL. See also PIXEL.

WATERSHED ANALYSIS SYSTEM - A software system used to characterize a watershed, drainage basin, or catchment. Basic functional capabilities often include the delineation of watershed boundaries, extraction of stream networks, and flow analyses. Watershed analysis systems are frequently parts of GEOGRAPHIC INFORMATION SYSTEMS.
WEEDING - Reducing the number of points defining a line while preserving the essential line shape. See also SMOOTHING, THINNING. Common weeding algorithms include:

- distance traversed algorithm,
- Nth point selection algorithm,
- moving average algorithm,
- angle selection algorithm,
- Williams' point relaxation algorithm,
- floating band algorithm,
- Douglas-Peucker algorithm,
- Walking dividers algorithm (31)(60).

WEIGHT - A scaling multiplicator expressing the importance of a variable in a particular operation.
WINDOW - 1. Rectangular portion of map selected for display.(7). 2. Portion of a screen of a terminal through which one may control an interactive session. 3. A band of the electromagnetic spectrum offering maximum transmission through a particular medium as in "atmospheric window" (42)
WINDOWING - The process of displaying that portion of a digital map lying within a WINDOW.
WORD - A group of bytes treated as a unit. See also LONG WORD, BYTE.
WORKSTATION - A single user, stand alone, micro or mini computer with both high processing power and large storage capacity.
WORM - (Write Once Read Many). A kind of mass storage device that is based on optical disk technology in which data are in permanent storage but can be accessed many times. See also CD-ROM.

X-WINDOW SYSTEM - Device-independent, network-transparent, WINDOW system allowing the simultaneous execution of multiple applications in windows,and/or the generation of text and graphics, in monochrome or color, on a bitmap display.(48)
ZENITH - The point in the celestial sphere that is directly overhead (21)
ZOOM - 1. A software or hardware function that allows for the display of progressively smaller (zoom in) or larger (zoom out) areas of an image on an interactive display device. 2. The process of using ZOOM.
Z-VALUE - Commonly, a value for a spatially distributed phenomenon (surface elevation, depth to groundwater) at a particular location.

REFERENCES

(1). Alagic, S., 1986, *Relational Data Base Technology*: Springer-Verlag, Berlin, 259 p.
(2). American National Standards Committee X3, 1983, *American National Dictionary for Information Processing Systems*: Information Processing Systems Technical Report, 149 p.
(3). Armando, J.A., 1983, *A Framework for the Analysis of Geographic Information Systems Procedures: the Polygon Overlay Problem, Computational Complexity, and Polygon Intersection*: PhD. Thesis, University of New York at Buffalo.
(4). Artwick, B.A., 1984, *Applied Concepts in Microcomputer Graphics*: Prentice-Hall International, Englewood Cliffs, New Jersey.
(5). Avery, T.E. and Berlin, G.L., 1985, *Interpretation of Aerial Photographs*: Burgess Publishing Company, Minneapolis, 554 p.
(6). Blahut, T.S., Chrzanowski, A. and Saastamoinen, J.U., 1979, *Urban Surveying and Mapping*: Springer-Verlag, Berlin, 372 p.
(7). Burrough, P.A., 1986, *Principles of Geographic Information Systems for Land Resources Assessment*: Clarendon Press, Oxford, 194 p.
(8). Cabay, S. and Vanzella, L., 1990, A comparison of partitioned and seamless organization of spatial data: In Proceedings of Symposium GIS for the 90's, March 13-16, Vancouver, B.C., p. 423-429.
(9). Colwell, R.N., (editor), 1983, *Manual of Remote Sensing:* Vol.1 and Vol. 2. American Society of Photogrammetry, Falls Church, VA, 2440 p.
(10). CPS-3 Users' Manual: February 1990 Edition, Radican Corp, Houston, Texas, 282 p.
(11). Darcy, L. and Boston, L., (editors), 1983, *Webster's New World Dictionary of Computer Terms*: Simon & Shuster, New York, 282 p.
(12). Feder, J., 1988, *Fractals*: Plenum Press, New York, 283 p.
(13). Folk, M.J. and Zoelick, B., 1987, *File Structures: A Conceptual Tool Kit:* Addison-Wesley, Reading, Massachussetts, 538 p.
(14). GIS 1990 Source book. GIS World Inc., Fort Collins, CO., 320 p.
(15). Grady, R.K., 1990, The lineage of data in Land and Geographic Information Systems (LIS/GIS): Journal of the Urban and Regional Information Systems Association, v.2(2), p. 2-6.

(16). Guptill, S., (editor), 1988, *A Process for Evaluating Geographic Information Systems*: Technology Exchange Working Group - Technical Report No. 1, Federal Interagency Coordinating Committee on Digital Cartography, United States Geological Survey Open-File Report 88-105, 57 p.
(17). Held, G., and Marshall, T.R., 1987, *Data Compression Techniques and Applications, Hardware and Software Considerations*: John Wiley & Sons, New York, 206 p.
(18). Hillier, F.S. and Lieberman, G.J., 1980, *Introduction to Operations Research*: Third Edition, Holden-Day Inc., 829 p.
(19). Hurn, F., 1989, *GPS: A Guide to the Next Utility*: Trimble Navigation, Sunnyvale, CA., 76 p.
(20). Jensen, J.R., 1986, *Introductory Digital Image Processing*: Prentice-Hall, Englewood Cliffs, New Jersey, 379 p.
(21). Johanssen, C.J. and Sanders, J.L., (editors), 1982, *Remote Sensing for Resources Management*: Soil Conservation Society of America, 665 p.
(22). Journel, A.G. and Huibregts, Ch.J., 1978, *Mining Geostatistics*: Academic Press, New York, 600 p.
(23). Kent, W., 1983, A simple guide to five normal forms in relational data base theory: Comm. ACM, v.28(2), p. 120-123.
(24). Khagendra, T., 1987, *Detection of Critical Points, First Step to Automatic Line Generalization*: Ph.D.Thesis, Ohio State University.
(25). Langran, G., 1990, Temporal GIS design tradeoffs: Journal of the Urban and Regional Information Systems Association, v.2(2), p. 16-25.
(26). Lynch, M.P. and Saalfeld, A.J., 1985, Conflation: automated map compilation - a video game approach: AUTOCARTO 7, Washington, D.C. March 11-14.
(27). Makower, J., Poff C., and Berhgeim,L., (editors), 1990, *The Map Catalog*: 2nd Edition, Tilden Press, 364 p.
(28). Marble, D.F., Calkins, H.W. and Peuquet, D.J., 1984, *Basic Readings in Geographic Information Systems*: SPAD Systems, Williamsville, N.Y., 384 p.
(29). Mark, R.W., (editor), 1990, The Census Bureau's TIGER system: Cartography and Geographic Information Systems, v.17(1), 133 p.
(30). Miller, W., 1984, *The Engineering of Numerical Software*: Prentice-Hall, Englewood Cliffs, New Jersey, 120 p.
(31). Monmonier, M.S., 1982, *Computer-Assisted Cartography: Principles and Prospects*: Prentice-Hall, Englewood Cliffs, New Jersey, 213 p.
(32). National Committee for Digital Cartographic Data Standards, 1988, Standards for digital cartographic data, American Cartographer, v.15(1), 140 p.
(33). National Research Council, 1980, *Need for a Multipurpose Cadastre*: Panel on a Multipurpose Cadastre, Committee on Geodesy, Assembly of Mathematical and Physical Sciences, National Academy Press, Washington, D.C., 112 p.
(34). National Transfer Format Secretariat, 1987, *The Assessment and Transfer of Data Quality*: Release 1.0, Maybush, Southampton, variously paged.
(35). NCD17 Terminal Users Manual, 1990.

(36). Peucker, T.K. and Douglas, H.D., 1975, Detection of surface specific points by local parallel processing of discrete terrain elevation data: Computer Graphics and Image Processing, v. 4. p. 375-387.
(37). Peuquet, D.J., 1984, A conceptual framework and comparison of spatial data models: Cartographica v.21(4), p. 66-113.
(38). Preparata, F.F. and Shamos, M.I., 1985, *Computational Geometry: An Introduction*: Springer-Verlag, Berlin, 330 p.
(39). Press, W.H., Flannery, B.P., Teukolsky, S.A. and Vetterling, W.T., 1986, *Numerical Recipes: The Art of Scientific Computing*: Cambridge University Press, Cambridge, 610 p.
(40). Pritsker, A.A.B. and Pegden, C.D., 1979, *Introduction to Simulation and SLAM*: John Wiley & Sons, New York, 588 p.
(41). Ralston, A. and Reilly, E.D., (editors), 1983, *Encyclopedia of Computer Science and Engineering*. IIth Edition, Van Nostrand Reinhold, New York, 1664 p.
(42). Robinson, A.H., Sale, R.D., Morrison, J.L. and Meuhrecke, P.C., 1984, *Elements of Cartography*: 5th Edition, John Wiley & Sons, New York, 554 p.
(43). Rock-Evans, R., 1981, *Data Analysis*: IPC Business Press Inc., 126 p.
(44). Rogers, D.F. and Adams, J.A., 1976, *Mathematical Elements for Computer Graphics*: McGraw-Hill, New York, 236 p.
(45). Ruiz, M.S. and Messersmith, J.M., 1990, *Cartographic Issues in the Development of a Digital GRASS Data Base*: Report No. SR. N-90/16, U.S.Army Construction Engineering Research Laboratory, Champaign, Il, variously paged.
(46). Saltor, G. and McGill,M.J., 1983, *Introduction to Modern Information Retrieval:* McGraw-Hill, New York, 448 p.
(47). Samet, H., 1990, *The Design and Analysis of Spatial Data Structures*: Addison-Wesley, Reading, Massachussetts, 493 p.
(48). Scheifler, R.W. and.Gettys, J., 1990, *X-Window System*: 2nd Edition, Digital Press, 738 p.
(49). Simpson, P.K., 1990, *Artificial Neural Systems: Foundations, Paradigms, Applications, and Implementations*: Pergamon Press, Oxford, 230 p.
(50). Smith, J.W.F., 1986, *Effects of Cartographic Error of Polygon Overlay with Geomorphic Data*: Unpublished M.Sc. Thesis, Department of Geography, University of Alberta, Edmonton, Alberta, 120 p.
(51). Snyder, J.P., 1987, *Map Projections - A Working Manual*: United States Geological Survey Professional Paper 1395, Government Printing Office, Washington, 386 p.
(52). Statistics Canada, 1988, *The Area Master File Users Guide*: Geographic Division, Statistics Canada, Ottawa, Ontario.
(53). Stefik, M. and Bobrow, D.G., 1984, *Object-Oriented Programming: Themes and Variations:* The AI Magazine, p. 40-62.
(54). Strahler, A.H., 1980, The use of prior probabilities in maximum likelihood classification of remotely sensed data: Remote Sensing of Environment, v.10, p. 135-163.

(55). Survey and Resource Branch, 1988, *Specifications and Guidelines 1:20000 Digital Mapping*: Release 3.0, Ministry of Environment and Parks, Province of British Columbia, 282 p.
(56). Tenenbaum, A.M. and Augenstein, M.J., 1981, *Data Structures Using Pascal*: Prentice-Hall, Englewood Cliffs, New Jersey, 545 p.
(57). Tomlin, C.D., 1983, A map algebra, In *Proceedings of Harvard Computer Conference*, July 31- Aug.4, Cambridge, Mass.
(58). Van der Lans, R.T., 1988, *Introduction to SQL*: Addison-Wesley, Reading, Massachussetts, 348 p.
(59). Vanzella, L., 1988, *Classification of Data Structures for Thematic Data*: Technical Report TR 88-14, Department of Computing Science, University of Alberta, Edmonton, Alberta, 145 p.
(60). Vanzella, L., 1988, *Map Generalization Project*: Report compiled for Land Information Services Divison, Alberta Forestry, Lands and Wildlife, Edmonton, Alberta, 230 p.
(61). Vanzella, L. and Cabay, S., 1988, Hybrid spatial data structures: *Proceedings 3rd. International GIS/LIS Conference*, San Antonio, Texas.
(62). Wegner, P., 1990, Concepts and paradigms of object-oriented programming: OOPS Messenger, *A Quarterly Publication of the Special Interest Group on Programming Languages, ACM Press*, v.1(2), August 1990.
(63). White, M.S. and Griffin, P.O., 1985, Piecewise linear rubber-sheet map transformation: *The American Cartographer*, v.12(2), p. 123-131.

APPENDIX IV

GIS Acronyms

AI-Artificial Intelligence
AM/FM-Automated Mapping/Facilities Management
AMF-Area Master File
ANSI-American National Standards Institute
ASCII-American Standard Code for Information Exchange
AVHRR-Advanced Very High Resolution Radiometer
BCD-Binary Coded Decimal
BIL-Band Interleaved by Line
BIP-Band Interleaved by Pixel
BIT-Binary Digit
BPI-Bytes Per Inch
BSQ-Band Sequential
CAD-Computer Assisted Design
CAD/CAM-Computer Aided Design/Computer Aided Mapping (or Manufacturing)
CASE-Computer Aided Software Engineering
CCT-Computer Compatible Tape
CD-Compact Disc
CD ROM-Compact Disk Read-Only Memory
CGA-Color Graphics Adapter
CGM-Computer Graphics Metafile
CIR-Color Infrared
CMY-Cyan Magenta Yellow
COGO-Coordinate Geometry
CPU-Central Processing Unit
CRT-Cathode Ray Tube
DBMS-Data Base Management System
DCDSTF-Digital Cartographic Data Standards Task Force
DEM-Digital Elevation Model
DIME-Dual Independent Map Encoding
DLG-Digital Line Graph
DLG-E-Digital Line Graph Enhanced
DN-Digital Number

DOS-Disk Operating System
DRAM-Dynamic RAM
DTED-Digital Terrain Elevation Data
DTM-Digital Terrain Model
DXF-Digital Exchange Format
EBCDIC-Extended Binary Coded Decimal Interchange Code
EDI-Electronic Data Interexchange
EDP-Electronic Data Processing
EGA-Enhanced Graphics Adapter
ERS-Earth Resource Satellite
ESD-Ecologically Sustainable Development
ETF-European Transfer Format
FTAM-File Transfer, Access and Management
GAM-Geographical Analysis MAchine
GB-Gigabyte
GBF-DIME-Geographic Base File/Dual Independent Map Encoding
GCEM-Geographical Correlates Exploration Machine
GCP-Ground Control Point
GDF-Geographic Data File
GIGO-Garbage In Garbage Out
GIS-Geographic Information System
GPC-Graphics Performance Characterization
GPS-Global (or Geodetic) Positioning System
IGES-International Graphics Exchange System
IHS-Intensity, Hue and Saturation (also HSI)
I/O-Input/Output
IQS-Intelligent Query System
IT-Information Technology
ITF-International Transfer Format
JERS-Japanese Earth Resources Satellite
KB-Kilobyte
LAN-Local Area Network
LIS-Land Information System
LUT-Lookup Table
MACDIF-Map and Chart Interchange Format
MB-Megabyte
MC&G-Mapping, Charting and Geodesy
MDIF-Mapping Data Interchange Format
MFLOPS-Million Floating Point Operations Per Second
MIPS-Million Instructions Per Second
MOS-Marine Observation Satellite
MSS-Multispectral Scanner

MTBF-Mean Time Between Failures
NAD-North American Datum
NAD27-North American Datum 1927
NAD83-North American Datum 1983
NTF-National Transfer Format
OO-Object Oriented
OODB-Object Oriented Database
OOP-Object Oriented Programming
OOPL-Object Oriented Programming Language
OOPS-Object Oriented Programming Systems
PC-Personal Computer
RAM-Random Access Memory
RDA-Remote Data Access
RDBMS-Relational Data Base Management System
RGB-Red Green Blue
RISC-Reduced Instruction Set Computer
ROM-Read Only Memory
RS-Remote Sensing
SAIF-Spatial Archive and Interchange Format
SAR-Synthetic Aperture Radar
SDSS-Spatial Decision Support System
SDTS-Spatial Data Transfer Standard
SGML-Standard Graphic Mark-up Language
SIM-Satellite Image Map
SOM-Satellite Orthophoto Map
SPOT-Systeme Pour l'Observation de la Terre
SQL-Structured Query Language
TB-Terrabyte
TCP/IP-Transmission Control Protocol/Internet Protocol
TIGER-Topologically Integrated Geographic Encoding and Referencing
TIN-Triangulated Irregular Network
TM-Thematic Mapper
UTM-Universal Transverse Mercator
VGA-Video Graphics Adapter
WAN-Wide Area Network
WORM-Write Once Read Many
WYSIWYG-What You See Is What You Get

Index

Numbers in italics refer to words in the Glossary

INDEX

INDEX

INDEX

WITHDRAWN